A Geography
of Australia

Landscapes
and Land
Uses

# A Geography of Australia

SECOND EDITION

NORMAN J SNELL

*The McGraw-Hill Companies, Inc.*

Sydney  New York  San Francisco  Auckland
Bangkok  Bogotá  Caracas  Hong Kong
Kuala Lumpur  Lisbon  London  Madrid
Mexico City  Milan  New Delhi  San Juan
Seoul  Singapore  Taipei  Toronto

Reprinted 2003

Text © 2003 Norman J. Snell & Geographical Association of WA
Illustrations and design © 2003 McGraw-Hill Australia Pty Ltd
Additional owners of copyright are acknowledged on the Acknowledgments page and in on-page credits.

Apart from any fair dealing for the purposes of study, research, criticism or review, as permitted under the *Copyright Act*, no part may be reproduced by any process without written permission. Enquiries should be made to the publisher, marked for the attention of the Permissions Editor, at the address below.

Every effort has been made to trace and acknowledge copyright material. Should any infringement have occurred accidentally the authors and publishers tender their apologies.

**Copying for educational purposes**
Under the copying provisions of the *Copyright Act*, copies of parts of this book may be made by an educational institution. An agreement exists between the Copyright Agency Limited (CAL) and the relevant educational authority (Department of Education, university, TAFE, etc.) to pay a licence fee for such copying. It is not necessary to keep records of copying except where the relevant educational authority has undertaken to do so by arrangement with the Copyright Agency Limited.

For further information on the CAL licence agreements with educational institutions, contact the Copyright Agency Limited, Level 19, 157 Liverpool Street, Sydney NSW 2000. Where no such agreement exists, the copyright owner is entitled to claim payment in respect of any copies made.

Enquiries concerning copyright in McGraw-Hill publications should be directed to the Permissions Editor at the address below.

---

**National Library of Australia Cataloguing-in-Publication data:**

Snell, Norman (Norman J.).
Landscapes and land uses 2/e.

2nd ed.
Includes index.
ISBN 0 074 71183 0.

1. Human geography – Australia. 2. Physical geography – Australia. I. Guy, Diane. Landscapes and land uses. II. Title.

304.20994

---

Published in Australia by
**McGraw-Hill Australia Pty Ltd**
**Level 2, 82 Waterloo Road, North Ryde NSW 2113, Australia**
Production Manager: Sybil Kesteven
Acquisitions Editor: Laura Whitton
Editor: Sharon Nevile
Permissions Editor: Jill Roebuck
Proofreader: Tim Learner
Indexer: Diane Harriman
Design (interior and cover): Jan Schmoeger
Mapmaker: Tony Fankhauser
Technical illustrator: Alan Laver
Cover image: see Acknowledgments page
Typeset in Utopia and News Gothic by Prototype Phototypesetting
Printed on 80 gsm woodfree by Pantech Limited, Hong Kong.

# Contents

| | |
|---|---|
| **Preface** | **vii** |
| **Acknowledgments** | **viii** |

## Chapter 1 The Biophysical Environment — 1

Introduction .................................................................... 1
Australia's physiographic regions ........................................... 1
Australia's climatic pattern ................................................. 13
Climatic controls ............................................................. 17
Vegetation patterns in Australia ............................................ 20
Natural factors influencing vegetation ..................................... 26

## Chapter 2 Biophysical Interrelationships in the South-West — 28

Topography of the South-West ............................................... 28
Soils of the South-West ...................................................... 33
The climate of the South-West ............................................... 34
Vegetation of the South-West ................................................ 39
The fauna of the South-West ................................................. 42
Biophysical interactions ..................................................... 44

## Chapter 3 Australian Agriculture — 47

Agricultural systems ......................................................... 47
Intensive and extensive agriculture ......................................... 49
Locational influences ........................................................ 50
The rural cultural landscape ................................................ 53
Current issues in Australian agriculture .................................. 55

## Chapter 4 Case Studies in Australian Agriculture — 59

Case study 1: Mixed crop and livestock farming in the South-West ....... 59
Case study 2: Extensive pastoralism in Central Australia ................ 71
Case study 3: Intensive pastoralism in Eastern Australia ................ 79

## Chapter 5 Australia's Mineral and Energy Resources — 94

The history of mining in Australia .......................................... 94
The cultural landscape ....................................................... 95
Case study 1: Gold mining in the central study area ...................... 97
Case study 2: Diamond mining in the northern study area ................ 108
Case study 3: Coal mining in the south-western study area .............. 114

# Contents

## Chapter 6 Australia's Population — **121**

Introduction .................................................................... 121
Population density and distribution ........................................... 121
Factors influencing Australia's population distribution ....................... 123
Changes in Australia's population ............................................. 127
Population patterns in the South-West ......................................... 134
Population characteristics of Perth ........................................... 135
Socioeconomic characteristics of Perth ........................................ 139

## Chapter 7 Settlement Patterns and Urban Networks — **145**

Settlement patterns ........................................................... 145
Urban networks and hierarchies ................................................ 148
Case study: The urban network of the Avon region .............................. 161

## Chapter 8 Urban Morphology — **174**

External morphology ........................................................... 174
Internal morphology ........................................................... 175
Models of city structure ...................................................... 178
Case study: Perth ............................................................. 179
Morphology of an Australian country town: Northam, Western Australia .......... 210

## Chapter 9 Practical Mapping and Research Skills — **219**

Introduction .................................................................. 219
Regionalism ................................................................... 219
Mapping skills ................................................................ 224
Interpreting topographic maps ................................................. 230
Interrelationships between the biophysical and cultural landscape ............. 234
Sketch mapping ................................................................ 235
Statistical analysis .......................................................... 236
Photographs ................................................................... 241
Fieldwork ..................................................................... 243
Web sites ..................................................................... 245

**Glossary** — **247**

**Credits** — **252**

**Index** — **253**

# Preface

This second edition of *Landscapes and Land Uses* has been written to meet the needs of students studying Geography in Western Australia and to reflect the significant changes that have occurred in the Year 12 syllabus since the publication of the original text. In addition, it contains a variety of material relevant to senior Geography courses in other Australian states.

The text engages the student in the study of Australia's biophysical environment and its inter-relationships. Elements of landform, soil, climate and vegetation are investigated at the national level as well as in the South-West of the continent. The human response to the environment is explored in the study of rural land uses, mining, settlement and demography.

Case studies in agriculture focus on the ways in which the biophysical and the cultural environments interact and on the factors and processes that influence this interaction. Studies include agricultural activities conducted in both humid and arid environments and which range from intensive to extensive land use. The focus of the case studies in mining is on the way in which this activity contributes to the economic wealth of the nation and how it impacts on the environment.

Demographic characteristics including population distribution and movements, and age and sex structures are investigated at national and local levels. These population studies also include analysis of the ways in which the population has changed over time and how it is projected to change in the future. Associated with issues of population is the nature of settlement. Australia is one of the world's most urbanised societies and in this sense the relative size and distribution of its towns and cities is an important element in this area of study. The case study of Perth in terms of its socioeconomic characteristics, patterns of urban land use and its influence on Western Australia's urban network is an important focus of the text.

A special emphasis has been placed on the development of mapping and practical skills with the inclusion of a chapter (Chapter 9) devoted to these. A comprehensive range of tasks and activities provides the student with the opportunity to develop the necessary degrees of proficiency in this area. Topics covered range from topographic mapping tasks through to fieldwork techniques and cover basic skills as well as higher level analysis.

The design of this edition is built on a thematic approach with information dealt with in a logical manner. The book has been structured to provide an easy-to-read format using various levels of subheading to organise information. The use of the most up-to-date information and statistical material gives the text a high degree of relevance and the provision of exercises at each stage within the chapters allows each topic to be studied and reviewed in an orderly manner. An extensive glossary is a welcome addition to the text and will be of great assistance to the student of Geography.

This text will not only assist senior students in the pursuit of their specific studies, but will also be of interest to the general reader with an interest in the unique character of the Australian continent.

# Acknowledgments

I would like to thank the following individuals and organisations for the assistance they have provided in the writing of this text:

- Beryl Snell
- Darren Lane
- Patrick Durack
- Alan May
- Emmy Terry
- City of Canning
- Shire of Goomalling
- West Australian Chamber of Minerals and Energy.

This text is dedicated to my wife, Beryl Snell, and my father, Ernest J. Snell (1911–1999, prospector, pastoralist, miner, bushman).

# Chapter 1

# The Biophysical Environment

> The purpose of this chapter is to locate and investigate the characteristics of Australia's landforms, climate and vegetation, and then to account for these.

## Introduction

Australia, along with Antarctica, India, Africa and South America made up the supercontinent of Gondwanaland, which was located on the equator 200 million years ago. As the supercontinent began to break up both Antarctica and Australia travelled south, reaching the present position of Antarctica about 70 million years ago. These two continents were the last to separate and, around 45–65 million years ago, the continental landmass of Australia broke away and moved northwards towards its present position.

Geologically, however, this was not the beginning of the landmass as we know it today. Many of the structures pre-date this event by millions of years. Approximately two-thirds of the continent had become stabilised by Precambrian times, with large sections such as the Western Shield, the Arnhem Block and the Kimberley Plateau already ancient and stable landforms. The oldest geologic region is the Western Shield (or Great Western Plateau), where rocks over 3500 million years old are found.

There has been little in the way of major upheavals in the last 150 million years, since the uplift of the Eastern Highlands. As a consequence of this long period of tectonic inactivity the landmass has been greatly worn down by the gradational processes of weathering and erosion. Compared to other continents Australia is very flat, with much of the landmass being less than 300 metres above sea level and the highest point being just over 2000 metres.

The minerals that make up the inorganic component of soils are closely linked to the exposure of new rock material to weathering forces. The absence of recent significant tectonic activity over much of Australia has resulted in soils that are relatively thin and infertile by world standards. Approximately 5 per cent of the continent's soils may be classified as high quality. The other 95 per cent exhibit moderate to severe chemical or physical limitations and hence require nutrient inputs to obtain a reasonable level of agricultural output.

Climatically the continent is very dry, with around two-thirds classified as either arid or semi-arid, but this has not always been the case. There is evidence of significant variations in the past. The laterites of south-western Australia were formed by a past warmer and wetter climatic period, while the iron ores of the Pilbara were laid down at a time when atmospheric oxygen levels were beginning to rise and trigger oxidation of different minerals. Remnant vegetations, such as the Livingstonia palms of Palm Valley in Central Australia and the pockets of rainforest on the western slopes of the Atherton Tableland, point to a wetter and warmer past. The impact of glacial and interglacial periods can be seen in the landforms of the Victorian and Tasmanian highlands and in the sea-level changes around the entire coastline.

Australia's vegetation reflects the impact of the soils and the climatic characteristics of the continent. Eucalypts dominate the flora, with over 70 per cent of the continent's plants belonging to this genus. They appear to have begun evolving around 34 million years ago, in response to the increasingly dry conditions. Eucalypts form part of a larger group of plants that have adapted to the infertile soils, lengthy periods of drought and the impact of fire—the **sclerophyllous** or hard, woody plants.

## Australia's physiographic regions

Landforms are the result of ongoing interactions between **gradational** and **tectonic** forces. These interactions can be complex and may occur over a very long time period. Landforms may also reflect interruptions and significant changes in the way that the two forces have operated during this time period.

Due to the relatively low level of tectonic activity over the last 80–90 million years, Australia's landforms tend to reflect the significant effects of the various gradational processes. Recent **diastrophic** movements have been the result of subsidence along earlier fault lines, rather than significant shifts in the tectonic plates. The location of the continent in the centre of the Indo-Australian Plate, away from the main plate boundaries, explains the absence of both volcanic and earthquake activity. Table 1.1 and figures 1.1 and 1.2 illustrate the major tectonic and gradational processes.

**Table 1.1** Gradational and tectonic forces

| Process | Meaning | Agencies/Forces | Examples of features produced |
|---|---|---|---|
| *Gradational processes* | | | |
| Weathering | Breakdown of **regolith** (rock) by physical, chemical and biotic means | | Soil, bare rock, fragments |
| • physical | • Mechanical breakdown of rock, e.g. rock cracking and peeling | • Water, temperature variations, ice | • Jointed boulders, sand, fragments (scree) |
| • chemical | • Decomposition ('rotting') of rock, e.g. oxidation ('rusting') | • Water, temperature variations, gases | • Clay, limestone caves |
| • biotic | • Breakdown of rock by plants and animals, e.g. rock cracking | • Burrowing, tree-root intrusion, acids | • Jointed rocks, scree |
| Erosion | The wearing away of the land by the gradational agents that move weathered materials | Wind, running water, waves, currents and ice | Peneplains, gibber deserts, coastal cliffs, glacial valleys, volcanic plugs |
| Transportation | Removal of weathered material by agents | Wind, running water, waves, currents and ice | — |
| Deposition | Laying down of material that has been transported | Wind, running water, waves, currents and ice | Sand dunes, river plains and deltas, beaches |
| Mass movement | Downhill movement of weathered materials under the influence of gravity, e.g. soil creep, mudflow, landslide, avalanche | Gravity, water, snow, earth tremors | Eroded slopes, alluvial fans, scree slopes |
| *Tectonic forces* | | | |
| Diastrophism | Movement of solid parts of the earth's crust (lithosphere), e.g. uplift, subsidence, horizontal movement | Internal pressure, heat | Mountain ranges, rift valleys, basins |
| • folding | • Bending of crustal rocks | • Compressional forces (→←) | • Fold mountains with anticlines and synclines |
| • faulting | • Fracturing of lithosphere; uplift, subsidence or horizontal movement along fault lines | • Compressional forces (→←), tensional forces (←→), shearing forces (↙↗) | • Horst blocks, rift valleys<br><br>surface fault lines |
| Vulcanism | Movement of molten earth materials upwards from the earth's mantle to produce surface and subsurface features | Heat, pressure | Volcanoes, lava plains and internal features, e.g. batholiths |

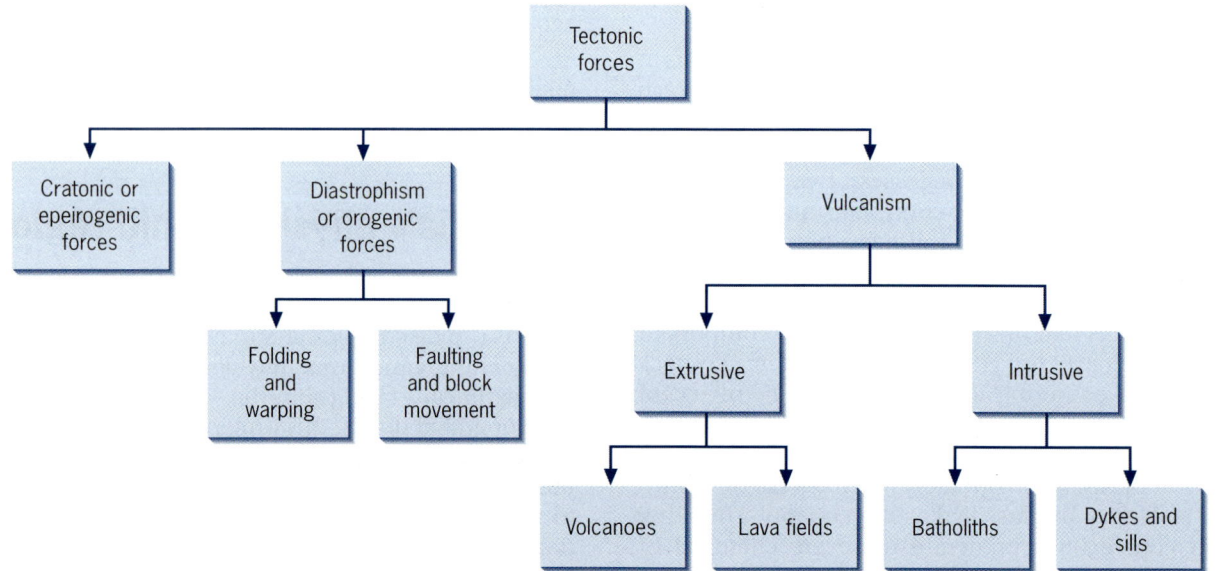

**Figure 1.1** Tectonic forces

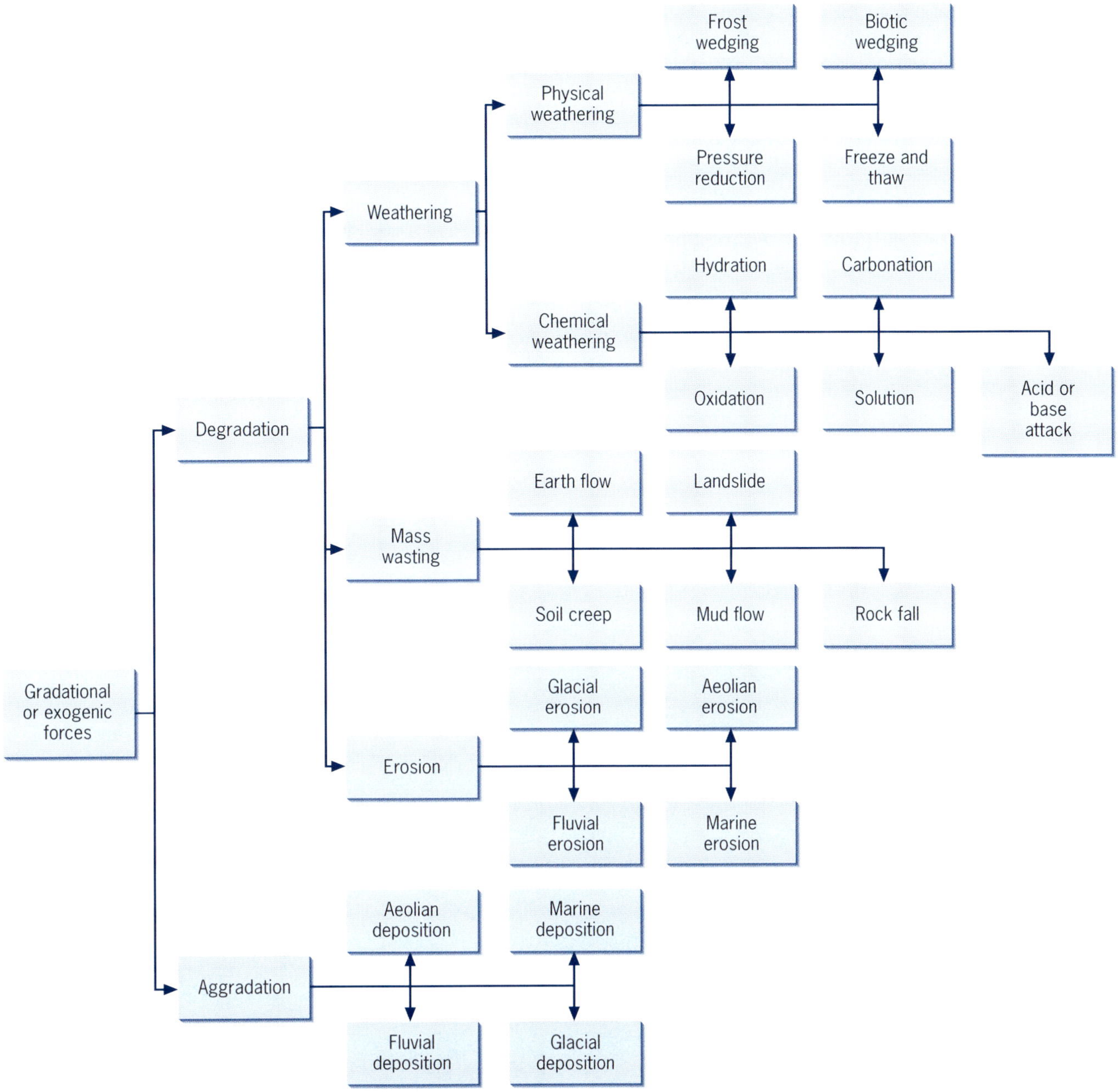

# THE BIOPHYSICAL ENVIRONMENT 3

**Figure 1.2** Gradational or exogenic forces

**Figure 1.3** Sedimentary sandstone overlain with marine limestone in Southern Victoria illustrates tectonic uplift and gradational deposition

Australia's landform regions fall into four main divisions (see figure 1.4):

- the Great Western Plateau, composed predominantly of the remains of the ancient rock shield of Gondwanaland;
- the Eastern Highlands, where the most recent tectonic uplift occurred;
- the Central Lowlands; and
- the Coastal Lowlands.

## The Great Western Plateau

The Great Western Plateau (Western Shield) became part of the earth's surface more than 3000 million years ago during the Archaean era. It has remained continuously above the surface of the oceans since the time when it was formed by outpourings of igneous material from the mantle. It has been subjected to long periods of weathering and erosion, along with several tectonic uplifts. At various times it has had massive outflows of volcanic lava—the remnants of one of the most extensive sequences of flood **basalts** in the world are located in the East Kimberley region. These are very old (530 million years) and outcrop over an area of 35 000 square kilometres, with another 115 000 square kilometres found buried beneath younger sedimentary layers.

The shield has been extensively eroded from its original height of some 1200 metres to its current average height of around 450 metres. This has exposed significant areas of **granite** and **gneiss**, which can be seen as rock domes or **monadnocks** in areas such as the Darling Plateau. The actions of wind and water combined with the various tectonic processes have produced a series of distinctive **fluvial** and arid landscapes on the plateau.

Recent evidence suggests that glaciers covering the Western Shield during ancient ice ages of the Permian period (some 280 million years ago) may have contributed to the general flatness of much of this region. As these massive continental glaciers moved outwards from the South Pole across much of Gondwanaland, they planed down the bedrock to form much of the present surface. It is believed that an ice sheet more than 5 kilometres thick once covered south-western Australia.

The distinctive sub-regions on the plateau will be discussed in more detail in the following sections.

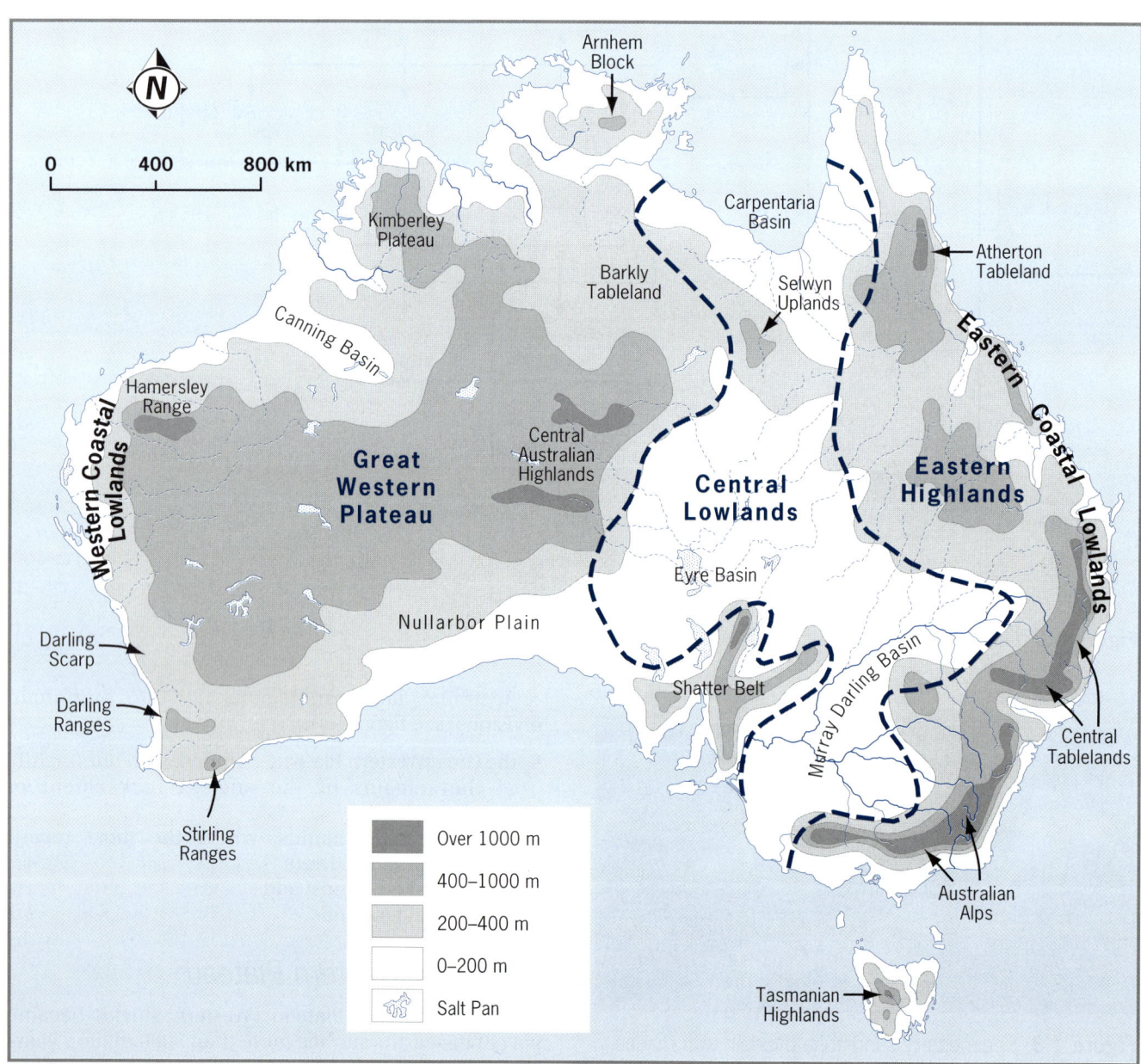

**Figure 1.4** Australia: physiography and drainage

**Figure 1.5** River red gums growing in dry river beds in Central Australia

## Sandy deserts, rocky deserts and salt lakes

Sand deposits, bare rock and gibber (weathered and polished pebbles) cover over half of the shield. Prevailing wind conditions have formed much of the sand into parallel dunes or longitudinal sand ridges that can extend unbroken for hundreds of kilometres. Some of these dunes are very old and are fossil remnants of earlier climatic conditions. They have been stabilised in many cases by a covering of desert scrub.

Where the sand and soil have been removed by wind and water, bare rock or stones are exposed. The stones accumulate as the sand is removed and can sometimes produce a barrier known as deflation armour to prevent further sand removal. The deserts of the Western Shield, which include the Great Sandy, Gibson, Great Victorian, Tanami and Simpson, provide many examples of these gradational processes.

A study of a map of the physical features of Australia will reveal significant areas of salt lakes (playas). They often form the basis of an internal or endoreic drainage system known as a bolson. These lakes will fill after rainfall, and may contain water for up to several years. They are, however, mostly dry. The pattern of these lakes indicates that, in times past, they were once part of major river systems that flowed across the plateau following the retreat of the Permian ice sheets, and when there is exceptional rainfall they can link up to form vast shallow inland seas.

## Ranges of the plateau

Standing out above the surrounding plateau are a number of highland areas dating from earlier geologic times. These include the Hamersley and Chichester ranges in the Pilbara, the Central Australian Highlands, and the Flinders and Mt Lofty ranges in South Australia.

The Hamersley and Chichester ranges, along with the Fortescue River, form a '**horst** and **graben**' (the result of faulting and movement of the crust). The Hamersley Range and the Pilbara Block were raised up, with the valley of the Fortescue slipping downwards to form a rift valley. During their uplift, both the Chichester and the Hamersley ranges were cut through by rivers to form steep-sided gorges. Wittenoom (figure 1.7), Weano, Hancock and Joffre gorges form spectacular canyons and waterfalls where water has cut though the layers of ironstone, **shale** and dolomite. Features such as box canyons, incised meanders and **alluvial** fans can be found throughout this region. The original plateau surface remains as flat-topped **mesas** and **buttes**. The striking bands of red and white iron oxide and **quartzite** sediments making up the Hamersley Range indicate a marine origin for this landform. Layers of iron-rich sediments were laid down in a shallow sea around 2000–2500 million years ago. (The deposit of iron at this time coincides with the formation of oxygen in the earth's atmosphere and the subsequent oxidation of this mineral.)

The Shatter Belt of South Australia is a series of roughly parallel step faults running north–south. These faults have created block mountain ranges and rift valleys. Their formation around 450 million years ago,

**Figure 1.6** Plateau surface, Hyden

# 6 LANDSCAPES AND LAND USES

**Figure 1.7** Wittenoom Gorge

during the Palaeozoic era, was probably associated with the same crustal upheavals which formed much of the Eastern Highlands. The dominant highland areas of the shatter belt are the Flinders and Mt Lofty ranges and, to a lesser extent, the Eyre and York peninsulas. Lake Torrens, and the St Vincent and Spencer gulfs, form rift valleys between these uplifted sections. The ranges are composed of crystalline rocks (granite) and sedimentary rocks (**sandstone**). These sediments were laid down around 500 million years ago, in a region known as the Adelaide Geosyncline, when an ancient sea that occupied much of the Central Lowlands split Australia into a western (Yilgarnia) and an eastern continental landmass.

The Central Australian Highlands are a series of roughly parallel ranges formed mainly from sediments laid down in the Amadeus Basin over 700 million years ago. Around 350 million years ago the sediments and older crystalline rocks were uplifted to produce the folded and faulted features of the Central Australian Highlands. Today, the eroded remnants of this uplift occur as the MacDonnell, Musgrave and Petermann ranges, separated by the Amadeus Trough. The main features of this region are shown in figure 1.9.

The Olgas (Kata Tjuta) and Ayers Rock (Uluru) represent sediments originally laid down as a result of erosion from the Petermann Ranges to the south. These sediments were in turn covered by more recent material when the area became a shallow sea. Uplift caused the layers to be folded and, in the case of Uluru, rotated nearly 90 degrees. As the newer and softer sediments wore away the more resistant older layers were exposed, which then became the Olgas and Uluru. A further significant landform in this region is Mt Conner—a mesa or tabletop mountain that is virtually surrounded by 100 metre cliffs. It is composed of relatively soft sandstone, capped with a resistant siliceous rock. This has allowed it to resist the erosional forces that have reduced the surrounding plain to its present level.

## Elevated plateaus

The Kimberley and Arnhem plateaus occur as elevated and eroded blocks on the northern margins of the continent.

The Kimberley Plateau has a complex geologic history. Over 1900 million years ago, during the Precambrian period, it was part of a continental landmass to the north of Gondwanaland. It separated and drifted south to where it collided with the rest of Northern Australia about 1800 million years ago. This created major upheavals within the earth's crust, causing both internal and external volcanic activity, and also compressed sedimentary material that lay in the ocean between the two landmasses. Subsequent movements caused further volcanic activity in the form of lava flows, as well as folding and faulting the landscape to produce spectacular sandstone ranges such as the King Leopold and Durack ranges. These form a perimeter to the plateau on its southern and eastern margins. Granite and quartzite rocks dominate the Mitchell and Gardner plateaus on the western side, where long fault and joint lines have been eroded into the narrow inlets and gorges that give the Kimberley coastline its unique appearance.

Purnululu (the Bungle Bungles) in the eastern Kimberleys is comprised of soft sandstone, deposited by fluvial action and subsequently raised up during one of the region's many earth movements. Its distinctive

**Figure 1.8** Flinders Range, SA

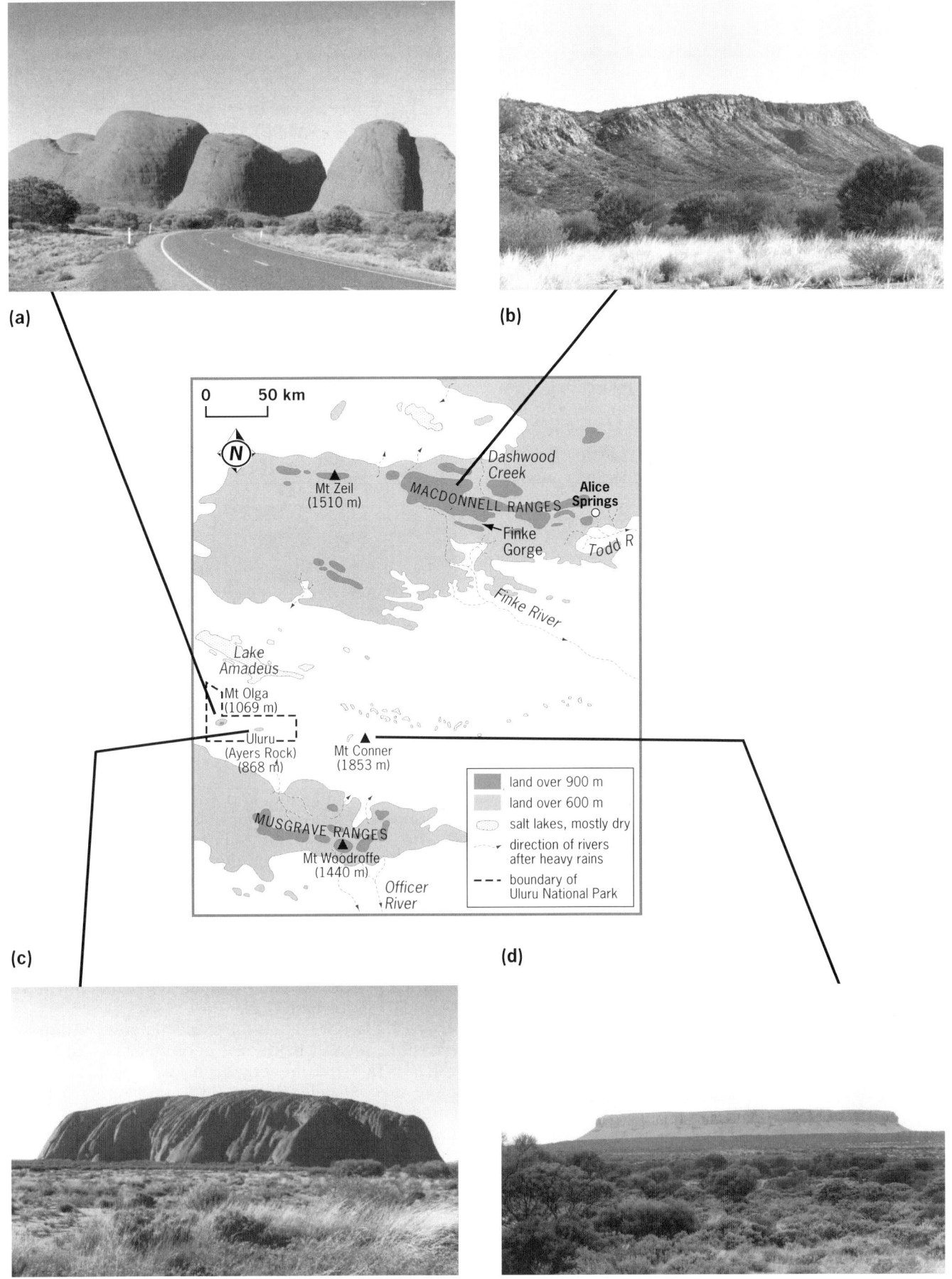

**Figure 1.9** The Central Australian Highlands: (a) the Olgas (b) the Macdonnell Ranges (c) Uluru (d) Mount Conner

banding is due to oxidation and the growth of black algae and is not the result of layers of sediment (figure 1.10).

To the south of the Kimberleys lies the Canning Basin. This region was submerged around 500 million years ago and during the Devonian period a series of barrier reefs began to form in this sea. These were subsequently uplifted and today can be seen as a chain of steep-sided ranges along the southern Kimberley region. Features such as Windjana Gorge, Tunnel Creek and Geikie Gorge can be found along this ancient reef.

The Arnhem Plateau is an uplifted sandstone block that rises up to 300 metres above the surrounding plains. This sandstone is over 1600 million years old and lies on top of a more ancient layer of igneous **dolerite**. It was laid down in distinctive layers (strata) and has been subjected to a long period of erosion that has produced many spectacular features, including the Katherine River Gorges—a series of 13 gorges cut into the edge of the Arnhem Plateau. The coastline of this region is composed of broad alluvial floodplains, built up with material eroded from the plateau. Kakadu National Park, on the eastern border of Arnhem Land, is typical of this type of landscape, with extensive wetlands along the South and East Alligator rivers during the monsoon season.

**Figure 1.10** The Bungle Bungles

**Figure 1.11** Uplifted Devonian limestone barrier reef, southern Kimberleys

**Figure 1.12** Katherine Gorge

## Karst landscapes

The Barkly Tableland and the Nullarbor Plain are featureless limestone plains of marine origin.

The Barkly Tableland is about 150 kilometres south-west of the Gulf of Carpentaria. It is composed predominantly of **dolomite**, a form of limestone deposited some 500 million years ago during the Cambrian period. While its surface is a relatively featureless and treeless plain, extensive caves have formed in the soluble dolomite, connected to the surface via sinkholes. The Selwyn Ranges mark the northern boundary of the tableland.

The Nullarbor Plain is the world's largest continuous limestone area. It is the bed of an ancient sea, where calcium-rich marine organisms were deposited for more than 60 million years. In some places the limestone deposits are up to 250 metres thick. Its uplift probably occurred around 25 million years ago, during the Miocene age. This movement happened without tilting or folding, thus creating the vast uniform plain that it is today. Cliffs ranging in height from 40 to 90 metres mark the southern edge of the plain and extend virtually unbroken for approximately 650 kilometres (see figure 1.13).

While there are no streams to mark its surface, an extensive network of subterranean caverns and streams lies below. It has the world's longest flooded cave system, estimated to extend for up to 450 kilometres through interconnecting chambers.

### The Darling Plateau and Escarpment

The south-western corner of Australia contains the Darling Fault, one of the major tectonic features of the earth's crust. It separates two distinct regions—the Yilgarn Block (core of the Western Shield) to the east and the Perth Basin to the west. The plateau is composed of very ancient crystalline rocks dating back to the Archaean era. Movement along the Darling Fault probably occurred during the Mesozoic era and was associated with the break-up of Gondwanaland. Due to erosion the original fault line lies around 1–3 kilometres west of today's Darling Escarpment and is covered by sediments washed down from the plateau or deposited by marine action.

The escarpment is most distinct between Perth and Bunbury. South of this it disappears as the Yilgarn Block slopes down to eventually form the southern coastline of the region. The edge of the escarpment is deeply dissected by the valleys of streams and rivers running westwards. These have in some cases followed the fault lines associated with its formation. East of this the plateau becomes more undulating and is relatively featureless except for the Stirling Ranges and the Porongerups in the south.

## The Central Lowlands

The Central Lowlands are composed of three distinct regions, stretching from the Gulf of Carpentaria in the north to the Coorong in South Australia. They are bounded by the Great Western Plateau and the Eastern Highlands. The Central Lowlands are not continuous, with the northern Carpentaria Lowlands being separated from the southern section by the Selwyn Ranges. These ranges are relatively low at around 300 metres and link the western plateau to the Eastern Highlands. They also form a watershed between the Carpentaria Basin and the Lake Eyre and Murray–Darling basins, which lie to the south.

The Central Lowlands have been subjected to significant tectonic forces over their geologic history. Recent evidence points to the whole of Eastern Australia being drawn downwards by as much as 350 metres below its present height by an underlying plate, which then allowed the seas to flood low-lying parts of Queensland, New South Wales, Victoria and Central Australia. As the continent drifted north, it gradually escaped from the influence of this plate and surged back up again, shrugging off the shallow seas that had covered it. Layers of sediment laid down by marine processes, as well as fluvial and glacial deposits from the Western Shield, made the land higher still and it remained dry even when the ocean levels rose following the last ice age.

### The Carpentaria Lowlands

The Gulf of Carpentaria is a shallow gulf that makes up most of the Carpentaria Lowlands Basin. A number of short, seasonal rivers flow into it from the eastern side of Arnhem Land around to the western edge of the Cape York Peninsula. These include the Norman, Gilbert, Leichhardt and Roper rivers. The coastline of much of the gulf is bordered by low-lying swampy land that extends up to 30 kilometres inland from the sea. It is an area of extensive saltwater creeks and wetlands produced by a daily rise and fall in the tide of around 6 metres. During the wet season the many rivers that empty into the gulf add to this inundation.

### The Lake Eyre Basin

The Lake Eyre Basin overlies the Great Artesian Aquifer and is crossed by a series of intermittent rivers in what is known as 'the Channel Country'. The waters of rivers such as the Georgina and the Diamantina, and Cooper's Creek, empty into Lake Eyre when they flood (infrequently).

The basin is the world's largest internal drainage system, covering approximately 1.2 million square kilometres of arid and semi-arid Central Australia. This is about one-sixth of the continent and, at 15 metres below sea level, Lake Eyre is Australia's lowest point. The area of the lake is about 10 000 square kilometres, that is, it would occupy a square measuring 100 kilometres by 100 kilometres.

**Figure 1.13** Nullarbor Plain and cliffs

### The Murray–Darling system

The Murray–Darling river system is Australia's biggest, both in length and in the volume of water carried. At just over 1 million square kilometres, its drainage basin comprises about 14 per cent of the continental landmass. It includes the Darling River (2740 kilometres long), the Murray River (2530 kilometres long) and the Murrumbidgee River (1690 kilometres long).

The waters of the Murray–Darling system are drawn from the snow and rain that falls on the Eastern Highlands. This water then runs westwards into the drier interior of the continent. Over most of its journey the river is typical of a lowland tract system with features such as broad floodplains, meanders, oxbow lakes, billabongs and natural levees. Near to its mouth in Lake Alexandrina and the Coorong, the Murray passes through a 200 kilometre gorge that was incised into the landscape as the region was gently uplifted during the Pliocene era.

## The Eastern Highlands

The Eastern Highlands occur as a series of tablelands separated by low passes along the eastern coastline of the continent. Their formation commenced during the Palaeozoic era, around 500 million years ago, with the development of the Tasman Fold (**geosyncline**). This formation was, in some respects, similar to the volcanically active island arcs that make up Indonesia today. This was a period of substantial volcanic activity that lasted until around 200 million years ago. Sandstone and limestone sediments laid down in the seas that covered the region prior to uplift now form a series of broad tablelands along the range.

The highlands were the last of Australia's geologic regions to be tectonically active, with the last significant tectonic activity occurring during the Pliocene era, in a period known as the Kosciuszko Uplift. During this time the sediments were folded and faulted, with the subsequent formation of areas of block mountains and rift valleys. This also produced new volcanic activity, which lasted up until 5000 years ago. This final period occurred around Mount Gambier in South Australia and Victoria and would have been witnessed by Aboriginal people. Indeed, their stone tools have been found buried beneath volcanic ash in this region of Victoria.

Significant deposits of black and brown coal can be found in the sedimentary layers of rock throughout the highlands, from Queensland to Victoria. The older black coal deposits, laid down during the Permian era, occur mainly in Queensland and New South Wales. Brown coal deposits occur in Victoria and are much younger, being formed during the Tertiary era.

The Eastern Highlands may be broken into four subregions: the Atherton Tableland, the Central Tablelands, the Australian Alps and the Tasmanian Highlands.

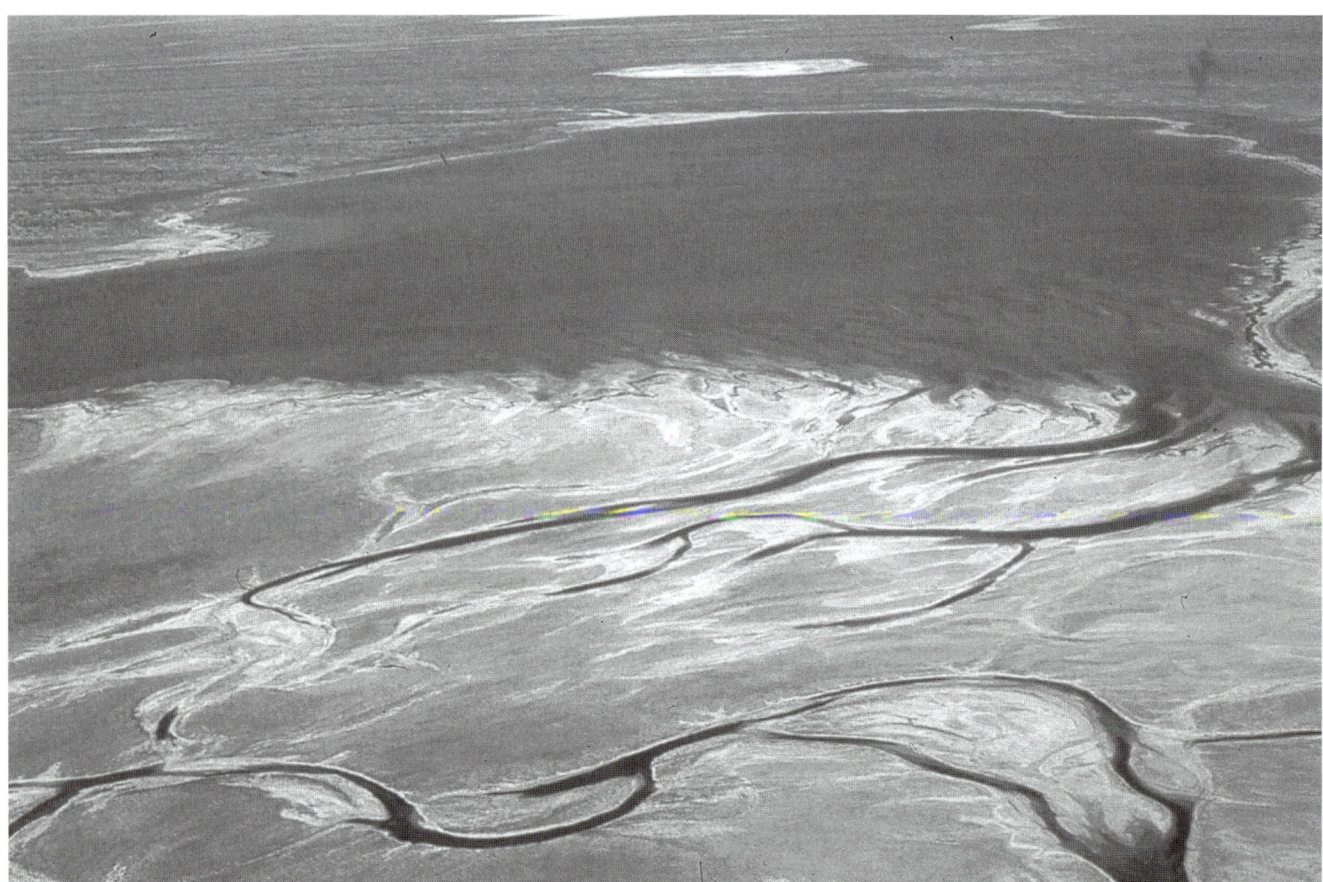

**Figure 1.14** Lake Eyre, SA

**Figure 1.15** The Blue Mountains

### The Atherton Tableland

Found to the west of Cairns, in Far North Queensland, the Atherton Tableland rises steeply from the coastal plain to a height of some 1500 metres with its eastern margin forming the Bellenden Ker Range. The surface of the tableland then slopes gently towards the west. It is covered by rich volcanic soils and contains a number of crater lakes and volcanic cones. The Undara Lava Tubes, on the western edge of the tableland, are unique landforms that were produced by lava flows from the Undara shield volcano around 190 000 years ago.

### The Central Tablelands

South of the Atherton Tableland lie the McPherson Range, the New England Tableland, the Liverpool Range and the Blue Mountains, together forming the Central Tablelands.

This region runs along the New South Wales coastline from southern Queensland to near the Victorian border. While the region is not high by world standards (around 1200 metres) it is geologically complex, with older granites on the western boundary and sedimentary sandstones on the eastern margin. The region generally slopes to the west, with the higher areas being along the eastern edge. The Blue Mountains are distinguished by deep canyons cut into the sandstone rock, ending in sheer cliffs. These proved to be a barrier to the westward movement of European explorers. Between the valleys the tablelands extend outwards like fingers, forming extensive highland ridges.

### The Australian Alps

The Australian Alps straddle the New South Wales–Victorian border and rise to an average height of 1600 metres. This region of the Eastern Highlands was subjected to significant stresses and tensions which generated lengthy faults, along which glacial and fluvial action produced deep, straight valleys. Outpourings of basalt in areas such as the Bogong High Plains in Victoria have helped the region to resist erosion and this in part accounts for its overall height. The Australian Alps contain Mt Kosciuszko which, at 2228 metres, is the highest point on the continent.

### The Tasmanian Highlands

Tasmania and the mainland were one continental landmass more than 12 000 years ago, separated at the end of the last major ice age by rising sea levels. The separation was aided by the frequent storms that occur at these latitudes—strong westerly winds (roaring forties) drive the waters of the Southern Ocean into the gap between Victoria and Tasmania, further eroding Bass Strait.

**Figure 1.16** Barron Falls, Atherton Tableland, QLD

**Figure 1.17** Mount Kosciuszko, NSW

Much of Tasmania is covered by dissected highlands associated with the same processes that produced the rest of the Eastern Highlands. The central and south-west regions of the island contain the highest landforms, including Cradle Mountain (1416 metres), Frenchmans Cap (1443 metres) and Mount Ossa (1617 metres). The highlands exhibit evidence of widespread volcanic and diastrophic activity. This, along with the heavy precipitation experienced over much of the region, has acted to produce a deeply dissected and rugged landscape. The Tasmanian Highlands also contain widespread evidence of glaciated landforms in the form of cirque and moraine lakes, glaciated valleys, arêtes and fiord coastlines.

## The Coastal Lowlands

These lowlands are comprised of a series of disconnected strips of land 0–200 metres in height, fringing the Australian continent. Where they occur they may vary in width up to 200 kilometres, though most are around 20 kilometres wide. In one sense the lowlands are not a separate geologic feature as they mainly reflect the characteristics of the adjacent landform regions. The Kimberley coastline, for example, is mainly made up of rocky headlands and long narrow inlets associated with the geology of this region. What the different coastal areas do share in common, however, is the impact of changing sea levels and the effects of fluvial and marine erosion, weathering and deposition. The coastal regions along the east coast are often narrow and discontinuous, with the areas of lowland separated by hills and ranges associated with the Eastern Highlands. This is illustrated by figure 1.20.

### Drowned coastlines

The continent's coastal regions have advanced and retreated with the changes in sea levels brought on by cooling and warming of the earth's climates. The ice ages associated with periods of falling temperatures produced lower sea levels and exposed areas of the continental shelf. With a warming of the earth's climates, sea levels rose and coastal landform features were drowned beneath the rising waters. Where rivers and glaciers once flowed they were replaced by marine estuaries, **Ria** and **fiord** coasts. Hills and mountains became islands and former coastal lowlands became part of the continental shelf. Other coastal areas which were land depressions formed bays and gulfs or wetlands. These are termed **sunklands**.

All of Australia's major rivers show some form of seawater inundation at their mouths and a number have offshore channels that mark their courses during earlier

**Figure 1.18** Glacial valley, Cradle Mountain

**Figure 1.19** Kimberley coastline in north-western Australia

**Figure 1.20** Coastal region in northern Queensland

times. In south-western Australia, the Swan and Canning estuaries and the Peel Estuary are just some of many examples. Sydney Harbour and Pittwater form part of an extensive system of ancient river valleys now drowned by the ocean (known as a Ria coastline). Long inlets such as Port Macquarie on the south-west coast of Tasmania were once glacial valleys and are now sounds or fiords.

### Depositional and erosional coastlines

Sheltered coastlines and those with shallow offshore areas are often regions where deposition advances the shore seawards. The gulf lands of Carpentaria are an area of substantial fluvial deposition. The silts and muds carried seawards by the rivers are trapped in the mangrove swamps that fringe the shore. Over time, this new land is stabilised and the mangroves are replaced by grasslands and trees. In areas of abundant sand, wave action can construct extensive beaches and spits. The Coorong in South Australia is a region of extensive advance. Evidence of earlier spits similar to the Younghusband Peninsula extend up to 65 kilometres inland as a series of parallel ridges. The combination of fluvial deposits from the Murray River and the movements of sand by longshore drifting have produced this deposition.

Exposed coastlines and those with deep offshore waters are often areas where erosion causes the shore to retreat landwards. The western entrance to Bass Strait funnels the Southern Ocean between the coasts of Victoria and Tasmania. Driven by westerly storms, the heavy seas attack the coastal rocks. The sandstone and limestone rocks along the Port Campbell coastline of Victoria have been eroded to produce spectacular cliffs, arches, stacks and sea caves. Some sections of this coast are eroding at rates of up to 50 centimetres per year. In other areas, where basalt or granite forms the coastline, the erosion can be measured in millimetres per thousand years.

### EXERCISES

1. Using Australian examples where appropriate, define the following terms:
   - tectonic activity
   - weathering and erosion
   - horst and graben
   - internal drainage
   - diastrophism
   - internal and external volcanism
   - drainage basin and watershed
   - igneous rocks
   - mesa and butte
   - drowned coastline
   - sedimentary rocks
   - Gondwanaland
   - continental shield
   - karst landscape
   - fluvial and glacial processes

2. Draw a transect from Perth to Brisbane and on it show the main physiographic regions. Include a brief summary of each.

3. Produce a table in which you find and list examples of Australian tectonic and gradational features. For example, the Hamersley and Chichester ranges are examples of tectonic diastrophism. They were uplifted blocks along fault lines, with the Fortescue River forming in a rift valley between the two.

4. Write a newspaper article of about 300 words entitled 'Australia—an Ancient Land' in which you describe the physiographic characteristics of the continent to an overseas readership. Use Publisher or a similar program to produce the finished product.

5. Produce a series of cards similar to the following examples, to be used in a class quiz.

   | Clues | Answers |
   |---|---|
   | • Contains the continent's oldest rocks<br>• Is also referred to as a shield<br>• Flat or undulating surface<br>• Is the largest physiographic region | • The Great Western Plateau |

6. Find a chart of the geologic time scale and then draw a timeline showing the points at which Australia's various geologic events occurred. Analyse the pattern revealed by the sequence of events and outline how the Australian continent gradually developed over time.

7. Discuss the evidence supporting the statement: 'Australia is an ancient and stable continent'.

## Australia's climatic pattern

The classification of localities into climatic types is an example of the application of formal regional criteria to spatial data. 'Formal' or 'uniform' regions group together areas with similar characteristics. The boundaries for these types of regions are set by the use of categories with pre-determined cut-off points. For example, a common division between arid and semi-arid climates is the 250 millimetre annual **isohyet**.

The greater the number of categories and variables used, the more complex will be the overall pattern of **formal regions**. In categorising climates it is common to use rainfall totals and distribution as well as temperature patterns. However, it would also be acceptable to include additional variables such as cloudiness, humidity and potential evaporation. In this respect it can be argued that there is no correct classification of climates. It is up to the classifier to decide how many regions there will be and what criteria will be used.

The classification system used in this text draws upon different examples, as well as using the Koppen system of classification. The rainfall data used is median rainfall totals rather than average totals, in line with Australian Bureau of Meteorology standards. Figure 1.23 illustrates the seven formal climatic regions (climates) to be discussed: tropical arid, tropical and warm temperate semi-arid, Mediterranean, monsoonal, tropical maritime, temperate maritime and temperate alpine.

### *Tropical arid*

The division between arid and semi-arid regions is often set by the use of the 250 millimetre annual rainfall isohyet. This is, however, a very basic means of

# 14 LANDSCAPES AND LAND USES

**Figure 1.21** The Twelve Apostles

**Figure 1.22** London Bridge

distinguishing between desert and semi-desert environments. Arid or desert regions are best determined by the following criteria:

- Actual and potential evapo-transpiration rates significantly exceed annual precipitation even during high rainfall years.
- Precipitation is irregular and unpredictable with extensive periods of drought.
- Low humidity and an absence of clouds contribute to high levels of solar energy reaching the ground during the day and a relatively rapid loss of heat during the night.
- Diurnal and seasonal temperature ranges tend to be greater than in other climatic regions.
- Tropical arid regions are dominated by high pressure systems throughout the year.

This climatic type occupies around one-third of Australia in a broad area extending from the west coast to the Eastern Highlands. The northern region occurs within the tropics and locations such as Marble Bar experience maximum temperature in excess of 40°C during summer. In the regions south of the Tropic of Capricorn, temperatures are generally lower and locations such as Coober Pedy and Kalgoorlie have average summer temperatures of 25–28°C. Where locations occur near the ocean, such as at Shark Bay, the movement of air from the sea moderates temperatures. A further variation in this climate may be seen in the Central Highlands region, where locations such as Uluru and Alice Springs have slightly cooler summer and winter temperatures and a higher rainfall compared to surrounding desert locations.

## Tropical and warm temperate semi-arid

Tropical semi-arid and warm temperate semi-arid climates occur as transitional zones between the arid and higher rainfall climates of Australia. They are sometimes identified as climatic regions falling between 250 and 500 millimetre average annual rainfall. Due to the transitional nature of the climate these regions share the characteristics of both adjoining climatic zones. The tropical semi-arid zone receives its rainfall in summer, while the warm temperate zone receives its rainfall either in winter or more evenly distributed over the year, but still with a winter maximum. Due to the variation in latitudinal positions the warm temperate zone is cooler in both summer and winter than the northern tropical zone. Both, however, share the following characteristics:

- a significant seasonal range in temperature due to their predominantly continental position;
- high levels of evapo-transpiration, particularly during the summer season;
- a seasonal rainfall pattern which is generally light, but highly variable;
- mainly influenced by the subtropical high pressure belt, but more likely to experience the effects of low pressure systems than in the arid climate. This is due to the seasonal shifts of the global pressure belts. It should be noted that the eastern sections of this climate are influenced by the rain shadow effect of the Eastern Highlands.

Like the arid zone, this climate occurs over about one-third of the Australian continent. It forms a semi-circle encompassing the arid region and passing through large areas of inland Australia. As indicated on the climate map, figure 1.23, it reaches the coastline in several areas.

## Mediterranean

The lower limit of this climate is generally defined by the 400–500 millimetre rainfall isohyet. Its winter rainfall and summer drought pattern set it apart from a number of other climates and on a world scale it occupies a relatively small area of the globe. The characteristics of the Mediterranean climate include the following:

- warm average summer temperatures and mild winter temperatures, with a moderate diurnal and seasonal range due to the predominantly maritime location;
- frontal rainfall associated with the northward movement of the sub-polar low pressure belt in winter;
- an extended period of drought from November to April (over summer) due to the southward movement and dominance of the subtropical high pressure belt;
- the formation of a low pressure trough associated with the northern heat lows in summer can bring heatwave conditions when it is offshore and a cool change when it crosses the coast in south-western Western Australia;
- a relatively predictable and reliable rainfall pattern (compared to other climates), and its occurrence during winter when evaporation rates are low enhances its effectiveness.

The Mediterranean climate occupies less than 5 per cent of the continent, and is found along the south-

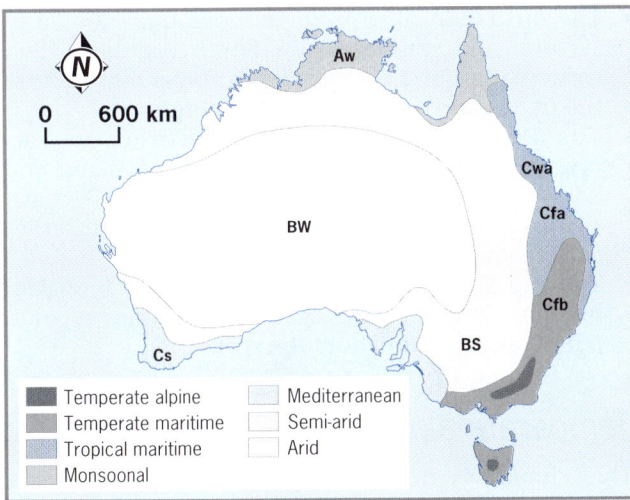

**Figure 1.23** Climatic regions of Australia

west coastline of Western Australia and around the gulf lands of South Australia. Rainfall is generally higher in the South-West compared to South Australia. Perth has a median rainfall of approximately 850 millimetres compared to Adelaide's total of 460 millimetres.

## Monsoonal

This climate is variously called 'tropical savanna', 'tropical wet-dry', 'tropical wet summer' and 'tropical monsoonal'. Rainfall totals in this zone can range from 500 to 1500 millimetres. While the rainfall pattern is predominantly one of summer rainfall, the northern margins of the climate exhibit more of a year-round pattern typical of an equatorial climate (some climate classifications identify the northern tips of Arnhem Land and the Cape York Peninsula as equatorial rather than savanna). This climatic zone occupies around 8–10 per cent of Australia. In addition to the characteristics already mentioned, the monsoonal climate also has the following features:

- Temperatures are relatively high throughout the year due to the region's proximity to the equator.
- The low latitude of the climate results in small diurnal and seasonal variations in temperature, with a location such as Darwin having a yearly range of 6°C.
- The seasonal reversal of the trade winds is associated with shift in the **thermal equator** and the equatorial low pressure belt, which brings onshore north-west **monsoons** in summer, and offshore south-east monsoons in winter.
- The summer wet season is a time of high humidity and heavy convectional rainfall, usually occurring in late afternoon. Monsoonal climate regions also experience a high level of thunderstorms (about 80 days in Darwin compared to 10 days in Perth and 20 days in Sydney).
- Along with the tropical maritime climate, the monsoonal regions are subject to the intense tropical low pressure cells (tropical cyclones) that develop over the Timor and Coral seas in summer. The cyclone season, which runs from November to March, has on average five cyclones per season.

## Tropical maritime

The tropical maritime climate also covers regions that are outside the tropics and it is sometimes referred to as 'humid subtropical' or 'subtropical wet summer'. This climate occurs in a narrow coastal strip from Cairns in North Queensland through to northern New South Wales. It is a climate in which precipitation occurs in all seasons and in which some of the highest rainfall areas in Australia are found (Babinda in North Queensland receives over 4.5 metres of rain per year on average). The tropical maritime climate is distinguished by the following characteristics:

- The summers are warm to hot and wet and the winters are mild to warm and wet, depending on the latitudinal position. The common element is the higher summer rainfall.
- Being found on the eastern slopes of the Eastern Highlands and along the eastern coastline it experiences a strong maritime effect, which acts to moderate temperature ranges.
- The Eastern Highlands create a significant **orographic** uplift, with enhanced precipitation along the windward slopes. Rainfall totals decline rapidly to the west pointing to the **rain-shadow** effect experienced on the western or leeward slopes.
- The northern region of the climate occurs within the zone of tropical cyclones, which will affect precipitation patterns during the months of November to March.
- Prevailing winds are onshore south-easterly trades associated with the movement of air from the subtropical highs to the region of equatorial lows. This onshore movement produces precipitation throughout the year. It should be noted that on the west coast of Australia these winds blow offshore, and therefore produce arid conditions.

## Temperate maritime

The temperate maritime climate is also referred to as 'marine west coast' or 'cool temperate marine'. This climate incorporates the southern highland regions and coastline of central New South Wales and extends southwards into Victoria and Tasmania. Like the tropical maritime climate, this zone experiences year-round rainfall. It is, however, distinguished from the former by the lower temperatures in both summer and winter and the greater temperature range caused by the more pronounced seasonal variations that occur at these higher latitudes. The following features characterise this climatic region:

- The frontal systems associated with the sub-polar low pressures and the mid-latitude westerly winds are important in bringing winter rainfall to the northern limits of the region and year-round rainfall in the southern regions of Victoria and Tasmania.
- The onshore influence of the south-east trades during summer results in rainfall in the northern section during this season.
- The Australian Alps produce orographic rainfall in Victoria and southern New South Wales on their western margins in winter, and summer rainfall from the trade winds on their eastern margins.
- The Tasmanian Highlands produce heavy year-round rainfall on the west coast of the island and have a rain-shadow effect on the eastern coast. Tasmania is dominated by the prevailing westerlies in both summer and winter.
- The mid-latitude position of this climate, along with its maritime location, generally produces mild temperatures throughout the year.

## Temperate alpine

This climate is not recognised in some references. However, the occurrence of precipitation in the form of snow during the winter season, and lower temperatures compared to the surrounding maritime climate, provide enough variation to allow for its classification as a separate climatic region. Temperate alpine regions occur in the Australian Alps and in the Central Highlands of Tasmania and occupy less than 1 per cent of Australia. The following characteristics help distinguish them from the temperate maritime climate:

- Altitude is an important influence on this climate, with below zero degree temperatures in winter allowing for seasonal snow cover.
- Annual precipitation is quite high, with a number of localities receiving in excess of 1500 millimetres in a year.
- Summers are mild to warm and any snow that does fall in this season does not remain long.
- As with all mountain climates, conditions may change in a very short period of time. They are often described as having four seasons in one day.

## Climatic controls

Climatic controls help to explain the characteristics and patterns of climatic regions. They are the causes of the climate, while the patterns of rainfall and temperature are the effects. Previous discussions of Australia's climates have identified some of these controlling influences and these discussions will be expanded upon under the following four headings.

### Shape and size

Australia's area is 7.7 million square kilometres and this size and its regular, almost rectangular, shape give rise to distinctive maritime and continental influences. Increasing distance from the sea produces a general decrease in the amount of precipitation and an increase in the extremes of temperature. This is not, however, as significant as the larger continental landmasses of the Northern Hemisphere, and the temperature ranges are therefore much smaller. The climatic difference between Perth and Giles illustrates this variation between coastal and inland locations (see table 1.2).

### Latitude

Australia has a generally low latitude position, with the continent being roughly divided by the Tropic of Capricorn. This places much of the continent within the tropical and subtropical regions of the Southern Hemisphere. Only Tasmania, at a latitude of around 40°S, falls well into the mid-latitude temperate regions.

The effect of this low latitude position is to produce limited seasonal variation in sunlight hours and to put much of the continent under the influence of a high angle of solar incidence, with its associated warming, throughout the year. The relatively small variation in temperature between the tropical and subtropical regions of Australia can be seen in the average annual temperatures for Cape Don and Bega (see table 1.2).

### Landforms

Landforms can shape the surface movements of air and the patterns of temperatures at varying heights. The presence of mountains can act to deflect air movements and thus cause air masses to affect different regions. Mountains can also create uplift and cooling of **air masses**, with a resulting decrease in temperatures and an increase in precipitation in processes known as adiabatic cooling and orographic uplift. Higher altitude positions are associated with lower atmospheric pressures and a reduced greenhouse effect, producing rapid heat loss at night. Even small variations in heights above sea level can bring about localised differences in both rainfall and temperature. Note the rainfall and temperature patterns for Thredbo in table 1.2.

### Air masses and pressure systems

The angle of inclination of the earth's axis produces a seasonal shift in the sun's energy between the Northern and Southern hemispheres. There is a corresponding movement of the thermal equator and the associated global pressure systems. The north–south shift in the equatorial low pressure, the subtropical high pressure and the sub-polar low pressure systems produces marked seasonal variations in precipitation between northern and southern Australia. There is also an associated change, from one season to the next, in the air masses related to these pressure systems.

During summer, the subtropical high pressure systems are positioned over the southern half of the continent, bringing fine conditions to the majority of the region. Tropical continental air masses associated with offshore winds produce hot and dry conditions over much of southern and western Australia. The northern regions of the continent are dominated by the presence of heat lows associated with the southward shift of the **intertropical convergence zone** and the thermal equator. Here, moist tropical maritime air masses are moved over northern Australia by the north-west monsoons. The only part of Australia that is influenced by the sub-polar lows in this season is Tasmania, which is affected by the westerly air stream all year round.

In winter, the subtropical highs are located about 15° (latitude) further north and bring fine, warm to hot conditions to the tropical north. Offshore, south-east monsoons direct tropical continental air masses over this region. In the southern areas the sub-polar low pressures replace the high pressure system. The associated cold fronts bring a convergence of tropical and polar maritime air masses, producing winter rainfall along the south-western and south-eastern regions of the continent. Figure 1.25 illustrates the seasonal shift in pressure systems as well as prevailing winds and air masses.

**Table 1.2** Climatic data for selected stations

| Giles, WA: Desert | Jan | Feb | Mar | Apr | May | Jun | Jul | Aug | Sep | Oct | Nov | Dec | Annual |
|---|---|---|---|---|---|---|---|---|---|---|---|---|---|
| Mean daily maximum temperature | 37.2 | 35.9 | 33.8 | 29.2 | 23.5 | 20.3 | 19.8 | 22.4 | 27.2 | 31.6 | 34.3 | 35.9 | 29.3 |
| Mean daily minimum temperature | 23.5 | 22.8 | 20.6 | 16.3 | 11.3 | 8.3 | 6.8 | 8.6 | 12.8 | 17 | 19.8 | 21.8 | 15.9 |
| Median (5th decile) monthly rainfall | 20.2 | 20.6 | 14.2 | 5.8 | 6.2 | 7.6 | 2.7 | 1.6 | 2.3 | 4.8 | 12.4 | 28.2 | 248 |
| 9th decile of monthly rainfall (mm) | 69.1 | 165.4 | 118.9 | 47.1 | 69.8 | 48.8 | 39.8 | 37.5 | 31.4 | 43.9 | 68.3 | 76.2 | 480.1 |
| 1st decile of monthly rainfall (mm) | 1.4 | 0.5 | 0 | 0 | 0 | 0 | 0 | 0 | 0 | 0 | 0.4 | 4.7 | 126.4 |
| Mean number of rain days | 5.5 | 5.3 | 4.1 | 3 | 3.9 | 3.4 | 2.2 | 2.4 | 2.7 | 3.3 | 4.8 | 6.4 | 47.1 |
| **Fitzroy Crossing, WA: Semi-arid** | Jan | Feb | Mar | Apr | May | Jun | Jul | Aug | Sep | Oct | Nov | Dec | Annual |
| Mean daily maximum temperature | 37.7 | 36.4 | 36.8 | 36.1 | 32.1 | 29.9 | 29.6 | 32.5 | 36.2 | 39.4 | 40.5 | 40 | 35.6 |
| Mean daily minimum temperature | 24.7 | 24.1 | 23.1 | 19.6 | 15.7 | 12.4 | 10.7 | 12.7 | 17 | 21.4 | 24.2 | 25.1 | 19.1 |
| Median (5th decile) monthly rainfall | 127 | 114.9 | 54.9 | 4.2 | 0.3 | 0 | 0 | 0 | 0 | 1.4 | 16 | 65.8 | 533.4 |
| 9th decile of monthly rainfall (mm) | 293.8 | 288.2 | 204 | 63.2 | 32.1 | 26.8 | 21.1 | 2.2 | 4.1 | 13.7 | 60.1 | 184.7 | 784.8 |
| 1st decile of monthly rainfall (mm) | 46.1 | 27.5 | 8.2 | 0 | 0 | 0 | 0 | 0 | 0 | 0 | 1.2 | 13.5 | 294.9 |
| Mean number of rain days | 12.4 | 10.7 | 7.2 | 2 | 1.5 | 1 | 0.7 | 0.3 | 0.4 | 1.3 | 4 | 8.3 | 49.8 |
| **Cape Don, NT: Monsoonal** | Jan | Feb | Mar | Apr | May | Jun | Jul | Aug | Sep | Oct | Nov | Dec | Annual |
| Mean daily maximum temperature | 31.3 | 30.9 | 31.1 | 31.4 | 30.6 | 28.8 | 28.2 | 29 | 30.4 | 31.7 | 32.5 | 32.2 | 30.7 |
| Mean daily minimum temperature | 25.3 | 25.2 | 25 | 25 | 24.1 | 22.5 | 21.6 | 22 | 23.2 | 24.8 | 25.9 | 25.9 | 24.2 |
| Median (5th decile) monthly rainfall | 264.5 | 239.3 | 233.5 | 85.6 | 7.3 | 0 | 0 | 0 | 0.2 | 15.5 | 86.2 | 191.1 | 1284.9 |
| 9th decile of monthly rainfall (mm) | 486.5 | 356.8 | 493.2 | 351.9 | 59.3 | 12.7 | 7.5 | 3.5 | 10.3 | 65.5 | 216.9 | 431 | 1701.6 |
| 1st decile of monthly rainfall (mm) | 115.4 | 123.8 | 121.8 | 28.6 | 0 | 0 | 0 | 0 | 0 | 0 | 18.5 | 78.7 | 1028.5 |
| Mean number of rain days | 17.9 | 17.5 | 18.1 | 10.1 | 3.3 | 0.8 | 0.7 | 0.4 | 0.9 | 3.1 | 8.4 | 14.2 | 95.4 |
| **Swan Hill, VIC: Semi-arid** | Jan | Feb | Mar | Apr | May | Jun | Jul | Aug | Sep | Oct | Nov | Dec | Annual |
| Mean daily maximum temperature | 31.4 | 31.2 | 27.9 | 22.8 | 18.4 | 14.9 | 14.4 | 16.3 | 19.3 | 22.9 | 26.8 | 29.8 | 23 |
| Mean daily minimum temperature | 15.4 | 15.5 | 13.2 | 9.8 | 7 | 4.9 | 4.1 | 5 | 6.7 | 9 | 11.6 | 13.9 | 9.7 |
| Median (5th decile) monthly rainfall | 13.4 | 11.5 | 16.5 | 17.9 | 29.6 | 31.4 | 29.8 | 34.6 | 27.8 | 26.6 | 21.8 | 16.8 | 347.3 |
| 9th decile of monthly rainfall (mm) | 60.4 | 69.6 | 58.6 | 57.3 | 70.9 | 67.4 | 58.4 | 65.6 | 62.5 | 81.1 | 58.7 | 57.8 | 473.1 |
| 1st decile of monthly rainfall (mm) | 0.8 | 0 | 0.2 | 2 | 4.3 | 9.1 | 9.4 | 9.6 | 8.2 | 6.5 | 2.4 | 1.4 | 196.1 |
| Mean number of rain days | 3 | 2.9 | 3.3 | 4.5 | 6.8 | 8.4 | 9.2 | 9.5 | 7.7 | 6.8 | 4.9 | 4 | 71.2 |
| **Bega, NSW: Temp. maritime** | Jan | Feb | Mar | Apr | May | Jun | Jul | Aug | Sep | Oct | Nov | Dec | Annual |
| Mean daily maximum temperature | 27 | 27 | 25.7 | 23 | 19.7 | 16.8 | 16.7 | 18.2 | 20.5 | 22.5 | 24 | 25.9 | 22.2 |
| Mean daily minimum temperature | 14.2 | 14.5 | 12.6 | 8.9 | 5.5 | 2.9 | 1.4 | 2.6 | 5.1 | 8.2 | 10.7 | 12.9 | 8.3 |
| Median (5th decile) monthly rainfall | 60.5 | 49 | 59.1 | 39.5 | 38 | 45.8 | 27.2 | 29.8 | 33.2 | 52.3 | 53.3 | 54.4 | 809.8 |
| 9th decile of monthly rainfall (mm) | 183.7 | 199.7 | 264.3 | 180.2 | 209.1 | 202.9 | 144.9 | 135.2 | 130 | 161.9 | 146.2 | 181.4 | 1342.2 |
| 1st decile of monthly rainfall (mm) | 16.6 | 9.4 | 8.2 | 6.2 | 4.8 | 6.6 | 1.5 | 2.8 | 10 | 16.9 | 9.7 | 14 | 516.7 |
| Mean number of rain days | 8 | 7.6 | 8.2 | 6.8 | 7.1 | 7 | 5.7 | 6.5 | 7.3 | 8.8 | 8.6 | 8.4 | 89.9 |
| **Thredbo, NSW: Temp. alpine** | Jan | Feb | Mar | Apr | May | Jun | Jul | Aug | Sep | Oct | Nov | Dec | Annual |
| Mean daily maximum temperature | 20.8 | 20.6 | 18 | 13.7 | 10 | 6.3 | 5.1 | 6.3 | 9.6 | 12.9 | 15.8 | 18.9 | 14 |
| Mean daily minimum temperature | 6.8 | 6.7 | 4.4 | 1.7 | −0.5 | −2.8 | −3.9 | −2.3 | −0.7 | 1.3 | 3.3 | 5.1 | 2 |
| Median (5th decile) monthly rainfall | 100.9 | 80.2 | 116.8 | 98.1 | 177.6 | 148.1 | 131.2 | 162.6 | 212.7 | 182.8 | 145.7 | 126.7 | 1863.8 |
| 9th decile of monthly rainfall (mm) | 217.8 | 144.1 | 202.9 | 230.2 | 298.4 | 263.1 | 322.6 | 304.8 | 353.1 | 380.8 | 260.6 | 183.9 | 2597.6 |
| 1st decile of monthly rainfall (mm) | 40.8 | 15.3 | 28.9 | 33.6 | 65.2 | 63.3 | 41 | 73.6 | 75.9 | 55.4 | 76.7 | 50 | 1247.7 |
| Mean number of rain days | 11.1 | 10 | 11.3 | 12.3 | 15 | 15.6 | 16.3 | 17.1 | 17.8 | 15.7 | 14.9 | 11.6 | 168.7 |
| **Innisfail, QLD: Tropical maritime** | Jan | Feb | Mar | Apr | May | Jun | Jul | Aug | Sep | Oct | Nov | Dec | Annual |
| Mean daily maximum temperature | 30.8 | 30.5 | 29.8 | 28.2 | 26.4 | 24.5 | 24.1 | 25.1 | 26.6 | 28.3 | 29.7 | 30.7 | 27.9 |
| Mean daily minimum temperature | 22.7 | 22.7 | 22 | 20.3 | 18.3 | 16 | 15.1 | 15.3 | 16.8 | 18.8 | 20.6 | 21.9 | 19.2 |
| Median (5th decile) monthly rainfall | 463.4 | 509.8 | 624.2 | 399.4 | 263 | 161.5 | 112.8 | 98 | 65.5 | 61.2 | 111 | 186.6 | 3537 |
| 9th decile of monthly rainfall (mm) | 1013 | 1071 | 1148 | 812.9 | 563.2 | 390.1 | 230.9 | 240.8 | 220.6 | 180.2 | 370.4 | 580.4 | 4783.1 |
| 1st decile of monthly rainfall (mm) | 132.4 | 266.4 | 211.4 | 153.5 | 93.3 | 41.5 | 36.4 | 20.4 | 3 | 5.6 | 16.2 | 38.8 | 2579.4 |
| Mean number of rain days | 16.4 | 17 | 19.4 | 18.8 | 16.9 | 12.7 | 11.7 | 10.4 | 8.6 | 7.9 | 9.7 | 12.1 | 161.6 |
| **Perth, WA: Mediterranean** | Jan | Feb | Mar | Apr | May | Jun | Jul | Aug | Sep | Oct | Nov | Dec | Annual |
| Mean daily maximum temperature | 29.7 | 30 | 28 | 24.6 | 20.9 | 18.3 | 17.4 | 18 | 19.5 | 21.4 | 24.6 | 27.4 | 23.3 |
| Mean daily minimum temperature | 17.9 | 18.1 | 16.8 | 14.3 | 11.7 | 10.1 | 9 | 9.2 | 10.3 | 11.7 | 14 | 16.3 | 13.3 |
| Median (5th decile) monthly rainfall | 4.1 | 5.4 | 11.4 | 38.6 | 122.1 | 177.5 | 168.4 | 135.7 | 70.2 | 49.3 | 17.3 | 9.6 | 854.2 |
| 9th decile of monthly rainfall (mm) | 21 | 33.3 | 48.3 | 89 | 200.8 | 285.4 | 263.2 | 216.1 | 134.4 | 94.4 | 44.4 | 31.6 | 1051.1 |
| 1st decile of monthly rainfall (mm) | 0.5 | 0 | 0.9 | 7.2 | 45.9 | 97.5 | 94.6 | 66.6 | 30.6 | 23.7 | 3.6 | 0.9 | 666.4 |
| Mean number of rain days | 2.9 | 2.7 | 4.3 | 7.6 | 13.8 | 17.2 | 18.2 | 17.2 | 14 | 11.1 | 6.5 | 4.2 | 119.6 |

THE BIOPHYSICAL ENVIRONMENT   19

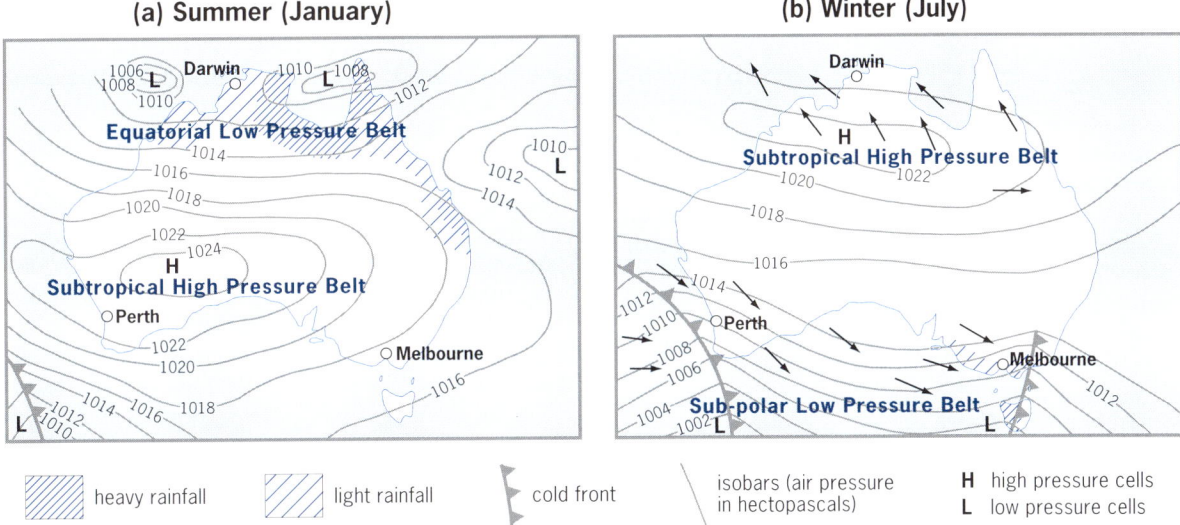

Figure 1.24 Common summer and winter weather patterns, Australia

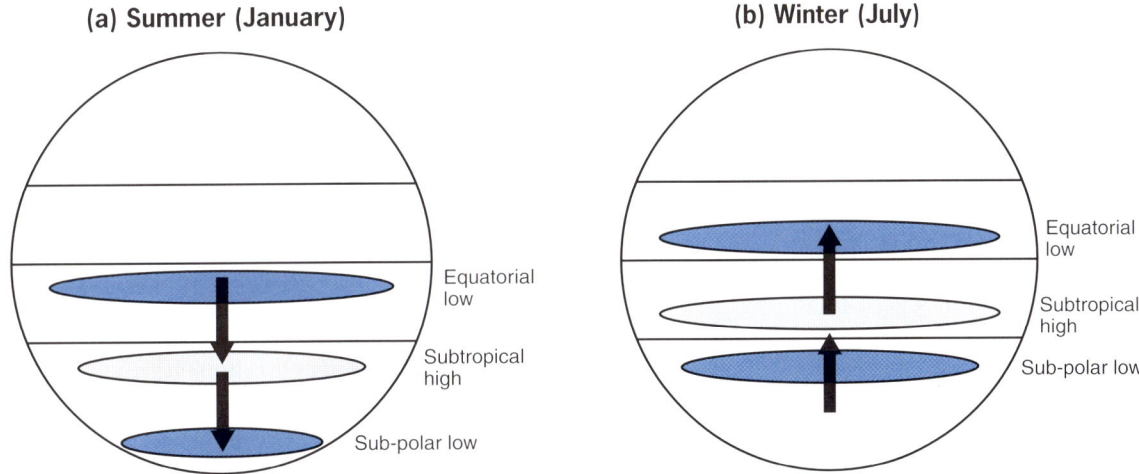

Figure 1.25 Summer and winter movements of the pressure systems in the Southern Hemisphere

### EXERCISES

1. Using Australian examples where appropriate, define or explain the following terms:

   - median and mean average rainfall
   - thermal equator and intertropical convergence zone
   - pressure systems
   - maritime and continental
   - arid and semi-arid
   - tropical cyclones
   - isohyet, **isotherm** and **isobar**
   - diurnal and seasonal temperature range
   - orographic rainfall and rain-shadow effect
   - prevailing winds
   - air masses
   - angle of incidence
   - monsoons
   - tropical and subtropical
   - humidity and evapo-transpiration

2. For each climatic zone prepare a summary using the following subheadings:

   Size and location
   Annual and seasonal temperature and rainfall patterns
   Climatic controls

3. Construct climatic graphs for each of the locations provided in table 1.2 and then compare and contrast each one using the following criteria:

   - average annual temperatures
   - seasonal and diurnal temperature ranges
   - total rainfall
   - seasonal rainfall patterns
   - rainfall intensity (monthly rainfall divided by number of rainy days)
   - rainfall variability (compare the first decile to the ninth decile as well as the median to determine the year-to-year differences).

4. Choose two climatic stations from table 1.2 and then explain how they are influenced by:

   (a) landforms
   (b) distance from or closeness to the sea
   (c) seasonal shifts in pressure systems and changes in prevailing winds.

# Vegetation patterns in Australia

Plants and animals form communities called **biomes**. Where groups of plants combine to form distinctive associations, these vegetation types are referred to as plant formations. There is a number of major **vegetation formations** in Australia and they can be grouped together in a variety of ways. The grouping in this text is only one way and other references will have their own methods of classifying vegetation. Distinctive vegetation formations evolve out of their response to a variety of natural factors that determine the relative success of individual plants within a community. The degree of success will then determine the extent to which the formation will spread. These natural influences do not always strictly limit a plant to a particular formation and many plants exhibit the ability to survive within a variety of conditions and may therefore be found within different vegetation formations. Hence it is possible to see the combination of plants from two or more distinctive formations over a wide zone of transition. Where the northern dry sclerophyllous forest of south-western Australia gives way to the wandoo woodlands there is a broad band of mixed vegetation that can extend up to 40 kilometres in width.

The classification of vegetation groups may be based on species, ground coverage, plant density, height and canopy. Other classification systems use climatic characteristics to identify different vegetation regions. The system used in this text to determine the vegetation formations of Australia is one of appearance and response to climate, soils and topography.

## Closed forests

A dense interlocking canopy and a deeply shaded environment on the forest floor characterise closed forests (rainforests). It is the most productive and diverse vegetation system in terms of its biomass and variety of species. The competition for light and space produces both distinctive horizontal and vertical characteristics in its various plant species.

### Tropical rainforests

The tropical rainforests of north-eastern Australia contain over 300 species of trees, epiphytes and climbers. Trees such as the hoop pine, cedar, kauri and Queensland maple may be found here. These trees in turn may be completely covered in vines, **epiphytes**, tropical orchids, climbing palms, mosses, fungi and lichens. The ground in the rainforest is relatively clear due to the low light intensity. However, shade-tolerant ferns, broad-leafed plants and fungi can be found here. The buttress roots of the tall trees dominate the view at this level, with their shallow root system extending outwards for as long as the tree is tall to gain the essential nutrients from the mat of decaying vegetative matter that covers the ground (see figure 1.28).

The best known of the tropical rainforests is the Daintree, north of Cooktown in Far North Queensland. This occupies an area of some 100 000 hectares. It forms part of the Wet Tropics world heritage site, an area of approximately 900 000 hectares representing the last significant area of lowland tropical rainforest in Australia

### Temperate rainforests

While rainforests tend to be associated with tropical regions, they also exist in temperate rainy areas. Here they lack some of the large, shade-loving plants of their tropical counterparts. However, they still contain a wide variety of plant species. Beneath their canopy, mosses are more dominant and form a thick carpet of green over branches and fallen logs. This forest type can be found in the Otway Ranges in southern Victoria, and in the south-west of Tasmania away from the coast. Here, trees such as myrtle, sassafras, leatherwood, blackwood

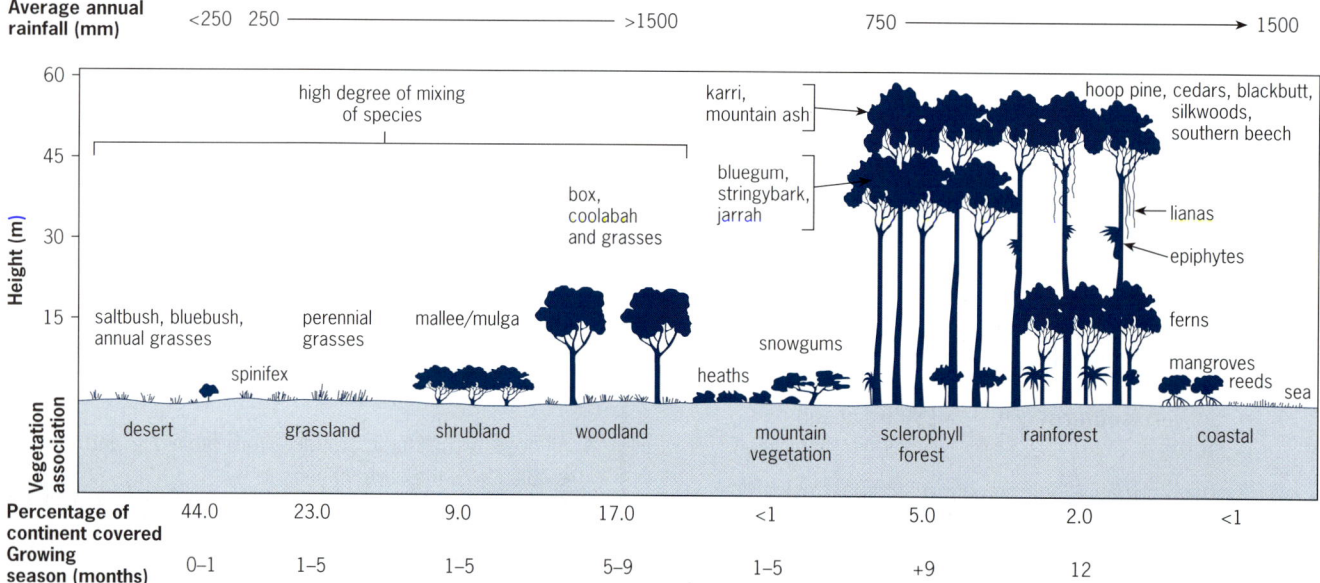

**Figure 1.26** The pattern of vegetation types in Australia

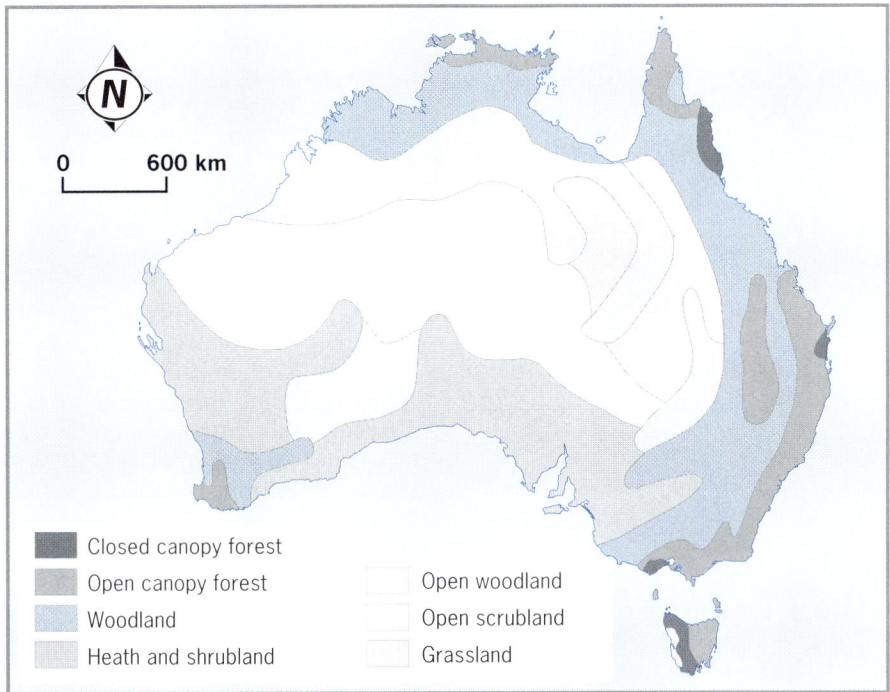

**Figure 1.27** Vegetation formations of Australia

and celery-top pine form the canopy with smaller epiphytes clinging to their branches and trunks.

Rainforests are one of the few plant communities not dominated by eucalypts. They depend on a continual supply of moisture, an absence of forest fires and an uninterrupted nutrient cycle. Should any of these conditions alter, they will quickly degenerate.

Australia originally had about 0.9 per cent of its landmass covered in rainforest. It tended to occur in discontinuous stands from Cape York Peninsula to

**Figure 1.28** Tree with buttress roots

**Figure 1.29** Tropical rainforest

# 22 LANDSCAPES AND LAND USES

**Figure 1.30** Temperate rainforest

Tasmania, running along the slopes of the Eastern Highlands and seldom extending more than 100 kilometres from the coastline. In many places it was limited to areas as small as 1 hectare. Today, rainforests occupy approximately 0.4 per cent of the continent. There is some evidence to suggest that the rainforests are remnants of earlier climatic periods and may once have been more widespread, when Australia was joined to the other continents of Gondwanaland.

## Open forests

Open forests are common throughout the humid regions of Australia, occurring mainly within 200 kilometres of the coast. They are lighter than the closed forests, because the foliage of the dominant trees is not as dense and the trees are more widely spaced. In Australia, open forests once covered about 5.2 per cent of the continent.

Most open forests are dominated by the eucalypts. Eucalypt open forests can be subdivided into two types according to their moisture requirements. Those found in the wetter areas form the wet sclerophyllous forests, while those in the drier areas are the dry sclerophyllous forests.

### Wet sclerophyllous forests

These are found throughout the humid regions of New South Wales, Victoria and south-east Queensland. They also occur in smaller areas in the extreme south-west of Western Australia and in parts of Tasmania. They are found on fertile soils in good rainfall areas, and in some locations may be considered to be a transition vegetation between the rainforest and the drier open forest types.

The wet sclerophyllous forest tends to be dominated by very tall tree species such as mountain ash, Tasmanian blue gum, blackbutt and karri. Like the rainforest, this formation also exhibits distinctive levels or strata. The understorey has a variety of medium sized trees that include wattles, she-oaks, peppermints and **banksias**. The ground cover may include a variety of ferns, including the bracken fern. This forest type has adapted to the impact of fire and requires intense fires at around 25 year intervals. If fires are more frequent, the species composition of the formation will change. Tree regeneration after fire is by means of **epicormic** shoots—a feature of all eucalypts.

### Dry sclerophyllous forests

This forest type is widespread throughout the humid margins of New South Wales, Victoria, South Australia, Western Australia and Tasmania. The 500–800 millimetre isohyets generally mark the upper and lower limits of the formation.

The drier open forests have adapted to cope with a lower nutrient supply than the wet sclerophyllous

**Figure 1.31** Tingle trees, unique to the wet sclerophyllous forests of the South-West

forests. Their soils are generally poorer, with lower moisture retaining capabilities. There is often a distinctive dry season, which means that the plants develop a variety of **xerophytic** characteristics. The lower and more seasonal rainfall pattern also results in a higher frequency of fire, with the plant species adapting to this.

As a consequence of these factors the forest tends to have shorter tree species, with a woody, hard-leafed understorey and a variety of nitrogen fixing plants. Dominant tree species include spotted and stringybark gums, peppermints, jarrah, marri, yate and tuart. The understorey includes grasstrees, banksias, wattles, she-oaks, hakeas and grevilleas. Grasses such as kangaroo grass may be found in some regions, while they are virtually non-existent in other areas. Ground cover in Western Australian dry forests is often in the form of spring flowering orchids and wildflowers.

## Woodlands

Woodlands have a park-like appearance and are dominated by a variety of eucalyptus trees and shrubs. Some of the common trees include ghost gums, box gums, grey gums and ironbarks. Wandoo and marri gums are common in the woodlands of south-western Australia.

This vegetation formation is generally found on the drier margins of the open forest, or in areas with a moderately low rainfall. It may also occur in higher rainfall areas where the soils are particularly sandy. As the rainfall becomes lower the woodlands are often replaced by shrub or grass lands. They may, however, penetrate some of these dry areas if there is a supply of water, such as along rivers or in areas of shallow groundwater supplies. The trees of the woodlands are often distinguished from the forest formation by the ratio of canopy to trunk. A typical woodland tree will have 50 per cent of its total height as the main trunk and 50 per cent as the spreading branches of the canopy. The generally wide spacing between the trees means that other trees do not inhibit the outward growth of the canopy. This allows sunlight to reach the ground easily and in a number of areas encourages the growth of seasonal grasses. It also means that the bushfires are less intense than those of the open forest. The distinction between different types of woodlands is often made on the nature of the understorey and ground cover. There are woodlands where various grasses are the dominant lower vegetation, while others have a variety of shrubs. The woodlands of Australia once occupied some 22.6 per cent of the continent.

### Shrub woodlands

These are found in the drier regions of New South Wales and northern Victoria, north-eastern Tasmania and southern Western Australia. The rainfall is around 400–600 millimetres in the south-eastern regions and as low as 200 millimetres in the south-western regions. This type of woodland is mainly subjected to hot to very hot summers and cool to mild winters.

The trees may grow up to 30 metres in height and in the eastern areas the dominant species are the white gum, stringybark, river red gum and box gum, while in the south-west the wandoo, York gum and jam (**acacia**) are common. The understorey may include a variety of wattles (acacia), she-oaks, sandalwood, kurrajongs and a variety of saltbushes in the more arid areas. The occurrence of this type of woodland has corresponded to the desirable areas for wheat production and in many cases it has been cleared for this purpose. There is also some evidence to suggest that the reduction in the frequency of fires in this vegetation zone, due to the introduction of European farming, may have

**Figure 1.32** Jarrah and marri trees, dry sclerophyllous forest

**Figure 1.33** Wandoo woodland

**Figure 1.34** Tropical paperbark (*Melaleuca*) trees fringing a billabong

encouraged the decline in grasses and the growth of a shrubby understorey.

**Savanna woodlands**

Located in the drier regions of northern New South Wales, southern and central Queensland, the Northern Territory and north-west Western Australia, the tropical savanna or grassy woodlands exhibit a wide variety of vegetation types. They may have short, widely-spaced trees and tall seasonal grasses, or alternatively the trees may be closer together and taller and the grasses shorter. The northern and north-eastern savanna woodlands consist of tall, coarse grasses such as Mitchell and Flinders grasses, spear grass and wild sorghum. The trees of the region include the Darwin box, the Darwin stringybark, the coolabahs and the baobab.

The temperate savanna woodlands are found in the cooler south-eastern areas of New South Wales, Victoria and South Australia. The tree species of this formation include the ironbark, coolabah, bimble box and grey box. Undergrowth may include a variety of wattle shrubs or mallee, with kangaroo and wallaby being the main grasses.

## Shrub or heathlands

The shrub and heathlands are formations composed of short, shrubby plants generally 2–5 metres in height. Where they are under 2 metres, the vegetation is classified as heath.

Many of the plants exhibit a mallee character—that is, a number of branches growing out of a large, woody, tuberous root and an absence of a main trunk. This **lignotuber** quickly produces shoots after fire or prolonged drought, allowing the shrub to regenerate. The number of branches appears to be influenced by the soil type and the number of bushfires. Poor or sandy soils and frequent fires produce mallee **eucalypts** with a number of thin branches, while better soils and less frequent fires result in fewer but larger branches. In all areas except for the driest the mallee shrublands are relatively dense, with the crowns of the bushes touching.

The majority of the shrub and heathlands are found in the southern arid and semi-arid regions of Australia. They extend in a continuous strip around the Great Australian Bight from Esperance to Adelaide as well as occurring in a large area in the mid-west of Western Australia. Their location suggests a relationship between the vegetation type and the winter rainfall pattern experienced in these areas. The shrublands and heathlands accounted for 9.5 per cent of the land area of Australia prior to European settlement.

A more open form of shrubland exists across much of Central Australia in large parts of the arid zone. This is a combination of mulga and spinifex. The mulga is a low, spreading tree with a short trunk and a number of xerophytic adaptations. It is generally found in areas which receive less than 300 millimetres of rain per year. The presence of open shrubland across much of the desert areas of Australia contrasts with the appearance of deserts in other parts of the world, where large areas are devoid of plants. The open shrublands once accounted for approximately 20 per cent of Australia's land area.

## Herblands or grasslands

This vegetation formation is dominated by the non-woody plants which are generally found within the arid

**Figure 1.35** Heathland near Esperance, WA

and semi-arid regions of Australia. In the main they have adapted to high summer temperatures, low precipitation, high evaporation and long periods of drought. The spinifex or hummock grass is the most widespread of this formation, occupying large areas in the northern arid and semi-arid regions. The structure and character of this grass shows a number of adaptations to the harsh climatic conditions. It is able to survive where even the hardiest shrubs would perish. The spikes of the grass are in fact leaves, reduced in area to prevent moisture loss. The spherical nature of the plant also decreases moisture loss by maintaining a slightly cooler and shaded interior.

The north-eastern grasslands of the Barkly Tableland and central Queensland, with their higher rainfall, are dominated by Mitchell and Flinders grasses, which are taller and more succulent than spinifex.

While the grasslands are generally associated with the drier regions of Australia, they may also be found in wetter regions. Here, factors such as poor soil drainage or exposure to wind or very cold conditions may prevent the growth of larger plants while allowing for the survival of grasses and herbs. The Bass Strait Islands provide examples of this.

It has been estimated that approximately 17 per cent of the continent is covered with the herbland and grassland formation.

## Desert formations

Low rainfall over much of Australia has resulted in many of the vegetation zones having plants with xerophytic adaptations. Thus it is possible to identify these characteristics in plants found in the more humid locations as well as those in the arid regions. In this sense it is not possible to identify a particular formation as being desert vegetation. As already stated much of the arid regions have shrubland or grassland formations or, in a few areas, no vegetation at all (the stony and sandy desert areas and the inland salt lakes). This section will therefore discuss the various adaptations to **aridity** found in a wide variety of Australian plants:

- mechanisms to maximise water intake
- mechanisms to minimise water loss
- mechanisms to maximise water storage
- mechanisms to avoid or escape droughts.

Infrequent or unpredictable rainfall can result in plants developing a variety of mechanisms to maximise the water intake when it does occur. For example, the mulga bush has two levels of roots. The near-surface roots absorb the light rainfall that only penetrates the first few centimetres of the soil, while the deeper roots tap underground reservoirs or make use of the water that penetrates deeper after heavy rainfall. The leaves and branches of the plant direct rainwater to the base of the trunk, thus maximising the available intake. Plants such as salt bush and blue bush have fine hairs on their leaves to trap and absorb dew.

The extended drought conditions of the arid and semi-arid regions encourage the survival of those plants that can minimise water loss. This is mainly achieved through the modification of leaf, bark and branch structures. Leaves are often small and are shaped to reduce their surface area. Stomata or transpiration pores are also reduced in number to prevent moisture loss. Many eucalyptus leaves have waxy, reflective surfaces and reduced chlorophyll to minimise the effects of strong sunlight. The bark of trees (such as the desert oak) can be thick and spongy with good insulating qualities or white (such as the ghost gum) to reflect sunlight. As a general rule, the more arid the conditions the smaller the plant. This, along with the habit of forming branches from the base, results in smaller total surface areas and maximum shading of the ground beneath. This is best illustrated by the mallee shrublands.

The capacity to store water in roots, branches and leaves is an important drought-resisting mechanism found in some plants. Plants that store water within fleshy leaf structures are called **succulents**. The lignotuber roots of mallee forms retain moisture and allow these plants to regenerate after extended droughts. While there are no indigenous cacti in Australia this species of plant has significant water storage capacity in the stems. The introduction of the prickly pear cactus in the nineteenth century resulted in this plant spreading rapidly in the inland areas of Queensland, where it was well adapted to the arid conditions.

The drought evaders or **ephemerals** are short-lived plants that germinate and produce seed stock after rainfall. They comprise a wide variety of herbs, grasses and wildflowers and are regulated by the occurrence of rainfall rather than the seasons. Some well-known flowering ephemerals of the arid regions include desert peas, everlasting daisies and pussy tails. These plants produce large quantities of seeds that might lie dormant in the soil for a number of years until the next period of sufficient rainfall.

## Minor vegetation formations

A number of minor vegetation formations that exist in various locations around Australia are distinguished from the major formations by their response to specific environmental factors. Such minor formations include wetland and riverine vegetation, coastal and alpine vegetation.

### Wetland and riverine vegetation

This type includes both marine and freshwater formations. **Hydrophytes** (water tolerant plants) dominate these environments, with distinctive zonation of plant species occurring as you move away from the edge of the wetland. **Riparian** or fringing vegetation occurs along the banks of streams and around the edges of wetlands. This location within the flood zone results in a variety of specialised plants including flooded gums, river red gums, bottlebrushes and swamp paperbarks (**melaleuca**). Undergrowth can include a variety of sedges and rushes. Freshwater and saltwater wetlands display differences based on the salt tolerance of the plants. Marine wetlands include samphires, mangroves, saltbush and salt grasses. The wetlands of Kakadu National Park are an important example of a monsoonal freshwater environment.

### Coastal vegetation

While this will vary significantly according to topography, soil and climate the sand dune environment is one that is sufficiently common along the coastline to warrant mention. This is a zone of dunes and swales with **calcareous** and **siliceous** sands. The poor water retention of these sands and the effect of salt and sand blasting on the vegetation results in the survival of only the hardiest species. Many exhibit a number of xerophytic characteristics designed to retain water and minimise moisture loss. The formations found in these coastal environments can include grasslands, heathlands and closed shrublands.

### Alpine vegetation

Australia's alpine regions are small in area. They cover southern Australia's highest mountains, including areas such as the Kosciuszko Plateau and the Tasmanian Highlands. In general the plants have adapted to cold winters and mild summers. The weather conditions can change rapidly, with cold snaps occurring even during summer. There are often marked differences in the vegetation type as a result of locational variations such as aspect, altitude, slope, exposure and soil moisture levels. Thus, alpine tussock grasses on exposed ridges may occur within a few metres of closed heath or woodland growing within a sheltered valley. Some adaptations by alpine plants include the ability to survive being buried under snow for the colder winter months or the ability to grow in waterlogged and permafrost soils. Common plants of the alpine regions include snow gums, hummock or tussock grasses, mosses, lichens and deciduous shrubs (see figure 1.37).

## Natural factors influencing vegetation

The natural factors that account for Australia's vegetation formations include climate, soils and topography.

**Figure 1.36** Mangrove trees, north Queensland

In addition to this is the element of time. Long-term climate change, the evolution of the landscape and the way the plants themselves have undergone change all contribute to the vegetation patterns that we can observe today.

### *Climate*

Climate has many different effects on vegetation. Rainfall amount, intensity and seasonal distribution can produce a range of distinctive plant types, from the water-loving hydrophytes through to the drought-resistant xerophytes. The amount of sunlight received and the seasonal variation can result in plants with a wide variety of photosynthetic abilities, as well as those that modify their growth patterns in response to the changes in day light hours. The change in the amount of daylight in high latitude regions triggers leaf shedding amongst deciduous plants.

Temperature patterns, often in combination with rainfall, will determine the length of annual growing seasons. Temperature extremes maintained for extended periods will cause wilting and death of plant species.

The effect of wind should not be overlooked. The transition of vegetation from coastal to inland locations will often be affected by the capacity of different plants to withstand the effect of strong prevailing winds that may sand-blast exposed plants.

**Figure 1.37** Tussock grasses, Cradle Mountain

## Topography

Factors such as altitude, slope and aspect will all influence the ways in which plant communities develop. The orographic uplift associated with the Eastern Highlands produces higher rainfall on the windward slopes and partly accounts for the predominance of forest formations in these areas. Altitude, with its effect on temperature, causes changes in vegetation associated with height above sea level. The alpine vegetation regions of the Australian and Tasmanian highlands illustrate this influence. Northward facing slopes in Australia will often show contrasting vegetation formations to those on the cooler and shadier southern slopes.

## Soils

Soils provide support, moisture and nutrients for plants. Their development reflects the way in which nutrients are cycled between the plants and the ground, and the way in which topography and climate affect their structure and characteristics. The degree of **acidity** or alkalinity (pH) of the soil will influence the types of plants that can grow. The predominance of tuart trees on the **alkaline** or calcium-rich soils of coastal south-western Australia illustrates the influence of soil pH on plant types. Soil structure and texture affect drainage and density. Soil depth, and the distribution of nutrients within the soil horizon, will influence root structures. For example, the shallow-rooted plants of the rainforests are a response to the shallow and relatively infertile nature of the tropical soils.

## Time

Climatic and geologic changes over time will alter plant communities. One plant species will be replaced by another, better suited to the new conditions. Sometimes ancient or remnant communities may become isolated by changing conditions. Individual plants remain as living fossils and evidence of past climates or geology. The evolution of the eucalypts was a response to the gradual movement of the Australian continent into warmer and drier latitudes. The presence of white mangrove plants in tidal creeks near Bunbury in south-western Australia indicates a past climatic period when this region was warmer. This is the most southerly location where this tropical plant is found (see figure 1.38).

**Figure 1.38** White mangroves in a tidal estuary at Bunbury

### EXERCISES

1. Using Australian examples, define or explain the following terms:

    - canopy
    - ground cover
    - hydrophytic
    - woodland
    - heath
    - sclerophyllous
    - understorey
    - xerophytic
    - forest
    - shrubland
    - grassland
    - Riparian

2. Describe the vegetation found along a transect from Perth to Cairns by drawing a cross-section and then placing notes and illustrations along it.

3. Using a grid system of calculating area from a map, as well as information provided in the text, determine the approximate percentage of land area occupied by each of the major vegetation formations shown in figure 1.27. Use this to produce a pie graph. Discuss the extent and location of each of the major formations.

4. Compare and contrast the major closed and open forest formations found in Australia, paying particular attention to strata, plant species and the influence of fire and climate.

5. Select two vegetation types from different study areas and then account for their similarities and differences by referring to the natural factors that influence vegetation.

# Chapter 2

# Biophysical Interrelationships in the South-West

The purpose of this chapter is to describe the biophysical environment of the South-West. This includes analysis of the ways in which the climatic conditions, the landforms, the soils, the vegetation and the animals have adapted and responded to each other to produce a variety of distinctive ecosystems.

## Topography of the South-West

The South-West (south-western Australia) contains some of the most ancient geologic landforms in Australia and has been subjected to a variety of tectonic and gradational processes over its long history.

Although there is a great variety of physical landscapes in the South-West, they can be grouped into three structural landform regions based on their appearance and geology. The Darling Plateau, the Darling Escarpment and the Coastal Lowlands are the most significant regions and are illustrated in figure 2.1.

### The Darling Plateau

The plateau forms part of the Yilgarn Continental Block (Craton), which dates back some 3500 million years BP (before present). It is composed of crystalline rocks

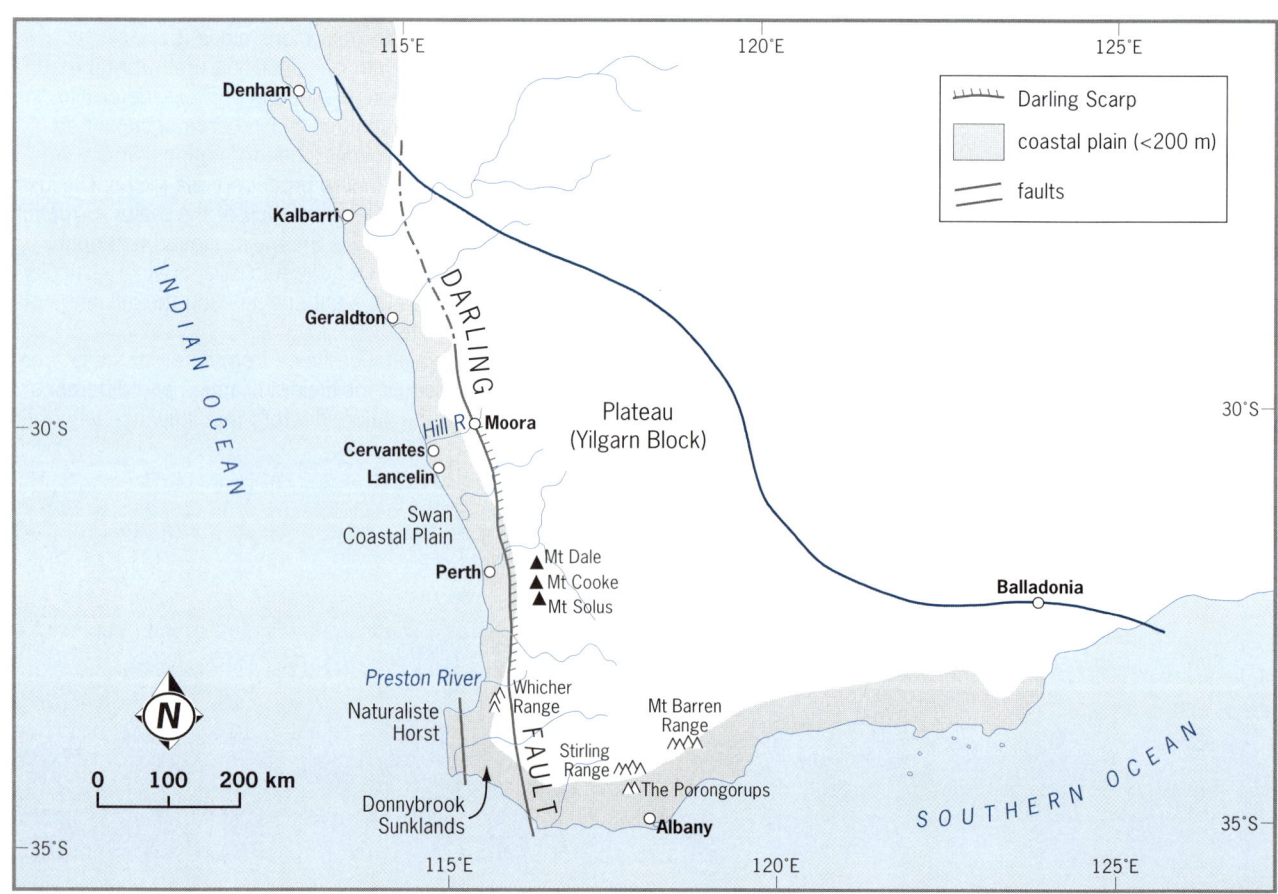

**Figure 2.1** Major landforms of the South-West

(such as granite) and metamorphosed igneous rocks (such as gneiss). These rocks were, in many cases, formed far beneath the original continental surface and have been exposed only after a long period of surface weathering and erosion. Granite outcrops and boulders can be found over much of the plateau and in a number of instances they occur as rounded hills or domes.

The Yilgarn Craton is bounded on its eastern side by the South Australian Craton and the collision of these blocks, around 1500 million years ago, produced significant mountain building or **orogeny** along this boundary. The remnants of this process can be seen in the Stirling, Barren and Porongerup ranges, on the southern part of the plateau.

The presence of extensive ice caps across much of the Darling Plateau during the Permian era, 250 million years ago, resulted in significant erosion of the plateau surface. This action was mainly responsible for the formation of the gently **undulating** hills that characterise much of the plateau surface. The presence of a wet tropical climate from around 65–30 million years BP produced much of the main valley and river systems that we see today, with fluvial erosion further exposing the underlying granites and gneiss. This period of climatic warming also produced the lateritic duricrust that occurs extensively along the western margins of the plateau.

Today, the plateau surface is relatively low lying and rises from 250 metres to 450 metres as you cross it from west to east. Its eastern regions have extensive salt lakes that are the remains of ancient river systems. These now form the basis for a series of internal drainage basins. The western and southern edges of the plateau have a series of rivers that meander across the surface until they reach the Darling Escarpment or the southern coastline, where they form steep dissected valleys. They then discharge their waters into the Indian and Southern oceans.

## The Darling Escarpment

The geologic history of the Darling Escarpment (or Darling Scarp) can be traced back to approximately 1100 million years BP, when this area was squeezed between the Yilgarn Craton and another **continental block**. The Darling Fault marked the edge of the mountain range formed by this process. At around 450 million years BP the fault again became active and formed an elongated trough along its western edge—the Perth Basin. Over about 300 million years this was filled with sediments to depths exceeding 10 kilometres.

When the continental landmass of Australia finally separated from Antarctica, about 65 million years ago, the separation did not occur along the main fault line and, as a consequence, part of the Perth Basin was also broken away and carried along with the landmass as it drifted northwards.

The Darling Fault is one of a series of roughly parallel faults found along the west coast, running from around Shark Bay in the north to Augusta in the south. The Darling Scarp marks the approximate location of this fault line. The Dunsborough Fault, between Cape Naturaliste and Cape Leeuwin, is another of these

**Figure 2.2** Bluff Knoll in the Stirling Ranges, composed of metamorphosed sedimentary rock including quartzite, **schist** and shale

faults. Along with the Darling Fault it is responsible for the low-lying area between Bunbury and Augusta known as the Donnybrook Sunkland. The raised section between Cape Leeuwin and Cape Naturaliste forms a block or horst, while the sunkland between this and

**Figure 2.3** Wave Rock at Hyden, an ancient granite outcrop

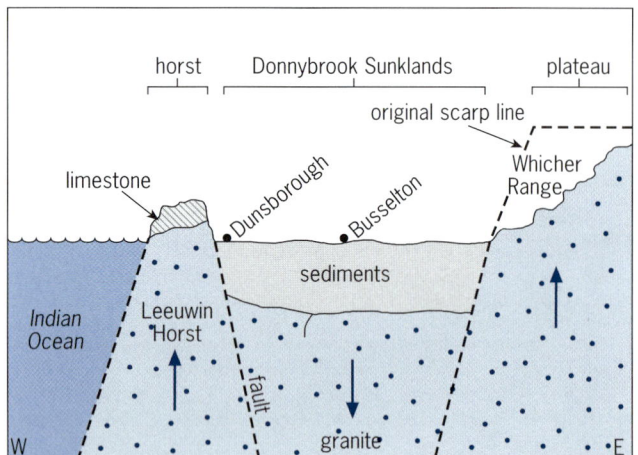

**Figure 2.4** Formation of the Naturaliste–Leeuwin Horst Block, WA

the Darling Ranges forms a **rift valley** or graben (see figure 2.4).

Today, the scarp is characterised by deeply dissected valleys formed by rivers that flow westward from the plateau. Remnants of ancient tectonic processes such as orogenisis and volcanism can be seen in the extensive granite and gneiss outcrops and rounded summits of Mt Cooke (582 metres), Mt Dale (543 metres) and Mt Saddleback (575 metres). Extensive erosion has caused the retreat of the original scarp face to the east. As a result, sediments that were carried down from the plateau now cover the main fault line. It lies buried on the coastal plain about 3 kilometres to the west of the present-day scarp face. The escarpment is most obvious from Bindoon to the north of Perth through to Manjimup in the south. Here its average height is between 250 and 300 metres.

## The Coastal Lowlands

The western and southern coastal lowlands show the impact of marine processes, with a series of changes in sea level having a significant influence on the geology and landforms of this region. In the last 2 million years, the polar ice caps contracted and expanded many times, often producing marked changes in the sea level. Coastal sand dunes developed in response to the consequent changes in the position of the coastline, and their locations can be seen in the limestone deposits that were left behind. These occur in the cliffs along the coastline or in the numerous offshore reefs.

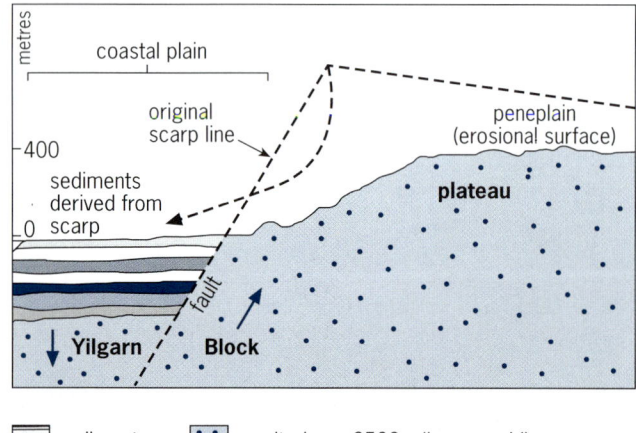

**Figure 2.5** Formation of the Swan Coastal Plain, WA

**Figure 2.6** A rare surface basalt flow at Back Beach, Bunbury

The sea level was approximately 130 metres below its present height 18 000 years ago, and the coastline was around 15 kilometres to the west of Rottnest. The Swan River passed by the northern end of today's island and its ancient channel can be seen in an undersea canyon found on the edge of the continental shelf. Since that time the sea level has gradually risen, reaching its present height some 6500 years ago. A number of river valleys were drowned in this process, forming a series of inlets and estuaries, for example the Swan and Canning estuaries, the Peel and Harvey estuaries, and Nornalup, Wilson and Walpole inlets. Subsequent dune and beach development in a number of instances closed off inlets and bays, which then formed coastal lakes. Lakes Preston and Clifton, south of Mandurah, were once connected to the ocean.

While the coastal lowlands of the South-West have been subjected to similar gradational processes, the variation in the geology of the west and south coasts has resulted in two distinctive types of topography.

## The west coast

The western coastal plain is dominated by extensive wind-blown coastal dunes and alluvial deposits washed down from the Darling Escarpment and the Darling Plateau. Its width varies from a few kilometres to around 80 kilometres at its broadest point. It is composed of two main areas. The largest is the Swan coastal system, running from Gingin in the north through to Dunsborough in the south. North of Gingin, where the climate is more arid, the absence of permanent rivers and the lower nature of the scarp has resulted in a dominance of Aeolian or wind-blown landforms. The sand dunes around Geraldton and Cervantes have developed to heights of 80 metres or more.

As you move inland from the coast towards the scarp you will first come across the most recent, wind-deposited sand dunes. These are the Quindalup dunes and they make up most of the shoreline, as it exists today. Beneath these recent deposits, and further east, extensive limestone deposits and yellow sands form the Spearwood dune system. The limestone deposits associated with these earlier dunes are known as Tamala limestone, formed from the movement of rainwater down through the dunes. This dissolved the fragments of shell found in the dunes and deposited the calcium as limestone. The Pinnacles, to the north of Perth at Cervantes, are an example of Tamala limestone and were originally formed by the action of plant acids and the leaching of calcium by rainwater. The oldest of the dune systems is known as the Bassendean formation and occurs to the east of the Spearwood system. It is a grey-white sand formation made up of heavily leached quartz or silica sands. It may sometimes have a yellow, oxidised subsoil and a hard layer of iron-rich silica where the groundwater meets the **watertable**.

**Figure 2.7** The Pinnacles near Cervantes, WA

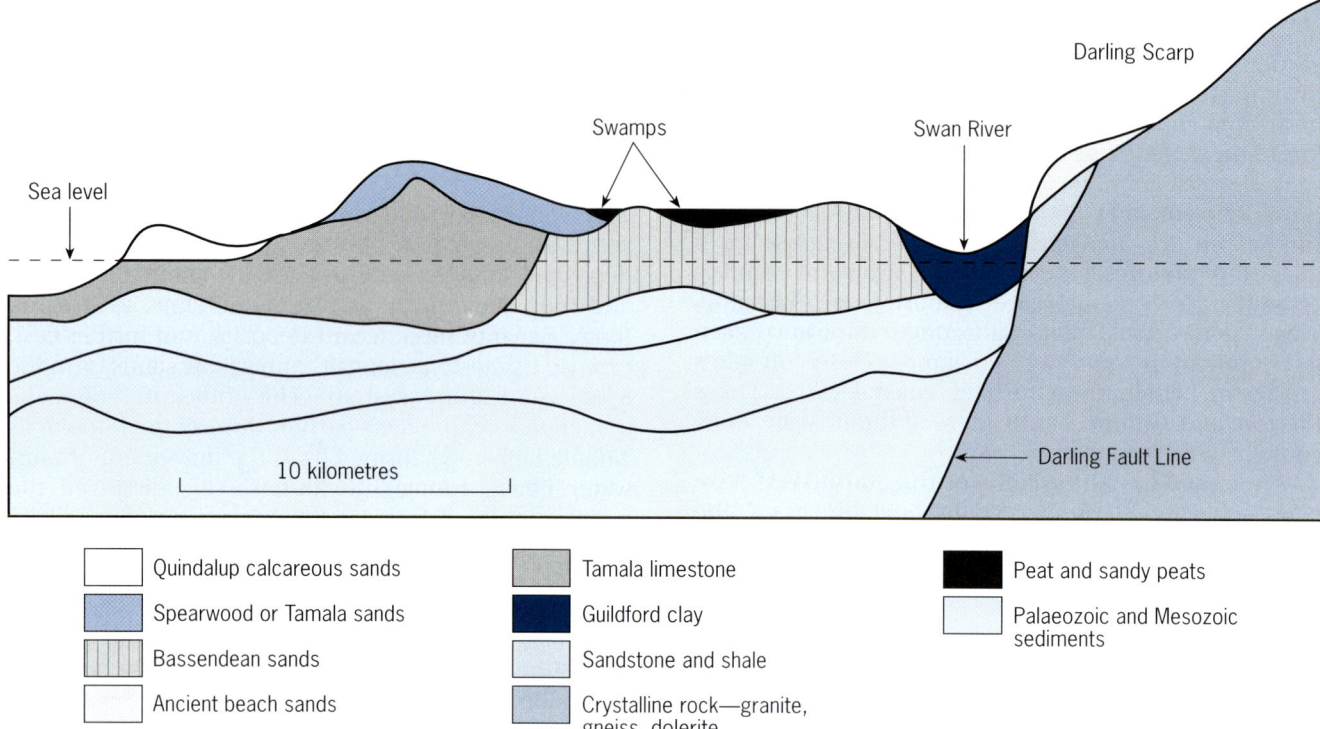

**Figure 2.8** Generalised soil transect, Perth Basin

The Bassendean dunes give way to the Guildford and Pinjarra alluvial soils on their eastern edge. These are composed of clay and lateritic material that has been washed down from the plateau and forms a westward-sloping plain or piedmont along the bottom of the scarp. The rivers that flow onto the coastal plain have tended to lay down the deposit in a series of alluvial fans, with the heaviest or coarsest material being deposited first and the finer alluvium being carried further westward to form a broad alluvial plain. Where major river systems such as the Swan and Murray have crossed the plain to reach the ocean alluvium has been carried into the areas of sand dune formation. Thus it is possible to find heavy alluvial loams and clays along the banks of the Canning River around Gosnells and Thornlie.

### The south coast

The south coast runs from Cape Leeuwin through to the east of Esperance, where it gives way to the cliffs of the Great Australian Bight. Unlike the west coastal plain, this region is broken up by outcrops of granite, gneiss and basalt associated with the southern edge of the Yilgarn craton. In addition, there are numerous rivers that flow southwards to enter inlets or estuaries. Limestone deposits and calcareous soils occur in low-lying areas along the coast between the granite headlands (see figure 2.9).

Geologically, the south coast is an area with a complex history. About 40 million years ago the sea level was some 300 metres higher than today and formed a huge embayment that stretched from Albany to Israelite Bay and inland to Norseman. Sediments rich with the skeletons of sponges were laid down in this region and gradually formed a sedimentary rock called spongelite. As the sea level fell, rivers cut into this material to form colourful gorges such as those within the Fitzgerald National Park.

### EXERCISES

1. Describe the landforms found along a transect from Perth to Merredin. In doing so refer to geomorphic processes, location and extent.
2. Match the following statements with the correct landform region.

   (a) Area of significant internal drainage and salt lake development
   (b) Coastal region with granite and limestone
   (c) Spearwood and Bassendean sand dune formations are found here
   (d) Area between Bunbury and Augusta, known as the Donnybrook Sunkland
   (e) One of a series of roughly parallel faults found along the west coast
   (f) The Pinjarra alluvial soils are found on the eastern edge

3. Draw a cross-section of the Perth Basin, from the coast to the plateau. Label and describe the significant landform features.
4. Identify and briefly explain the main geomorphic processes responsible for the development and appearance of the Darling Escarpment.
5. Analyse and discuss the relationships between landform development and sea level changes using examples from the South-West.

## Soils of the South-West

Soil is the combination of weathered rock material and decaying organic matter. The type of parent rock, the type and effect of the different weathering processes, the type and quantity of the organic matter, the period of time over which the soil formation has occurred and the landforms on which the soils have formed, determine its characteristics.

Soils that have developed in a given region in response to a particular climatic regime are called zonal soils. Most of the world's soils are classified as zonal. Soils that are transported by wind, water or ice to a new location are called azonal. Alluvial soils on river floodplains fall into this category.

The climate of an area is the single most important factor influencing the development of soils. The combination of different rainfall and temperature conditions can produce distinctly different soils, even when the base or parent rock material is similar. Hot wet climates produce deeply weathered and leached soils. Hot dry climates produce thin, granular soils with mineralisation near the surface due to the upward movement of water as a result of evaporation and capillary action. Neither of these soils is particularly fertile. The most fertile soils are those associated with the temperate grassland regions of the world. Moderate seasonal rainfall in these areas allows for the soils to develop, while at the same time reducing the impact of leaching (downward movement of minerals) or surface evaporation and capillary action (upward movement of minerals). The availability of new minerals from freshly exposed or developed parent rock material will also affect the overall fertility of the soil. The soils of the South-West are very old and lack phosphorus, potassium and nitrogen and are therefore generally infertile. Away from the coast they are mainly red and yellow in colour, indicating a high level of iron.

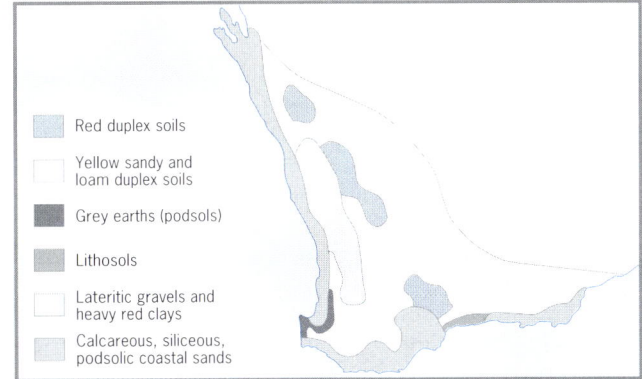

**Figure 2.10** Generalised soils, WA

### Coastal soils

These soils are predominantly sandy, with grains of silica quartz being the dominant inorganic material. Near the coast these sands are high in calcium derived from the dissolving shells of marine organisms. In some cases this is leached downwards to be deposited as limestone. These soils are classified as calcareous sands and are often associated with the Quindalup dune formation. Away from the coastline the calcareous sands can give way to siliceous sands. These sands are yellow in appearance, due to oxidation of minerals such as haematite (iron oxide) and alumina. Beneath the surface there are often ridges of limestone. This soil type relates to the older Spearwood dune formation and, like the calcareous sands, is also alkaline in nature. Further inland on the coastal plain the siliceous soils give way to the grey acidic soils of the Bassendean dune formation. These soils are referred to as **podsolic** and are of low fertility. Leaching of this soil has, in many cases, resulted in a layer of iron rich rock, called 'coffee rock', which has been deposited at the depth of the watertable.

As the land increases in height along the inland margin of the coastal plain, clays and loams replace the sandy soils. These are the red earth alluvias that are the result of fluvial deposition from the adjacent scarp and plateau. They form a narrow alluvial plain, which includes the Swan Valley and Ellen Brook to the north of Perth, and the Pinjarra plains to the south. Compared to the other soils of the coastal region they are quite fertile, with good water retention (see figure 2.8).

The coastal soils are, in summary, the product of marine erosion and transportation by wind. The

**Figure 2.9** Esperance beach with granite boulders

moderate rainfall associated with the Mediterranean climate of the region has leached them. They range from alkaline sands near the coast to acidic in the near coastal areas and where swamps and wetlands originally dominated low-lying sections. Further inland, these sands are replaced by heavier loam and clay-type soils influenced more by fluvial transportation and deposition. As you can see from figure 2.10, this is not continuous along the entire coastal zone of the South-West. The southern coast has the sandy soil types broken in places by the heavier clay or red duplex soils associated with the soils of the plateau.

## Plateau soils

The soils that cover most of the plateau are the result of the erosion of lateritic soils that developed some 40 million years ago and the breakdown of the underlying granites. Remnants of the lateritic soils can be seen in the gravel and ironstone ridges found along the eastern edge of the escarpment in areas such as Wundowie, Boddington and Dwellingup. Laterites are found extensively in the wet tropical regions of the world and are deeply weathered soils. They are red to yellow in colour, with a concentration of iron oxides and alumina in the top layers or horizons. They are generally high in clay and will harden to a brick-like consistency when dry. Between 40 and 25 million years ago, during the Eocene era, a warmer and wetter tropical period influenced the South-West, resulting in deep weathering of the rocks. The rainfall leached out most of the soluble minerals, leaving behind greatly weathered granites and a hard, iron-rich surface known as duricrust. This then is the origin of the laterite soils of today.

## Duplex soils

Inland from the Darling Scarp the most common soil is the duplex type. Duplex soils have two distinctive layers. The surface layer (A horizon) is generally yellow or red sands or gravelly loam. Beneath this, the B horizon has a higher concentration of clay. This layer prevents further downward leaching of soluble minerals such as sodium chloride and gypsum or calcium sulphate, which has become a problem in areas where rising watertables bring these back to the surface, causing salinity. Due to their long process of weathering, these soils are relatively infertile and need the introduction of a variety of trace elements, nitrates and phosphates to be suitable for farming. The duplex nature of the soil has given it good water retention characteristics, as the sandy or loam top layer allows the water to penetrate the soil while the clay subsurface layer prevents it from draining away. This is beneficial for the growth of cereal crops in the region.

## Saline loams

These occur in the salt lakes and clay pans that occupy the river channels of ancient drainage systems. They are alluvial in nature, being deposited in these broad, flat valley regions after heavy rainfall. High levels of evaporation during the summer period encourage the concentration of salt and gypsum on the surface, making these soils generally unsuitable for any form of agriculture.

## Lithosols

These soils are classified as azonal or recently developed soils. They form on steep surfaces and lack a well-developed soil profile. As the name suggests, they are stony or rocky in their overall composition. Located on the southern coastline of the South-West, they correspond to the steeper sections of the Yilgarn Block where it meets the Southern Ocean. It is not until the slope angle is reduced that they can weather and accumulate organic matter to fully develop in response to climatic and other conditions.

### EXERCISES

1. Identify and explain the processes by which climate influences soil development, using examples from south-western Australia.
2. Explain or define the following:
   - laterite
   - calcareous sands
   - alkaline and acidic soils
   - saline soils
   - zonal soils
   - **pedalfers**
   - soil profile
   - podsolic soils
   - leaching
   - duplex soils
   - **pedocals**
3. What factors contribute to the level of soil fertility and why are South-West soils generally considered to be of moderate to low fertility?
4. Describe and explain the changes in soils found along a transect from Bunbury to Merredin.

## The climate of the South-West

While the South-West region may be identified as having a Mediterranean or warm temperate western maritime climate there are observable differences in rainfall, temperature, humidity and evaporation levels from north to south and from west to east.

As the global pressure systems shift north and south with the changing seasons this brings a significant seasonal variation in rainfall over the South-West. With the onset of summer in the Southern Hemisphere, the pressure systems move southwards. A broad low pressure belt forms over the northern part of Australia and a high pressure system dominates the southern portion. This is the subtropical high pressure belt and it directs easterly offshore tropical continental (cT) air over the South-West for much of the season. The cloudless skies and the flow of dry continental air associated with this pressure system keep the daily maximum temperatures of the region high and result in little or no rainfall.

In winter, the pressure systems move northwards and the subtropical high pressure system is located over northern Australia. In the south it is replaced by lower pressures associated with the Sub-polar Low. The centre of this low tends to be south of the south coast of Australia, with cold fronts sweeping up in a westerly airstream to bring rainfall to the southern edge of the continent. The cold fronts form a region of contact between cold stable polar maritime (mP) air and warmer moist tropical maritime (mT) air. The colder and denser air produces uplift of the tropical air along a front causing the formation of cumulus and cumulostratus clouds. Depending on the intensity and speed of the advancing front, showers of rain may be brief and light or longer and heavier. The increased cloud cover, onshore winds and shorter daylight periods of winter result in cool to mild daily maximum temperatures.

## Temperature variations

Variations in temperature occur as the result of distance from the equator, distance from the ocean, and seasonal changes in the earth's axis, and the effects of day and night, altitude, humidity levels and wind direction. These factors all play a part in shaping the patterns of temperature experienced in the South-West.

### North to south coastal

There is approximately eight degrees difference in latitude from Shark Bay in the north to Walpole in the south. The South-West therefore ranges from near tropical (25°S) through to mid-latitude temperate (35°S). The annual average temperature for Denham at Shark Bay is 22°C, while at Walpole it is 15.8°C. This variation can be partially explained by the sun's angle of incidence associated with the curvature of the globe. Other contributing factors include the difference in ocean temperature and the influence of the landmass on the winds that pass over it. Shark Bay on the west coast has higher temperatures produced by the easterlies blowing from a hot interior during summer. Walpole, on the other hand, receives relief from high temperatures as the winds tend to come from the south-east and therefore blow over water.

Another factor contributing to higher summer temperatures along the west coast is the formation of low pressure troughs along the coastline. When these are positioned offshore they direct hot, north-easterly winds over the region, bringing a period of heatwave conditions to the areas from Shark Bay through to Busselton. When the trough moves across the coast and is positioned inland there is a break in the high temperatures and the heatwave is replaced by a cool change and onshore winds. The effect of this is much less pronounced along the southern coastline and, as a result, temperatures are significantly lower there.

### West to east inland

As you move inland from the west coast the temperatures become more extreme, both higher in summer and lower in winter. Perth has an average January temperature of 23.8°C and a July temperature of 13.2°C. Southern Cross, which is approximately 400 kilometres inland, has a January temperature of 26°C and a July temperature of 10.3°C. This is known as the continental effect—locations close to the ocean will have day and night temperatures moderated by the movement of air to and from the sea. Water temperatures vary little between summer and winter and this influence is transferred to the land by sea breezes. Away from the sea the moderating effect decreases. As the land has a greater capacity to absorb and release heat than the sea, this is reflected in the greater **diurnal** as well as seasonal variations in temperature. Temperature ranges therefore increase with increasing distance inland. Table 2.1 illustrates these variations. It should be noted that the general decrease in monthly temperature ranges towards the middle of the year corresponds to the increase in rainfall and cloud cover associated with the Mediterranean climate.

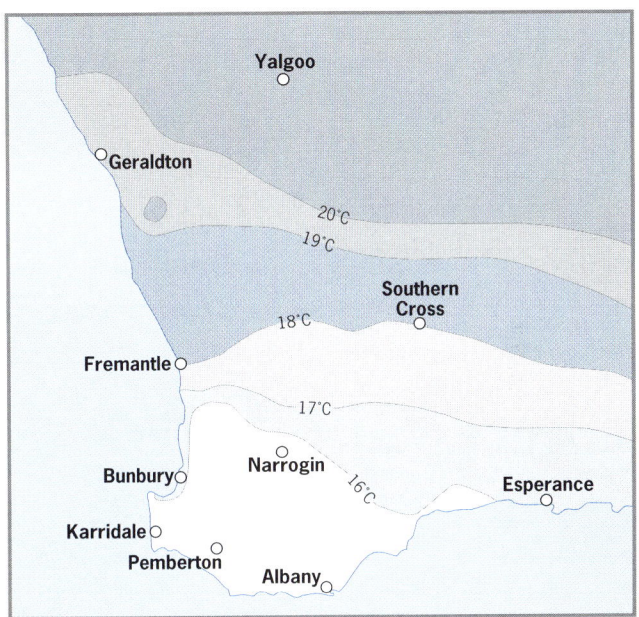

**Figure 2.11** Annual average temperature

## Rainfall patterns

While the rainfall pattern for the South-West conforms to the characteristic summer drought and winter rainfall of the Mediterranean climate, there are distinctive differences from the north to the south and from the west to the east. The position and movement of cold fronts across the South-West can generally explain this. Topographic features can also account for variations in rainfall patterns. Figures 2.13–2.17 illustrate these variations. In addition to the spatial variation of rainfall distribution there is also the element of seasonal rainfall variability. The **Southern Oscillation Index** (SOI), and the intensity of the pressure systems associated with this index, is a pointer to the differences in the amount and duration of rainfall experienced by the study area. While the Mediterranean climate is generally one of fairly reliable and predictable rainfall, there are periods of time when drier winter seasons are experienced as well as times when summer rainfall can vary.

### North to south variation

The movement of the sub-polar low pressure systems north and south with the change of season results in lower rainfall and a shorter wet season on the northern

**Table 2.1** Temperature data for selected stations, south-western Australia

| Geraldton | Jan | Feb | Mar | Apr | May | Jun | Jul | Aug | Sep | Oct | Nov | Dec | Annual |
|---|---|---|---|---|---|---|---|---|---|---|---|---|---|
| Mean daily maximum temperature | 31.7 | 32.6 | 30.8 | 27.6 | 23.8 | 20.7 | 19.4 | 20 | 22 | 24.4 | 27 | 29.5 | 25.8 |
| Mean daily minimum temperature | 18.3 | 19.1 | 17.9 | 15.4 | 12.8 | 11 | 9.4 | 8.9 | 9.2 | 10.9 | 13.8 | 16.3 | 13.6 |
| Mean monthly average temperature | 25 | 25.85 | 24.35 | 21.5 | 18.3 | 15.85 | 14.4 | 14.45 | 15.6 | 17.65 | 20.4 | 22.9 | 19.7 |
| Range | 13.4 | 13.5 | 12.9 | 12.2 | 11 | 9.7 | 10 | 11.1 | 12.8 | 13.5 | 13.2 | 13.2 | 12.2 |
| **Yalgoo** | Jan | Feb | Mar | Apr | May | Jun | Jul | Aug | Sep | Oct | Nov | Dec | Annual |
| Mean daily maximum temperature | 37.2 | 36.3 | 33.5 | 28.5 | 23 | 19.2 | 18.2 | 20 | 24 | 27.4 | 32.1 | 35.5 | 27.9 |
| Mean daily minimum temperature | 20.7 | 20.7 | 18.6 | 14.4 | 10.1 | 7.7 | 6.2 | 6.8 | 8.7 | 11.4 | 15.2 | 18.4 | 13.2 |
| Mean monthly average temperature | 28.95 | 28.5 | 26.05 | 21.45 | 16.55 | 13.45 | 12.2 | 13.4 | 16.35 | 19.4 | 23.65 | 26.95 | 20.55 |
| Range | 16.5 | 15.6 | 14.9 | 14.1 | 12.9 | 11.5 | 12 | 13.2 | 15.3 | 16 | 16.9 | 17.1 | 14.7 |
| **Fremantle** | Jan | Feb | Mar | Apr | May | Jun | Jul | Aug | Sep | Oct | Nov | Dec | Annual |
| Mean daily maximum temperature | 27.3 | 27.9 | 26.4 | 23.6 | 20.3 | 18.1 | 17.1 | 17.3 | 18.5 | 20.1 | 23 | 25.4 | 22.1 |
| Mean daily minimum temperature | 17.8 | 18.1 | 17 | 14.9 | 12.7 | 11.1 | 10 | 10.2 | 11 | 12.3 | 14.5 | 16.5 | 13.9 |
| Mean monthly average temperature | 22.55 | 23 | 21.7 | 19.25 | 16.5 | 14.6 | 13.55 | 13.75 | 14.75 | 16.2 | 18.75 | 20.95 | 18 |
| Range | 9.5 | 9.8 | 9.4 | 8.7 | 7.6 | 7 | 7.1 | 7.1 | 7.5 | 7.8 | 8.5 | 8.9 | 8.2 |
| **Narrogin** | Jan | Feb | Mar | Apr | May | Jun | Jul | Aug | Sep | Oct | Nov | Dec | Annual |
| Mean daily maximum temperature | 30.9 | 30.1 | 27.2 | 22.9 | 18.1 | 15.1 | 14.3 | 15.2 | 17.8 | 20.5 | 25.6 | 28.8 | 22.2 |
| Mean daily minimum temperature | 13.6 | 13.6 | 12.5 | 10 | 7.6 | 6.1 | 5.1 | 5 | 5.7 | 6.9 | 9.5 | 11.7 | 9 |
| Mean monthly average temperature | 22.25 | 21.85 | 19.85 | 16.45 | 12.85 | 10.6 | 9.7 | 10.1 | 11.75 | 13.7 | 17.55 | 20.25 | 15.6 |
| Range | 17.3 | 16.5 | 14.7 | 12.9 | 10.5 | 9 | 9.2 | 10.2 | 12.1 | 13.6 | 16.1 | 17.1 | 13.2 |
| **Bunbury** | Jan | Feb | Mar | Apr | May | Jun | Jul | Aug | Sep | Oct | Nov | Dec | Annual |
| Mean daily maximum temperature | 27.6 | 27.8 | 25.9 | 22.9 | 19.8 | 17.6 | 16.8 | 17.1 | 18.1 | 19.9 | 22.9 | 25.5 | 21.8 |
| Mean daily minimum temperature | 15.1 | 15.4 | 14.3 | 12.2 | 10.4 | 9.2 | 8.4 | 8.4 | 9.2 | 10.3 | 12.2 | 13.9 | 11.6 |
| Mean monthly average temperature | 21.35 | 21.6 | 20.1 | 17.55 | 15.1 | 13.4 | 12.6 | 12.75 | 13.65 | 15.1 | 17.55 | 19.7 | 16.7 |
| Range | 12.5 | 12.4 | 11.6 | 10.7 | 9.4 | 8.4 | 8.4 | 8.7 | 8.9 | 9.6 | 10.7 | 11.6 | 10.2 |
| **Karridale** | Jan | Feb | Mar | Apr | May | Jun | Jul | Aug | Sep | Oct | Nov | Dec | Annual |
| Mean daily maximum temperature | 24.6 | 24.7 | 23.7 | 22 | 19.1 | 16.9 | 16.1 | 16.4 | 17.3 | 18.5 | 21.3 | 23.4 | 20.3 |
| Mean daily minimum temperature | 13.6 | 13.4 | 12.5 | 10.8 | 9.8 | 8.9 | 8.1 | 8.1 | 8.8 | 9.4 | 10.9 | 12.5 | 10.6 |
| Mean monthly average temperature | 19.1 | 19.05 | 18.1 | 16.4 | 14.45 | 12.9 | 12.1 | 12.25 | 13.05 | 13.95 | 16.1 | 17.95 | 15.45 |
| Range | 11 | 11.3 | 11.2 | 11.2 | 9.3 | 8 | 8 | 8.3 | 8.5 | 9.1 | 10.4 | 10.9 | 9.7 |
| **Pemberton** | Jan | Feb | Mar | Apr | May | Jun | Jul | Aug | Sep | Oct | Nov | Dec | Annual |
| Mean daily maximum temperature | 26.1 | 26.3 | 24.2 | 21 | 17.8 | 15.8 | 14.9 | 15.3 | 16.6 | 18.6 | 21.1 | 23.8 | 20.2 |
| Mean daily minimum temperature | 13.1 | 13.5 | 12.6 | 10.8 | 9.3 | 8.1 | 7.2 | 6.9 | 7.5 | 8.5 | 10.2 | 11.9 | 10 |
| Mean monthly average temperature | 19.6 | 19.9 | 18.4 | 15.9 | 13.55 | 11.95 | 11.05 | 11.1 | 12.05 | 13.55 | 15.65 | 17.85 | 15.1 |
| Range | 13 | 12.8 | 11.6 | 10.2 | 8.5 | 7.7 | 7.7 | 8.4 | 9.1 | 10.1 | 10.9 | 11.9 | 10.2 |
| **Albany** | Jan | Feb | Mar | Apr | May | Jun | Jul | Aug | Sep | Oct | Nov | Dec | Annual |
| Mean daily maximum temperature | 22.8 | 22.9 | 22.2 | 20.8 | 18.5 | 16.5 | 15.6 | 16.2 | 17.2 | 18.4 | 20.4 | 21.9 | 19.5 |
| Mean daily minimum temperature | 15 | 15.3 | 14.5 | 12.6 | 10.6 | 9 | 8 | 8.3 | 9.2 | 10.2 | 12.3 | 13.9 | 11.6 |
| Mean monthly average temperature | 18.9 | 19.1 | 18.35 | 16.7 | 14.55 | 12.75 | 11.8 | 12.25 | 13.2 | 14.3 | 16.35 | 17.9 | 15.55 |
| Range | 7.8 | 7.6 | 7.7 | 8.2 | 7.9 | 7.5 | 7.6 | 7.9 | 8 | 8.2 | 8.1 | 8 | 7.9 |
| **Esperance** | Jan | Feb | Mar | Apr | May | Jun | Jul | Aug | Sep | Oct | Nov | Dec | Annual |
| Mean daily maximum temperature | 26.3 | 26.2 | 25.1 | 23.1 | 20.3 | 17.9 | 17.1 | 17.7 | 19.3 | 21.1 | 23 | 24.6 | 21.8 |
| Mean daily minimum temperature | 15.6 | 16 | 15 | 13.1 | 11 | 9 | 8.2 | 8.6 | 9.5 | 10.6 | 12.6 | 14.4 | 12 |
| Mean monthly average temperature | 20.95 | 21.1 | 20.05 | 18.1 | 15.65 | 13.45 | 12.65 | 13.15 | 14.4 | 15.85 | 17.8 | 19.5 | 16.9 |
| Range | 10.7 | 10.2 | 10.1 | 10 | 9.3 | 8.9 | 8.9 | 9.1 | 9.8 | 10.5 | 10.4 | 10.2 | 9.8 |
| **Southern Cross** | Jan | Feb | Mar | Apr | May | Jun | Jul | Aug | Sep | Oct | Nov | Dec | Annual |
| Mean daily maximum temperature | 34.6 | 33.6 | 30.5 | 25.7 | 20.5 | 17.1 | 16.3 | 18 | 21.8 | 25.4 | 29.6 | 33 | 25.5 |
| Mean daily minimum temperature | 17.2 | 17.1 | 15.1 | 11.5 | 7.6 | 5.7 | 4.4 | 4.8 | 6.5 | 9.2 | 12.8 | 15.5 | 10.6 |
| Mean monthly average temperature | 25.9 | 25.35 | 22.8 | 18.6 | 14.05 | 11.4 | 10.35 | 11.4 | 14.15 | 17.3 | 21.2 | 24.25 | 18.05 |
| Range | 17.4 | 16.5 | 15.4 | 14.2 | 12.9 | 11.4 | 11.9 | 13.2 | 15.3 | 16.2 | 16.8 | 17.5 | 14.9 |

perimeter of the region. Geraldton has a total rainfall of 449 millimetres, with 70 per cent of this falling between May and August. Pemberton, on the other hand, has 1204 millimetres and its wet season extends from April through to September. As the fronts extend their influence northwards in winter they affect the southern areas first. The northern areas receive the bulk of their rainfall when the frontal systems have become well established, with only the strongest extending up into the semi-arid regions that border the northern edge of the study area. As the rainy season comes to an end it is the northern areas that are affected first as the low pressure belt shifts southwards.

### West to east variation

When the frontal systems cross the west coast and move inland there is a decline in the amount of rainfall they produce. The general pattern is one of highest rainfall on the coast, with progressively smaller amounts inland. By the time the eastern margin of the study area is reached the annual rainfall is somewhere between 250 and 300 millimetres. This corresponds to the temperate semi-arid climatic zone. The decline in rainfall can be explained by the change in the character of the air masses associated with the frontal systems. Initially, a cold front forms when warm tropical maritime air is drawn into contact with cold polar maritime air in a low pressure cell. The effect of this is to cause the warmer air to rise and cool. Moisture is condensed and precipitated in this process. This continues as the front crosses the coastline. Once over land the air masses begin to merge with the warm tropical air, becoming cooler and drier. Further uplift is lessened and the rainfall declines. The initial differences between the two air masses, and the speed at which they are drawn into the low pressure cell, determine the degree to which the front will carry the rainfall inland. In addition to this there is a decrease in the available moisture within the air mass, given the fact that significant amounts have already fallen as rain.

## Topographic variations

The Darling Escarpment, while only a few hundred metres high, has an orographic effect on the eastward-moving fronts. As the scarp forces the frontal system to rise, rainfall increases. The slope is in effect enhancing the uplift caused by the cold air mass within the frontal system. Perth, which is on the coastal plain, receives about 850 millimetres of rain per year, while locations such as Karnet and Dwellingup receive about 1200 millimetres. This increased rainfall along the scarp face and in the Darling Ranges also has the effect of reducing the amount of rain that is carried inland. Figure 2.12 showing annual median rainfall in the South-West demonstrates this decline quite clearly,

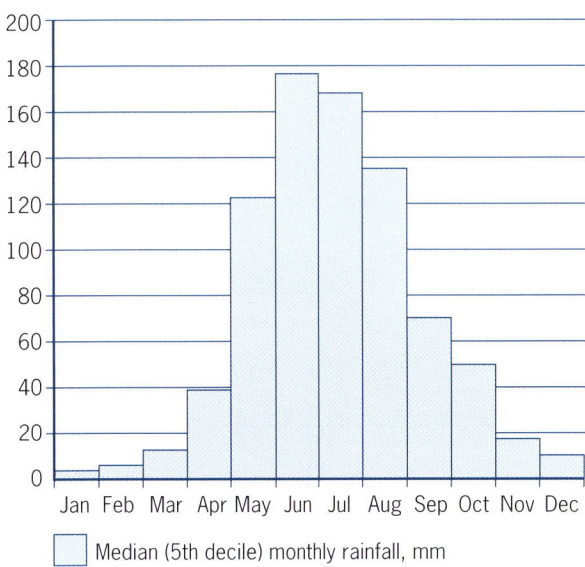

**Figure 2.13** Median rainfall, Perth

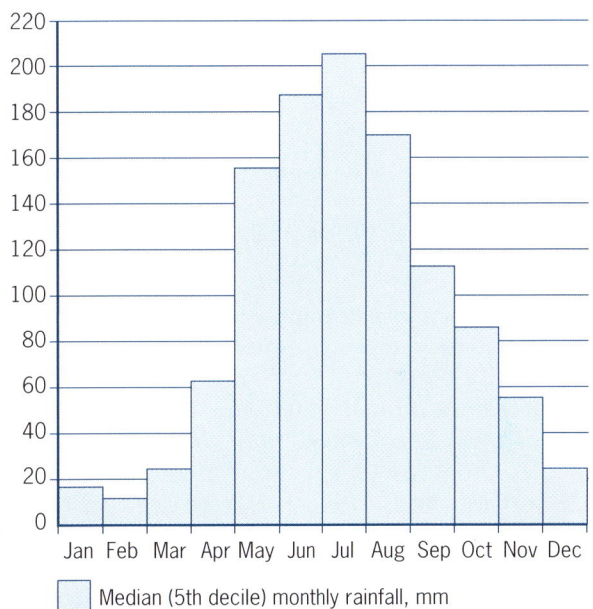

**Figure 2.14** Median rainfall, Pemberton

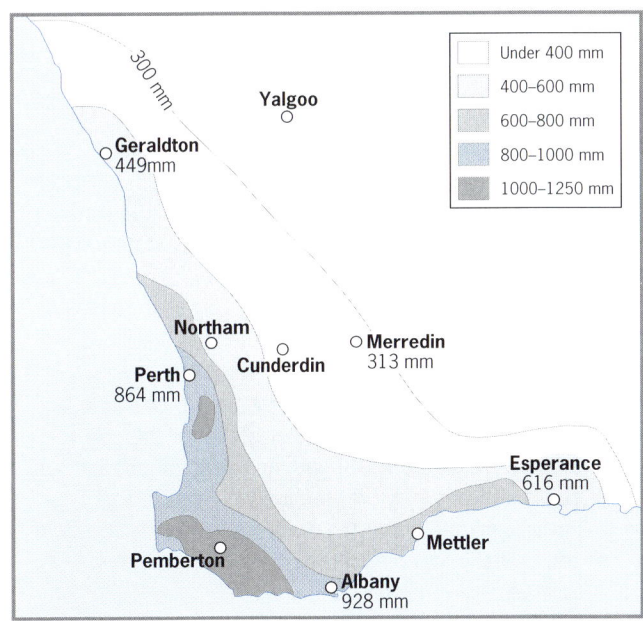

**Figure 2.12** Annual median rainfall

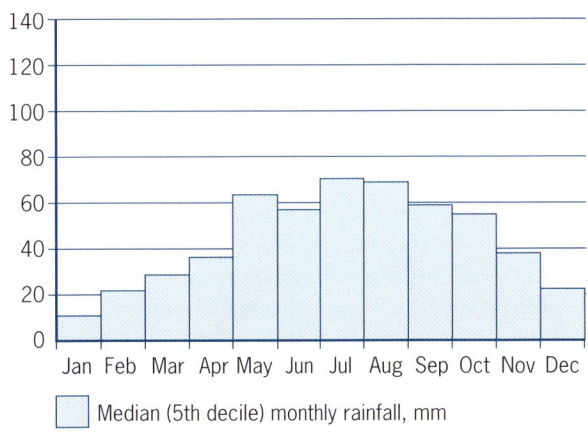

**Figure 2.15** Median rainfall, Mettler

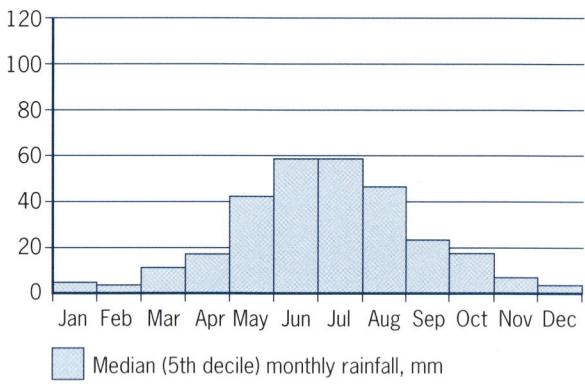

**Figure 2.16** Median rainfall, Cunderin

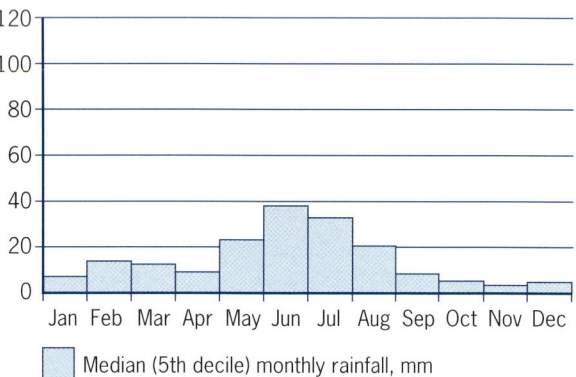

**Figure 2.17** Median rainfall, Yalgoo

with towns such as Northam, less than 100 kilometres from the coast, receiving around 436 millimetres per year.

Note that the rainfall pattern for Mettler, with higher falls in May, is due to the periodic incursions of moist tropical air that comes across the interior of the South-West from the north. This occurs from late April to early May. The falls then taper off in June and increase again from July to October due to a prevailing southerly air stream. This feature occurs through to Esperance but is not a feature of locations to the west of Mettler (see figure 2.15).

### Rainfall variability and reliability

The variation of rainfall from the mean average for the South-West, shown by figure 2.19 (see page 40), indicates a long period of below-average rainfall since the 1970s. In addition, the maps illustrate a pattern of alternating rainfall variability between the South-West and the rest of the continent. The explanation for the change in rainfall amounts in Western Australia is twofold: the change in sea temperature associated with the periods of time when the warm Leeuwin Counter Current is dominant; and the operation of the Indian Ocean equivalent of the El Nino and La Nina sea temperature anomalies.

The west coast generally has a cool current (West Australian Current) that flows northwards. However, there are times when the current nearest to the coast reverses and brings warm water from near Shark Bay as far south as Cape Leeuwin. This counter current produces increased evaporation, lower air pressure and higher rainfall. During years when the Indian Ocean in the tropical areas becomes warmer, tropical air may reach the South-West in summer and produce increased rainfall during what is normally a period of drought. In addition, recent CSIRO evidence points towards a 20 per cent decline in rainfall in the South-West which some have attributed to the effects of global warming.

### EXERCISES

1. Using examples from south-western Australia, define or explain the following climatic factors:

   (a) median and average rainfall
   (b) rainfall reliability
   (c) average monthly minimum temperature
   (d) average monthly maximum temperature
   (e) diurnal temperature range
   (f) seasonal or annual temperature range.

2. Construct a series of line graphs for temperature for the locations provided in table 2.1 and then analyse the annual and monthly temperature ranges. When accounting for the differences, consider the possible effects of distance from the sea and changing levels of atmospheric moisture or humidity.

3. Describe the changes in temperature patterns that occur as you move from Geraldton to Pemberton and from Bunbury to Southern Cross.

4. Study the rainfall graphs provided and then describe each according to the following criteria:

   - total rainfall
   - wettest month
   - driest month
   - wettest 3 months
   - driest 3 months

5. With reference to factors such as orographic effects, air pressure and prevailing winds, latitude and distance from the sea, account for the differences noted in the rainfall graphs.

6. Study the maps provided showing long-term rainfall patterns and then describe the changes that have occurred in the rainfall patterns in the South-West. What factors affect a region's rainfall variability?

7. Study the sequence of weather maps for the South-West (figure 2.18) and then describe the passage of a cold front as it crosses the west coast. Pay particular attention to the air pressures, wind direction and probable speeds, air masses involved in the formation of the front and the changes that would be experienced in temperatures and rainfall over the time period.

# BIOPHYSICAL INTERRELATIONSHIPS IN THE SOUTH-WEST 39

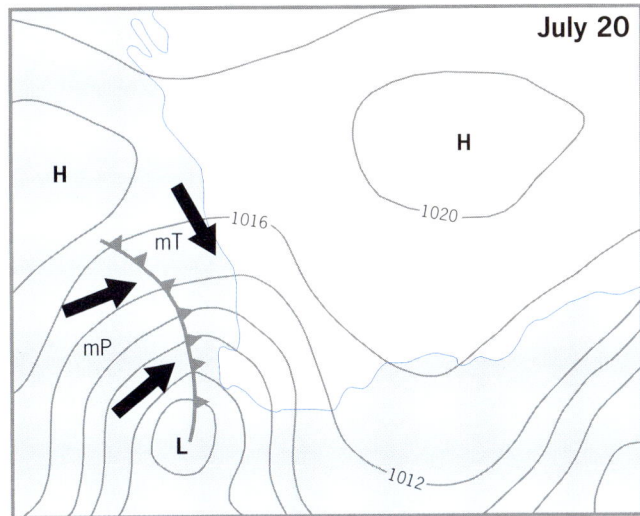

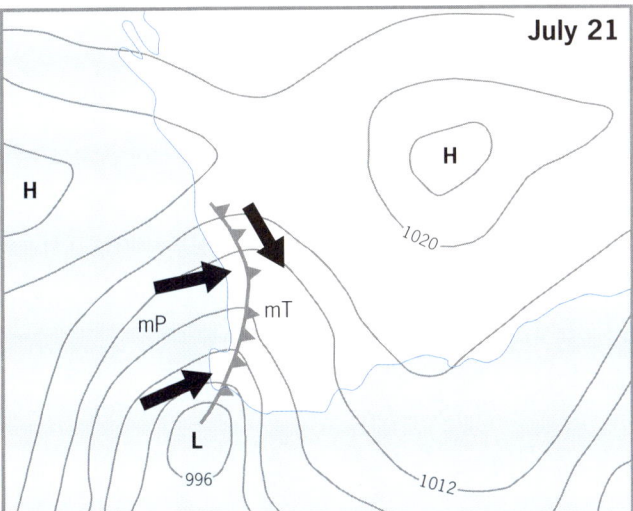

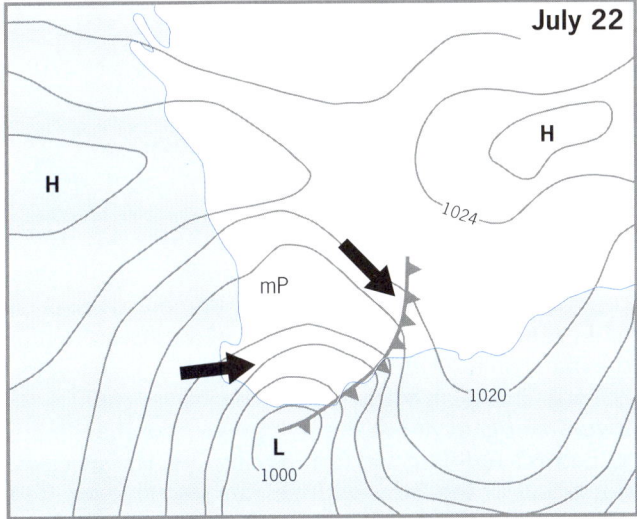

**Figure 2.18**

## Vegetation of the South-West

Differences in temperature, rainfall, topography and soils have created a corresponding difference in the type and pattern of vegetation found in the South-West—from open forest through to shrubland and heathland. The main forest formations are situated in the wetter south-western corner, separated from the west coast by the tuart and marri woodlands. Here, the combination of soils and rainfall has created a distinctive zone of forest vegetation. The further north you go the less dense the vegetation becomes, as the annual rainfall totals decline. Changes in rainfall also account for the change from forest to woodland and then scrub as you move from west to east, illustrated in figure 2.20. In some areas the original vegetation has been significantly modified by the use of the land for agriculture, while in other areas the range of native plant species is substantially intact, preserved in state forests or national parks. The South-West region has been isolated from the rest of Australia by the surrounding expanse of desert and semi-desert environments. This isolation has produced distinctive variations in the plant species, with some 70 per cent of the plant species in the South-West found nowhere else in the world. When a plant or animal becomes isolated from the broader population by some physical barrier such as mountains, oceans or deserts then they may follow a separate evolutionary path to their counterparts in other locations.

### *Open forests*

#### Wet sclerophyllous forest

This vegetation formation (also known as 'shrubby tall open forest') occurs in the extreme south of the study area and is dominated by the karri. Other tall trees include the yellow and red tingle, marri and blackbutt. It is limited to a small region extending from Nannup through to Denmark on the south coast and northwards to Pemberton, in an area where the annual rainfall exceeds 1100 millimetres and the dry season is only four months long. Competition for space and light has resulted in the dominant karri species growing tall and straight to heights exceeding 30 metres. The best examples of the species are found in areas of deep red loam soils, where heights can exceed 80 metres. Branching occurs high up in the canopy which, unlike the rainforest, shades less than 50 per cent of the ground. Where the karri is planted in open areas the tree has a wide spreading appearance, with the branches forming much lower and the main trunk being shorter. Unlike that of other forests, the ground cover here is very thick with a variety of acacia, banksia and melaleuca shrubbery.

#### Dry sclerophyllous forest

This forest formation borders the wet sclerophyllous forest on the western, northern and eastern edges. It extends northwards in the 1000–600 millimetres rainfall zone to an area just north of Perth. While it reaches the southern coastline, it tends to follow the

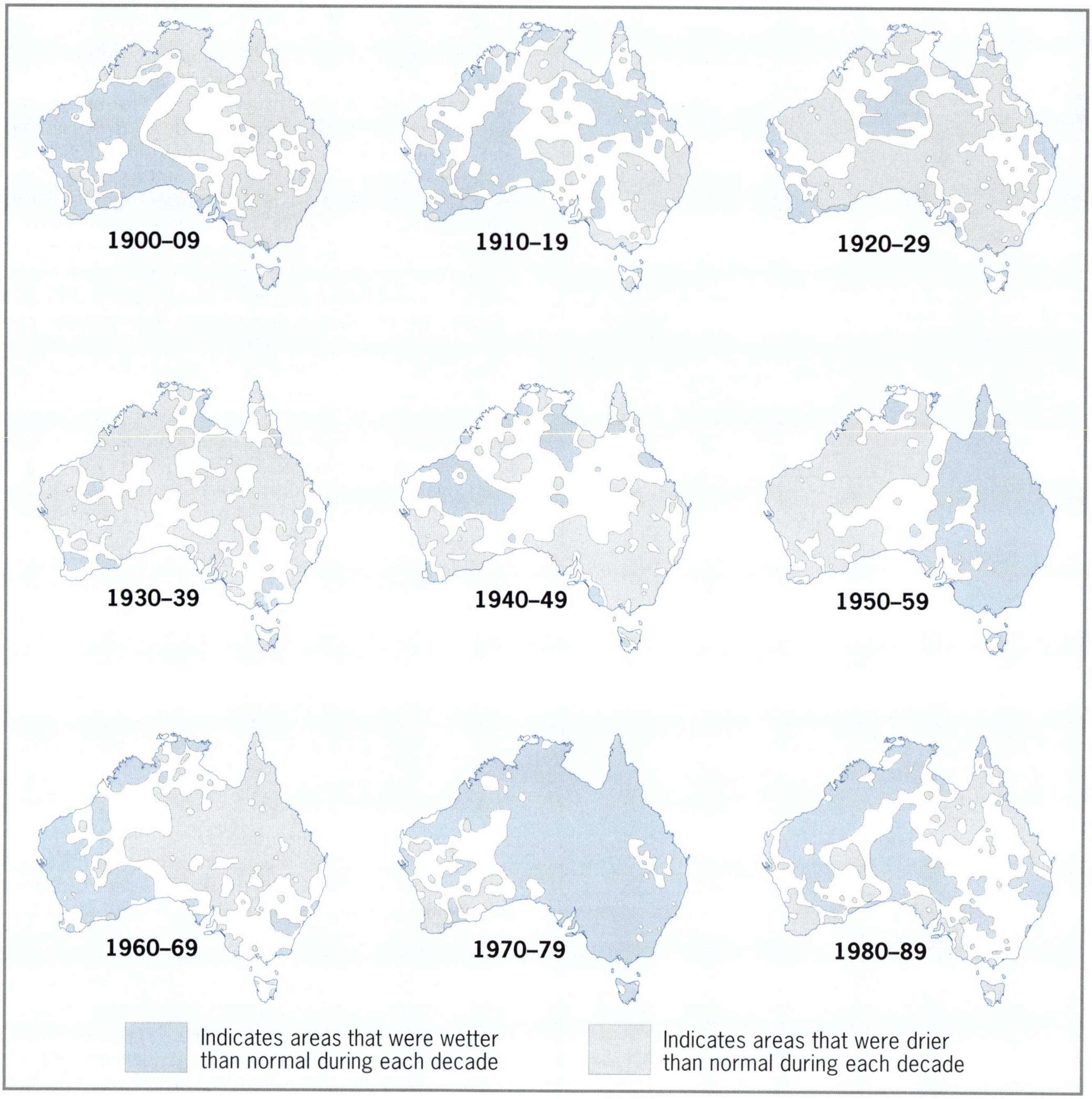

# 40 LANDSCAPES AND LAND USES

| 1900–09 | 1910–19 | 1920–29 |
| 1930–39 | 1940–49 | 1950–59 |
| 1960–69 | 1970–79 | 1980–89 |

Indicates areas that were wetter than normal during each decade

Indicates areas that were drier than normal during each decade

**Figure 2.19** Australia's long-term rainfall variations

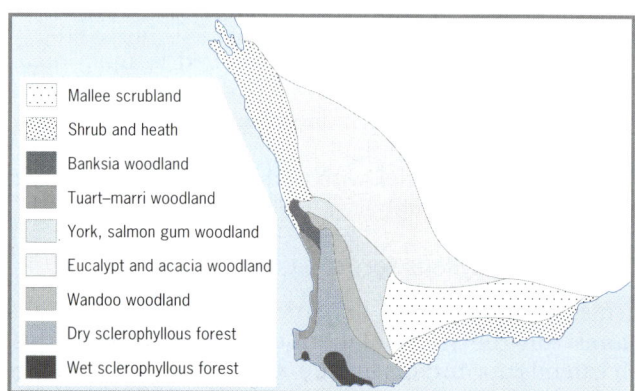

- Mallee scrubland
- Shrub and heath
- Banksia woodland
- Tuart–marri woodland
- York, salmon gum woodland
- Eucalypt and acacia woodland
- Wandoo woodland
- Dry sclerophyllous forest
- Wet sclerophyllous forest

**Figure 2.20** Vegetation formations

escarpment to the north and is separated from the west coast by coastal woodlands. The dominant tree specie of the forest is the jarrah. Other significant trees include the marri, blackbutt and wandoo.

Early European settlers knew the jarrah as the Swan River mahogany. It is a dense hardwood timber, highly valued as a building and furniture timber. The species is not as tall as the karri, with heights seldom exceeding 40 metres, and the trees are more widely spaced, with the canopy providing shade of around 15–30 per cent. The jarrah tree has an extensive, deep root system that can penetrate the hard lateritic duricrust and soils of the scarp to reach ground water, making it well adapted to the drier conditions in which it is located. The ground cover is less dense than that found in the wet sclerophyllous forest and the undergrowth species exhibit xerophytic adaptations to the longer dry period. In the northern section, around Mundaring, a variety of shrubs including the grass tree, zamia palm, she-oak, banksias and acacia can be found.

Jarrah can also be found growing on the coastal plain, particularly on the Guildford (alluvial) and Bassendean (podsolic) soils. These trees are generally shorter than their escarpment counterparts, with a more spreading canopy. Here they form an association with the banksias and marri that make up this woodland formation.

## Woodlands

Where soil, topography and climatic conditions are restrictive the open forest gives way to woodlands. This formation is characterised by a parkland appearance. One means of classifying this formation is to compare the length of the canopy to the bole (trunk). Where the canopy is 50 per cent or more of the bole for the majority of the tree types, then it is a woodland formation.

### Coastal woodlands

The two main woodland formations in the South-West occur along the lower west coast, stretching from Busselton to Lancelin. These include the tuart and peppermint woodlands near Busselton and the banksia–marri woodlands that once covered much of the Perth metropolitan region.

The tuart woodlands have also been classed as open forest in some references. This vegetation type has adapted to the sandy calcareous soils of near coastal locations. It is limited to areas where the annual rainfall generally exceeds 600 millimetres. The tuart's root system extends down to tap the shallow ground water supplies beneath the sandy ridges and plains on which it grows. The best examples of this species are found in the Vasse district near Busselton, where the trees can exceed 40 metres with a girth of some 7 metres at the base. Here in the Ludlow State Forest they are the dominant type and the lack of undergrowth gives the area a park-like appearance. Other examples of tuarts can be found as far north as Yanchep National Park, where they are shorter than that their southern counterparts (see figure 2.21).

The banksia–marri woodland formation is inland from the tuart and extends further north, to around Lancelin. It is mainly found on the Bassendean and Guildford–Pinjarra soil formations. The banksia is a plant well adapted to the low fertility sandy soils of the coastal regions. It takes on a variety of forms, ranging from small shrubs through to medium-sized trees. Its hard-leafed structure and numerous xerophytic adaptations allow it to withstand extended periods of drought. The often large well protected seed pods common to the protea family protect the seeds during fire and allow for seedling regeneration of the species once the fire has passed.

**Figure 2.21** Tuart woodlands

The coastal woodlands have a variety of melaleuca and acacia shrubs within the undergrowth, often with small hard leaves and, in some cases, thorns. The widespread grass tree is also a common feature of the woodland. In addition this formation often contains a range of spring-flowering wildflowers and ground orchids.

### Plateau woodlands

To the east of the dry sclerophyllous forest the lower rainfall results in a variety of woodland formations. These include the wandoo woodlands, York and salmon gum woodlands and the dryandra dry woodlands. In many cases these original formations have been cleared to make way for mixed farming activities and it is not until the eastern edges of the study area is reached that the remaining semi-arid woodlands are encountered.

The wandoo woodland borders the eastern edge of the dry sclerophyllous forest, occupying a band about 50 kilometres wide stretching from Gingin in the north to Boyup Brook in the south. The trees are generally widely spaced with limited undergrowth and less than 20 per cent canopy cover. The wandoo is a medium-sized tree growing to around 15 metres in height. It has a white and tan mottled bark, which sheds in long stringy pieces to reveal a smooth trunk. Its timber is very dense and hard, making it suitable for some building purposes. Like many other inland plants it has a number of xerophytic characteristics including hard leaves, an extensive root system and a protective bark. The undergrowth contains grass trees, zamia palms, acacias and banksias.

## Shrublands and heathlands

North of Lancelin the coastal woodlands begin to give way to shrub and heathlands. Near the coast the harsh environment (strong winds and salt spray) prunes the vegetation to a relatively uniform height and produces a predominantly closed heath formation. The shrubs are densely packed, with little open ground. This is similar on the south coast to the east of the open forest. The plants of the heathlands are from the melaleuca, acacia, grevillea and banksia families.

Where the vegetation ranges from around 2–4 metres in height it forms the shrublands. Eucalyptus shrubs in this formation take on the mallee characteristic, forming multiple branches from a single tuberous root system. Much of the southern interior of the South-West was covered in this type of vegetation. In this area it tends to be closely spaced, with limited opportunity for the development of significant undergrowth. On the eastern edge of the study area the mallee shrublands are more open, allowing plants such as saltbush and bluebush to develop between the eucalypts.

In the 1960s significant areas of the southern mallee formations were cleared to open up the region to wheat farming and grazing.

### EXERCISES

1. Using examples where appropriate explain the following vegetation features:
   (a) woodland
   (b) wet and dry sclerophyllous forest
   (c) mallee
   (d) bole
   (e) heath.

2. Identify and discuss the factors that have influenced the vegetation formations in the South-West.

3. Produce a table similar to that illustrated below in which you summarise the main characteristics of the vegetation formations of the South-West.

| Vegetation formation | Extent and location | Main tree species | Appearance (canopy, ground cover, density and height) | Main environmental controls |
|---|---|---|---|---|
|  |  |  |  |  |

4. Describe and explain the changes in vegetation that occurs along a transect from Pemberton in the south-west through to Merredin in the east.

5. Find and draw illustrations of the following vegetation types:
   - karri
   - marri
   - peppermint
   - mallee acacia

## The fauna of the South-West

While the native animals of Australia reflect the isolation of the continent from other landmasses, particularly following its separation from Gondwanaland, the animals of the South-West have distinctive characteristics resulting from their location in a region surrounded by ocean and desert barriers. Hence, while the marsupials dominate the continent, the South-West contains some species which are unique to the region.

Native animals have generally adapted to the vegetation and climatic conditions of the region, which have affected their dietary habits, populations, reproductive cycles, habitat and distributions. The capacity of these animals to survive since European settlement has depended on their adaptability to changes in the vegetation of the region and their ability to react to the introduction of new animal species. In some cases native animals have not responded well and are endangered or have become extinct. In other cases they have survived and prospered in the new environment. This section looks at the unique characteristics of the fauna of the South-West and their adaptability to the changes brought about by European settlement as well as their response to the climate, soils and vegetation of the region.

## Birds

There are about 500 species of birds found in Western Australia, of which some 300 live in the South-West. Many of these are common to the whole of Australia, with only about 12 being unique to the South-West. The latter include the white-tailed black cockatoo, the noisy scrub bird and the western thornbill. Of the 12, five are classified as threatened or in need of special protection, primarily due to the loss of habitat or the impact of feral animals. Some birds, including the magpie, pink and grey galah, silver gull and Australian shelduck have expanded their populations. These species have been favoured by the introduction of new food sources associated with European agriculture or, in the case of the silver gull, the availability of food scraps in open rubbish disposal sites. Two examples of birds found in the South-West are provided to illustrate the relationship between these animals and the wider environment.

### The white-tailed black cockatoo

The white-tailed black cockatoo is **endemic** to the south-west of Western Australia. These birds can often be seen flying over central parts of Perth or feeding in suburban gardens. Flocks of several hundred arrive in late autumn searching for feeding areas. State pine forests planted in the metropolitan area after World War II gave them access to a new food source—pine cones. Their natural food is predominantly the seed cones of the marri and banksias, which they tear open with their strong beaks to extract the seeds within. They will also strip bark off trees to feed on the grubs beneath. They pair for life and nest mainly in tree hollows in the wandoo, jarrah and marri trees of the northern Darling Range. The removal of large, older trees with suitable nesting hollows has had an effect on their capacity to reproduce, as has increasing distances between the nesting sites and the location of food supplies (the longer the parents are away from the nest, the greater the likelihood of the chick dying). They will generally produce one successful offspring per year. Given their lifespan of around 35 years it is difficult to know if the population is in decline, or the rate at which this might be occurring.

### The black swan

The black swan is found over much of Australia even though it is often associated with the south-west of Western Australia. It is the largest of the freshwater birds, with a height of about 1–1.5 metres and a wingspan of some 2 metres. Immature swans are greyish-brown in colour for about two years, then they mature and their feathers change to shiny black. During moulting the birds are flightless. Black swans' flattened bills are red and white with fine grooves that help grip the underwater plants; their legs and webbed feet are black. They do not dive for food. Their long necks allow them to reach down and feed on plants in deeper water than can geese or ducks.

Nesting season occurs from February through to September. Black swans mate for life and build nests in the wetlands, usually on islands or on water plants. These nests are large and built of sticks and vegetation by both parents. While most species of swans will not allow another swan into their territory during nesting season, black swans are the exception and will often build their nests in colonies.

Once the female lays four to eight pale green eggs, both parents incubate the eggs for 35–40 days, with the female doing most of the nest-sitting while the male does most of the defending and chasing of 'enemies'. The young swans, called cygnets, are covered with a fluffy grey-coloured down that they keep for 3–4 weeks. Shortly after hatching, the cygnets can swim and they quickly learn to feed themselves. They also may ride on their parents' backs for longer trips into deeper water. This family unit stays together for about nine months until the next breeding season, when the young swans go off to find their own territory.

While the birds have adapted to European settlement they have been affected by reduced habitat. The in-filling of swamps and wetlands and the development of farms and urban centres around the major river systems of the South-West has had an impact on their overall numbers.

## *Mammals*

Western Australia's native mammal numbers have declined significantly since European settlement. Some 10 species have become extinct and another 36 are threatened or endangered. These are the medium-sized, non-flying animals, ranging in size from the honey possum through to the native cat or chudich. The loss of habitat, introduction of predators such as foxes and cats and competitors such as the rabbit, and changing fire regimes have all had an effect on the animal populations. Animals with low reproduction rates, specialised diets, short lifespans and limited ability to run, hide or defend themselves are those that are most vulnerable.

Two mammals with significantly different requirements and responses are the numbat and the ash grey mouse.

### The numbat

The numbat was once distributed over 25 per cent of the Australian continent, but its numbers have declined sharply since the beginning of European settlement. Its range had shrunk to the south-west of Western Australia by the 1960s. By the late 1980s, only two colonies were left in the wild (west of Ongerup and south of York). In 1990 it was found in the Dryandra State Forest and the Perup Management Priority Area (east of Manjimup), and had been discovered in several suburbs south of Perth, all in the south-western portion of Western Australia.

There are several reasons for the numbat's decline, including the effects of introduced predators such as foxes and the clearing of land for agriculture. Land clearing eliminates dead and fallen trees, which can be used for shelter and as a source of termites (an adult numbat will consume up to 20 000 termites per day and requires an area of around 100 hectares from which to obtain this food). Another factor contributing to their decline could have been the cessation of Aboriginal fire stick farming practices. The smaller, controlled fires started by Aborigines reduced the frequency of larger, more devastating bushfires, which could wipe out the numbat's habitat, food source and shelter.

### The ash grey mouse

This native member of the rodent family has a body length around 7 centimetres and weighs some 30 grams. It inhabits low heath and shrub vegetation in south-western Australia, avoiding the higher rainfall, forested areas of the region and preferring instead an area corresponding to the 600–300 millimetre isohyet zone. The habitat of the mouse stretches from near Shark Bay through to Esperance. Much of its range

**Figure 2.22** Black swans

corresponds to the wheat and sheep belt of the South-West and in these cleared areas it has been restricted to islands of remnant vegetation.

The ash grey mouse is nocturnal, spending its days in burrows some 60 centimetres below the surface. It prefers sandy to gravelly soils in which to construct these burrows. During winter it obtains its moisture requirements from a diet of seeds and bark. In summer, when the plant moisture is lower, it changes to a variety of insects to meet its moisture needs. In most areas breeding occurs in spring; the animals reach maturity in August. In the eastern extremities of its range, around Southern Cross, breeding is in response to the periods of rainfall and increased food supply rather than the season. The animals have a population density of about two to four per hectare and a life expectancy of one year, thus replacing their population annually. With the introduction of grain farming through most of its territory the mouse has benefited from this new food source and maintains a significant population.

## Reptiles

Of Australia's 750 known species of reptile, more than 440 can be found in Western Australia. Few reptiles in the state are declared rare or threatened and none are known to have become extinct since European settlement. A variety of reptiles can be found in the South-West including lizards, tortoises and snakes. The short-neck western swamp tortoise is the best known of the endangered species and there is a captive breeding program under way to ensure its survival. Its habitat includes one or two swamps near Bullsbrook, just to the north of Perth. Lizards are common throughout the region and include monitors, skinks, slinks and geckos. The snakes of the South-West include the dugite, western tiger snake, death adder, western brown or gwardar, crowned snake, hooded snake and the carpet python.

### The dugite

This species is found across the southern half of the South-West, from Perth to Augusta, and eastwards to the Nullarbor Plain. It is around 2 metres in length when fully grown and generally brown or copper in colour. Its habitat ranges from coastal plains to the Darling Plateau. It prefers areas of sandy soil and can be found in shrubland, heath, woodland and forest environments. Since European settlement its numbers have increased due to the growth in house and field mice populations associated with the development of agriculture. It is sometimes confused with the gwardar, a close relative.

### The ornate crevice dragon

This lizard is found on granite outcrops and in the surrounding woodlands of the semi-arid regions of the South-West. Its colouration blends with the rock surface of the granite on which it is found and can vary from blue and black through to red and white. It is extremely well camouflaged and difficult to detect when stationary. Food varies with the seasons, from insects such as ants in summer through to flowers in spring. While their water intake is limited, they make use of the pools of water that form on the granite surfaces after rainfall. The adult is around 29 centimetres from head to tail and it is very common within its habitat.

> **EXERCISES**
>
> 1  Identify and explain the factors that influence the distribution of fauna within the South-West.
> 2  Identify and explain the factors that determine the ability of an animal species to survive changes in their habitat.
> 3  Discuss some of the ways in which the animals of the South-West have adapted to the climatic pattern of the region.

## Biophysical interactions

Biophysical interactions between climate, landforms, vegetation, soils and fauna can occur in a number of possible combinationss, including two-way or multiple interactions. Thus it is possible to establish links between climate and landforms and also between climate, landforms and vegetation. In reality, all biophysical elements combine to influence each other to a greater or lesser extent through the establishment of complex interactions. For example, the autumn breeding season for the white-tailed black cockatoo is affected by vegetation, landforms and climate. These factors influence the availability of food and breeding sites, and therefore the survival of chicks.

For purposes of clarity, this section has been limited to two-way interactions between the different biophysical elements and these have been identified in point form. Examples for each point may be found in earlier sections of the text or by using other sources of information.

### Climate and landforms

- Coastal regions have atmospheric temperatures moderated by the movements of air between land and sea. Sea water's small temperature change between seasons means that onshore winds cool the land in summer and warm it in winter. Coastal areas also generally have higher rainfall regimes than locations further inland.
- Mountains or hills produce orographic uplift and can increase rainfall on their windward side while reducing it on their leeward side.
- The weathering and erosion of landforms is closely linked to climatic patterns. Humid and arid landforms illustrate this relationship.
- Factors such as terrain and aspect help produce micro-climatic variations. Northward-facing slopes in the study area will experience slightly higher temperatures.

### Vegetation and climate

- Transpiration from vegetation influences precipitation and humidity. Forested areas can produce micro-climatic variations, which result in higher

humidity, moderate temperatures and higher rainfall.
- As the amount and seasonal duration of rainfall increases, so does the density and the height of the vegetation. The variation between the wet and dry sclerophyllous forests illustrates this relationship.
- The seasonal patterns of rainfall affect the overall characteristics of the plants. The summer drought period in the South-West produces vegetation with a variety of xerophytic responses.
- The combination of rainfall and temperature determines the length of the growing season for plants. This ranges from nine months on the south-west coast through to four months on the eastern edge of the study area.

## Climate and soils

- Soils result from physical and chemical weathering of rock. Climate is a significant force shaping the character of soils—it accounts for the formation of the world's major zonal soils. The soils of the higher rainfall areas tend to be more deeply weathered and leached than those in the drier areas of the South-West. Soils in arid regions with high levels of surface evaporation have moisture drawn upwards, which results in calcium and salts being deposited in the surface layer. These soils are termed pedocals. Soils that develop in hot and wet conditions are termed pedalfers. They have concentrations of **aluminium** and iron and are subjected to leaching.
- The erosive agents of wind and running water can transport soils to new locations. Soils moved by rivers are termed alluvial, while soils moved by the winds are termed Aeolian. The humid nature of the South-West climate has allowed for the development of river systems with sufficient discharge to create extensive areas of alluvial soils at the foot of the Darling Escarpment. Strong onshore winds have resulted in an extensive pattern of dunes along the west coast of the study area.
- The colour and texture of soils can produce micro-climatic changes, particularly with respect to local temperatures. This is influenced by the albedo or reflectivity of the soil surface. Lighter soil types reflect more light than darker soils, which tend to absorb the sun's energy (light energy is converted to heat energy when it is absorbed by solid objects). This heat can then be given off at night and will influence minimum temperatures.

## Vegetation and soil

- The organic component of soil is closely linked to the biomass or amount and type of vegetation that it supports. Dead vegetation produces decomposing matter called humus. This breaks down in the soil, releasing nutrients and influencing the profile or structure of the soil.
- The root system of plants can also break up the soil and allow the penetration of moisture and atmospheric gases, both of which play a part in continued weathering of the inorganic components.
- Vegetation also produces acidic compounds, which can aid in the further breakdown of parent rock material and in the production of new soil.
- Soil pH and composition can affect the types of plants that occupy certain areas. The alkaline nature of calcareous sands supports tuarts, while the deeper clay and gravel soils of the escarpment support jarrah and marri. The permeability of the soils (their drainage characteristics) can also influence the types of vegetation found on them. Swamp melaleuca and flooded gums are successful in areas of waterlogged soils. Wandoo can grow on gravelly clay soils, which are often thin and overlay subsurface granite. Their spreading root system can penetrate the hard soils and anchor the tree in the crevices and cracks of the base rock.
- Where plant nutrients are in limited supply, shrubs such as the acacia adapt by developing nitrogen fixing root nodules. Others, such as the nuytsia or christmas tree, survive as parasites by drawing nutrients from host trees.

## Vegetation and landforms

- The angle of slopes and their aspect can, in conjunction with other factors such as soil and climate, influence vegetation patterns and species within an area. North-facing slopes in the South-West have a higher exposure to sunlight and therefore a higher temperature. Plants in these locations have generally adapted to these hotter and drier conditions and exhibit a range of xerophytic characteristics.
- Topography (relief) can affect the degree of shelter from, or exposure to, the elements. Coastal areas demonstrate this influence, with stunted heath being common on the tops of sand dunes and larger shrubs or small trees being found in the swales between the dunes.
- Vegetation plays a part in shaping the landforms of an area in several ways. Plants act as an agent of physical weathering, with roots penetrating the crevices of rocks. In addition, decomposing organic material produces acidic conditions which can act on the land surface. Plants will also bind soil together and influence the shapes created by erosional agents such as wind and water. Heavy vegetation cover produces a rounded landscape, while areas devoid of vegetation have a more angular appearance. This is often referred to as humid and arid landscapes.

## Topography and soils

- As soil forms on slopes it may be moved by **mass wasting** to the base of the slope. Here it can accumulate as alluvial outwash or flood plain alluvium. This soil may also be graded, with the coarser material being deposited on the upper slope while the finer colloidal material is carried further down.
- Where water accumulates in hollows soils may be poorly aerated and low in aerobic bacteria. Their development is more acidic and lacking in nitrogen.

- The weathering of soils in humid locations influences the shape of the landscape. Hills tend to be rounded with incised valleys. The soil contains more organic matter and is broken down more by chemical processes. Thus the combination of the soil and vegetation cover influences the gradational appearance of the landforms. In arid environments with thinner and coarser soils the landforms are more angular, with the infrequent flooding moving large quantities of unprotected soils and exposing the underlying rock.
- The type of landforms that may develop can be influenced by the predominant soils of an area. The lack of surface drainage features or streams and valleys on the coastal plain of Perth is in part a result of the sandy permeable soils, which encourage infiltration rather than surface water movement. In contrast, hilly areas with less permeable soils tend to have well developed surface drainage features.

## Fauna and climate

- Where adverse climatic conditions, such as aridity and high temperatures, prevail animals may respond by living in insulated burrows and being nocturnal. This is the case with small marsupials and other mammals in the South-West.
- Breeding or reproduction is often timed to coincide with the optimum period of food supply and the most favourable climatic conditions. Thus many animals in the study area breed between autumn and spring in response to the flowering pattern of the vegetation.
- The prolonged dry summer season has led a number of animals in the semi-arid eastern region of the study area to reduce water loss through concentrated excretion and low moisture **scats**.

## Fauna and vegetation

- Most native herbivores have developed a tolerance to the toxicity of the natural vegetation. In addition these animals have the capacity to digest the hard, fibrous leaves that are common in the Australian bush. They have evolved to extract nutrients from the predominantly eucalypt vegetation.
- Vegetation also offers many animals nesting and breeding habitats. In a mature forest or woodland, trees are a variety of ages and conditions and provide hollows, high and low canopies, a mix of fruits and flowers and a wide range of insect life. They can therefore accommodate a variety of arboreal and terrestrial animals by providing food and shelter.
- Fauna can assist in the spread of plant seeds and in the fertilisation of flowers. Thus they introduce plant species into new areas.
- Insects such as sawfly lavae can have significant impacts on native vegetation by eating the leaves of the plants. In recent years this has placed the tuarts in the South-West under a degree of pressure, with plants in some areas having much of the foliage removed. Insects also play a vital role in the pollination of a range of native plants.

## Fauna and soils

- Insects play an important role in breaking down organic material and releasing the nutrients for reuse. In addition, insects such as ants carry organic material below ground where it decomposes. Numerous small subterranean insects help break up the soil and maintain its friability.
- Larger animals and birds produce nutrient-rich excreta which contains phosphates and nitrates, thus enriching the soils.
- All animals through burrowing and mechanical action can play a part in the breakdown of regolith to produce soil.
- Some of the burrowing mammals and reptiles will be influenced in their choice of habitat by the nature of the soil. They will avoid territories where the soil is too hard or too sandy and less suitable for the formation and maintenance of burrows.

## Topography and fauna

- Animals may assist in the exposure and subsequent erosion of the landscape through the removal of vegetation or the digging of burrows.
- The calcium-rich shells of a myriad of sea creatures have been decomposed by water action to form extensive limestone deposits in coastal regions, thus shaping the landforms of these regions.
- Landforms such as granite outcrops or steep cliffs can provide a variety of habitats for specialised animals. Offshore islands have allowed some animals to survive after their mainland relatives have died out.

### EXERCISES

1 For each of the 10 combinations of geographic interrelationships outlined provide two examples from the South-West.
2 Identify the likely interrelationship that is occurring with regards to the following situations. For example, in situation (a), the high concentration of lime in the coastal sands and the low levels of fertility have favoured the tuart over other tree species. Hence the interrelationship is one of soil and vegetation.

(a) The tuart is a tree found almost exclusively on the coastal strip from Busselton to Perth.
(b) The northern limit of the wet sclerophyllous forest is near Manjimup.
(c) Both plants and animals play a part in the breakdown of regolith.
(d) Locations such as Karnet and Dwellingup have a higher rainfall than surrounding areas.
(e) Soils on the ridges and slopes of the escarpment are thin and stony.
(f) Alluvial soils are deposited along the banks of rivers and streams.
(g) Extensive limestone deposits occur along the west coast of the South-West.
(h) Many vegetation species in the South-West have extensive root systems, protective bark and tough waxy leaves.

# Chapter 3
# Australian Agriculture

The purpose of this chapter is to identify characteristics, processes and issues associated with agriculture in Australia.

## Agricultural systems

Agriculture can be defined as the organised and planned cultivation of plants and tending of animals in order to produce food and other useful products for eventual human consumption. Where these products are mainly consumed by the producer then the agricultural activity is subsistence in nature. Where the objective is to produce a surplus for a market then the activity is commercial in nature. While agriculture is the predominant activity for over 50 per cent of the world's population, the percentage of people directly employed in Australian farming is less than 5 per cent.

Agricultural systems manipulate or modify **ecosystems** to a greater or lesser extent, depending on the type of farming undertaken. Aboriginal Australians practised various types of farming that produced changes in the environment. Fire stick farming involved the periodic burning of woodland and grasslands to encourage new vegetation growth and bring animals into these areas, which could then be hunted. They may have also planted various fruit-bearing trees in order to increase the abundance of these seasonal food sources. In comparison to the farming methods introduced by Europeans these forms of agriculture had a limited impact on the environment.

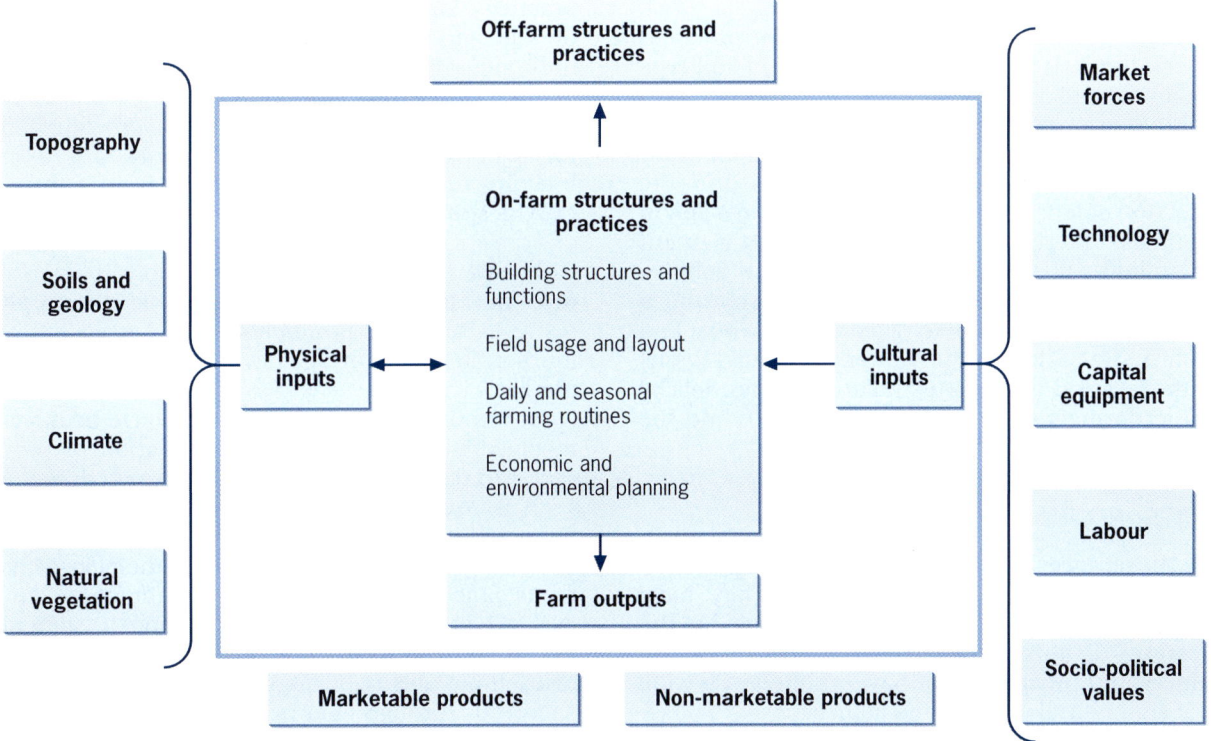

**Figure 3.1** The farming system

Any farming system can be viewed as a series of processes or practices in which the natural environment is utilised or modified by the use of labour and technology in order to meet the needs of the farmer or the consumer. The major elements of a farming system are shown in figure 3.1.

## Physical inputs

While farmers have the opportunity to modify elements of the physical environment to optimise their agricultural productivity, natural advantages still play an important role in determining the type and success of the agricultural enterprise. Climatic factors such as rainfall amount and distribution and temperatures affect the length of growing seasons and the type of farming that can be undertaken. The mild temperatures and absence of extended dry periods in southern Victoria are, in part, the reason why this region has a natural advantage in dairying.

Soils may be heavy or sandy, with shallow or deep profiles. Their structure and fertility play a part in the type of agriculture being practised and in the productivity of the land being farmed. Many soils in the south-west of Western Australia lacked suitable trace elements or minerals and were of low productivity until soil technologies made them commercially viable.

Topography has played an important role in broad acre farming. Large-scale farm machinery such as that used in wheat or dryland grain farming is best suited to flat or undulating topography. Valleys have long been favoured locations for many types of irrigated farming. These activities have also taken advantage of the alluvial soils associated with this landform type. The location of citrus and stone fruit orchards in the valleys along the Darling Escarpment and the vineyards of the Middle Swan Valley illustrate this preference.

Natural vegetation can influence farming practices on several levels. It provided an indication of soil type and suitability to pioneer farmers. It was seen as a natural fodder or pasture for grazing animals. It could act as a barrier to the development of certain areas for farming. It provided building materials for the construction of fences and other farm structures and in recent times it has been seen as an important element in controlling soil erosion, waterlogging and salinity. The mallee-type woodlands of much of southern Australia were easily cleared to make way for wheat and sheep farming, whereas the wetter, forested regions of south-west and eastern Australia were often left by the earlier farmers due to their dense growth and the difficulty in clearing it.

## Cultural inputs

People have, over time, observed and developed responses to the natural environment. They have manipulated ecosystems to their advantage by the application of this knowledge and by inventing machinery and processes that have resulted in a wide array of agricultural systems and landscapes.

The application of technology takes many forms. Mechanical technologies include the development of preparation, planting and harvesting equipment. Biotechnologies can apply to the processes of animal breeding and husbandry. The development of hybrid plant species and methods of cloning both plants and animals are examples of recent technologies in this field. Biotechnologies including GMO (genetically modified organism) research are aimed at modifying plants and animals to withstand a range of environmental stresses as well as reducing the need to use agricultural chemicals, water and soil nutrients. Other technologies include chemical means of enhancing soils and controlling unwanted plant or animal species in order to maximise yield and quality. Computers and agricultural software have revolutionised farm management. Predictive models, where variables such as weather forecasts, market analysis, optimum levels of production and cropping/livestock rotations are applied to determine the best use of the natural and human resources, are now used in commercial farming.

There is a close relationship between the use of technology and the need for rural labour. A major emphasis in Australian agriculture has been the drive to reduce labour requirements by the use of large-scale labour-saving equipment. The substitution of labour with capital equipment has seen the rural workforce of Australia decline from 7 per cent in 1980 to 4 per cent in 1996. There has also been a shift in the workforce from full time to part time, with a significant proportion of workers engaged in seasonal rather than full-time continuous work. This decline in the rural workforce, combined with the amalgamation of farm units, has seen an overall rural depopulation as well as the loss of services to rural communities.

Where the primary objective of the farmer is to produce for consumption by a designated market, then the expectations of that market will affect the farming practices and processes. The growing popularity of organic agricultural produce is one illustration of this influence. The ability to produce pesticide-free or 'natural' foodstuffs is the end result of this changing market force. Another example is the growing demand for lean meat products. The result has been the breeding of low fat or lean varieties of cattle and pigs.

Widespread changes in markets may see the demise of some agricultural systems and their replacement with new farming activities as supply is adjusted to meet new demands. Global or regional oversupply will result in marginal producers being forced out of the market, while shortages will bring them into the market.

Governments in many countries have been actively involved in the structure and operation of the rural sector of the economy. The use of production controls such as **quotas**, marketing boards, land allocation, financial aid, and the provision of infrastructure and advisory services are all examples of their close involvement. Since 1990 the federal government has enacted over 80 pieces of legislation relating to agriculture. These range from laws to enforce quarantine restrictions and controls over the use of agricultural chemicals through to laws to deregulate the marketing of rural products. The *Wool International Privatisation Act 1999* and the *Dairy Industry Adjustment Act 2000* are examples of market deregulation. The introduction of the federal government's national competition policy in the late 1990s has influenced these and other laws affecting production and competition in Australia.

Cultural influences on farm products and practices may be seen as an input into the farming system. The growth in the dairy industry in Australia (milk and milk-related products) illustrates the significant influence of northern European dietary habits in the population. The changing nature of market garden produce corresponds to the growth in migrants from southern Europe and South-East Asia. For example, South-East Asian vegetables now grown in Australia include snake bean, okra, lemon grass, Chinese cabbage and burdock. The dietary preferences of cultural groups within a population will therefore influence the type of market that exists as well as the way in which agricultural land will be used.

## Farm throughputs or practices

The farming enterprise combines physical and cultural inputs to produce the desired agricultural outputs. The combination of the different inputs involves the application of farming practices and routines to produce the desired products. This process is termed throughput. It involves setting up buildings and other structures to assist in rural land use, the establishment of farming routines and practices, and management and modification of the natural environment in which the farming activities occur. Over time, these practices will change in response to changes in the environment, technologies, markets and a variety of socio-political forces. The nature of the farming activity will be reflected in the observable **cultural landscapes** that develop both on the farm and in the farming region in which the land use is found.

In some respects farming practices are as individual as each farmer. Therefore, while it is possible to observe similar seasonal routines, equipment, buildings and patterns on wheat and sheep farms in Australia, each operator will manage the land in response to their own beliefs regarding best practice for the land or the way to achieve optimum output.

Farms do not operate in isolation from the surrounding region. Each farm's practices and activities will impact on other farms as well as any urban areas and natural ecosystems. Thus, when a farmer decides to dam a stream or clear land they can change the conditions under which their neighbours operate. When they use fertilisers or pesticides they can affect waterways and other natural environments. Their economic decisions may have far reaching impacts on the urban centres that rely on the purchases of the farmer. Thus, on-farm structures and practices will influence the way in which off-farm structures and practices operate and develop. This in turn will influence the appearance of the regional cultural landscape.

## Farm outputs

Farm outputs include both marketable and non-marketable products. The main objective of the farmer is to grow products that will meet the demands of the market while at the same time using both natural and capital resources in an efficient and sustainable manner. To this end products may vary depending on the season, and changes in consumer demand, and in accordance with longer-term planting cycles. A wheat farmer, for example, may produce wheat one year, canola in the second year, barley in the third and graze sheep in the fourth.

In 1999–2000, agricultural outputs in Australia were worth approximately $30 billion, representing 4.8 per cent of the nation's gross domestic product (GDP). The growth in the value of agricultural products in recent years is indicated in figure 3.2.

The largest commodities in terms of gross value of production in 1999–2000 were: cattle and calves slaughtered ($5.1 billion); wheat ($4.8 billion); milk ($2.8 billion); and wool ($2.1 billion).

Non-marketable products include waste matter produced as a result of the farming process as well as environmental outputs. For example, crop residues such as stubble may be burnt by the farmer, thus releasing carbon dioxide and particulates into the atmosphere. Nutrient runoff into waterways is another example of these types of outputs.

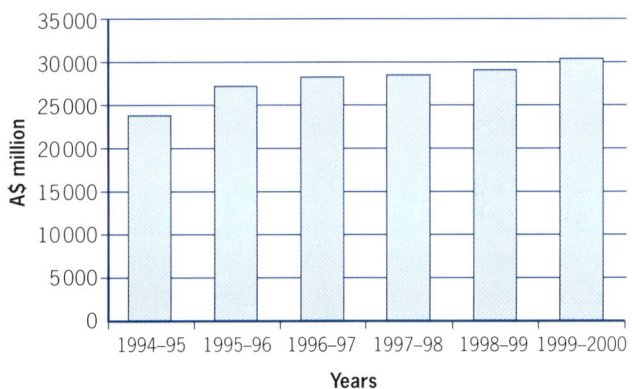

**Figure 3.2** Value of agricultural production

## Intensive and extensive agriculture

The degree to which farmers apply capital to the farming activity, and the extent to which the environment is modified as a result of the agricultural practices, indicate how intensive or extensive the land use is.

## Intensive agriculture

Intensive agriculture is characterised by the following features:
- small farm sizes, generally less than 100 **hectares**;
- amount of capital equipment used per hectare is high;

- labour input per hectare is high;
- significant environmental modification to the farm site;
- farm practices are closely monitored and controlled;
- overall levels of investment per hectare are high;
- outputs or returns per hectare are high.

In 1997, potato farming was carried out on a total of 3021 hectares in the south-west of Western Australia. Average production per hectare was 38 tonnes, while the Australian average production was 32 tonnes. In 1997–98 the average return to farmers in Western Australia was $310 per tonne ($11 780 per hectare). The average farm is around 60 hectares and is intensively cropped. Significant capital investment in irrigation systems, planting and harvesting equipment and on-farm storage contributes to a high cost of production per hectare. Figure 3.3 illustrates the intensive nature of this type of farming activity.

**Figure 3.3** Potato harvesting equipment

Other examples of intensive agriculture include market gardening, fruit growing, viticulture, tropical fruit farming, dairying and fat lamb or intensive grazing. These activities are generally located in regions with extended growing seasons that are close to major urban markets, or with well-developed transport links to them.

**Figure 3.4** Intensive farming

## Extensive agriculture

Extensive agriculture is characterised by the following features:

- large farm size, generally over 1500 hectares;
- low level of capital improvements and usage per hectare;
- labour input per hectare is low;
- often limited environmental modification to the farm site;
- overall levels of investment per hectare are low;
- output or returns per hectare are low, however total returns may be high given the large area being farmed;
- farm size makes it difficult to closely monitor the operation of the property.

In Northern Australia cattle are grazed over large properties, known as stations. The cattle graze on native pastures at very low stocking densities, often one head per 50 hectares, compared to the intensive grazing of southern areas, where stocking rates of three head per hectare are common. Cattle stations can exceed 1500 square kilometres (150 000 hectares), with some being the size of a small country. The largest cattle station in Northern Australia is more than 1.6 million hectares. The annual return per hectare from these very large properties is often around $2, given the amount of land required to support one animal.

The raising of cattle for beef in Northern Australia and the grazing of sheep for wool in the semi-arid regions of the continent are two examples of extensive agriculture. The pattern of intensity (scale) of rural operations in Australia is illustrated by the map in figure 3.5.

## Locational influences

The distribution and type of land use practised within a region will reflect a number of locational decisions made by farmers, governments, markets and the community. Why a region develops as a producer of wine or wheat or beef cattle can be seen in the way in which these decisions have been influenced by environmental conditions, commercial considerations and a range of cultural, historical and political factors. Figure 3.6 illustrates some of these influences.

### Environmental controls

The influence of environmental controls depends on the physical requirements of the agricultural activity, the intensity of production and the capacity of the farmer to alter the environment. Where the activity involves significant modification of the environment and the capacity to solve problems using technology, environmental controls become less significant. Market gardeners, through the use of fertilisers, irrigation and greenhouses, are less reliant on the environmental conditions than large-scale pastoralists, where natural pastures are relied upon and factors such as drought and flood are very difficult to avoid.

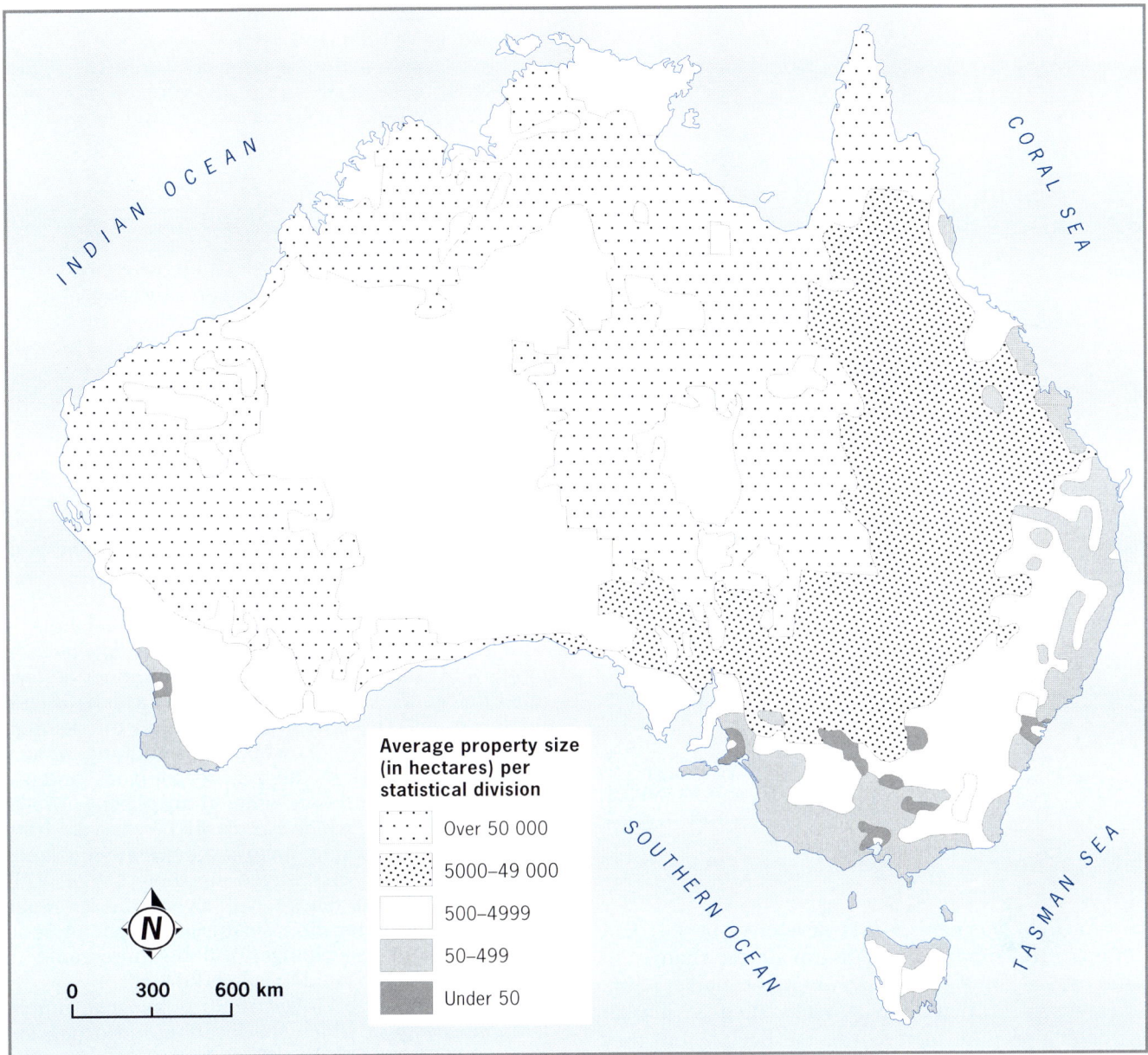

Figure 3.5 Pattern of intensity of agricultural land uses in Australia (blank areas are non-agricultural lands or urban areas)

Despite the capacity of people to manipulate physical conditions, the locational influence of the environment will still play a part in providing regions with a natural advantage for certain types of agriculture. It will ensure that one location has greater levels of productivity when compared to other, less desirable locations. Dairying and whole milk-production in Victoria illustrate this. The mild climate and even distribution of rainfall, combined with the fertile alluvial and volcanic soils of the dairy zone, help the Victorian farmer achieve greater outputs per hectare than in other Australian states.

## Commercial considerations

Land values for rural properties can vary from $100 000 per hectare for market garden properties near large urban markets to $10 per hectare for large pastoral leases in Northern Australia. 'Moola Bulla' pastoral lease, near Halls Creek in the Kimberleys, changed hands in 2002 for a record price of $18 million. It covers an area of 660 000 hectares—twice the size of Singapore—so this represented a sale price of $28 per hectare. The need for the farmer to receive an income that will justify the value of the land partly explains the variation in land uses that occurs at different distances from the market point. High yielding intensive farming is more successful in near-market locations, while the low yield extensive activities occur in the more distant locations.

The variation in transport costs associated with different distances to the market point also affects the land use decision and the return per hectare. These patterns of land use are illustrated by the economic rent

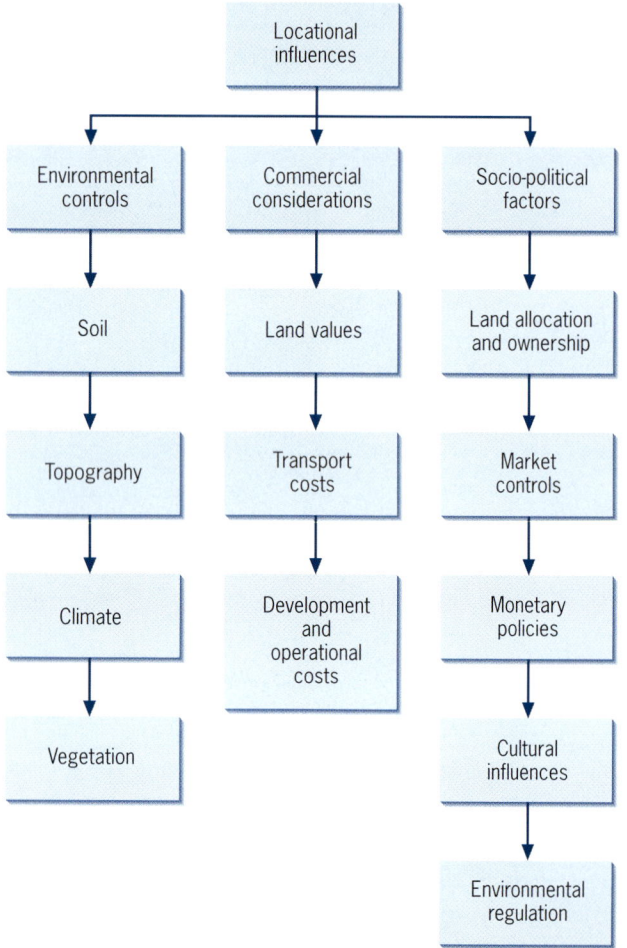

**Figure 3.6** Locational influences

or land rent mechanism first developed by J H Von Thunen. This model is used to explain the changes in land use seen as you move away from the market point.

Intensive horticultural activities such as market gardening produce high returns per hectare, but they also incur high transport costs and this significantly reduces the farmer's income if they carry out production on land that is away from the market. Other less intensive activities such as dairying generate less income per hectare, but their transport costs are also lower and therefore this activity can be successfully carried out in more distant locations. Dairying in turn gives way to wheat and sheep farming, with lower returns per hectare and the need for large areas of land to produce an income. Fortunately produce only needs to be transported once or twice per year, making the transport cost for wheat and sheep farmers relatively low. Figure 3.7 illustrates the theoretical effects of economic rent on land use patterns.

A further cost involved in determining the location and type of agriculture practised is the cost associated with developing the land. The advantage of relatively level land for intensive-irrigation farming encourages the use of coastal and river plains for this type of agriculture. Undulating or rugged topography presents higher development costs and thus tends to favour more extensive rural land uses.

Other factors such as labour costs and the higher maintenance costs of capital equipment may also discourage intensive forms of farming in isolated locations.

## Socio-political factors

There is a range of historical, political, social and cultural factors that can affect when, where and what type of farming might take place on a piece of land.

In Australia, there has been a long history of distributing Crown lands for rural uses. The allocation of **freehold** and **leasehold** land is one such example. Close-area farming land was distributed or sold with freehold title while large grazing or pastoral properties remained the property of the Crown (government) and were leased for set periods of time. This difference in tenure resulted in distinctions in how much capital investment the occupier was willing to put into improving the land and the ways in which the land could be used.

In Western Australia, the *Land Act 1933* sets a number of restrictions on the use of leasehold land. The land is leased on the condition that it may only be used to graze livestock. The Act contains a number of penalties that control the alternative uses to which the land might be put. These include restrictions on clearing, sowing or sale of introduced pastures, running of stock that is not approved by the minister for lands, and use of the land for purposes other than grazing. Unlike freehold title land, which can easily be changed from one type of rural production to another, leasehold lands remain relatively unchanged in their usage. This difference in land tenure will, over time, produce differences in the locational patterns of agriculture in Australia. Land use changes will be more rapid in freehold areas compared to leasehold lands.

The provision of infrastructure by government is another way in which the location and expansion of rural land use can be affected. Between 1905 and 1916, the Western Australian Government financed

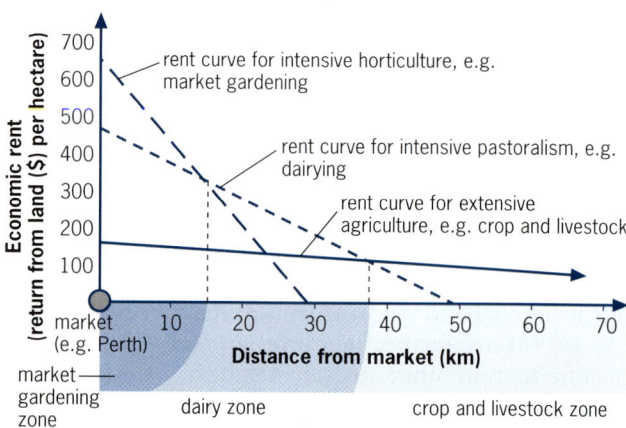

**Figure 3.7** The effect of economic rent on the location of agricultural land use

expansion of the railway system into the wheat and sheep belt. As the system expanded, new farms were developed along the routes and the wheat and sheep belt expanded north and south along lines to Geraldton and to Albany. Added to this, grain cooperatives and storage silos were set up, and rural roads and ports were improved—all part of the support given to develop rural industry in the state.

Since the beginning of the 1970s there has been a change in political attitudes towards the expansion of agricultural areas into the remaining natural ecosystems. Farmland is now seen as a finite resource that needs to be managed and conserved. Conservation controls restrict further clearing of existing properties and funds are provided for the rehabilitation of degraded lands, mainly through the replanting of vegetation. Landcare is one such initiative. Farmers within local regions form Landcare groups or cooperatives. These groups plan and implement programs aimed at rehabilitating waterlogged, eroded or salt affected land. The federal government supports these measures through tax deductions for money expended to combat land degradation. Activities that attract full deductions include tree replanting, fencing of remnant vegetation, feral animal and weed control and the development of management plans. In 1995–96, approximately 25 000 hectares of land in the agricultural areas of the south-west of Western Australia was revegetated.

Marketing and production controls on rural produce initiated by government legislation can influence the location and extent of various rural land uses. For example, the existence of a whole-milk quota system in most states in Australia up until 2000 allowed for the control and protection of milk production. As part of the deregulation process, which commenced in the late 1980s, all states have repealed legislation governing pricing and sourcing of drinking milk, and the state milk authorities which administered these controls are being wound up. Since that time a number of farmers have been forced to leave the industry and former dairy farms have been purchased for plantation timber production, vineyards and intensive cattle grazing. Thus, the locational patterns of rural land use are changing in response to these new challenges.

Monetary and **fiscal policy** can and does exert a marked influence on the success or otherwise of the Australian rural industry. The floating of the dollar against other currencies has had some success in making the country's rural exports competitive overseas and has encouraged the development of new rural activities. As overseas markets develop, land uses will adjust to meet new demands and the locational patterns of agriculture will change. Fiscal policy with regards to the availability of credit and the interest charged on loans also impacts on farmers. A wheat farmer may carry a line of credit up to $100 000 per year in order to finance the planting of a crop, which may not be sold and paid for until the start of the next season. If interest rates climbed this could lead to severe hardship and the loss of the land to this type of agriculture.

## The rural cultural landscape

A cultural landscape reflects the degree of change brought about by human modification to the original natural environment. It will be determined by a number of factors combining in such a way as to produce distinctive rural landscapes. In studying a landscape, the elements that give it its appearance will include the transport systems and linkages that connect farms with towns, and towns with other towns within and beyond the region. Roads and road hierarchies, railway and seaway linkages, transport nodes and transhipment points, airports and storage facilities all interact to a greater or lesser extent and help to provide a framework around which the cultural landscape develops.

The intensity of the dominant agricultural system also plays a role in shaping the landscape. Ranging from intensive horticulture through to the extensive rangeland grazing systems, the landscape reflects the degree to which the environment has been modified and the spacing between the farm units as well as the rural and urban settlements. Associated with the intensity of land use is the overall productivity and wealth that is generated. Thus, highly commercial and successful agricultural pursuits will be reflected in the landscape by the quality and extent of the infrastructure and other constructions, while areas in decline or those with limited commercial development will produce a different cultural landscape.

Over time cultural landscapes will change in response to changes in environmental, social, economic, technological and political conditions. **Farm amalgamations** and increases in the economic scale of operations will produce a landscape in which fewer people live and where derelict and deserted farmhouses indicate the joining together of small holdings to produce larger properties. The loss of fertile soil through such factors as erosion and salinity changes the appearance of the farms and the patterns of grazing or cropping within a region.

Type of land ownership and land use controls will also influence the appearance of the landscape. This is both a political and an economic response to the natural landscape. If land is surveyed and then identified by a formal title which can be owned or leased and bought and sold, then its extent will be clearly defined by fencing or markers and a distinction between public and private land will act to shape the layout of different land uses. Within an Australian or European landscape the locations of the transport linkages, the rural land holdings and the urban settlements have been based on a tradition of individual ownership and governmental planning policies.

It is also possible to observe the patterns of the cultural landscape by looking at the overall appearance of the farming region as well as the appearance of individual farms. Within the farm boundaries, the types and locations of buildings, fences, watering points or irrigation facilities as well as the degree of clearing of natural vegetation and changes to topography and soils all provide indications of the nature of the cultural landscape.

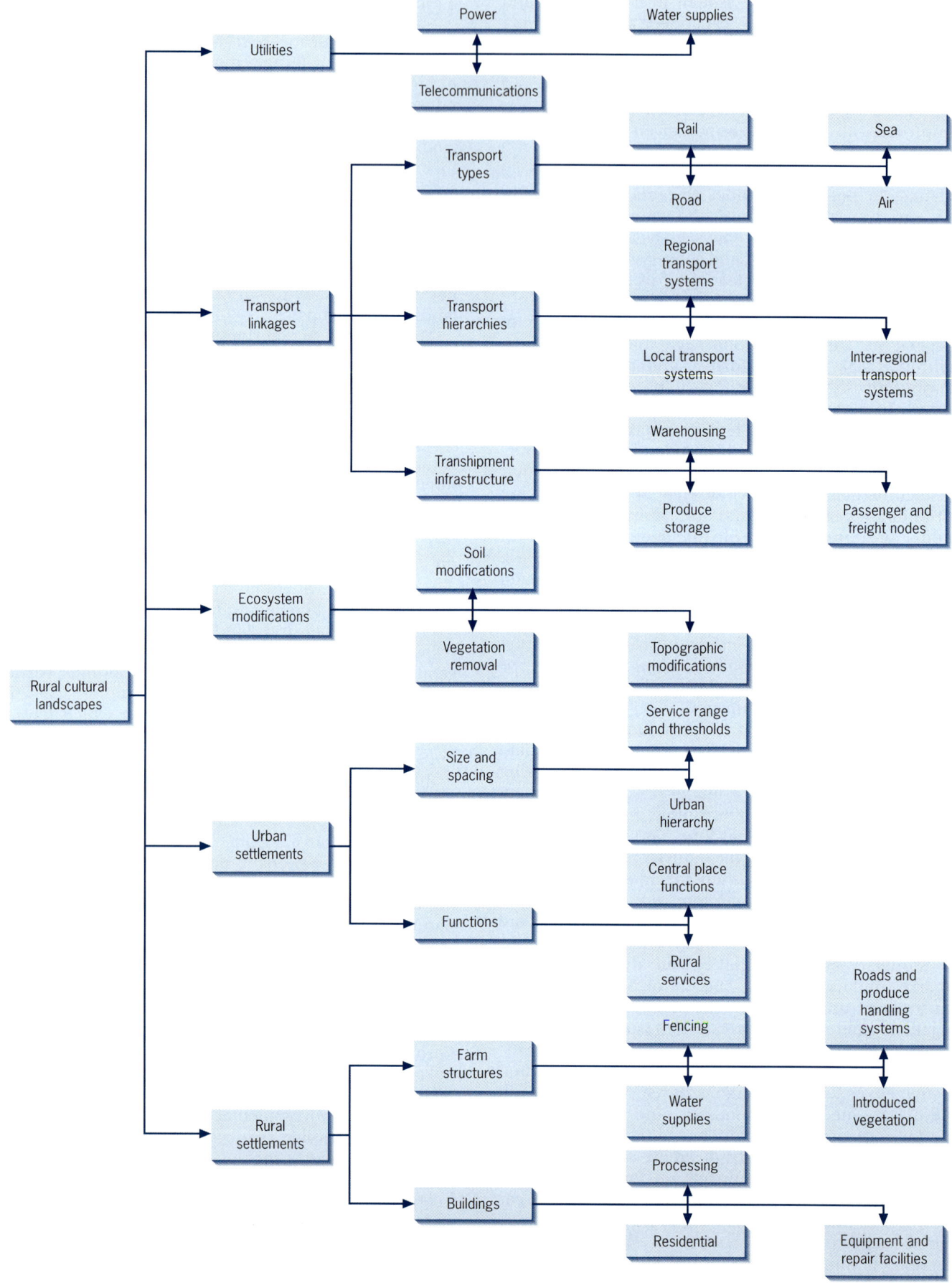

**Figure 3.8** Rural cultural landscapes

Beyond the farm gate how the individual properties link together and how they are served will produce the regional or off-farm landscape. Thus, a region in which vineyards and orchards are predominant will have a different appearance to one where mixed crop and livestock farming is practised. In figure 3.8 the elements that make up a cultural landscape have been identified.

Within the rural cultural landscape a variety of shapes or spatial patterns will develop in response to the interaction of different influences. These may include site factors, the development of transport systems, the effects of government planning, the evolution of the landscape over time, and the level of economic activity within the region. The pattern may be one in which cultural features are clustered (nucleated) or dispersed. If the pattern is one of dispersal then the intensity of the land use will influence the distance between the scattered features. The dispersal may also be regular or irregular in its appearance, depending on factors such as topography, soils or vegetation as well as the methods of surveying used. Regular dispersal patterns include **linear**, **radial** and grid patterns. Irregular patterns include dendritic and random dispersion. The irregularity may be due to features such as rivers, coastlines and variations in the land's shape or relief.

Clustered patterns occur where cultural features within the landscape have aggregated or grouped around a desirable point. This may be a transport node or a favourable physical site. The pattern may be one of clustering along a road (a linear pattern), or around a central point (a circular pattern). The shape of the clustering may also radiate out along transport routes. Figure 3.9 illustrates the possible appearances of the cultural landscapes within agricultural regions.

## Current issues in Australian agriculture

There are many issues in Australian agriculture associated with social, cultural, economic and environmental events and trends. These will be dealt with further in the chapters on agricultural case studies, population and urban networks. In this section a summary of some of these issues has been included to illustrate the scope of the concerns facing Australian farmers today.

### Ecological issues

- Nutrients have increased to five times their natural levels in agricultural landscapes, thus increasing the potential for leakage from land to rivers and estuaries. The increasing of nutrient loads in rivers and estuaries is termed **eutrophication**. The high nutrient levels encourage algal growth, which in turn de-oxygenates the water, causing stress and death to aquatic life forms.
- Biomass productivity from agricultural landscapes has doubled compared to natural levels supported by the original ecosystem. This is a reflection of the role of fertilisers and farm management systems in delivering profitable agricultural products. A

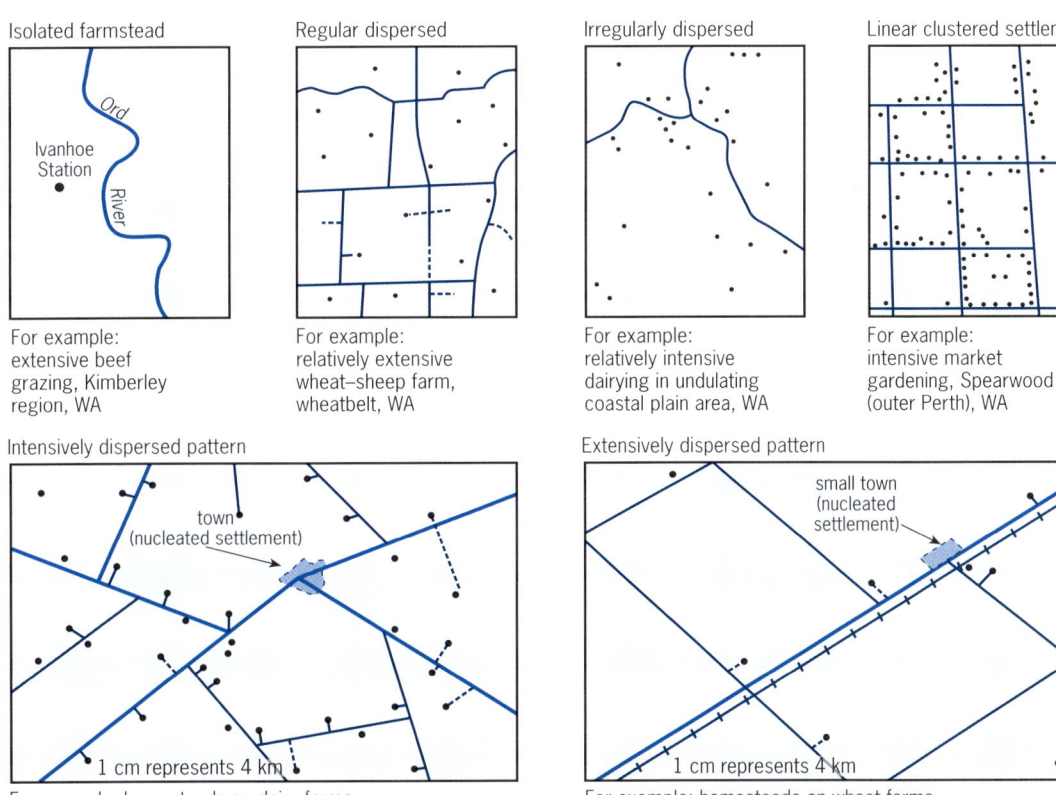

**Figure 3.9** Rural settlement patterns

consequence of this increased plant material is the associated growth in animal and insect populations that benefit from the new food source.
- Almost 1200 million tonnes of soil potentially moves on agricultural hill slopes each year. Much of this soil is transported to lower areas or into streams or rivers. This situation points to the need to put in place soil management strategies that will keep this soil in place.
- As the soils are moved down slopes, about 50 million tonnes of sediment enters the rivers with about 50 000 tonnes of **phosphorus** attached to the sediment. This indicates the need for management practices that will keep the soils on the farms. The use of tree belts and permanent grass cover on slopes can help reduce sheet and gully erosion.
- About 44 million tonnes of sediment each year enters the rivers from gullies, including about 11 000 tonnes of attached phosphorus, demonstrating that even following a long period of government and industry attention to this most obvious of soil erosion activities, much still remains to be done to improve land use practice.
- A further source of sediment is the soil and rocks that are removed from unprotected river banks. Each year some 33 million tonnes of sediment is eroded from banks and carried downstream along with sediments from other sources. This includes some 9000 tonnes of attached phosphorus. The planting and preservation of riparian vegetation is an important means of preventing this erosion.
- It is predicted that nearly 19 000 tonnes of total phosphorus and 141 000 tonnes of total nitrogen will be transported down rivers to the coast each year. This is particularly significant on the east coast of Australia, where valuable agricultural soils are being washed into the ocean or estuaries. In the long term this is likely to have an impact on areas such as the Great Barrier Reef National Park, as well as producing algal blooms that will affect both fisheries and recreational areas.
- Fifty per cent of Australia's agricultural (cropping) soils (44 million hectares) have soil **pH** values below optimal levels (that is, less than 5.5). This will place pressure on the continued production of acid-sensitive crops. The reduction in soil acidification by the use of lime on a regional scale is currently inadequate. It is estimated that the rate of lime application to **acid soils** needs to increase by some 80-fold to solve this problem. Surface and subsoil acidity exists in all Australian states and it has been estimated that the total area of acidity is eight to nine times that affected by dryland salinity. Soil acidification is a more serious problem than salinity, in terms of area affected and cost to the economy.

The largest areas of acid soils are in New South Wales, Western Australia, Victoria and Queensland. Soil acidity is caused by the removal from the fields of crops or pastures which are slightly alkaline, the leaching of **nitrogenous** fertilisers below the root zone, overuse of nitrogenous fertiliser, and a build-up and decay of organic matter. It should be noted that

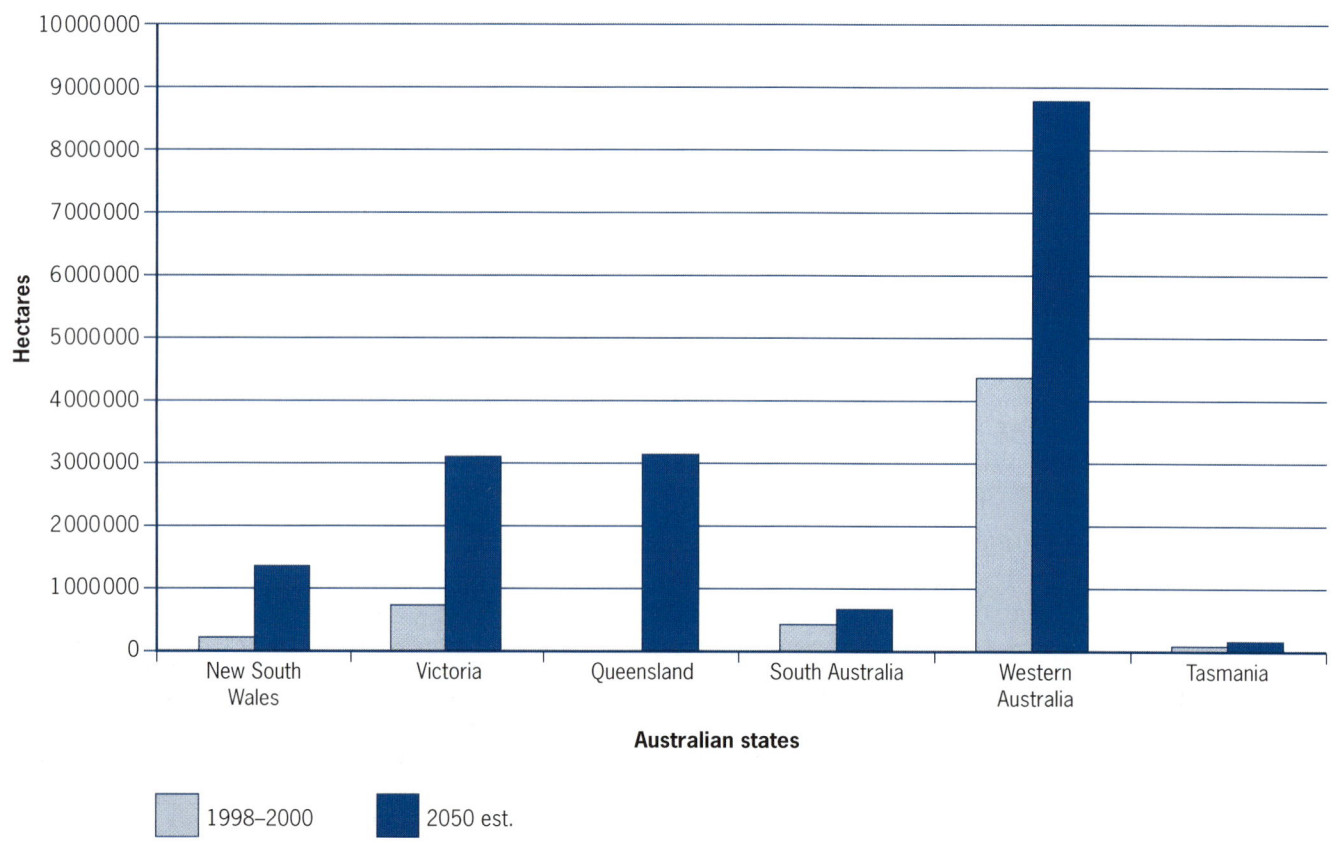

**Figure 3.10** Dryland salinity forecasts

soil acidity is a natural process that occurs as soils age, however farming practices can speed up this change. The major production losses come when acidity increases to the point where toxic elements in the soil, notably aluminium and manganese, dissolve. The aluminium damages root growth, restricting the ability of plants to take up nutrients and water, and excess manganese uptake is toxic to shoots.

- The widespread clearing of woodland and mallee scrublands for dryland farming, particularly the cultivation of cereal crops, has had a significant effect on the soil salinity levels. Susceptible areas have been lost, as the productive agricultural soils have become too salty to farm. The extent of this is illustrated by figure 3.10.

Salinity is a natural feature of the Australian landscape. Large areas of inland salt lakes as well as coastal regions contain high levels of salts. This is identified as primary salinity. Secondary salinity is the result of introducing European farming practices into the landscape. The way in which the spread of salinity occurred in response to the clearing of land and the growth of annual shallow-rooted crops is illustrated in figure 3.11.

At first the soils remain unaffected by the clearing and planting of crops. Gradually, **ground water** levels increase and the watertable rises towards the surface, bringing salt with it. Low-lying areas are subjected to increasing levels of salinity. Ground water flow through the soils also increases and salts are flushed into rivers and streams, causing them to become increasingly **brackish**.

Western Australia has the largest area of dryland salinity in Australia and the highest risk of increased salinity in the next 50 years. Currently salinity affects some 9 per cent of the agricultural land in the South-West, however an estimated 4.3 million hectares (16 per cent) of the region have a high potential of developing salinity due to shallow watertables. This is predicted to rise to 8.8 million hectares (33 per cent) by 2050.

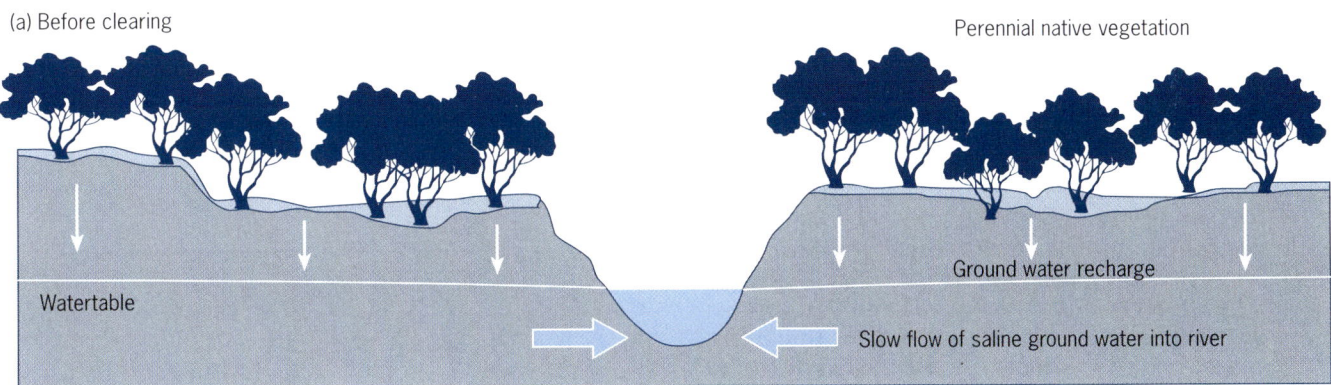

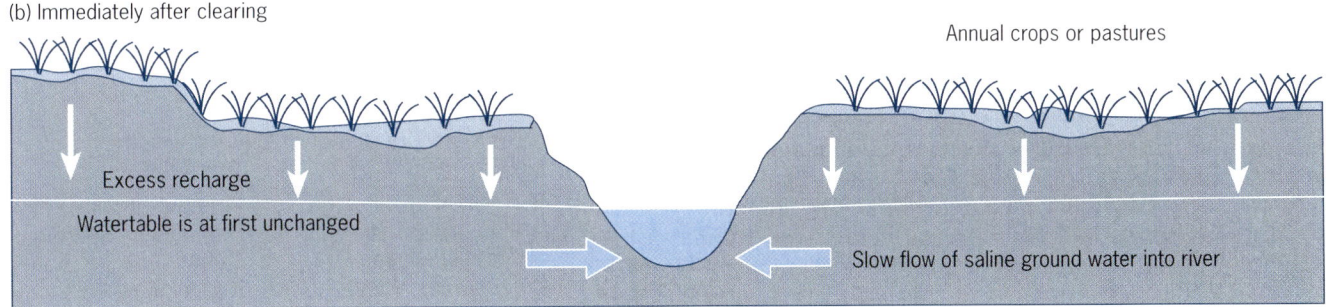

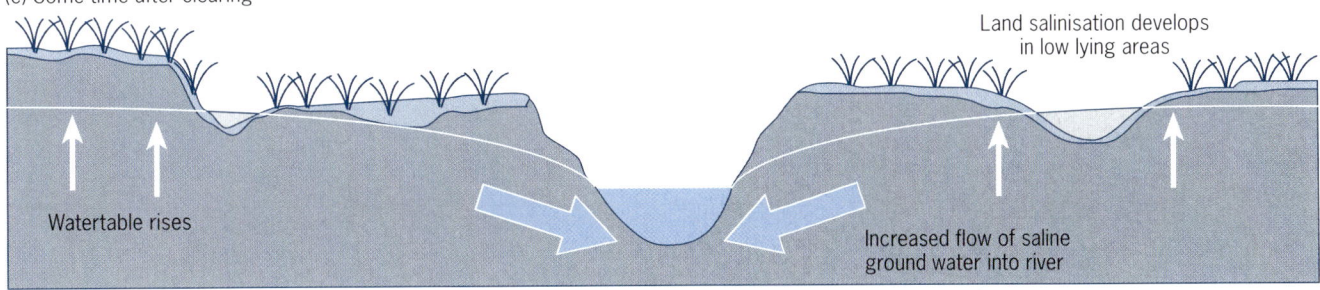

**Figure 3.11** Land clearing and soil salinity

In 2000, the risk of salinity was predominantly in the eastern wheat belt in valley floors and adjacent areas. Eastern sections of the northern wheat belt also exhibit high risk. There are also some coastal areas at high risk around Bunbury and Donnybrook Sunkland. Salinity expansion by 2050 will be mainly in the Great Southern and south coast regions.

## Changing rural landscapes

- Farm numbers have decreased from 178 000 in 1982 to 145 000 in 1996–97. Average property size has increased in the cropping and grazing industries. This trend points towards an ongoing process of farm amalgamation and the adoption of economies of large-scale agricultural production.
- Land use and farming systems will continue to evolve in response to changes in commodity prices, market arrangements and natural resource conditions and opportunities. In the South-West the changing land use is illustrated by the decline in dairying and the expansion of vineyards and tree plantations.
- Livestock industries have reached a plateau. Areas under cotton, sugar cane, potato, rice and horticulture have all increased since 1983. Viticulture is also expanding in many regions.
- Between 1982 and 1997, cereal grain yields per hectare have improved in many regions, notably where crops are more diversified in regions of more reliable rainfall. In the drier areas increased productivity has been less spectacular. Improved nitrogen management was associated with strong productivity gains in several regions of Australia. This increase, however, should be contrasted with the increasing areas of land that are being lost to production through the effects of acidification, **salinity** and erosion.
- Annual variations in yields of Australia's dominant crop, wheat, due to climate have been reduced through development of drought-tolerant species and disease control.
- The area of irrigated agricultural land in Australia has increased by 26 per cent in the last 20 years. Two-thirds of all irrigated land is in the Murray–Darling Basin where nearly half is used for pasture. Irrigated areas Australia-wide under cotton, sugar cane, pasture and fruit have also increased. Often this increase has been at the expense of natural environments where rivers are dammed to meet expanding water needs or where bigger demands are placed on underground water supplies.

### EXERCISES

1. What are the essential elements in any farming system?
2. What are the main characteristics of intensive agriculture and what factors account for the location of intensive farming in Australia?
3. Identify, using examples, how extensive agriculture differs from intensive agriculture.
4. Calculate the approximate area of New South Wales, Victoria and Western Australia that will be subject to salinity in 2050 and suggest ways in which this expansion might be restricted.
5. Using diagrams, explain why the impact of salinity on agricultural land may be described as a 'time bomb, waiting to explode'.
6. Identify the cultural and physical factors that might be responsible for the change in the rural land use of an area. For example, the decline in dairy farming and the growth in tree plantations.
7. Study the following diagram of the land rent mechanism and then explain the points marked from 1 to 6.

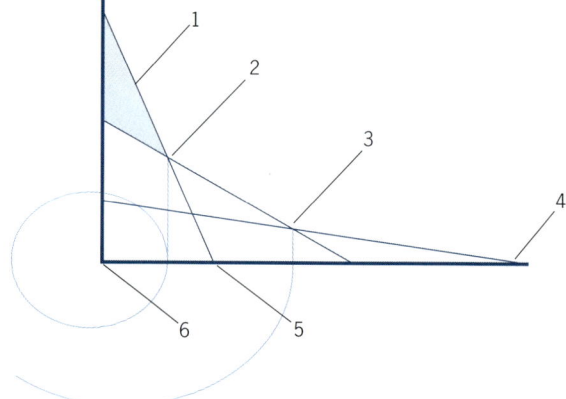

8. Using the model of the rural cultural landscape (figure 3.8), classify the following features according to their place within the structure. Find and include examples of your own by referring to topographic maps of rural landscapes:

- planted windbreak
- hedge
- farm tracks
- railway siding
- district high school
- regional agricultural department office
- vines and fruit trees
- large country town
- bare field
- windmill
- stone wall
- farm machinery sales
- grain silo at railway station
- fruit packing shed
- airport
- small village
- road train

# Chapter 4

# Case Studies in Australian Agriculture

The purpose of this chapter is to focus on three agricultural case studies from the south-west, central and eastern regions of Australia. The operation of farms as a system and the factors influencing the location of the different land uses are investigated. In addition, the economic significance and environmental impacts of each farming system are analysed.

Figure 4.1 illustrates the major rural land uses in Australia and their location. Significant areas of non-agricultural land include urban lands, national parks, state forests, Aboriginal lands and areas that are too harsh to support farming. In this chapter three case studies of agricultural systems will be discussed covering the south-west, central and eastern regions of Australia.

### EXERCISES

1. Investigate and briefly describe the land use types shown on the map of Australian agricultural regions.
2. Outline the factors that might account for the presence of vacant land.
3. Rank the agricultural land uses from intensive to extensive and suggest approximate sizes of the average property in hectares.
4. Essay: To what extent does the land rent mechanism determine the location of Australian agricultural regions? In answering this question you need to consider how other factors may have shaped the regions. These would include topography, climate, soils, vegetation, cultural influences and political decisions.

## CASE STUDY 1

## Mixed crop and livestock farming in the South-West

Wheat and barley are the main cereal crops grown in the South-West. In 1999, Western Australia's wheat harvest of 8.2 million tonnes was Australia's largest and represented 38 per cent of total wheat production, while the 1999 barley harvest of 1.5 million tonnes was Australia's second biggest (after South Australia with 2 million tonnes) and represented 25 per cent of total barley production. Out of a total value of all agricultural production for the state of $4.3 billion, wheat and barley contributed approximately $1.6 billion or 47 per cent. The comparative values of all major rural produce are illustrated in figure 4.2. Recent declines in the value of wool and meat have seen a shift in the output of mixed crop and livestock farms towards more profitable wheat production, with sheep and lambs declining from some 38 million head in 1990 to 27 million in 1999. During the same period, wheat output rose from around 5 million to over 8 million tonnes per annum. In 1999 there were approximately 2600 farms producing wheat and other cereal crops and 3700 mixed grain and sheep or beef cattle properties. These represented 46 per cent of all agricultural properties within Western Australia. In the 10 years from 1989 to 1999 there has been a decline in the number of rural properties by approximately 10 per cent as a consequence of farm amalgamations and changes in land uses.

### Location of mixed crop and livestock

The wheat and sheep belt occupies an area of some 25 000 million hectares, extending from Northampton in the north through to Esperance in the south. As it extends southwards, the region broadens in line with the rainfall distribution pattern. It is roughly limited on the western side by the 750 millimetre isohyet and on the eastern margin by the 270 millimetre line. Its general location is shown on figure 4.3 and on the map of Australian agriculture (figure 4.1).

#### Influence of the natural environment on location

*Climate*

Where rainfall exceeds 750 millimetres wheat crops may be prone to disease, especially where the soil is prone to waterlogging. This can affect the roots systems by causing them to rot. Moisture retention within the leaves and seed head of the plant can also cause a fungal growth known as rust. It should be noted, however, that recent developments by the CSIRO in the breeding of new wet-resistant species plants might see a shift into higher rainfall areas in the future. While not as dependent on rainfall patterns, sheep can also be affected by climatic extremes. Sustained rainfall with

# 60 LANDSCAPES AND LAND USES

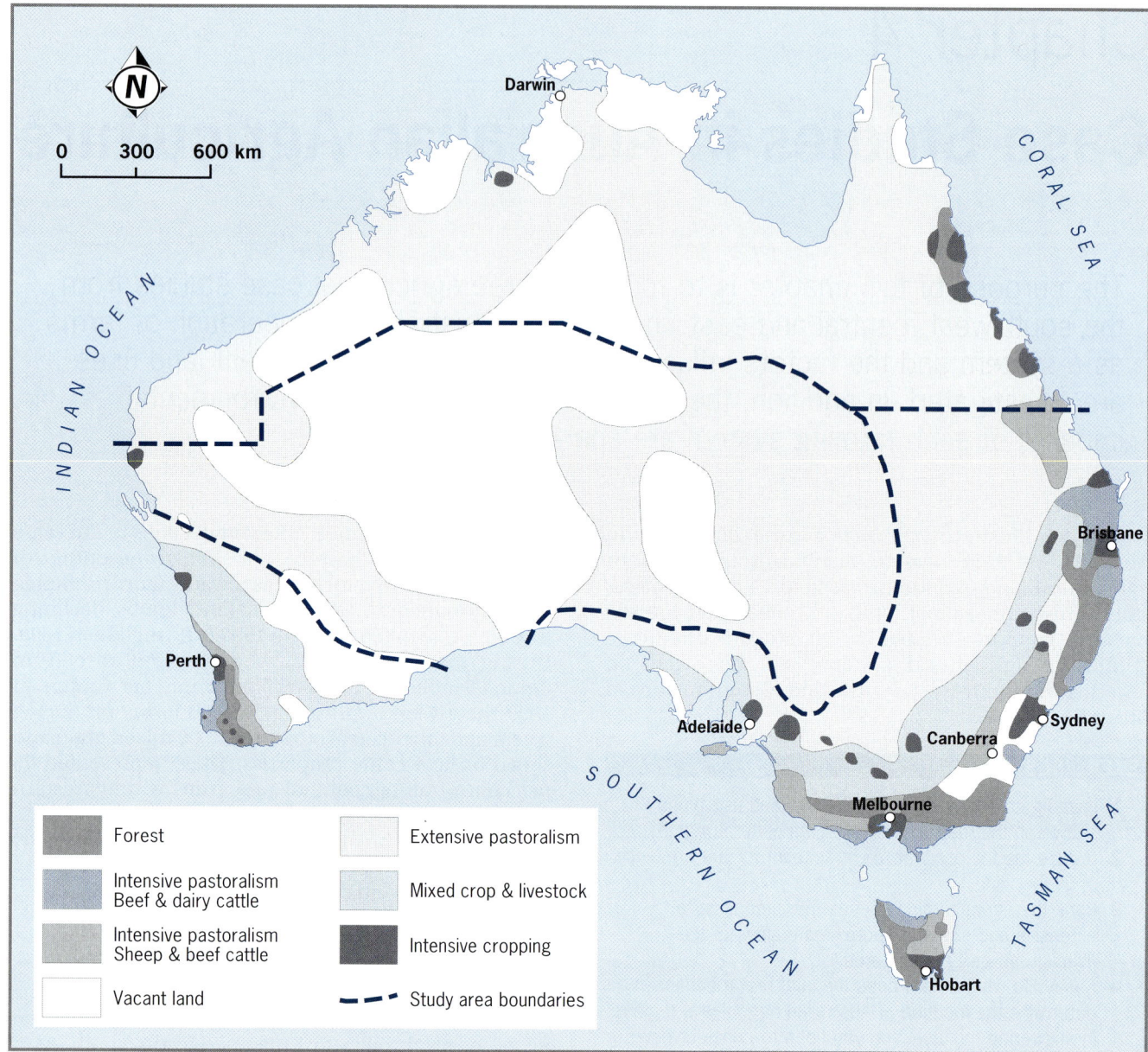

**Figure 4.1** Australian agricultural regions

waterlogging of the soil can affect the feet of the animal, allowing fungal and bacterial growth within the hoof. The result can be a disease within the horny material, known as footrot. The combination of rain and temperature can also impact on the animals. This is especially apparent during spring lambing, when sudden cold and rainy conditions can lead to stock losses.

Temperatures within the South-West are generally mild to warm and do not have a major impact on the location of wheat farming. However, it is desirable to have hot and dry conditions in the months of November to January, when the wheat needs to ripen and dry out to permit harvesting. The effects of low, unreliable rainfall and extremes of temperature on the eastern boundary of the wheat–sheep belt act to limit further expansion into this region.

## Topography and soils

Topographical features can influence the location and development of broad acre cropping to a degree. In general the area is flat to undulating, allowing mechanised farming activities to be conducted with ease. Some areas of the western sub-region (wheat–sheep belt of the South-West) are characterised by steep slopes and rugged terrain, which makes the use of large-scale machinery difficult. In the eastern areas lateritic breakaways, granite outcrops and saltpans further disrupt the pattern of farming.

The soils of the wheat–sheep belt are derived from the geologically ancient Great Western Plateau and are infertile in their natural state. Two main types of soils exist: the light, sandy duplex soils, which are located on sand plains, and the heavy gravel and clay red duplex soils which are found in valley areas. Areas of sandy

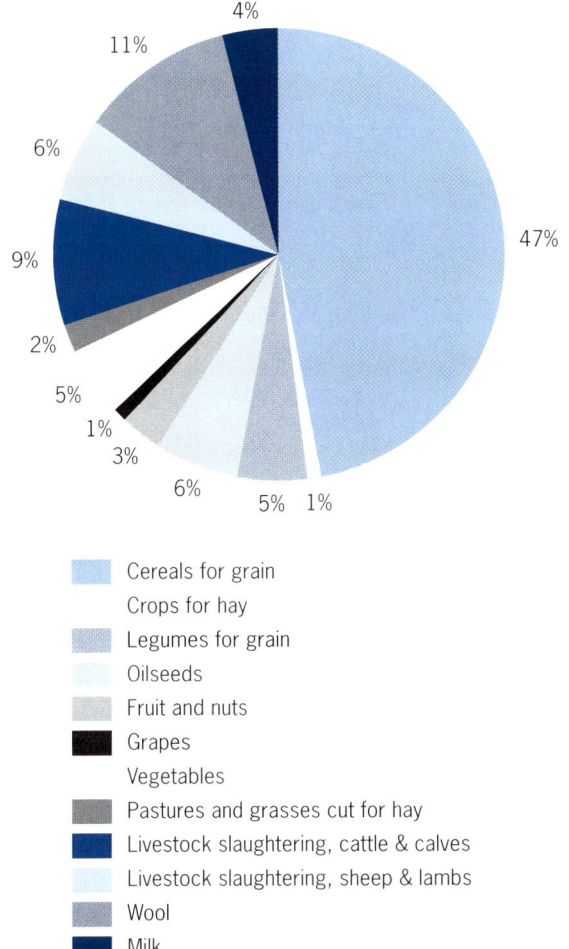

**Figure 4.2** Value of Western Australian products, 1998–99

Legend:
- Cereals for grain
- Crops for hay
- Legumes for grain
- Oilseeds
- Fruit and nuts
- Grapes
- Vegetables
- Pastures and grasses cut for hay
- Livestock slaughtering, cattle & calves
- Livestock slaughtering, sheep & lambs
- Wool
- Milk

soils account for about two-thirds of the wheat–sheep belt. Many of these areas have been developed since World War II. They have become more productive with the addition of phosphates and trace elements such as copper, zinc, molybdenum and manganese in various fertiliser mixes. Growing leguminous pastures and crops such as subterranean clover and lupins also assists in overcoming problems of nutrient deficiencies. To the east of the wheat–sheep belt, desert soils are generally stony or thin and coarse and offer little opportunity for cultivation. To the west the coastal soils are infertile, with poor water retention, and would require substantial improvements to make them suitable for the cultivation of wheat.

*Natural vegetation*

While the location of the wheat–sheep belt was not strongly influenced by the natural cover of vegetation, as this cover was easy to clear, it should be noted that the early selection of land for wheat farming often used the natural vegetation cover as an indicator of the best soil types for wheat. In developing their properties farmers used large teams of horses to remove the mallee scrub which existed in the area. On the wetter western margins, clearing was more difficult because farmers had to cope with eucalypt woodland of marri or wandoo. Similarly, York and salmon gum woodland in the drier south-east presented clearing problems. In later years, with the assistance of large-scale machinery, farmers cleared the sandier areas of scrub very quickly.

### Economic factors affecting location

The location pattern of wheat–sheep farming in the South-West can be partly explained by the competition between various agricultural users for farmland close to the major coastal (urban) areas (the markets). A location close to the market is desirable because it increases accessibility and therefore reduces the farmer's transport costs. The consequent competition for near-city agricultural land has the effect of keeping prices for that land high. (Land prices generally decrease with increasing distance from the market. See the economic rent mechanism explanation, chapter 3.) Land uses generally change from intensive to extensive as the distance from the market point increases. The nature of wheat and sheep farming is such that since the 1930s farm sizes have increased from less than 400 hectares to a current average of 3000 hectares, with some exceeding 5000 hectares in order to be economically viable. In addition, the infrequent movement of produce to the market point using bulk transport methods results in a lower transport cost than the more intensive forms of farming, such as market gardening and dairying. As the map of Australian agriculture illustrates, the location of the different farming areas does in part show the effect of competition for the land.

### Historical and political factors affecting location

For almost a century, the government of Western Australia was directly involved in promoting the growth and development of the wheat–sheep belt, as part of its efforts to encourage settlement and increase productivity. John Forrest, the Commissioner for Crown Lands and later premier of Western Australia from 1890 to 1901, believed it was the duty of the state to do everything in its power to encourage agricultural development. He pointed out that, without agriculture, the state could not support its population, and that without sufficient population it would never prosper. During the years when Forrest was the dominant figure in Western Australian politics, he worked hard to foster the development of agriculture in general, and the wheat–sheep belt in particular. His 1887 Land Regulations Act authorised the establishment of 'agricultural areas'. Lands were surveyed and their fertility established before they were offered for sale on very attractive terms. Almost 50 such areas were proclaimed, most of which were in the wheat–sheep belt in an area extending from Northampton to Tenterden (north of Albany) and from York to Southern Cross. The area contained almost 300 000 hectares of land and absorbed most of the new rural settlement between 1890 and 1915. A shortage of funds prevented the government of the day from constructing the railway network required throughout the new agricultural areas. Private investors were encouraged to form companies and construct the lines in exchange for

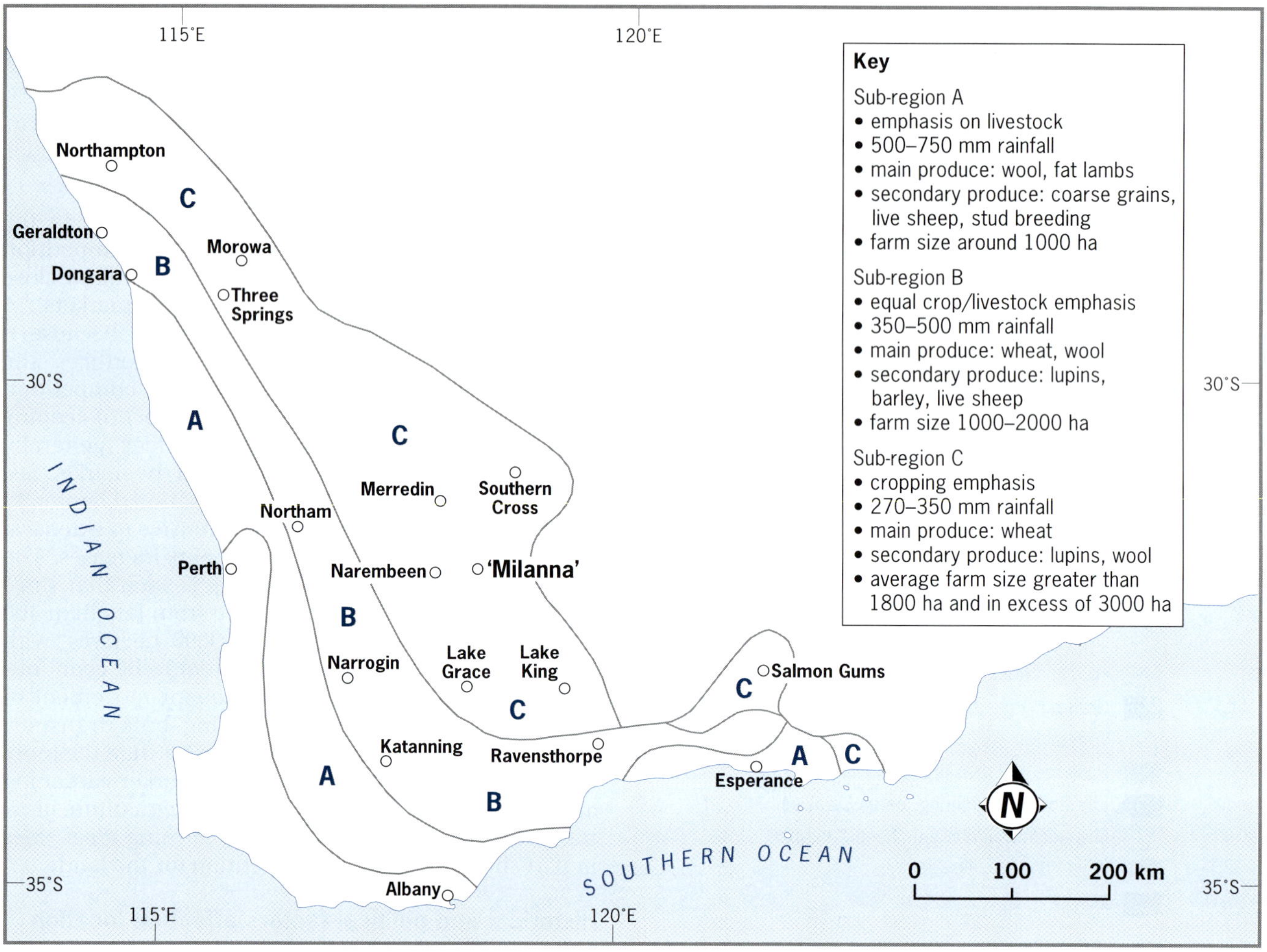

# LANDSCAPES AND LAND USES

**Figure 4.3** Wheat–sheep belt of the South-West

**Key**

Sub-region A
- emphasis on livestock
- 500–750 mm rainfall
- main produce: wool, fat lambs
- secondary produce: coarse grains, live sheep, stud breeding
- farm size around 1000 ha

Sub-region B
- equal crop/livestock emphasis
- 350–500 mm rainfall
- main produce: wheat, wool
- secondary produce: lupins, barley, live sheep
- farm size 1000–2000 ha

Sub-region C
- cropping emphasis
- 270–350 mm rainfall
- main produce: wheat
- secondary produce: lupins, wool
- average farm size greater than 1800 ha and in excess of 3000 ha

3000 hectares of land beside the railways for every kilometre completed. The Great Southern Railway between Beverley and Albany and the line between Midland Junction and Walkaway were built on this basis.

Responsible government was granted in 1891, and at about this time the financial resources of the state were boosted by the gold rushes. The government directed much of its newly acquired wealth into rural development. The Agricultural Bank (later to become the Rural and Industries Bank and then, more recently, BankWest) was established in 1894, in order to lend money to settlers on very easy terms. The Bureau (later Department) of Agriculture came into being in the same year. In the following decades, William Lowrie and George Sutton, who were two of the bureau's most capable officers, used their expertise to improve mixed crop and livestock production in the state. Between 1905 and 1916 the government borrowed heavily to finance the expansion of the railway system into the wheat–sheep belt. During that time over 3200 kilometres of loop and spur lines were laid. No matter which political party was in power, the railway system continued to develop, as did the area under wheat.

Following World War I, the politicians returned to their task of agricultural development. Ex-servicemen and immigrants, most of whom lacked farming experience, were encouraged to take up the lighter country found on the eastern margins of the wheat–sheep belt and in the timbered areas near its south-western boundary. The government intended that this phase of rapid expansion would culminate in the '3500 Farms Scheme'. The plan was the result of an

**Figure 4.4** Diesel train and grain carriages

agreement with the British Government whereby 3500 English migrant farmers would pioneer new lands beyond the existing margins of cultivation. Because of the onset of the Great Depression, the scheme was never put in place. The Depression caused other profound changes in the farming of the wheat–sheep belt. Hundreds of farmers were ruined by the collapse of world wheat prices. They had no alternative but to walk off their farms. The farms were restructured into larger units and a greater emphasis was placed on sheep farming.

Recovery in the 1930s and stagnation during World War II were followed by the renewed growth of the wheat–sheep belt after 1945. Once again, expansion came in response to healthy wheat and wool prices and to a belief by politicians that closer settlement was desirable. Eleven hundred war-service farms were made available in the wheat–sheep belt, including farms in new land areas near Eneabba and to the north-west and north-east of Albany. The government also signed a land grant with the American-based Chase Syndicate for the development of 600 000 hectares of sand plain in the Esperance area. Although the syndicate failed, the Esperance Land Development Company, which replaced it, succeeded in creating a major sheep, cattle and cereal-growing region. Most of the land was held by individuals or family farming enterprises. Meanwhile, with the encouragement of politicians, the eastern frontier of the wheat–sheep belt continued to advance into areas where soil structures were dangerously fragile and pasture development virtually impossible.

Since the 1970s the growing realisation that the wheat belt faces major environmental problems has seen a shift in attitudes about agricultural expansion. A range of tough environmental laws now carefully protects remaining forest and woodland areas and the emphasis is on consolidation on and conservation of existing farmland. Many farmers are now looking to rehabilitate land affected by erosion, waterlogging, acidity and salinity.

### EXERCISES

1. Draw a map of the South-West and show the location of the mixed crop and livestock region. With reference to chapter 2 as well as the information provided in chapter 4, place annotations on the map to show the way in which physical factors influence the location of this agricultural region.
2. Which of the physical factors do you think has had the greatest influences on the location of the mixed crop and livestock region? You will need to justify your selection by providing appropriate evidence.
3. Identify and explain the main economic and political factors that influenced the location and development of the mixed crop and livestock zone.

## Crop and livestock farming as a system

The size, appearance, farm practices and outputs of wheat and sheep farms in the South-West are a reflection of the inputs from the physical and cultural environments (see figure 4.5). The farm, Milanna, which is referred to in this case study along with the surrounding farms, forms a distinctive landscape in the lower rainfall region of the eastern wheat belt of the South-West. If the inputs substantially alter then this will have an impact on farm outputs, appearance and practices. Sometimes these variations are beyond the control of farmers. Seasonal variations in rainfall are one such input. The late arrival of rain will reduce the yield per hectare while the occurrence of thunderstorms during the harvest may destroy or damage the crop.

### Physical inputs

*Climate*

Wheat requires a minimum of 250 millimetres of rainfall during the growing season, which runs from April–May through to October. Opening rains of around 25 millimetres are needed to promote germination of the seeds, with follow-up rains during the initial growth period (May to July). Rainfall has to be reliable, with less than 20 per cent variation above and below the average. The Bureau of Meteorology classifies the South-West region as having a low to moderate rainfall variability and it is a comparatively good location for the cultivation of wheat. The development of drought-tolerant wheat varieties by the CSIRO and other agencies has influenced the location of the wheat growing region of the South-West, with most farms falling between the 750 millimetre and 270 millimetre rainfall isohyets.

*Soils*

The soils on which the wheat–sheep belt is based may be classified as either heavy or light. The heavier soils are represented by the red duplex soils or clays. Often these occur in the western margin of the wheat belt as gravel and loam over clay. The lighter soils are the yellow duplex soils, which occur as a sandy loam over clay. Both have good water retention characteristics, though by world standards they are low in fertility and require regular applications of fertilisers. In the South-West the ancient soils that are found in the mixed crop and livestock zone are lacking in important nutrients and trace elements and are unable to support agriculture without fertilisers and mineral supplements. Each year about 1.5 million tonnes of artificial fertiliser, valued at some $400 million, is applied to grain producing farms. In Western Australia, farmers apply fertiliser to some 80 per cent of their cropped areas. In contrast, the application in Victoria is about 50 to 80 per cent and in New South Wales about 50 per cent. Commonly used fertilisers include phosphate, nitrogen, zinc and sulphur based products.

The impact of factors such as soil erosion, water logging, acidification and salinisation also makes it important for the farmer to implement careful soil management practices.

*Topography*

Generally flat to undulating land is favoured for this type of farming. With the increase in the size of harvesting and planting equipment the need to have large, relatively flat fields is an important factor. Where outcrops of granite or laterites occur they have been generally left uncleared, with the farming of the land taking place around these areas. In the more undulating land closer to the Darling Ranges farm operations are generally smaller in scale, with less of the farm being cropped and more emphasis on pasture and livestock.

## Natural vegetation

While significant amounts of natural vegetation are cleared in the process of developing a crop and livestock farming property, the natural vegetation can be seen as an input into the farming system. It has been a source of fuel as well as building and fencing material during the period in which the wheat belt was developed. Areas of vegetation that were left standing provided shelter for livestock as well as acting as windbreaks against soil erosion. The reintroduction of native vegetation to farms as part of a campaign against increasing soil salinity and erosion is also an important element of this physical input.

## Human inputs

Crop and livestock farmers in the South-West are required to make a heavy and ongoing investment in machinery, equipment, fuel and fertilisers. As in other Australian agricultural systems, farmers develop very good management skills and benefit from research at both a government and private level. Because of the high costs involved in farming, and the uncertainties of world markets, farmers have to be efficient in order to be economically viable.

## Capital investment

The growing and cropping of cereals on crop and livestock farms requires a heavy capital investment, as illustrated by the machinery shown in figures 4.6 and 4.7. Combine harvesters range in price from $100 000 to $330 000, while the large 4WD tractors used on the larger properties may cost in excess of $220 000.

Many farms will carry equipment worth in excess of a million dollars. The type and extent of this equipment is illustrated by what is found on 'Milanna'. As indicated in table 4.1, the farm has an extensive range of machinery, sheds, silos, water and fuel tanks, windmills and many other smaller items. The family began farming in the area in 1933 on a newly released 400 hectare block, and since that time has constantly invested in a wide range of equipment in response to changes in farming methods, crop types and technology.

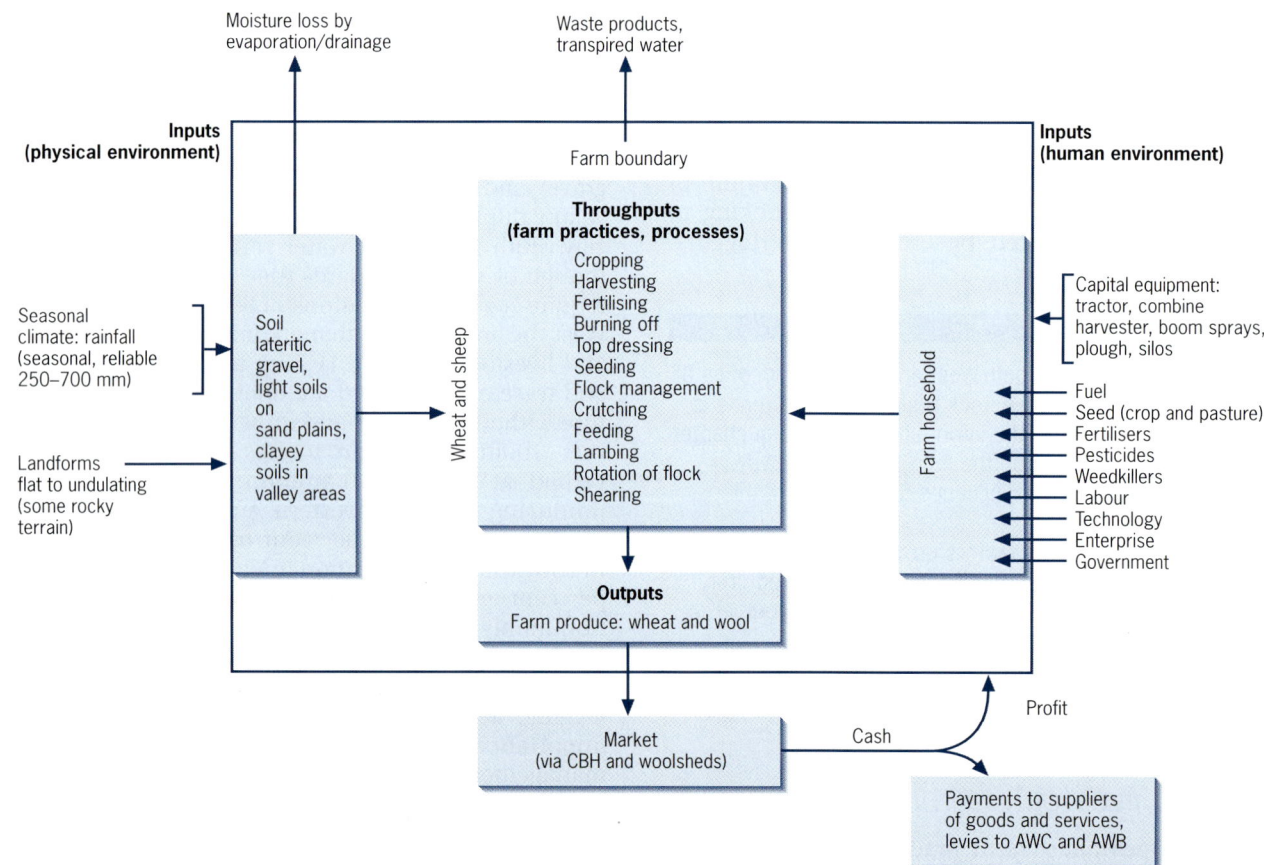

**Figure 4.5** The wheat–sheep farming system

**Figure 4.6** Combine harvester

**Table 4.1** Selected capital equipment on 'Milanna'

| Number | Item |
| --- | --- |
| 1 | Steiger 4WD tractor |
| 1 | Ford 5000 tractor (front-end loader) |
| 1 | Massey Ferguson combine harvester |
| 1 | John Shearer 4-150 air seeder and cultivator |
| 2 | 8 tonne trucks |
| 2 | Toyota Landcruisers 4WD |
| 1 | 18 metre boom spray |
| 2 | Chamberlain 24 disc ploughs |
| 1 | 24 run Chamberlain combine seeder |
| 1 | 5 tonne super spreader |
| 1 | 12 tonne mobile field bin |
| 1 | Grain auger (series of smaller units—hay mower, rake and baler, bale loader, post hole digger, dram mud scoop) |
| 8 | Grain silos (5 x 21 820 litre, 2 x 29 095 litre and 1 x 54 555 litre) |
| 3 | Machinery sheds |
| 1 | Sliding roof fertiliser shed |
| 3 | Diesel tanks (1 x 2700 litre, 2 x 4500 litre) |
| 2 | 2250 litre petrol tanks |

## Changing technology

Since the 1970s, machinery used on crop and livestock farms in the South-West has increased in size and complexity. Powerful 4WD tractors have the capacity to cover large areas quickly while towing equipment such as cultivators up to 20 metres in width. Australia is also a world leader in hand-free tractors, which use satellite navigation systems (Global Positioning Systems or GPS) to guide the tractor and plough within a field.

Other new technologies include the use of precision farming strategies via the introduction of GPS, geographic information system (GIS), variable (application) rate technology (VRT) and digital information technology. This development will allow farmers to map and respond to variations in the quality

**Figure 4.7** 4WD tractor

and quantity of a crop within a single field or paddock. They will able to relay information to a tractor-mounted computer, which will adjust applications of such inputs as fertilisers, lime and pesticides as it travels over the field.

The agricultural sector is one of the largest users of chemicals, with many of the productivity gains made over the last 30 years being achieved with the aid of agricultural chemicals. However, the focus for rural industries, over the last 10 years in particular, has been to achieve lower chemical application rates, softer chemicals and more specifically targeted chemicals. Initiatives such as Integrated Pest Management (IPM), the Farm Chemical User Training Program, and quality assurance are all examples on the part of industry to use fewer chemicals, and to use them in an environmentally sustainable and safe manner.

## Labour requirements

Because of the high input of technology, labour requirements on crop and livestock farms have been progressively reduced. The combined effects of replacing labour with machinery, and farm amalgamation have resulted in a substantial increase in property sizes while at the same time significant reductions in rural labour forces. Whereas it would take one person a day to prepare 5 hectares of land for seeding in the early 1900s, using horse drawn machinery, in the 1990s they could prepare 100 hectares using modern tractors and ploughs. On many farm operations such as 'Milanna', family members manage most of the operations, although seasonal labour may have to be used at peak work times such as harvesting, seeding and shearing. Specialist contractors are also employed by farmers for such tasks as fencing and mechanical repairs.

## Government inputs

Historically, government inputs into the crop and livestock industry have been most important. Various governments provided infrastructure such as railways, roads, water supplies, grain receival depots and vermin proof fences. Today a comprehensive road, rail and port network provides for the efficient and speedy delivery of bulk wheat to metropolitan and overseas markets. It should be noted that the recent relaxation of regulations covering the movement of grain via road and rail systems, as a consequence of national competition policy, has seen a decline in the movement of wheat and livestock by rail and an increase in the transportation by road using B-doubles and pocket road trains.

**Figure 4.8** Road train

With the movement in the 1990s towards market deregulation, the wheat industry saw the end of the Australian Wheat Board in 1999, along with government underwriting of wheat prices. The board now operates as a private company owned by the growers and is allowed special monopoly rights to promote and market wheat overseas. While individual growers can seek to export wheat privately, the AWB Ltd has the right to veto such sales. Domestically individual growers have been free to seek buyers in Australia since 1989.

In the late 1990s the federal government commenced industry restructuring and deregulation in the Australian wool industry. The Australian Wool Corporation was replaced with an interim growers' advisory board. They worked with the government to implement a new marketing and development structure. The result is a two-tier structure. One tier will manage the 2 per cent wool tax for research and development, the other will be a stand-alone commercial business aimed at making profits for woolgrowers, with shareholders free to trade shares. Thus the Australian Wool Research and Promotion Organisation and the Australian Wool Exchange Ltd have replaced the AWC.

## Throughputs

### The seasonal pattern for wheat

Table 4.2 illustrates the main work routine for 'Milanna'. With minor variations, this can be taken as typical for crop and livestock farms in the South-West.

**Table 4.2** Major farm practices and seasonal routine on 'Milanna'

| Months | Cropping | Livestock |
|---|---|---|
| January | Fertiliser carting | Water carting |
| February | Stubble burning, top dressing (superphosphate) | Crutching/Drenching |
| March | Initial cultivation | Hand feeding |
| April | Cultivation | Hand feeding |
| May | Pre-emergent spraying, final preparation and seeding | Changing paddocks |
| June | Completion of seeding | Tailing, ear marking, culling |
| July | Post-emergent spraying | General maintenance |
| August | Making firebreaks | Fencing, shearing and dipping |
| September | Hay cutting | Culling (old stock) |
| October | Hay raking/baling | Purchasing new rams |
| November | Beginning of harvest | Mating |
| December | Harvesting, Grading seed grain | Removal of rams, Changing paddocks |

Wheat growing in the region can be classified as winter wheat production, with planting commencing in autumn and harvesting in early to middle summer. The main periods of activity occur in these seasons. Preparation of the field for planting involves either the ploughing of the soil or, as is more common, the use of chemical weedicide and a system known as minimum tillage. This reduces the impact on the soil profile and reduces the amount of compaction caused by heavy machinery. It also reduces fuel costs associated with the cultivation of the soil.

Once the fields have been seeded the only task that may be undertaken is the further spraying of weeds to remove any that may have germinated after the planting of the crop. The farmer now relies on the winter rains to ensure a satisfactory yield.

Around December and January the attention turns to harvesting and the crop is removed using combine harvesters, which remove the head of wheat or barley, separate the chaff and store the grain in hoppers, from where it is placed into trucks for transportation to rail or port silos. The value of wheat and other grain production from crop and livestock farms in 1999 is illustrated in figure 4.9.

### The seasonal pattern for livestock

With the completion of seeding the farmer's attention focuses on livestock management. Shearing usually occurs in winter because it is a convenient time in the overall work program. This illustrates the **work complementary** nature of the crop and livestock routines. Other livestock-related activities include tailing, mulesing, ear marking, drenching, dipping and culling of old or inferior animals.

The mating of rams and ewes occurs in either May or November. May mating produces spring lambs in September for meat consumption, while November mating allows for births in autumn and early winter,

## CASE STUDIES IN AUSTRALIAN AGRICULTURE

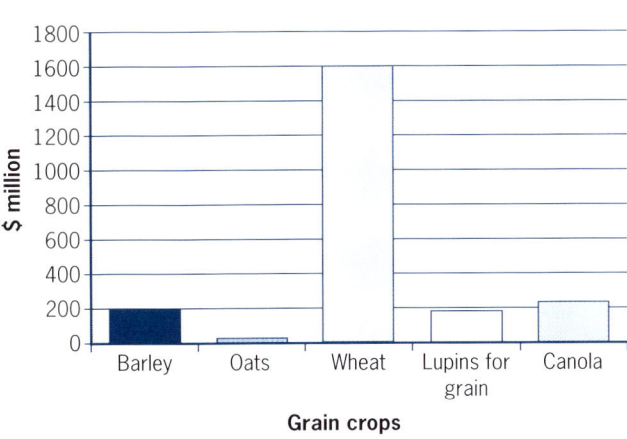

**Figure 4.9** Value of grain production in Western Australia, 1999

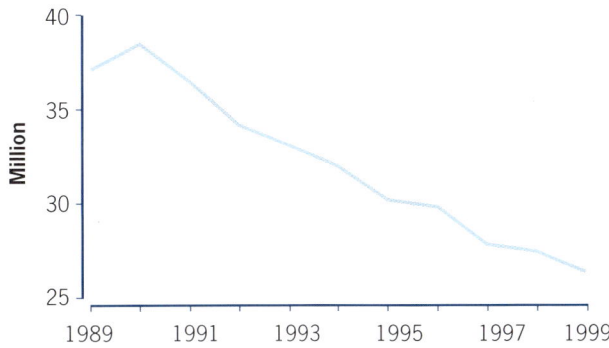

**Figure 4.11** Sheep and lamb numbers on properties in Western Australia, 1989–99

which makes good use of the pasture which germinates during this time. In recent years there has been a shift away from sheep grazing on crop and livestock farms in response to the poor prices received for wool. This has seen a corresponding increase in various types of grains being planted. In addition a shift to beef cattle production has seen a significant increase in herds in 2000.

### Outputs

Farm outputs in the crop and livestock region will vary according to the location of the farm, crop and livestock rotation practices, seasonal and other environmental influences and changes in markets. Figure 4.2 illustrates the relative values of the major rural commodities. It can be clearly seen that wheat is a major product within the total, with Western Australia producing 29 per cent of the nation's field crops. The changing nature of rural outputs from the crop and livestock region can be seen in figures 4.10 and 4.11. The dramatic fall in wheat production in 1994 reflected the failure of the growing season arising out of poor winter rains in much of the wheat belt. The sustained fall in sheep and lamb numbers reflects poor wool prices as well as reduced demand for meat over this time. In contrast meat cattle numbers have grown with the herd increasing by 14 per cent between 1989 and 1999.

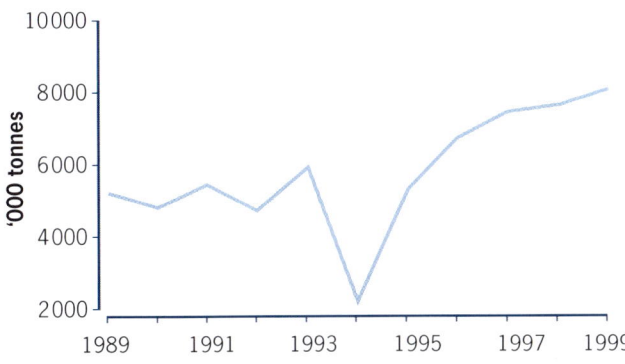

**Figure 4.10** Wheat production, 1989–99

### EXERCISES

1. Using an A3 sheet of paper draw a large diagram of the mixed crop and livestock farming system and then place on it summaries of the inputs, throughputs and outputs.
2. Identify and discuss the advantages of combining grain and livestock production within a farming system. In doing so consider both the environmental and the economic advantages.
3. Identify and discuss the factors that influence the variation in outputs from a mixed crop and livestock farming system.
4. Using information on climate from chapters 1 and 2 construct a chart (like the one following) to show the relationship between the seasonal routine for wheat and the pattern of temperatures and rainfall in the South-West. Discuss the relationships that appear between the climate and the farm routine.

| Month | Rainfall | Temperature | Routine |
|---|---|---|---|
| January | | | |
| February | | | |
| March | | | |

## The cultural landscape

### The regional cultural landscape

Because of the surveying system used in the early days of settlement, the overall cultural landscape of the wheat–sheep belt displays a regular grid pattern. The transport network that has been established throughout the region reinforces the rectangular layout of farms. Railways were constructed to enable large volumes of grain to be transported efficiently to major ports prior to export. The design of the system was based on the criterion that no farmer would have to produce further than 24 kilometres from a railway siding, which resulted in a regular spacing of the lines. The flat topography allowed for the development of a grid-shaped road network. Well-maintained but largely unsealed roads permit farms to be serviced and wheat to be transported by trucks to cooperative bulk handling bins sited on railway lines. In recent times there has been a shift towards road transport, with increasing volumes of commodities such as wheat, wool, livestock and general farm requirements being moved by road.

The transport network links the towns that are dispersed throughout the wheat–sheep belt. These towns are primarily service centres rather than residential and market towns, and they vary substantially in size—from tiny settlements that perhaps comprise a wheat bin and general store, to bustling multifunctional centres such as Northam, Narrogin, York, Merredin and Katanning. The primary function of the towns is to service the rural community, offering facilities such as the grain receival bin (often located well out of town), livestock saleyards and agricultural machinery agencies, fuel depots and other associated rural requirements. Depending on their size, the towns also provide a range of general goods and services and are the focus of social and recreational activities in the district.

The development of the eastern and south-eastern wheat belt regions in the 1950s and 1960s occurred at a time when road transport was beginning to dominate and the emphasis on rail networks was declining. Unlike the closer settlement patterns found in the central and western wheat–sheep regions, these more recent areas are characterised by larger farms and service centres based on the **functional distances** associated with greater mobility afforded by the motor vehicle. Towns such as Lake Grace, Newdegate, Lake King and Ravensthorpe illustrate this spacing. In contrast, towns established during the expansion of the railway system, such as Narrogin, Wagin, Dumbleyung and Katanning are much closer together and the average farm size is smaller.

### The farm site cultural landscape

Due to the size of crop and livestock farms and the nature of the farming practices and traditions, the majority of farming families live on the farms, thus producing a pattern of dispersed rural settlement. See figure 4.13, which illustrates the cultural landscape of the Katanning district in the southern wheat belt of Western Australia. As well as houses, there are likely to be outbuildings such as machinery sheds, a shearing shed and livestock yards, grain silos, fuel tanks and perhaps living quarters for farm and seasonal labour. This relatively isolated lifestyle generates the need for facilities such as an extensive network of power and telephone lines, services such as school bus routes, and regular trips to the nearest town for farm and domestic supplies.

A striking feature of the landscape surrounding the farmhouses is the extent to which it has been cleared of natural vegetation. Isolated clumps remain on rocky outcrops or along watercourses, providing some shade for stock. Windbreaks of original woodland or planted eucalypts are common. These assist in cutting wind speeds over paddocks which, when freshly cultivated, are subject to erosion by the strong winds (particularly easterlies) that occur frequently in the wheat–sheep belt. The growing emphasis in recent times on tree replanting programs, such as those initiated by local Landcare groups, has seen widespread planting of trees and shrubs in response to problems of soil waterlogging, erosion and salinity. This initiative is now beginning to have an effect on the farming landscape. In addition to the features of the cultural landscape associated with grain growing, there are also elements that indicate the changes associated with the raising of livestock. Clay-walled dams, dug to supply drinking water to sheep or cattle, have been constructed in the paddocks and either trap water from sheet flow or streams, or are filled by seepage from ground water. The construction of fences that subdivide the farm into paddocks, as well as gates and grids, is further evidence of livestock raising.

### Changes in the cultural landscape

The cultural landscape of the wheat–sheep belt is undergoing change. Tough economic times have forced some farmers to sell their properties, leave country districts and settle in large urban areas in Western Australia. This rural depopulation has resulted in an increasing number of vacant farmhouses. As a consequence, businesses in farming towns have suffered because of the loss of customers. Once population figures in the district fall below what is required to sustain a business, proprietors have no option but to close.

The trend among consumers to bypass businesses in smaller centres is accentuated by the mobility provided by private transport. Farmers often choose to drive to larger centres to obtain goods and services, a practice that has led to centralisation in the settlement pattern. With the increasing use of road transport rather than rail, roads are continually being upgraded, while some railway lines have been abandoned or are only operated on a seasonal basis.

Recent trends to reverse the general decline in the wheat belt towns can be seen in the ways in which local communities and local governments are working to retain industries as well as developing new ones. The larger more successful centres such as Narrogin and Wongan Hills illustrate this effort, with businesses and services becoming centralised within these towns.

**Figure 4.12** Vacant buildings, Greenhills, WA

### EXERCISES

1. The main features within a cultural landscape include transport systems and rural and urban settlements. Study the cultural landscape provided for Katanning (figure 4.13) and then describe its overall appearance. Refer to shapes or patterns as well as spacing and density.
2. Describe the effects that a decline in rural population will have on the cultural landscape of the mixed crop and livestock zone.
3. Identify and discuss the changes in the cultural landscape of the mixed crop and livestock zone that you might expect to see as you moved from Northam to Southern Cross along the Great Eastern Highway. In answering this question consider the changes in farm sizes, the spacing of urban settlements, the clearing of vegetation and the expected proportions of cropping and livestock raising. Also consider how physical factors as well as distance form the market may have influenced the changes in the cultural landscape.
4. Study the layout of 'Milanna' (figure 4.14) and then describe its overall appearance. In doing so refer to size, shape and distribution of various cultural features.

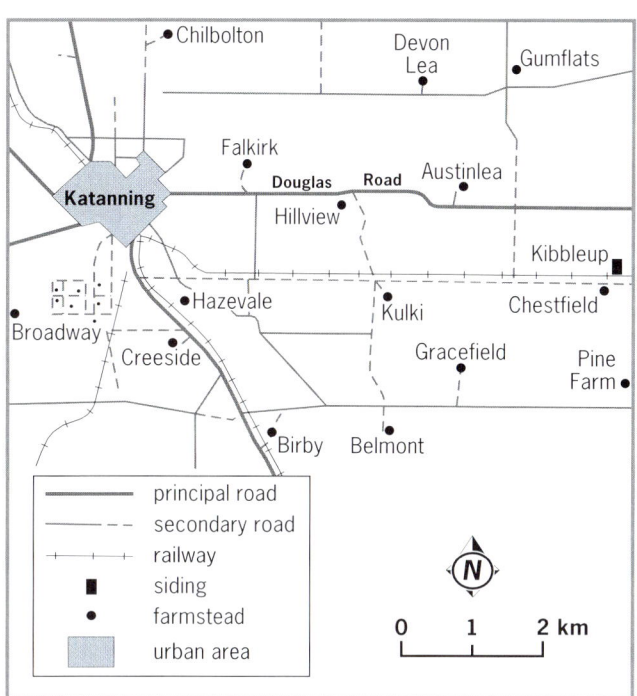

**Figure 4.13** The cultural landscape of the wheat–sheep belt in the Katanning district, WA

## Environmental impact of crop and livestock farming

The rapid development, over the last 80 years, of some 25 million hectares of land devoted to crop and livestock farming has dramatically changed the natural environment of this part of the South-West. Clearing the natural vegetation and substituting for it bare fallow and annual crops and pasture has disrupted ground water levels. The roots of the original mallee and shrub vegetation deeply penetrated the soil and subsoil to keep ground water and dissolved salts below the roots of annual crop and pasture plants. In addition, permanent vegetation cover continually transpired water into the atmosphere. After clearing, this cycle is broken and the ground water is renewed annually by rainwater, which is slightly salty and causes watertables to rise and bring past salt deposition to the surface. It is estimated that at least 1.6 million hectares (about 9 per cent) of once productive land is now salt affected and that, by 2010, this will increase to 2.9 million hectares (16 per cent). In addition less than half of the potential surface water supplies within the region are still fresh, with many farm soaks and dams no longer useable for stock water. Coupled with this, rising water and salinity are affecting the buildings in many rural towns by eating away at the brickwork. The impact is also felt in areas of remnant vegetation, where the effect of salinity from surrounding cleared land is destroying the native vegetation.

The most widely practised method of rehabilitating salt-affected soils is the planting of salt tolerant species of vegetation that have the capacity to lower the watertable and stabilise the movement of the salt. Another method is the use of drainage channels and contoured banks to either move the ground water to evaporative basins or slow the movement of water into lower-lying sections of farm land. The introduction of salt-tolerant crops and fodder trees to salt-affected land also allows unproductive soils to be brought back into use.

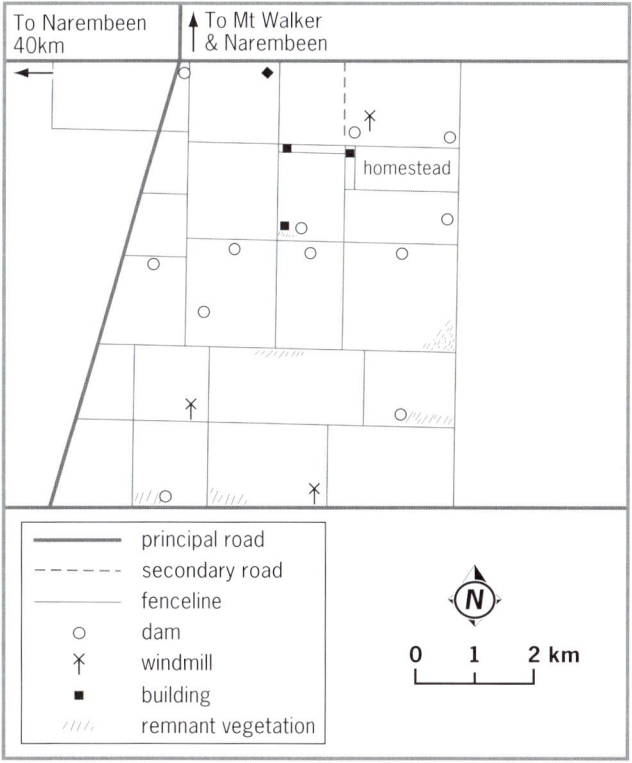

**Figure 4.14** The layout of 'Milanna'

Large-scale tree planting is a long-term solution to the serious problem of soil erosion caused by wind and water. This problem has developed because of the removal of natural vegetation, and because of the heavy stocking rates on some farms. Deep-rooted natural vegetation binds the soil profile, but clearing prevents this binding effect and heavy cultivation frees up the surface of the soil, allowing strong winds to carry away the fine topsoil. Such erosion is particularly evident during autumn, when ploughing is in progress in preparation for seeding, and during periods of drought, when paddocks are overgrazed. Soil instability also occurs around watering points, where sheep congregate in large numbers.

Soils are eroded by surface runoff during thunderstorms, which frequently leads to deep gullying. Farmers have long been aware of the deterioration of their land. In the 1970s a severe drought, followed by cyclone Alby, motivated people to commence tree planting. In subsequent years the numbers of trees planted increased rapidly. The use of chemicals has also had a serious impact on the environment. Insects, weeds and diseases represent an increasingly difficult problem for farmers, which is made worse when natural controls are affected by the use of chemicals. Farmers have no choice but to use a wider range of chemicals to protect their cereal crops, pastures and livestock. Unfortunately, there have been side effects as a result of this, including contamination of produce by protectorants, chemical residues in soil and livestock meat, and the occasional misuse of chemicals. The farming community is aware of the impact of its activities on the environment and farming practices have been altered to ensure that land degradation is minimised. The measures taken by farmers include establishment of windbreaks, contour ploughing on slopes, introduction of minimum tillage techniques, conservation grazing and pasture management. While farmers are urged to be environmentally conscious, the steps needed to minimise land degradation can be expensive. For example, tree planting and contour work are sound long-term practices, but do not produce economic returns in the short term. At the same time, economic pressures are forcing farmers to be more productive. Because many farmers are unwilling or unable to meet the costs involved, government is taking more direct responsibility with respect to the environment. Legislation and incentives are intended to encourage farmers to adopt more environmentally acceptable farming practices in the wheat–sheep belt.

## Problems and prospects

### Environmental issues

South-West wheat farmers contend with a variety of environmental issues including the uncertainties of weather conditions, the effects of various pests and diseases and the longer-term concerns regarding soil quality. The following examples, taken from the 2001 season, illustrate the types of problems that affect the farmer's ability to achieve good yields.

- Winter frosts in the Great Southern region of the wheat belt reduced the annual output by 450 000 tonnes. (The impact of frost when the seed head is forming damages the plant and reduces the quality and quantity of grain produced.)
- Late rains, particularly in northern agricultural regions, led to several interruptions to harvesting. The rain also caused sprouting in wheat (when the grain within the seed head germinates and sends out sprouts). The bulk handling of wheat makes it critical that this wheat is not mixed in with grain that is unaffected.
- The fungal disease powdery mildew affected wheat crops across the northern grain growing regions of the state, as well as some crops in the Esperance area. The cause of the outbreak appeared to be fungus that had survived on the stubble from the previous year's crop. Wheat varieties such as Brookton and Cunderdin are among the most susceptible to powdery mildew.

Other environmental problems facing mixed crop and livestock farmers include the invasion of crops by a variety of weeds such as skeleton weed, cape weed, wild oats and rye grass. These compete for nutrients and space and reduce crop yields. They also cause problems during harvesting. Insect pests have long been a problem in regions of broad acre farming and grasshoppers and locusts, which pupate in pastoral areas periodically, cause problems in the wheat producing regions of the state.

Also, the issues associated with soil management that were discussed in chapter 3 present another significant environmental impact in the wheat–sheep belt of Western Australia.

### Economic issues

The heavy reliance of the Australian wheat producer on export markets makes this commodity particularly sensitive to changes in world supply and demand, as well as changes in international exchange rates. Issues such as the removal of farm subsidies in Europe and the United States and unrestricted access into these markets have wide-ranging consequences for farmers in the South-West. The impact of farm subsidies in the United States during the mid 1990s saw prices for Australian producers drop by some $60 a tonne, with prices in 1991 reaching $135 a tonne. While prices have now recovered to about $270 a tonne in 2001, the uncertainty of fluctuations in supply and demand could see this trend reversed.

The following events in 2001 illustrate the effects of world economic conditions on the demand for Australian wheat.

- The largest pasta-making company in the world, Barilla, believes more durum wheat from Australia could be exported to Italy. In 2000, the company bought two boatloads of Australian durum. Currently there is a shortage of durum wheat in the world and sales such as this illustrate the capacity for markets to grow.

- Australia sells 35 per cent of its wheat to the Middle East and any deterioration in conditions within this region could have a major impact on Australia's access to this market. The impact of the terrorist attack on the World Trade Center in New York in September 2001 could bring economic or trade sanctions and the possibility that the Australian Government would support this action.
- The United States Department of Agriculture released its official estimate of the 2001 US winter wheat crop. The report outlined a significant reduction in anticipated yield. The crop was estimated at 19 million tonnes, down from 23.5 million in 2000. This decline provided good news for Australian farmers, with increased sales and a small rise in price of about $4 per tonne.

Other domestic issues such as the level of debt carried by farmers and the effect of changes in interest rates can also have an impact on the economic viability of the farming enterprise. The average debt carried by wheat and sheep farmers in 1999 was $147 000, rising to $500 000 for some. Such a situation requires farmers to manage their finances carefully and it places restrictions on their ability to invest in new areas of production.

### EXERCISES

1. Identify how mixed crop and livestock farming has impacted on the environment and the ways in which these problems are being dealt with.
2. Identify and analyse some of the economic uncertainties that can impact on the mixed crop and livestock farmer.

## CASE STUDY 2

# Extensive pastoralism in Central Australia

Extensive grazing or pastoralism occurs on very large leasehold properties found mainly within the semi-arid regions of Central Australia. It is dominated by the grazing of sheep within the temperate or southern semi-arid areas and by beef cattle ranching within the northern or tropical semi-arid regions. The carrying capacity of this land is very low, with cattle properties achieving a stock density of one animal for every 50 hectares, while sheep stations may operate with a density of one animal for every 10 hectares. The reliance on natural pastures that may be slow to regenerate, and the problems of an unreliable rainfall pattern are the main forces limiting the carrying capacity of these types of agricultural systems. In Western Australia the temperate semi-arid regions account for less than 6 per cent of the state's total sheep population even though they represent about 25 per cent of the land area.

Unlike other agricultural systems, extensive pastoralism is based on the leasing of government-owned land (Crown land). In addition to the purchase of a lease, the property owner pays an annual rent for each hectare of land (currently this is around 50c per hectare). Thus, a property of 300 000 hectares would have an annual rental bill of $150 000. This system of land tenure places a number of restrictions on the way in which the owner can use the land, as the primary purpose is the allocation of grazing rights. Thus, a mining company may enter onto the property for the purpose of searching for and extracting minerals. In recent times the issue of native title rights over Crown lands has brought a degree of uncertainty to the occupiers of pastoral leases.

While extensive grazing relates to both sheep and cattle, the focus of study in this agricultural system will be extensive sheep pastoralism carried out on large properties (called sheep stations) within the central study area.

**Figure 4.15** Wheat field and harvester

**Figure 4.16** Merino sheep flock on extensive grazing pastoral property

## The location of extensive sheep pastoralism

Sheep stations occupy two significant areas within the central study area. The eastern region stretches from southern Queensland, through central and western New South Wales into the drier regions of Victoria and South Australia. The western section covers a vast area of Western Australia, stretching from the north-west coast through the Murchison and Eastern Goldfields districts and into the Nullarbor Plain. It is an area that covers some 1.5 million square kilometres. Almost 900 000 square kilometres occur in Western Australia, where the number of pastoral leases currently stands at 500. Most of the region falls within the 200–400 millimetre isohyets and is an area where rainfall is too low or unpredictable to sustain intensive agriculture. Pastoralism plays a vital role in the management of the three-quarters of Australia that is classified as rangeland. As well as producing food and wool, the pastoral industries assist in the management of feral animals, vegetation change and fire over vast areas.

In 1997, the statistics on sheep and wool production in this zone were: 65.3 million head of sheep, producing 314 000 tonnes of wool at an average stocking rate of 0.1 head per hectare (or 10 hectares per head).

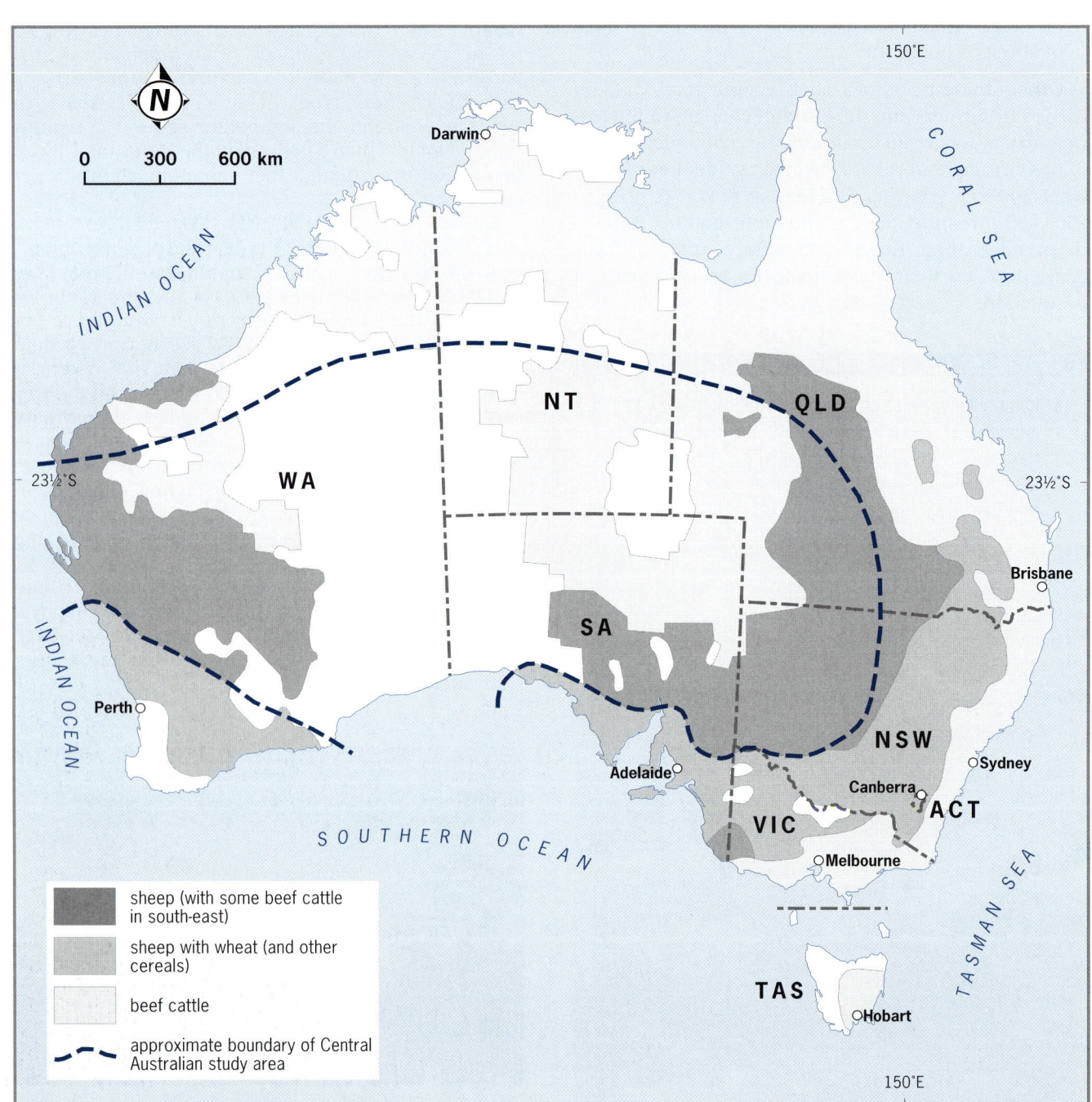

**Figure 4.17** Sheep and beef grazing area

## Influence of the natural environment on location

### Climatic influences

The physical environment in which extensive sheep grazing occurs has had an impact on both the location and the agricultural practices of the farm enterprise. This system occupies areas that are generally considered to be marginal lands because of the low and erratic rainfall. Annual rainfall can occur within a few days, followed by long dry periods or extended droughts. Periods of six to eight months without rain are not uncommon.

Evaporation rates are significantly high during summer and surface water supplies quickly evaporate. In addition the relatively flat nature of the Western Plateau and the Central Lowlands, on which the pastoral activities occur, encourages the formation of large salt lake systems or playas, where rainfall accumulates and concentrates soil salts and gypsum. This has the effect of rendering these areas unsuitable for grazing.

Offsetting the lack of rainfall in a number of areas are artesian and subartesian water supplies that are tapped by pastoralists in order to provide a reliable supply of water for the livestock. Significant deposits of artesian water occur on the west coast, Nullarbor Plain and Central Lowlands. These water supplies, trapped under pressure in porous aquifers, reach the surface when tapped. The windmill is a common sight on sheep stations, pumping water from shallow subartesian **aquifers** into large tanks, where it is stored and released into troughs through float operated valves.

### Natural pastures

Sheep in the **pastoral** lands of Queensland and north-central New South Wales are grazed on the native perennial Mitchell grass, a hard tussocky plant that produces new growth in summer. The annual Flinders grass is also highly **palatable** to sheep. Stocking rates in this savanna grassland can be as high as one sheep per hectare, compared with the grassland/saltbush country of Western Australia where one sheep to around 10 hectares is the normal stocking rate. Towards the drier interior the carrying capacity decreases markedly, with as much as 20 hectares of land required for one sheep.

### Soil and topography

In general the soils of these regions are highly weathered and very infertile. Soil fertility is predominantly linked to the underlying rock type. However, the redistribution of soil and nutrients by wind and water results in areas of better soils. Even in a landscape that may appear flat and featureless, water and nutrients concentrate in patches, around bushes and trees, or in gentle depressions.

These water and nutrient rich sites are the key to the productivity of an area. They are very important for native plants and animals, as well as for the pastoral industry. Thus the distribution of animals over the landscape is not even, but focused on the main watering points and better areas of pasture. Each station will have areas that are not suitable for grazing due to poor quality or poisonous plants and these are left unstocked.

Topography and soil characteristics have affected the location of sheep grazing only indirectly when compared with the overwhelming importance of water and the availability of palatable shrubs. Low-relief topography characterises much of the extensive sheep-grazing areas. This varies from the flat to undulating character of the Yilgarn Block, with its breakaway country, occasional granite outcrops, sand plain country and salt lakes, to the featureless limestone landscape of the Nullarbor and the arid and harsh regions of South Australia and western New South Wales.

The outcome of the generally low carrying capacities in sheep-grazing areas has been the development of very large properties. Although the average size of stations in the semi-arid area is 150 000 hectares, some are much smaller and some very much larger. Properties exceeding 400 000 hectares are found in the Murchison and Gascoyne regions of Western Australia and stations of 250 000 hectares are typical around the Eastern Goldfields area. The largest properties in semi-arid lands (some of well over 15 000 square kilometres) occur in the less favourable parts of Western Australia, south-western Queensland and in western New South Wales. The low carrying capacity, productivity and extensive nature of the pastoral zone are illustrated in figure 4.18. When compared to other sheep grazing zones the pastoral zone is the largest, and yet has the lowest flock size and wool yield.

## The influence of economic factors on location

Because of the constraints imposed by the physical environment of the pastoral region, and the leasehold restrictions that limit the pastoralist's ability to change or modify the environment, it is virtually impossible for any other form of agriculture to be economically

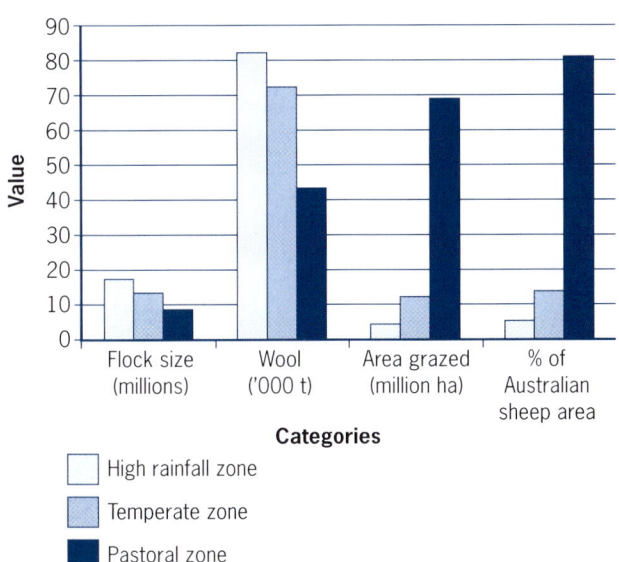

**Figure 4.18** Sheep and wool production, 1999

productive in this area. Closer to the coast, where conditions are suited to a range of land uses, relative economic profitability is largely responsible for the pattern of land uses.

Those land uses with higher returns per unit area are located close to markets. While sheep can be raised for wool in slightly wetter areas, the higher profitability of other land use activities (those with a higher economic rent per unit area) means that sheep grazing is largely concentrated in drier areas. Profitability decreases with increasing distance from markets, due to increasing transport costs. More intensive users of agricultural land have high transport costs, particularly those users that produce bulky, perishable goods that have to be transported frequently to the market. Costs associated with market gardening, for instance, would rise dramatically with increasing distance from the market. At a point still relatively close to the market this form of agriculture would cease to be a profitable enterprise. Wool, however, is not as costly to transport. It is valuable in relation to its weight, is not perishable and only needs to be transported once a year. There is only a gradual increase in costs with increasing distance from markets, so pastoralism can be undertaken in more remote areas.

The economic profitability of pastoralism in the marginal areas has been hard hit by market fluctuations, financial depression and drought, all of which affect a farmer's income and the ability to invest in and improve the property. When fluctuations in wool prices are coupled with climatic uncertainty and increasing costs of transportation to markets and shipping ports, there is limited scope for the profitable investment of capital into the farming system. Only through restricting investment in poor seasons, operating on a very large scale or, in some cases, diversifying into other activities, has pastoralism in these marginal areas remained viable.

### The influence of historical and political factors on location

The location of extensive sheep grazing in the semi-arid lands can partly be explained by decisions that were made in the past relating to the expansion and operation of pastoral systems. Pastoral landscapes evolved largely as a result of the demand by England for wool fibre, a demand created by the Industrial Revolution and by a growing population. The crossing of the Blue Mountains opened up the interior's natural pastures. Fine-woolled sheep were introduced into Tasmania, Western Australia, Victoria and South Australia (and finally into Queensland in 1940). In each case the better-watered grassland areas were taken up first, followed by the shrubland and poorer grassland as farmers in the better areas began to diversify into wheat.

The early pastoralists (squatters) in the eastern region ran sheep in open rangeland fashion, grazing their animals on unfenced land to which they had no legal title. The temporary 14 year lease issued by the government led to a lack of capital investment by pastoralists and a general deterioration of the range-land until leases were finally lengthened to 99 years. In Western Australia, temporary occupation leases had first been issued to pastoralists in south-western forested areas. The major push by pastoralists to the drier areas of the east and north-west, in the 1860s, occurred largely in response to favourable reports by explorers, the shortage of grazing land in the eastern colonies and the wealth brought by gold finds.

Pastoralists occupied the Murchison and Ashburton lands in the 1870s, and in the 1880s land as far east as the Eucla area was released to pastoralists. The gold rushes of the 1890s led to improved transport links (for example, the railway line from Perth to Kalgoorlie) and, therefore, to the spread of pastoralism even further inland. Over time more liberal leasing laws came into effect, with the push of grazing into more arid areas and with increasing awareness of the importance of the pastoral industry to the economy. The graziers were only able to graze sheep on the rangeland because of the government-instigated system of leasehold tenure for these enormous land areas. Because production achieved is low per unit area, no truly realistic value can be placed on the land—the amount of land required would be too expensive for the farmer to purchase. Graziers are therefore essentially caretakers on the land that provides them with their living. In return for the opportunity to graze sheep for a long period of time, they are required to invest in watering points, buildings and fences and to ensure that environmental degradation does not occur. This is necessary to maintain and increase the value of the leases.

### EXERCISES

1. On a map of Australia show the location of extensive pastoralism and then describe its extent and distribution within the central study area.
2. Outline how the physical environment has influenced the location of extensive pastoralism.
3. To what extent has distance from the major Australian markets influenced the type of agriculture practised within the extensive pastoral regions of the central study area?
4. Using the graph in figure 4.18 compare and contrast the three zones.

## Extensive sheep grazing as a system

Figure 4.19 shows the manner in which sheep stations operate as a system in terms of inputs (both natural and cultural), throughputs and outputs. A sheep station is a very large farm unit in which the farming activity is devoted almost entirely to the production of wool from sheep that graze within very large fenced paddocks. The number of sheep carried varies from property to property, and is usually dictated by the carrying capacity of the land at any time. To ensure constant levels, sufficient ewes are run so that natural increase covers losses that occur. The system is one in which environmental factors dictate the structural characteristics of the farm system and the nature of the processes that occur within it.

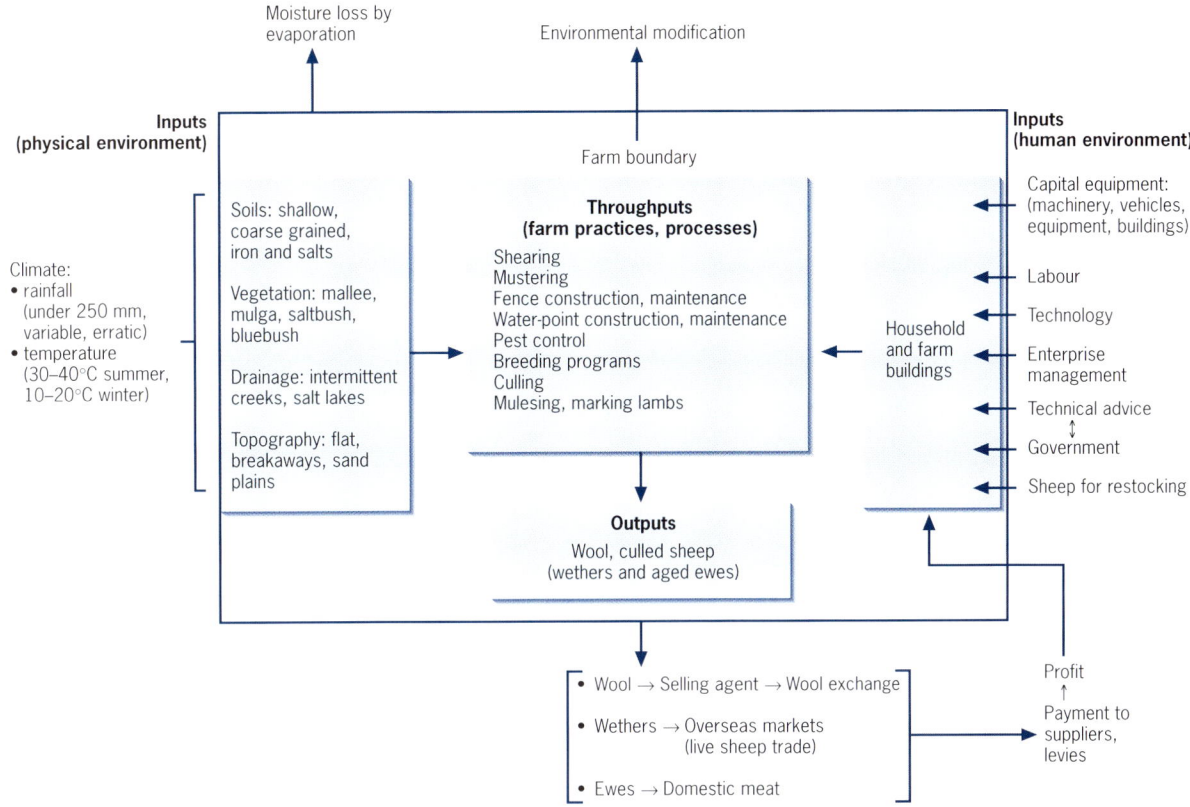

**Figure 4.19** Model of an extensive pastoralism (sheep) farming system

## Physical inputs

The physical inputs into the system are basically those that affect the location of extensive sheep-grazing activities. Rainfall is low (generally less than 250 millimetres), variable (sometimes as little as 150 millimetres and sometimes over 400 millimetres) and erratic. Temperatures are hot in summer, and winters are cool to cold (locations further from the sea have more extreme temperatures). Soils are coarse-grained, rocky and shallow and the topography is generally flat. As a result of these environmental conditions, particularly rainfall, the vegetation consists of saltbush, bluebush and other perennial shrubs, various types of acacia, mulga, and short-lived and unreliable winter grasses. The only significant modification of the physical inputs occurs with the tapping of subsurface water to supply stock water during the dry season. Because sheep cannot range too far from water, watering facilities have to be supplied at frequent intervals (mostly 7–10 kilometres apart) over the grazing lands. The water tends to vary in quality, sometimes containing levels of salt that are only just acceptable for stock use. Because too much salt causes problems, particularly with milking ewes and their lambs and young weaners, it is important to give sheep access to the best-quality water.

## Human (cultural) inputs

### Capital equipment and investment

A large amount of capital investment is required by the operator of a sheep station for the enterprise to remain economically viable. Although the total amount of capital invested is large, the amount invested per hectare is actually small because of the huge property sizes. While the cost of machinery and equipment is substantial, most capital investment programs today are aimed at increasing water availability and upgrading fencing. Properties often have numerous windmills, bores and stock dams and the installation of new watering points and the checking and maintenance of these points are usually the most time-consuming processes undertaken on a pastoral property.

The construction and maintenance of fencing around the paddocks also requires large amounts of capital, with new fencing costs running at $1500 to $2000 per kilometre. These fences have often been constructed at the expense of the natural vegetation, with posts originally cut from the mulga trees found on the properties. New fencing now relies more on the use of steel pickets. Capital items such as shearing sheds and equipment and shearers quarters and facilities also require substantial capital investment.

Sheep stations invest heavily in vehicles, machinery and equipment. This can include stock trucks and trailers, utilities and 4WDs, motorbikes, front-end loaders and graders. Larger properties may include fixed-wing aircraft and helicopters for use in mustering and property inspections. Many properties have fully equipped workshops as well as fuel depots and machinery storage facilities.

*Labour inputs*

When calculated on a per hectare basis, the labour input on sheep stations is extremely low. Often properties are operated by three or four people and additional labour is hired-in during the peak seasonal times, such as mustering and shearing. Capital improvements requiring extra labour, such as new fencing or new bores, are often undertaken by specialist contractors rather than by the manager or owner.

Labour shortages are looming as an issue in the pastoral industry, with the average age of Australian shearers now exceeding 40 years. Coupled with the increasing rates of retirement is the lack of interest shown by younger workers in this field of employment. In the period 1992 to 2002, the number of Australian shearers decreased from 30 000 to 8000. With the recent clearing of the national wool stockpile and some improvement in prices, any recovery in the industry may well be hampered by a lack of shearers, shed hands and wool classers.

*Technological inputs*

Over time technological inputs, while they have not changed the basic system of grazing animals on natural pastures, have addressed the issues of distance, stock control and water supplies. The advent of fencing in the 1850s played a significant part in herd management and the grazing of the natural pastures. It also allowed the pastoralist a degree of security in preventing stock losses. The sinking of wells and the development of water supplies by the use of artesian bores has also played an important role in the maintenance of herds during the extended dry seasons. The introduction of the diamond drill, with rigs capable of drilling over 1000 metres, has brought this valuable resource into production.

Improvements in transportation have played a vital part in the management of very large properties and in the movement of stock and wool to market. The on-station use of motorcycles and helicopters during mustering and in moving livestock, as well as the use of road trains to move animals or bales of wool to the saleyards, shows the importance of improved transport technologies.

## Throughputs

Table 4.3 shows two examples of a typical work calendar for a sheep station. The busy period on a station is the mustering and shearing of the flocks. Animals are progressively moved closer to the **homestead** paddocks in readiness for this task. It is also a time when the animals are dipped to remove external parasites and drenched to control intestinal worms. Old or weaker animals are culled, while others are classed and organised into suitable herd sizes on the basis of age and sex. They are then moved to paddocks according to the size of the flock, the number of watering points within the paddock and the quality of the pastures available. One of the largest environmental concerns on a sheep station is erosion of bare soils by wind and water and so care is taken to ensure that areas are not overgrazed.

**Table 4.3** Work calendar: 'Hampton Hill' and 'Oudabunna' stations

| Month | Hampton Hill | Oudabunna |
|---|---|---|
| January | General maintenance including dam and fence construction | Reallocation of sheep to paddocks after shearing, windmill checks |
| February | Mustering | Windmill checks, trapping of straggler sheep, removal of rams from ewes |
| March | Mustering and shearing | Windmill checks, trapping and shearing of straggler sheep |
| April | Shearing | Windmill runs |
| May | Shearing | Lambing |
| June | Culling/breeding program | |
| July | General maintenance including weed eradication, fencing, checking of watering points | Maintenance of windmills, fencing, buildings; regeneration of overgrazed areas |
| August | | |
| September | Lambing, lamb marking | |
| October | Lamb mulesing | Marking of lambs, checking and treatment of fly strike |
| November | General maintenance including tank construction, road upgrading, weed eradication | Trapping and mustering prior to shearing |
| December | | Shearing, breeding program, marking lambs, culling |

Throughout the year fences, bores and windmills must be checked at least every three to four days, and maintained regularly. The routine servicing of the machinery as well as repairs is also a regular task. In addition to these activities, other tasks include the maintenance of vehicles and some rehabilitation of degraded areas. During times of drought some properties import hay to provide supplementary feed or in some cases move stock to other properties where pastures are still in good condition or where rain has fallen.

## Outputs

Sheep stations are primarily concerned with the production of wool and this is their main output. This tends to be coarse wools (finer wools are produced on smaller farms in higher rainfall regions). Where pastoralists practise split shearing, the wool is transported twice a year to the auction houses located in the main port cities. Overseas and domestic buyers inspect the wool clip and then bid on the individual bales. There are several major companies involved in the auctioning of wool in Australia. These include Elders and Westfarmers. In March 2002 coarse (22–24 micron) wool was being auctioned at approximately $9.80 a kilo, an increase of some $3.00 over the same time in 2001.

While the prime purpose is to maintain a flock of wool producing sheep, pastoral properties do supply animals for meat as a consequence of culling and herd management. These animals may be sold to domestic abattoirs or exported as part of the live sheep trade. The major market for the latter are the Middle Eastern countries of Saudi Arabia, Kuwait and the United Arab Emirates. Merino wethers were sold for approximately $40 per head in 2002.

> **EXERCISES**
>
> 1 Using an A3 sheet of paper draw a large diagram of the extensive grazing (pastoral) system and then on it place summaries of the inputs, throughputs and outputs.
> 2 Discuss the relationship between the physical environment and the farming activities practised in the extensive pastoral zone. In completing this activity consider the seasonal routine, the needs of the livestock, the infrastructure found on the grazing properties, the impacts of climatic uncertainties and the effects of vegetation, soil and topography on the pastoralist's use of the land.

## The cultural landscape

### The regional cultural landscape

Due to the extensive nature of sheep grazing the most obvious feature of the cultural landscape is the very disperse settlement and transport patterns. There are large distances between urban centres and few sealed roads. Minor, unsealed roads connect properties to each other and the landscape often has little evidence of human occupation. Western Australia's extensive sheep-grazing lands contain very low population densities, with a rural population spread of less than one person for every 64 kilometres. The population therefore does not generate sufficient demand for the development of central places. Instead, many of the towns that have developed in the arid and semi-arid regions have done so in response to the demands of the mining industry rather than pastoralism.

The appearance of the natural vegetation suggests that there has been little or no obvious modification. It is not until it is compared to undisturbed sites that the changes become apparent. The grazing lands have had significant numbers of trees removed to provide fencing material and the grazing of the rangelands has reduced the variety of shrubs, with some of the slower growing species being removed altogether. Figure 4.20 illustrates the general appearance of the pastoral landscape near Kalgoorlie in Western Australia.

### The farm site cultural landscape

The large, isolated pastoral leases are generally surrounded by other stations and are divided into large paddocks by wire fencing. The paddocks may vary from several hundred to several thousand hectares in size and their most obvious internal features are the watering points. These can be in the form of windmills, bores, tanks, stock troughs or small dams. Often trapyards are constructed at these locations, as they become important gathering points for the livestock. Each paddock and watering point is connected by unsealed tracks similar to that shown in figure 4.20.

The homestead complex is in many cases a small, self-contained community with numerous buildings.

**Figure 4.20** Pastoral landscape

These include the main house, shearers' and stockmen's quarters, shearing sheds, stockyards, machinery sheds and repair shops, cookhouse or mess, fuel depots, water tanks and generator or power house. Some properties may also have unsealed airstrips, which provide emergency landing facilities for the Royal Flying Doctor Service or are used by the pastoral company. Figure 4.22 illustrates some of the structures associated with the homestead. The extensive shearing shed is used once or twice per year and must be able to cater for the large flocks of sheep which will be held in the stockyards or pens prior to shearing.

## Problems and prospects

### Environmental issues

On an average pastoral property of 50 000 hectares, approximately 5500 hectares is affected by some form of degradation. The main environmental challenges facing the pastoral industry include water erosion, weed infestation and wind erosion.

An example of a joint livestock industry response to a common regional issue is in the Western Division of New South Wales. In this region, beef and wool producers are confronted with similar environmental challenges that result from the integration of their production systems into landscape systems. The populations of native grazing animals, along with the pressures exerted by introduced grazing animals such as rabbits and goats, affect the capacity of the leasehold land to provide adequate feed supplies to the pastoralists' sheep. In response to these problems, the leaseholders met and produced a management plan that was based on the limitations imposed by the environment. Some of the key elements of the plan included establishing the appropriate level of stocking density (number of hectares per animal) to achieve a sustainable rate of grazing and allowing adequate time intervals for the regeneration of the native shrubs and grasses. In addition to sustainable stocking rates, other strategies include prescribed burning to encourage new pasture growth, the sowing of introduced pasture species, reintroducing native pasture species and controlling feral animal populations.

**Figure 4.21** Watering point and trapyards, 'Oudabunna' Station

## Economic issues

Producers in the pastoral regions of Australia work longer hours, have farm debt levels higher than the industry average, have negative business profit and lower than industry average off-farm income. In 1998–99 the average price for clean wool was $5.50 per kilo, with fine wools fetching around $7.00 per kilo and coarse blends about $3.00 per kilo (note that in 2002, sales have produced returns of almost $10 per kilo). An average sheep produces about 5–6 kilos of wool per year, with shearing costs in this period being about $3.15 per head. Other costs including jetting, dipping, mulesing, marking, supplementary feeding and transporting the wool in the same period ran out at $4.00 per head. Thus, one animal generates wool to the value of $33.00 with a net return to the producer after costs of

**Figure 4.22** Shearing shed and mustering on a sheep station

**Figure 4.23** Feral animals in the pastoral zone

$25.85. On a property of 100 000 hectares and running a flock of 10 000 sheep, the return would be $258 500. However, from this the cost of repairing fences, bores, windmills and machinery, as well as purchasing new equipment and paying the annual rent on the lease, has to be deducted. In 1998–99 the average net income on pastoral leases was negative $55 218. This meant that nearly all pastoral leases operated at a loss. The consequence of such a situation is a gradual depletion of savings and a deterioration of equipment and capital. The economic conditions faced by pastoralists, as well as the issues of environmental management and land tenure, all point to the marginal nature of this type of land use.

### EXERCISES

1. Refer to the characteristics of extensive agriculture found in chapter 3 and then, using examples, outline the ways in which extensive pastoralism conforms to these.
2. Extensive pastoralism may be described as a marginal land use. Using examples, discuss the ways in which this type of agricultural activity could be seen as economically and environmentally unsustainable.
3. Describe and account for the cultural landscape associated with extensive pastoralism.

## CASE STUDY 3

## Intensive pastoralism in Eastern Australia

Dairying is an intensive form of pastoralism in which cows are primarily raised for their milk. This type of agriculture is also called advanced grassland farming. It is one in which the animals are treated as biological factories, converting grass and other herbage into milk to meet the needs of large urban markets. With a farm-gate value of production exceeding $3 billion, it ranks third behind the wheat and beef industries. Dairying is also one of Australia's leading rural industries in terms of adding value through further processing. Much of this processing occurs close to farm areas, generating industry and employment in country regions.

Dairying has a long history in many areas of Australia. While the bulk of milk production occurs in Victoria, all states have viable, productive industries, supplying fresh milk to nearby cities and towns. In addition, a wide range of high quality manufactured products, from fresh lines such as yoghurt to bulk and specialised powders, to a variety of cheese types, are produced in most Australian states. Milk production can be broken into two main categories. The fresh milk supplied daily to the market is classed as **whole milk**, while the milk that is processed to produce a variety of milk-based products is **manufacturing milk** supplies.

The majority of Australian dairy farms are family owned and operated. The number of dairy farms has declined over the last two decades, from over 30 000 in 1975 to just under 12 000 currently. Figures 4.24 and 4.25 demonstrate this change in farm numbers.

The rationalisation in farm numbers is largely attributable to a reduction in government support given to the dairy industry over this period, and an accompanying increase in exposure to market forces. In order to keep pace with rising costs and falling or static returns, Australian farms have generally had to become larger and more efficient.

Average herd size has increased from 77 cows in 1975, to an estimated 190 in 2001. However, there are many farms with herds greater than 250 cows. Improved herd genetics, as well as advances in pasture management and supplementary feeding regimes, have seen average annual yield per cow increase from 2750 litres to 4624 litres over the same period.

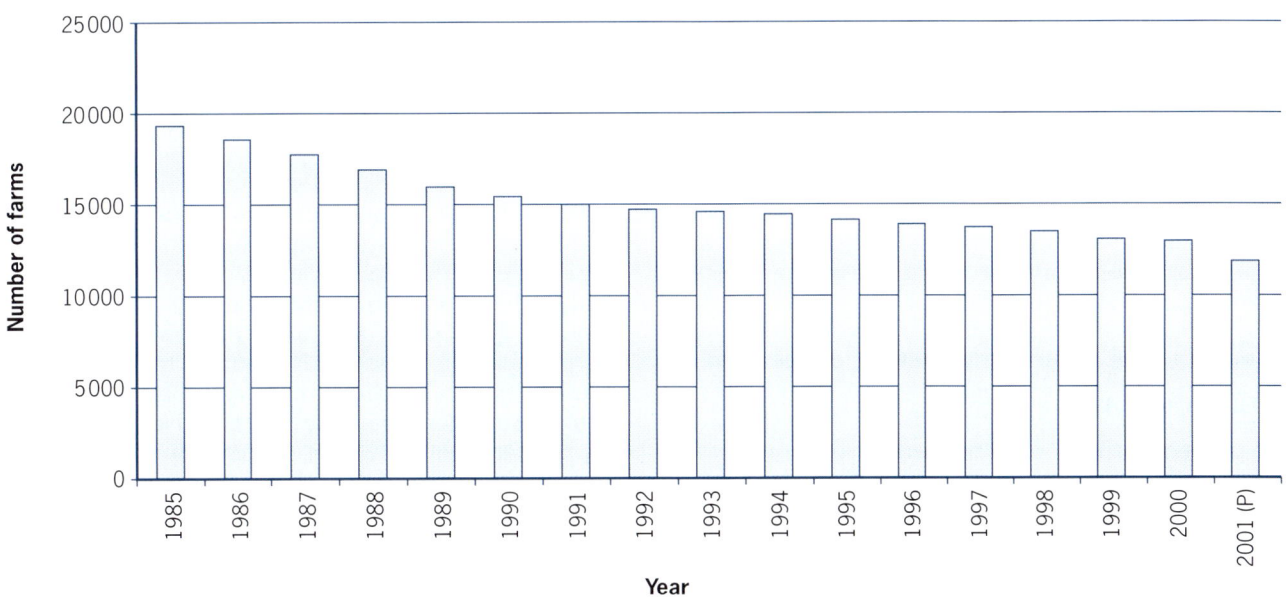

**Figure 4.24** Australian dairy farms

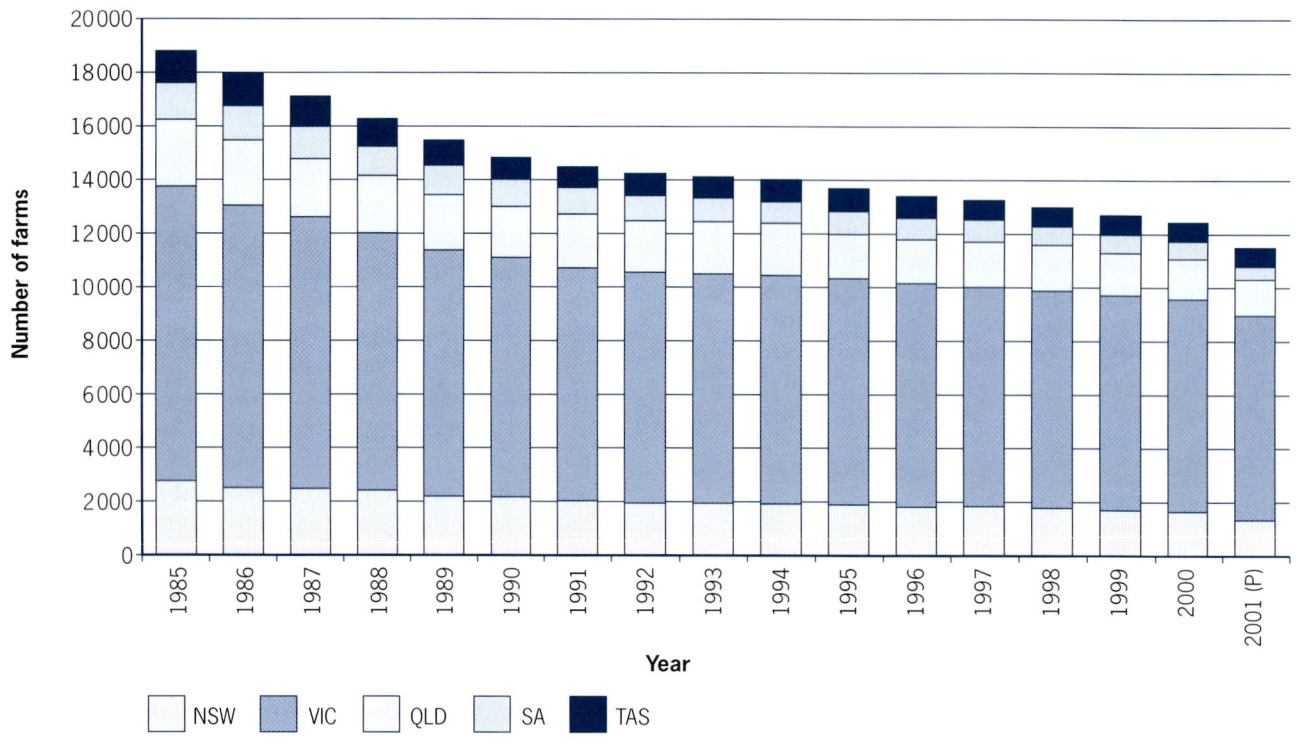

**Figure 4.25** Farm distribution in Eastern Australia

## The location of intensive pastoralism

Dairy farms in Eastern Australia are nearly all in coastal and near-coastal locations, with the majority of farms and production being found in Victoria. As the graph in figure 4.25 shows, in 2001 approximately 65 per cent of the farms in the eastern study area were located in this state. These farms are evenly distributed within the three main dairy zones: Gippsland to the east of Melbourne; the South-West Districts to the west; and the North Central District on the Murray River. Victoria is Australia's largest milk-producing state, accounting for more than 60 per cent of the national milk production and more than 70 per cent of manufacturing milk production. The average Victorian dairy herd has increased from 150 head in the late-1970s to around 250 head per farm in 2000.

The only significant inland location within the eastern study area occurs on the New South Wales–Victorian border, in the Riverina and the Murray–Goulburn Valley where irrigated pasture supplied by water from the Murray River has allowed intensive pastoralism to develop.

### Influence of the natural environment on location

#### Climatic influences

One of the most important reasons for the concentration of dairying in coastal areas is the relatively high all-season rainfall of over 750 millimetres, in areas which experience moderate temperatures. To be an efficient producer of milk, a dairy cow requires a constant supply of nutritious grasses (approximately 68 kilograms of fresh fodder per day). Climatic conditions which are too cold and dry would produce coarse, unpalatable pastures. The Victorian coastal districts are climatically the best in Australia, receiving an average yearly rainfall of over 750 millimetres, with the best dairying districts receiving more than 1000 millimetres. The cool, moist climate of north-west coastal Tasmania is also particularly suited to dairying. Some drier dairying areas depend on irrigation to supplement natural rainfall, and dairy farmers consider it essential for maximum production even in some high rainfall areas. In the Gippsland area of Victoria, dairying is mostly carried out under natural rainfall conditions. In the drier areas of northern Victoria, however, pasture production depends on flood irrigation via pipes or open channels leading from the large dams located on the major rivers of the Murray River system. Although the rainfall average for the Bega Valley (in New South Wales) is 900 millimetres per year, the region suffers from regular droughts and short dry periods (as shown in figure 4.27). Consequently, two-thirds of the farmers in this area use water from the Brogo and Bega rivers to spray-irrigate their pastures. In many dairying areas that experience unreliable rainfall and periodic drought, hay is cut and fodder crops are grown to supplement pastures. South Australia, for example, has a less-than-ideal climate for dairying as it experiences hot, dry summers. Meadow hay is used in this season to supplement relatively poor pastures. In the Bega River area low temperatures limit pasture and crop growth from June to late August, and high temperatures combined with low effective rainfall limit growth in summer. Pastures are therefore supplemented with hay (cut and baled when dry in summer) and silage (a green-fodder concentrate, cut and compressed in

CASE STUDIES IN AUSTRALIAN AGRICULTURE  81

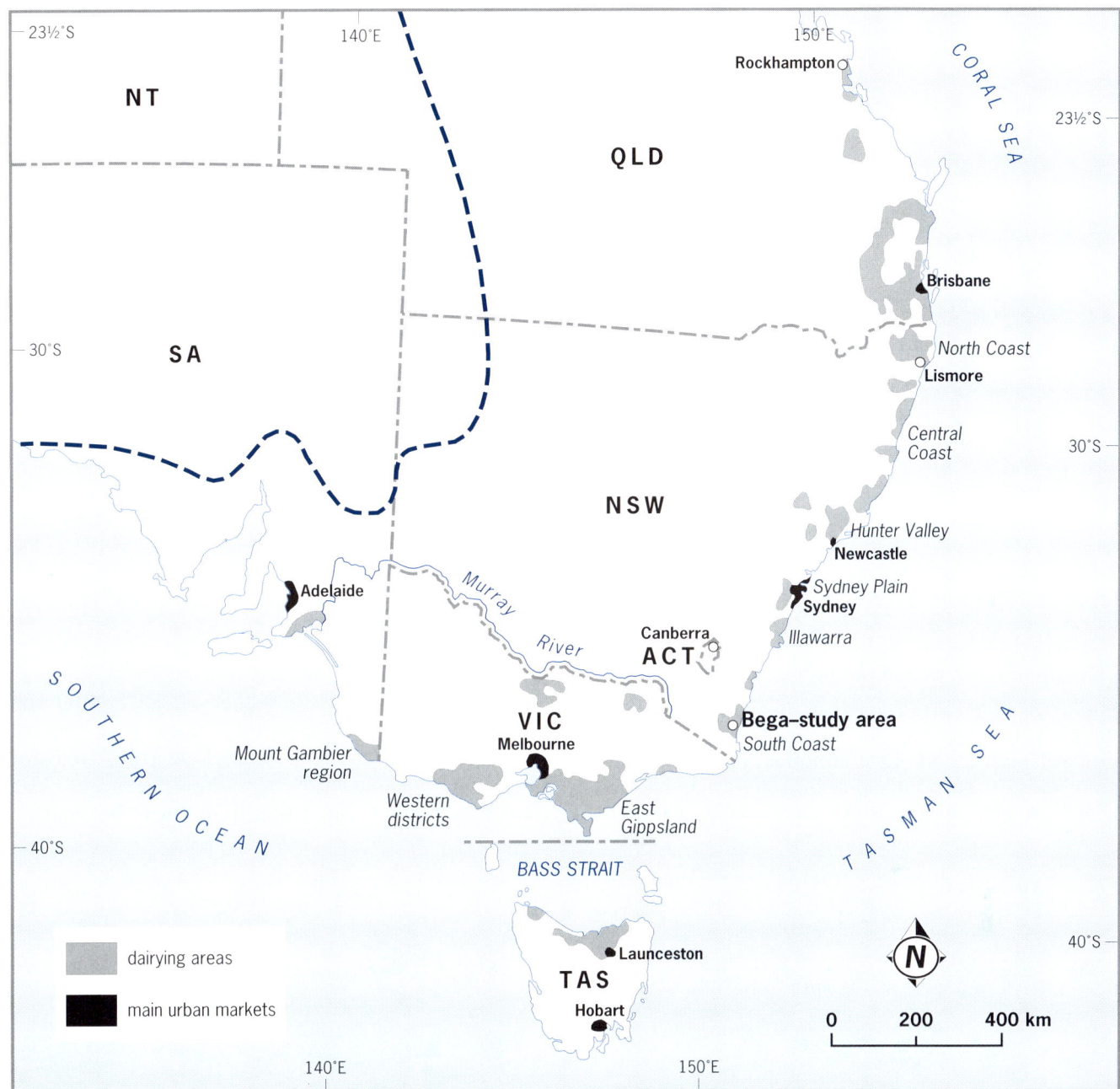

**Figure 4.26** The location of dairying in Eastern Australia

October, when pasture growth is at its best). Fodder crops such as maize, oats and sorghum are also grown in the area.

*Soil and topography*

Soil fertility and slope are other physical factors that help to explain the fragmentary nature of dairy farming, particularly along the coast of New South Wales (see figure 4.28). The fertile alluvial soils found in the river flats and plains of the valleys of the eastward-flowing rivers (and the lower hill slopes) are ideal for dairying. By contrast, the slopes of the rugged ridges and spurs which separate the valleys are often too steep to be cleared, or their soils are too thin and infertile to grow pasture. The steep slopes and relatively infertile granitic soils of the Bega River Valley catchment area, for example, are easily eroded and therefore not suitable for large-scale clearing. Dairying is largely restricted to the well-drained, relatively fertile, loamy soils of the river plains. Coastal areas of Victoria have larger areas of unrestricted coastal lowland than does the eastern margin. The volcanic soils of areas of the Western Districts and the fertile loamy and alluvial soils of Gippsland are particularly productive, containing Australia's heaviest concentrations of dairy cattle. The interior plains in the state's central-northern area have also been developed for large-scale dairying.

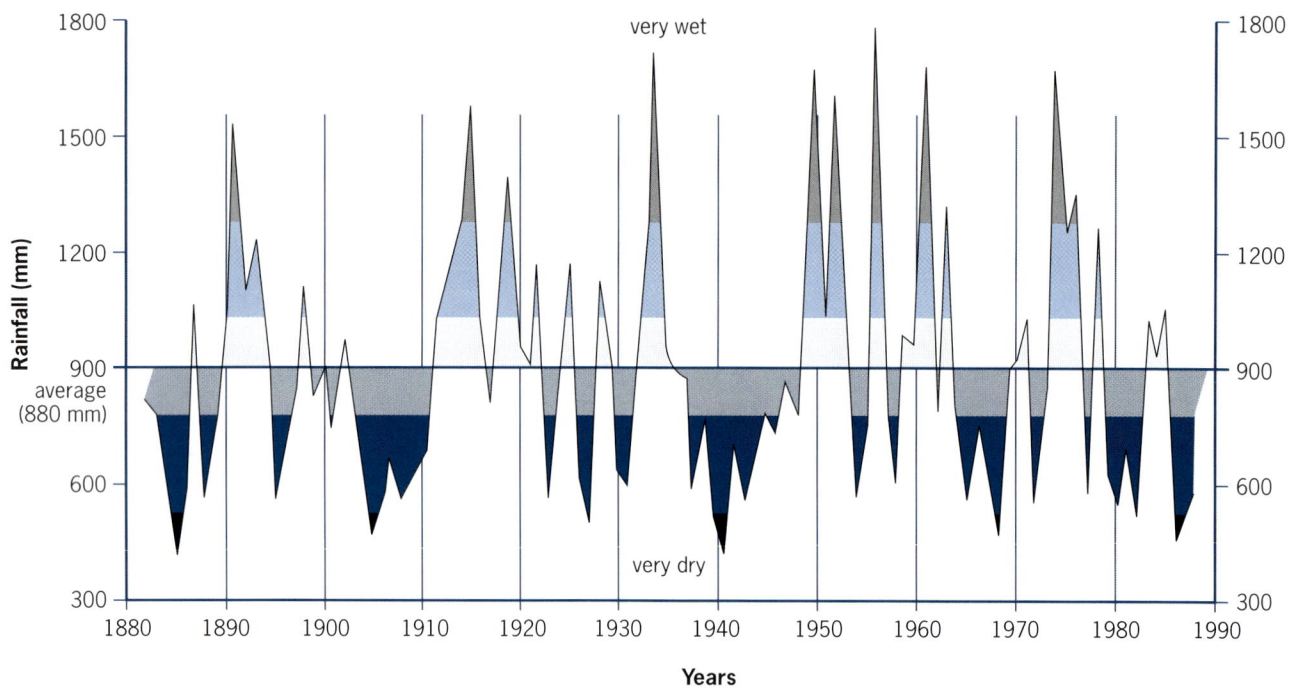

**Figure 4.27** Annual rainfall, Bega (from the 1880s)

### The influence of economic factors on location

Important economic considerations have led to the location of dairying in some areas considered physically unsuitable. Most important of these considerations is the need for access to major urban markets which require dairy products on a daily basis. The pattern of dairying in Eastern Australia consists essentially of a series of major nuclei around the various state capitals, clearly reflecting their 'pull' as major markets. Dairy farms that produce fresh milk for the whole-milk markets have traditionally located close to these markets in order to reduce the cost of

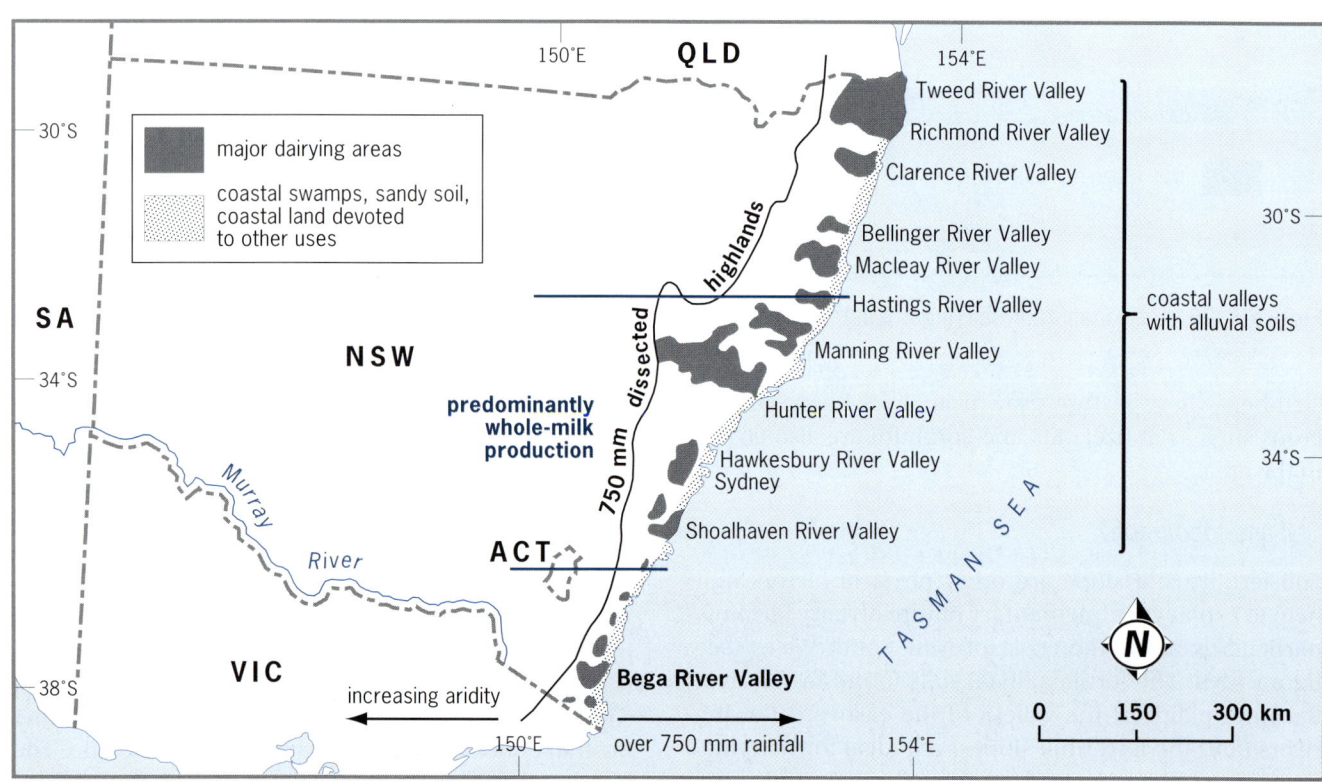

**Figure 4.28** The effect of physical factors on the location of dairying in New South Wales

constantly transporting a relatively bulky and perishable product in expensive refrigerated transport.

Dairy farms that supply milk for butter and cheese production have tended to be located further away from urban markets. The processing of milk into reduced milk products and cheese involves a weight reduction. For example, 9 litres of milk weighing 10 500 grams are used to produce 500 grams of butter. Processing also involves a decrease in perishability and an increase in value of products such as cheese and butter. Because they are now more concentrated and less perishable they are able to bear the higher costs of transportation to city markets, unlike the fresh milk from which they were produced. Dairy farms that are located further from major markets therefore tend to specialise in the production of milk for manufacturing. In these outer areas, too, unfavourable climatic conditions may restrict farmers' output of milk, so that they are unable to produce adequate fresh milk to satisfy the demands of a whole-milk market (examples include some farms in northern New South Wales and southern Queensland).

The processing factories associated with the dairy industry are generally found scattered throughout the outer, 'manufacturing milk' areas in relatively large towns. This 'raw materials orientation' results largely from the need to reduce weight and perishability. A number of dairy factories, usually run by farmers' cooperatives, are found throughout the manufacturing milk zones in the far north and south coasts of New South Wales, in east Gippsland and the Western Districts of Victoria, and in the Mount Gambier district of South Australia. In north-east Tasmania, for example, butter and cheese are produced at several factories in centres such as Smithton, Devonport, Burnie and Ulverstone. In many cases a proportion of the output of a dairy farm within the 'manufacturing milk' zone is directed to the whole-milk market. Some of the milk that arrives at the Bega cheese factory, for example, is transported in bulk to the whole-milk processors of Sydney, Wollongong and Canberra, from where it is distributed in consumer packs. Some is packaged and sent by refrigerated van to Queanbeyan and is distributed throughout various towns from Bodalla in the north to Eden in the south. Milk sent in bulk to Canberra is packaged and distributed in Canberra and south-eastern New South Wales. The remaining milk is made into cheese at the factory (see figures 4.29 and 4.30). In broad terms, dairying and its various processing factories tend to be located closer to larger cities. The reason for this is related to the results of competition among various agricultural land users for near-market locations.

Competition for land close to cities arises because farmers wish to minimise the costs of transporting products to market. Competition for restricted land near the market forces near-market land prices up, so that only intensive users of the land (those that can produce high yields, and therefore high profits, per hectare) can afford the near-market sites. Dairying can provide a greater return per hectare than a more extensive form of agriculture such as wheat farming, so it will be found close to the market. It is not, however, able to outbid intensive horticultural enterprises (such as market gardening and flower growing) as these enterprises are able to earn an even higher profit (economic rent) per unit area. Market gardening is, therefore, generally found closer to the market than dairying. The relative profitability of land uses at certain distances from the market also helps to explain the near-market location of whole-milk dairy farms and the more remote locations of the farms that produce

**Figure 4.29** The Bega cheese factory

**Figure 4.30** Loading tankers for delivery to milk markets

mainly manufacturing milk. Because economic returns to land decline with distance from the market, the highest returns in dairying come from whole-milk production, so they are able to outbid those farmers who produce manufacturing milk for the near market locations. The higher value, processed dairy products are profitable further out because they can withstand higher transport costs. Butter and cheese are sent to overseas markets because their higher value means that even higher transport costs can be absorbed.

### Market deregulation and location

Commencing in the early 1980s there has been a progressive movement by federal and state governments, in consultation with dairy farmers groups and other dairying authorities, towards a deregulation of the industry and a progressive removal of **tariffs** and market control mechanisms. The final stages of this action were achieved in 1999.

Historically, state governments were responsible for ensuring year-round supplies of fresh milk to consumers. Individual state governments achieved this using either pooling or quota systems to source milk from farms, and by controlling prices and distribution from the farm gate to the consumer. During the 1990s, the various states phased out post-farm gate controls, until only farm pricing and sourcing regulations remained.

In the late 1990s, all Australian state and federal governments signed on to a national competition policy (NCP). The guiding principle for this process is that there should be no regulatory restrictions on competition unless it can be clearly demonstrated to be in the public interest.

In early 1999, the industry approached the federal government with a plan for an orderly, national approach to the deregulation of the whole milk sector. On 28 September 1999, the federal government announced it would implement the Dairy Structural Adjustment Program (DSAP). This program places an 11 cents per litre tax on drinking milk products, with the funds raised being used to fund the smaller and less efficient producers to either grow or move out of the market. As part of the deregulation process, all states have repealed legislation governing pricing and sourcing of drinking milk, and the state milk authorities that administered these controls have been wound up.

The impact of this change on the locational pattern of dairy farming will, over time, be quite dramatic. As more and more whole milk products from the more productive regions appear on the supermarket shelves, the areas in which production is less efficient or more expensive will lose many dairy farms. It is a system that may well see the dominance of the Victorian producer as other areas lose farmers who take advantage of the various packages funded by the levy. The Dairy Exit Program provides an optional tax-free exit payment of up to $45 000 for eligible dairy producers wishing to leave the industry, while other restructuring packages are aimed at assisting farmers to upgrade or increase their efficiency and production levels.

As the distinction between whole and manufacturing milk zones disappears, the traditional patterns of location will also decline as the remaining zones produce for a combined market. Thus, farmers in areas traditionally focusing on milk for cheese, butter and processed milk products may supply whole milk when they can compete by producing cheaper supplies at certain favourable times.

## The influence of historical and political factors on location

The location and functioning of the dairying areas in Eastern Australia reflect the historical development of dairying and improvements in technology over time. Modern dairy farming was established in the early days of European settlement and its growth corresponded with Eastern Australia's increasing urban population and its increasing demand for milk. The first European settlers brought with them a taste for butter and cheese. Consequently some areas of coastal New South Wales, such as the Illawarra region, were developed in the 1800s to produce dairy products for the Sydney market. Development of the fresh-milk industry, however, had to await the coming of refrigerated transport in the 1880s. The replacement of the milk truck (which collected milk in cans from the nearby farms every day) by the milk tanker meant that the sale of liquid milk was no longer limited to areas within a day's journey of the farm. Because dairy products could be stored safely and transported over long distances, more remote areas of forest were cleared for dairying and a new export industry in butter became possible. The butter export industry of New South Wales and Victoria was aided by the introduction of milk separators, which enabled the farmers to separate the milk themselves, sending cream to the butter factory and retaining skim milk to feed to pigs or calves. A decade later the Babcock Tester was introduced, allowing the butterfat content of milk to be tested and farmers to be paid fairly for their produce.

In 1889, following the discoveries of Louis Pasteur, milk was pasteurised, adding further to its keeping properties and qualities. The net result of these inventions and improvements was the establishment of cooperative butter and cheese factories in areas quite remote from major urban settlements. The location of the dairying industry in Eastern Australia can also be explained by the role of government planning and intervention. Both federal and state governments have at times recognised the need to expand dairying for various economic, political and social reasons. The spread of dairying into regions of often marginal suitability has been encouraged in the past. For example, the spread of dairying along the far north coast of New South Wales was closely related to policies of promoting a closer settlement pattern of small family-run farms. Government considered dairying to be a good method of land settlement, as it would give one of the highest densities of rural population. This in turn would encourage the establishment of country towns to provide essential goods and services.

Some of these schemes (known as closer settlement schemes) involved resettling soldiers returning from World War I. The introduction of dairying on irrigated blocks in northern Victoria and the Riverina, for example, was the result of the deliberate 'soldier settlement' policy at the time. Dairying began in the Bega Valley in New South Wales in the 1870s in response to the continuing demand for milk by urban populations. Prior to this time, sheep had been grazed in the area. Sheep gradually gave way to dairy cows and by 1880 a third of the farms on the 'estate' were supplying milk to a local cheese factory. The Bega Co-operative Creamery Company was formed in 1899 and in 1900 the original butter factory opened. Important developments in the cooperative's history since 1900 are shown in table 4.4, which illustrates the influence of changing demands and improved technology on the area's dairy industry.

**Table 4.4** Significant events in the history of the Bega dairying industry

| Year | Event |
| --- | --- |
| 1890 | Original butter factory opened. |
| 1924 | Butter factory constructed on present site. |
| 1944 | Name changed to The Bega Co-operative Society Ltd. |
| 1954 | Commencement of milk receivals, town deliveries and cheese manufacture. |
| 1956 | Commencement of powder production. |
| 1960 | Opening of Canberra branch. |
| 1966 | Implementation of bulk-milk collection and refrigerated farm-milk vats. |
| 1969 | New cheese factory commissioned. |
| 1971 | Cheese production began. |
| 1972 | Factory ceased butter manufacture. (Factory is now a tourist attraction.) Joined with Tilba Co-operative. |
| 1975 | Opening of Queanbeyan branch. |
| 1976 | Successful campaign to gain share of the New South Wales market for fresh milk. Closed production at Tilba. |
| 1982 | Despatch tankers converted from 13 500 litre capacity to 22 500 litre capacity. |
| 1983 | Capacity of Bega cheese factory doubled. Complex occupies 1.5 hectares. Milk packaging in Bega upgraded. Closure of Queanbeyan factory. |
| 1985 | Arrangements made with large Victorian co-operative Bonlac for Bonlac to market Bega cheese in states other than New South Wales and the Australian Capital Territory. |
| 1986 | Milk packed in 2 litre plastic bottles. |

### EXERCISES

1 With reference to farm sizes and location, identify and discuss the changes that have taken place in dairy farming over the last 10 years.
2 With reference to the map showing location identify at least three features of the distribution of dairy farms in the eastern study area and briefly account for these.
3 Identify the two environmental influences that you consider to have had the most significant effect on the location of dairying in Eastern Australia and justify your selection.

4. Discuss the extent to which the land rent mechanism has influenced the location of the dairy producing regions in Eastern Australia.
5. Identify and describe the types of government regulations that have applied to dairy farming and how these have altered in recent times.

## Intensive pastoralism as a system

Figure 4.31 demonstrates the way in which the dairy farm operates as a system. The farm receives inputs from the physical environment, which are manipulated by the farmer to ensure a regular supply of high-quality milk from the milking herd. Compared with some other forms of pastoralism dairying requires a high level of investment per hectare in capital equipment and technology.

### Physical inputs

*Climate*

The most important physical input of the dairying system is climate, as sufficient rainfall and moderate temperatures are required to produce a constant supply of high-quality green pastures. When combined with adequately drained and reasonably fertile soils, and relatively flat land, a mild, moist climate provides highly satisfactory conditions for dairying. Water supplies for supplementary irrigation during the drier periods have been developed by constructing dams and irrigation channels. In some cases this has required the levelling of the land to permit flood irrigation of the pasture, while in other instances spray irrigation of the land is practised.

In the Bega Valley region, over 70 per cent of farmers irrigate from the waters of the Bega and Brogo rivers. The farm operator on 'Sunnybank', a case study farm in the Bega Valley (the location of which is shown in figures 4.32 and 4.37) has an irrigation licence for 60 hectares. A low-pressure travelling spray irrigator running off a 60 horsepower pump is used on the river flats, and lateral sprinklers (five sprinklers for every 2 hectares) are used on the foothills region of the farm. In 14 hours the irrigation system puts 50 millimetres of water on the pastures. The dairy herd is kept on these irrigated pastures day and night.

*Soils*

The high rainfall of many dairying areas has been responsible for the leaching of nutrients, leaving relatively infertile soils. Soils in the Bega area, with the exception of near-river alluvial soils and some derived from volcanic rock, are of low to moderate fertility. The major nutrient deficiencies are phosphorus, sulphur and molybdenum. On 'Sunnybank', fertilisers are applied to the soil twice a year. Relatively large amounts of lime (32 kilograms per hectare to combat acidification) and potassium (27 kilograms per hectare) are used, along with smaller amounts of nitrogen, phosphorus and sulphur. Effluent from the milking sheds is also pumped onto the poorly drained clay soils of the granite hills area.

*Natural vegetation*

Unlike more extensive systems of beef or sheep grazing, dairying is entirely dependent on introduced imported species of grasses, principally paspalum, perennial rye,

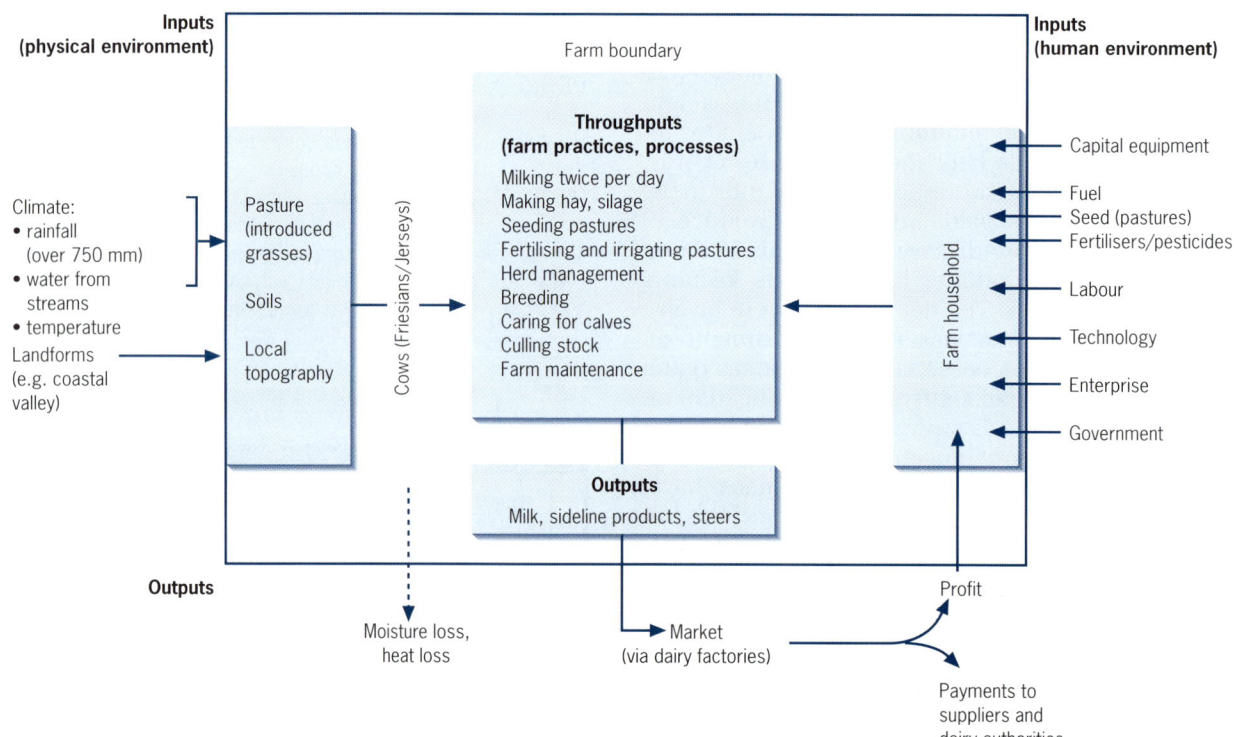

**Figure 4.31** Model of a dairy farming system

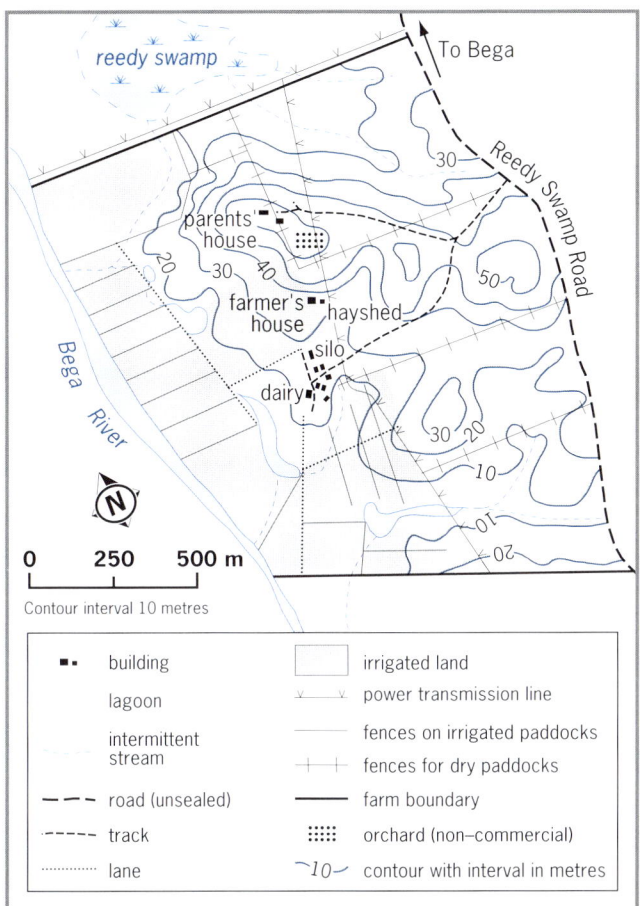

**Figure 4.32** 'Sunnybank' dairy farm, Bega

clovers, cocksfoot, couch and kikuyu. The original forest vegetation has been almost entirely removed from the land. It may remain as small stands along streams, or individual trees may have been retained as shade for the stock.

In dairy farming the land is used and worked relatively intensively. Farms vary in size from 80 hectares to over 200 hectares. The average farm size in the Bega Valley is over 200 hectares. Farms that have access to fertile river flats in the valley tend to be smaller than those farms on less fertile hill country. Herd sizes range from 80 cows to over 400, the average being about 150 cows per farm. Jerseys and Friesians are the dominant breeds of cattle. 'Sunnybank' is 191 hectares in size and carries a dairy herd of 130 Holstein–Friesian cows. The stocking rate of 0.7 cows per hectare is generally above average for this district and is considered by the farmer to be the farm's maximum capacity with the current area of land under irrigation.

Dairy farms that concentrate on the production of fresh milk for urban markets tend to be smaller in size than farms that produce milk for processing. All paddocks on the whole-milk producing farms are sown with improved pastures and fodder crops and the farms tend to be more productive. There is little time or need to run sideline animals or activities, so farms are stocked only with cattle that have high milk yields. Farms in the manufacturing-milk zone are less well developed and pastures are not improved to the same degree. Milk production tends to be more seasonal and farmers usually engage in sideline activities such as beef and veal production and pig raising. Hay and silage are cut and special fodder crops (maize, lucerne, sorghum and oats) are grown as supplementary feed at times when pasture growth is restricted. The rye grass and clover pastures on 'Sunnybank' are supplemented by hay and silage up to four times per week if pastures are poor. Up to 3000 bales of hay are made during summer, and silage is made in October and then stored in ground pits. Some grain is brought into the valley from the Riverina area. Most farmers also give their cows additional measured supplementary feed while they are in the milking bails. Cows on 'Sunnybank' are fed grains such as barley, the quantity of which depends on the amount of milk the cows are producing at any time.

## Human (cultural) inputs

A considerable investment in machinery, equipment, fuel and fertilisers is required in dairy farming. There is a great dependence on high levels of scientific research, management skills and specialist advisory services. The farmer regulates the inputs from the physical and human environments in order to create the most favourable conditions for the dairy cows.

### Capital equipment and investment

Dairy farming is a capital-intensive enterprise. Added to the initial cost of the land purchase is the high cost of providing capital and equipment such as the dairy and the associated yards. In addition there are the costs of milking machines, automatic grain feeding systems, milk vats, general farm machinery and fences. The maintenance and improvement of the herd of dairy cows is also a significant area of expenditure. When calculated out over the number of hectares, the high capital investment clearly shows that dairying is an intensive form of agriculture.

### Labour inputs

Dairying is the most labour-intensive pastoral system in the east. The demanding twice-daily milking regimen requires that the farmer adhere to a rigid work schedule. Most farms are worked as family units (there is generally a low level of employed labour and share farming). A husband and wife team runs 'Sunnybank', with the farmers working approximately 70 hours per week. A farmhand is employed for 40 hours per week and an occasional relief milker is employed on weekends. Contract workers are also employed occasionally to make silage and hay.

### Technological inputs

Technology and scientific research have contributed significantly to greater production levels in dairying. Farms are highly mechanised. Extremely efficient herringbone or rotary milking machines extract milk automatically via an electronic pulsating system. Milk is carried to a receiving tank, where it is filtered to remove impurities and then piped to a vat to reduce its temperature to 4°C. Modern rotary milking machines, such as the rotolactors shown in figure 4.33, are capable of milking a herd of 140 cows in approximately 75 minutes. Computers are now being used to record details of herd health, production and feeding requirements.

Technological advances have also been made in fodder production, conservation and storage. Considerable scientific research has been responsible for improved efficiency and production. For example, research has determined what type of introduced pasture is best suited to a particular locality, and which combination of grasses will yield the best year-round pasture growth. Irrigation has also undergone considerable technological improvement over the years. Scientific management of the herd and appropriate animal husbandry techniques are critical to the effective functioning of the dairy system. Cows have to be kept healthy, contented, safe and comfortable to ensure that energy is not wasted, that milk production is maximised and that the butterfat (cream) content of the milk is kept high. The maintenance and improvement of herd quality is given top priority. Most of the cows in the milking herd on 'Sunnybank', for example, are artificially inseminated, and records are kept every month of the amount of milk produced by each cow and of the butterfat and protein levels it contains. The herd is also culled to rid it of low-yielding and poor-quality milk producers. Cows are checked for diseases such as mastitis, and a vet visits 'Sunnybank' once a month.

In the 1990s, many farmers replaced or modified their dairy sheds, vats and milking machines. Adoption of new dairy shed technology has enabled dairy farmers to increase milk production per week of labour from an estimated average of 3950 litres in 1991–92 to 6057 litres in 1999–2000. Farmers have also substantially increased stocking rates by producing and using more supplementary feed, including silage, concentrates and grain. These increased stocking rates have been accompanied in many cases by management practices aimed at ameliorating their adverse environmental effects. For example, the number of farms that allow effluent to run off into the paddock without any treatment has fallen by more than half since 1991–92, to fewer than 25 per cent of farms in 1999–2000. Dairy farmers are also increasingly adopting computer technology. Around 46 per cent of dairy farms were using computers in 1999–2000, compared with only 38 per cent in 1997–98.

**Figure 4.33** A rotary milking machine

*Government involvement*

Traditionally, governments at both the state and federal level have always been closely involved with the dairy industry. This involvement ranged from land allocation through to marketing controls. Dairy farming and processing was subsidised until the 1990s, because of the need to keep milk production high for the local urban markets. Because dairy farming is not an ideal form of land use given Australia's climatic constraints (and in comparison with the more favourable conditions in New Zealand and Western European countries), various schemes had been put in place to keep the dairy industry viable. This included setting the domestic prices of dairy products at an artificially high level. Tariffs were placed on imported products and quotas or limits had been imposed on substitutes such as margarine. All of this was done in the belief that it would protect the industry from the highly competitive international industry and at the same time satisfy the needs of the consumers. As previously mentioned, the implementation of the national competition policy in the mid-1990s has had a dramatic effect on the way in which the dairy industry operates. The removal of whole-milk quotas has changed the protected status of the producers who owned these licences to provide whole milk to the urban markets. It has opened up state markets to interstate competition and, with the removal of tariffs, has also exposed Australian producers to international competition. The result is a strong belief in the industry that its longer-term survival depends on the ability of individual farmers to become larger and more efficient. This is a trend that has already become apparent with rising herd numbers and declining farm numbers. The size of the average dairy farm has increased by more than 40 per cent in the past 15 years. The number of cows milked per farm has increased by nearly 75 per cent, while milk yield per cow has risen by more than 40 per cent. The net result has been a 160 per cent increase in milk production per farm over this time.

## Throughputs

The work routine associated with dairying is a daily, rather than distinctly seasonal, one. During the milking season the farmer, who is usually on a motorbike, herds the cows from their paddocks twice a day. Small groups of the cows go into the shed to be milked and the rest wait in a fenced area. The cows return to the paddock after they have been milked and it may take between 70 and 80 minutes to milk the whole herd. When the daily milking process is complete, the farmer spends time hosing down the yard and cleaning the machinery with hot water and special detergents to make sure it is spotless. Milk is sampled each day before being transported, to ensure that it is free of antibiotics and bacteria, and a sample is taken so that the amount of fat and protein can be determined. Sick animals are tended to and after-calving attention is given to cows that require it. The farmer milks the cows every day until they are 'dried off' around May (note that this is not the case in Western Australia, where winter rainfall patterns results in the period May to October being the peak production period). Cows are dried off by not being milked. This allows the animal to build up condition in order to give birth and to commence further milk production.

From the time they are dried off until August when their calves are born, the cows do nothing but graze. The farmer normally uses these months to carry out maintenance work around the farm such as painting, fixing machinery and fences, building new sheds or barns. Many farmers and their families also use this less busy time to take a holiday before calving starts in spring and the whole process begins all over again.

On 'Sunnybank', eight hours a week are set aside for artificially inseminating cows, and two hours per day are spent shifting the travelling irrigator and the lateral sprinklers. The maintenance of fences, pumps, machinery and irrigation equipment is an ongoing chore. Other more occasional jobs include hay and silage making, weed removal, sod-seeding and fertilising of pastures, and culling of the herd.

## Outputs

Dairy farm outputs include milk destined for whole-milk or drinking milk consumption, and milk for the manufacture of cheeses, butter, dried and condensed milk and yoghurts. Other outputs include the sale of older animals for meat as well as the sale of some calves as vealers. In addition to these intended outputs, the movement of nutrients via various drainage systems off the farm can be considered as an output.

In Eastern Australia the trend has been for increased productivity or yield per cow and this is reflected in figure 4.34, which shows the amount produced on average in each of the states of the study area. The high output achieved by South Australian farmers is due to the emphasis on whole-milk production and the quota system which required year round production. The industry in this state primarily operated to serve Adelaide. When we look at total milk production, a very different picture emerges. Figures 4.35 and 4.36 show the dominance of two states—New South Wales and, in particular, Victoria.

In recent times Australian milk output has expanded at a faster rate than domestic consumption. As a result an increasing proportion is exported. Australia exports around 50 per cent of its annual milk production and more than 60 per cent of manufactured products. These exports are concentrated in Asia, with Japan and South-East Asia accounting for more than half of Australia's exports by value. In the Bega Valley where Sunnybank farm is located output has grown strongly from $40 million in 1995, to around $70 million in year 2000. This equates to 140 million litres of milk. Fifty-five per cent of this production was for liquid consumption with the balance for cheese.

## EXERCISES

1. Using an A3 sheet of paper draw up a large diagram of the intensive pastoral dairying system and then on it place summaries of the inputs, throughputs and outputs.
2. Refer to the characteristics of intensive agriculture found in chapter 3 and outline how closely dairy farming conforms to these.
3. Explain how physical and economic factors have influenced the location of dairy farms in Eastern Australia.
4. Compare and contrast the dairy industry in New South Wales and Victoria.

## The cultural landscape

### The regional cultural landscape

The cultural landscape associated with dairy farming results from the interplay of the system's inputs, throughputs and outputs. The dairying areas of Eastern Australia are neat and attractive landscapes of relatively small, highly developed farms. The small size of the farm units results in a relatively intensive, regular distribution of farmhouses and settlements connected by 'feeder' roads to minor roads which are connected in turn to major highways. A cooperative is often an

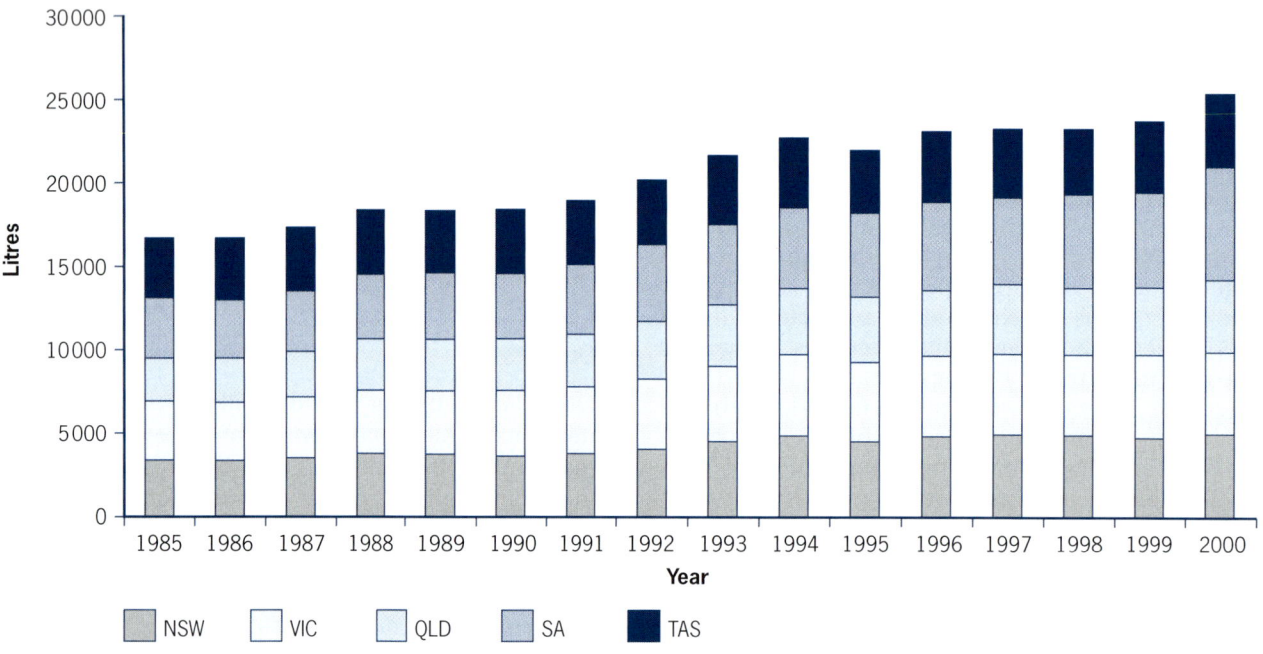

Figure 4.34 Output per cow per year

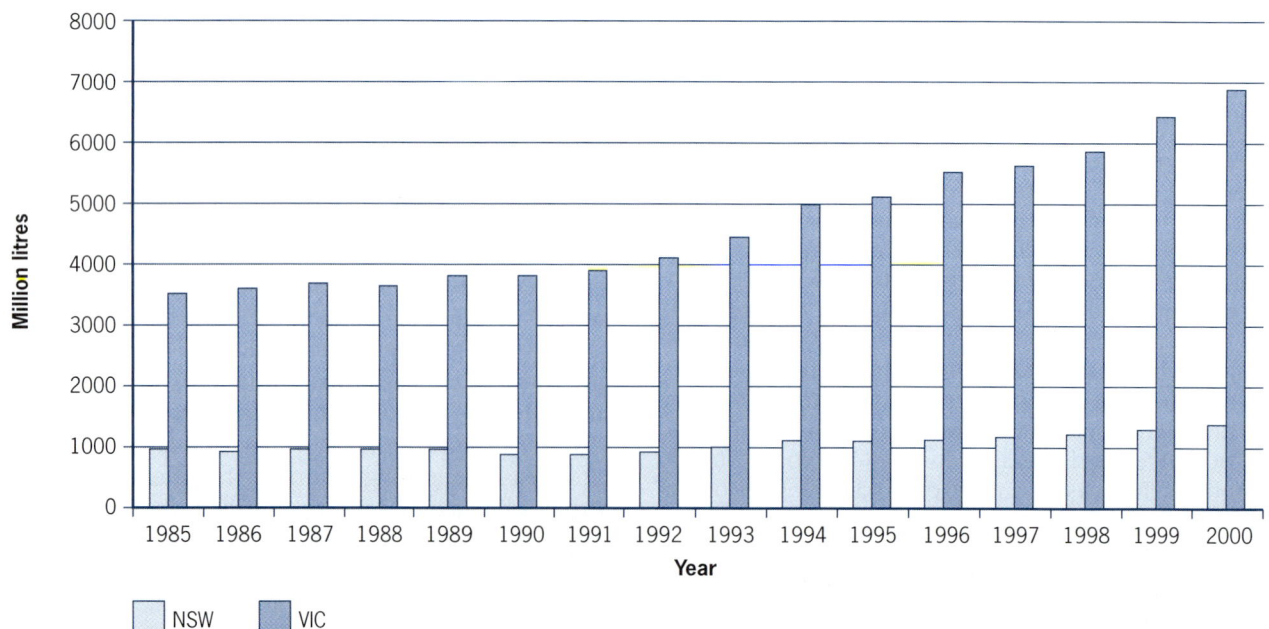

Figure 4.35 Milk production in New South Wales and Victoria

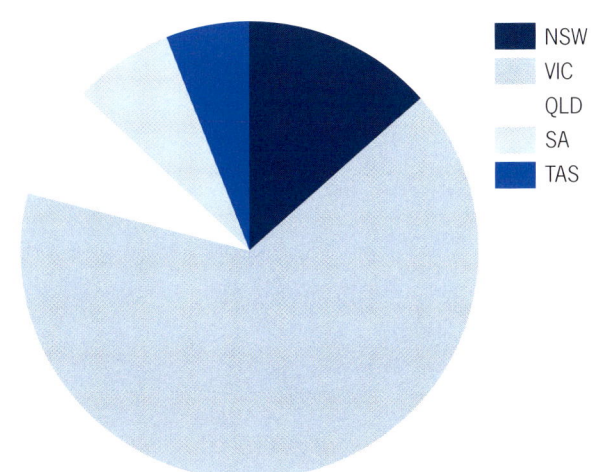

Figure 4.36 Milk production, 2000

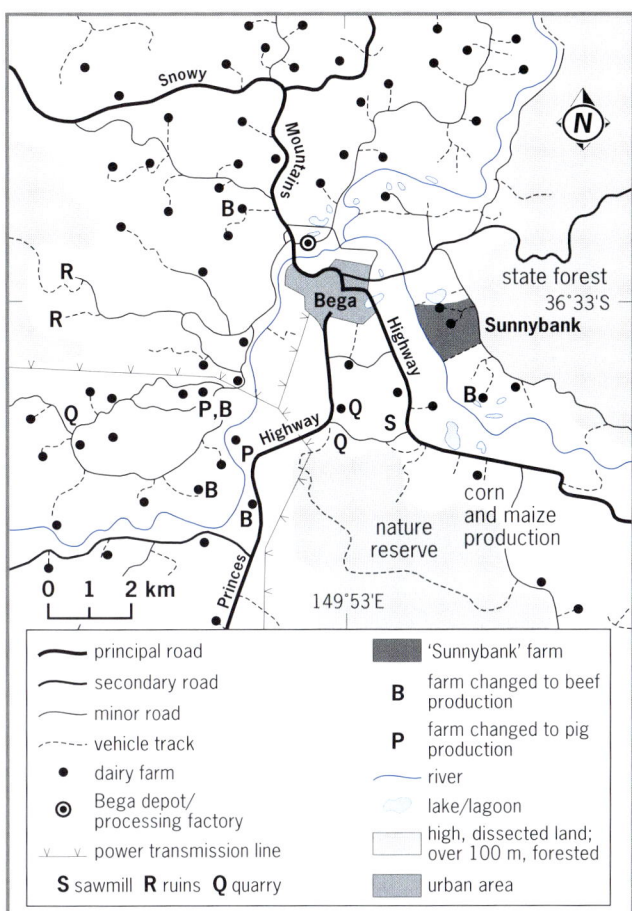

Figure 4.37 The cultural landscape, Bega area

integral part of the dairying landscape. Landforms play a part in shaping the layout and appearance of dairying regions, with flat to undulating land conforming to a regular dispersed pattern while hillier areas are influenced by the contour and drainage patterns.

Figures 4.37 and 4.38 illustrate the cultural features of the Bega dairying region. The river flats and lower slopes of the catchment area have been cleared. Farmhouses are regularly dispersed across the landscape. Small feeder roads connect the farms to a network of all-weather roads, which fan out across the cleared area, linking farms to the Bega collecting depot and processing factory located on the northern banks of the Bega River. Although not identified individually, numerous dairies, haysheds, small rural dams and silos are found scattered throughout the area. The environmental impact of pastoralism is subtler and less immediately obvious than that of intensive cultivation. However, the need to produce large quantities of dairy produce for a large urban market, with its expectations of uniform quality and supply, has resulted in the dramatic alteration of the pre-existing natural ecosystem by the dairy farming system. The natural ecosystem's forest and woodland have been cleared and replaced by improved pastures and fodder crops. Only on steeper slopes and hilltops has the natural vegetation been left standing. The natural water cycle has been altered in many areas by irrigation systems and the provision of watering points.

### The farm site cultural landscape

Farms are divided into well-cleared, fenced paddocks of various sizes, which contain shelter trees and, in some cases, dams. Paddocks are generally fairly small. 'Sunnybank', for example, has 65 per cent of its pastures as 2 hectare paddocks. The farm also has some paddocks of 8–9 hectares, which are strip grazed (with the aid of electric fencing). Paddocks are grazed in rotation, every 18–21 days. The dominant farm fixture is the dairy complex, which contains herringbone or rotary milking machines, vats and feeding systems, washing equipment and holding yards. Other farm fixtures include irrigation equipment (pumps, travelling irrigators and sprinklers), storage facilities for hay and silage, the farmhouse and the farm machinery shed.

## Problems and prospects

### Environmental issues

Although the degree of environmental modification is very high in the dairying system, farmers are now more aware of potential negative environmental effects created by dairying. Considerable care is exercised by farmers to minimise environmental damage. Of particular concern to farmers and agronomists are the problems of soil erosion, flooding, soil salinisation, eutrophication (the growth of algae and weeds in wetlands) and decreased soil fertility caused by overuse of the land. Many dairy farmers consider that the spread and control of noxious weeds is one of the most costly and time-consuming tasks that they face. The larger farms rarely have the resources to control the entire farm. One treatment of chemicals can cost as much as $250 per hectare, and two treatments are usually required to make any impact.

In the past, soil erosion has been created in some dairying areas by over clearing of steeper slopes of valley sides. Sheet erosion (the removal of a thin layer of soil by surface water) and gully erosion are initiated by

**Figure 4.38** The dairying landscape, Bega

the impact of raindrops on bare soil. Gully erosion is often made worse by water flowing down the paths worn by the herd of cows. Pasture depletion by overgrazing is also a contributing factor and is often the result of increases in herd sizes. Soil conservation measures have been important in reducing the amount of soil erosion. Farmers now practise rotational grazing to ensure that pastures are not eaten too low.

Over clearing of catchment areas by various users of the land and generally poor farming practices can result in increased surface runoff from the sides of valleys, leading to greater siltation of the river channels and subsequent flooding downstream. Flooding actually extends the area of erosion by undercutting the banks of the creeks, damaging the river channel and restricting the stock's access to their water supply. An extensive program of tree planting in the Hunter River's catchment area was undertaken as a result of the problems of soil erosion and flooding in the valley. Severe flooding in 1955, for example, saw the river at Singleton rise 13 metres in 24 hours, resulting in damage to property, transport and communications links, stock and crops. Surveys completed as early as 1943 had shown that about 58 per cent of the valley was showing signs of erosion.

A number of problems can arise because of the overuse of irrigation and fertilisers. Waterlogging of the ground, for example, can lead to acidification and the breakdown in soil structure. Farmers occupying the river flats in the Bega area are aware of potential problems and work closely with an agronomist from the Department of Agriculture to ensure that correct watering techniques are used in irrigation. Eutrophication can also occur in estuaries downstream from farms that apply too much phosphate to soils (particularly where soils are relatively sandy and cannot retain nutrients). The use of drains also aids phosphorus transportation. Slow-release fertilisers, which do not leach easily, are currently being developed. Efforts are made in the Bega region to reduce the problems of waste disposal. The Bega cheese factory, for example, produces 500 000 litres of whey (liquid waste from milk) per week from cheese production. It is pumped directly onto three nearby farms, where it is applied to pastures via irrigation. This represents a far better alternative than pumping it down the Bega River. At 'Sunnybank', effluent from the dairy is pumped onto the relatively infertile hillier area of the farm and is distributed with a moveable sprinkler.

## Economic issues

As previously mentioned, the most significant economic event to confront dairy farmers in recent years has been the deregulation of the industry. The impact of this has been felt both on the farm and in the towns where the industry is a major contributor to the local economy. It is an impact that has varied according to the focus of the farm output. Where the emphasis is on manufactured milk products, the effect has been less obvious. Such is the case in the Victorian dairying regions, where most of the output was used in the production of cheeses and other milk products. However, where the farmers held milk production quotas and supplied whole milk to metropolitan markets the effect has been more dramatic. In New South Wales the average farm-gate price of milk has fallen substantially. At the time of deregulation, the price for market milk was a little over 53 cents per litre (cpl) and this returned a net farm-gate price of around 48 cpl for the average farm (it varied depending on the distance from Sydney and the volumes of market milk transported to Sydney or consumed locally). With deregulation, market milk farm-gate prices payable by processors reduced to around 28 cpl, meaning that farmers saw a 20 cpl fall for market milk, which on average comprised 50 per cent of their production (this varied considerably from farm to farm, depending on quota holdings). In the first six months of deregulation, 193 dairy farms ceased dairying in New South Wales. The second six months saw another 78 cease dairying. In 2000–01 the total number of dairy farms supplying New South Wales processors reduced from 1725 to 1454, representing a decrease of 15.7 per cent.

Deregulation has in part delivered what it promised. While states such as New South Wales, Queensland and South Australia have had to undergo significant restructuring of the industry, given their reliance on quota-regulated production of whole milk for the domestic market, the resulting efficiencies forced on the industry by greater competition have resulted in an increase in exports to a world market. In the 12 month period from July 2000 to June 2001, total Australian dairy exports were valued at $3.006 billion. This compares to $2.418 billion for the same period in 1999–2000. This growth in export earnings was the result of substantial gains in cheese and powdered milk sales into the Asian markets.

### EXERCISES

1. With reference to the cultural landscape model found in chapter 3, describe the characteristics of the typical cultural landscape found in the dairying region of Eastern Australia.
2. Based on the information provided, draw a map of a typical dairy farm in Eastern Australia showing the locations of the buildings, paddocks, remnant vegetation and any other significant features.
3. Discuss the impact that industry deregulation has had on the location and operation of the dairy industry in the eastern study area.

# Chapter 5

# Australia's Mineral and Energy Resources

> The purpose of this chapter is to investigate the location of selected mineral and energy resource activities, to describe and account for the cultural landscape associated with these activities, and to outline the environmental impacts associated with mining processes.

Australia has the world's largest known economic reserves of a number of minerals including lead, mineral sands (ilmenite, rutile and zircon), nickel, tantalum, uranium and zinc. In addition, it is in the top six world-wide for reserves of bauxite, black and brown coal, cobalt, copper, diamond (gem and industrial quality), gold, iron ore, manganese ore and rare earth oxides. Australia's mining industry is of great economic importance to the nation. In 1998–99 minerals and energy contributed 8.8 per cent to the nation's GDP. At the same time the industry employed some 80 000 workers at the mine site and another 325 000 in associated manufacturing and processing industries. The contribution to the nation's exports in the same period amounted to almost $40 billion, which represented 35 per cent of total exports. In addition the industry makes an ongoing contribution to the areas of research and development, training and higher education.

The mining industry is of special significance to Western Australia, where it earns in excess of 75 per cent of the state's export income and in 1999–2000 contributed over $800 million to the state government in the form of royalties and lease rents. On a global scale the state's mining enterprises contribute about 12 per cent of the world's iron ore, 8 per cent of the gold, 18 per cent of the **alumina** and 41 per cent of the diamonds. In 1965–66 the value of agricultural production was more than seven times that of mining, but by 1999–2000 the value of production by the mining industry was almost four times that of agriculture.

Compared to agriculture, which accounts for some 60 per cent of Australia's land use, the area from which minerals are extracted is less than 1 per cent. The environmental impact of mining is therefore relatively localised. The type and degree of the impact of mining will vary from location to location depending on the methods of extraction used, the nature of the ore body and the type of on-site processing carried out.

## The history of mining in Australia

The history of the Australian mining industry began thousands of years before the first Europeans settled in the continent. Aboriginal people mined ochres and collected specialised rock for axe and spearheads. At Wilgie Mia, near Cue in Western Australia, red ochre was mined and traded across much of the central region of the state. European mineral exploitation began at the end of the eighteenth century with the discovery of small coal deposits. Since then mining has had its booms and depressions, but from the time of the first substantial discoveries of gold, it has been a major contributor to Australia's economy and infrastructure. Mining has provided the nation's basic industrial requirements including construction materials, fuel and industrial raw materials, and it has been a major contributor to the nation's exports. The discovery of minerals in often remote areas has encouraged the expansion of settlement frontiers and the decentralisation of both population and industry, with towns, railways and ports established to serve the mines and **smelters**. Mining has also encouraged technological advancement, both in its own and other fields.

It was the discovery of alluvial gold in 1851 near Bathurst in New South Wales and, soon after, the rich Victorian fields that provided the first major period of growth in the mining industry. As search and discovery quickly spread, people from many lands migrated to the eastern states to look for gold. The infrastructure built to support this new industry, and the realisation of the mineral potential of the continent, greatly influenced the development of Australia from the 1850s onwards.

By the beginning of the twentieth century, the mining industry, with associated smelters and refineries, was well established. Gold was still dominant, accounting for three-quarters of the total value of metallic mineral mine production, with copper, lead and silver accounting for most of the remaining quarter.

In the late 1930s the mining industry, although well established, played only a minor role in the Australian economy. The need for new ore reserves of many minerals was the major concern of the industry in the late 1930s and early 1940s. It was generally believed that the growth in the mining industry and the discovery of new deposits was over. The Commonwealth Government placed an embargo on the export of iron ore in 1938, when reserves of high-grade ore were believed to be no more than 260 million tonnes.

In the 1950s the mainstays of the industry were lead, zinc, copper, gold and coal, with only the first four being exported in any quantity. In Western Australia the 1950s saw the discovery and development of mineral sands at Capel, and oil at Exmouth Gulf. In the mid-1960s, the Australian mining industry began to expand, with growth in both production and exports combined with a change in the relative importance of the various commodities. Gold and base metals declined, while coal, iron ore and 'other minerals' increased in relative terms. By the late 1960s Australia was a world force in black coal, bauxite, iron ore, nickel, manganese, titanium and zirconium. In addition, the first major uranium deposits had been found. It was during this time that iron ore mines were developed in the Pilbara, the first bauxite mining was commenced in the Darling Ranges and nickel was discovered at Kambalda in Western Australia.

Australia's first commercial oil field was discovered at Moonie in Queensland in 1961. In 1964–65 there was a series of important discoveries. Oil was found at Barrow Island in the north of Western Australia and gas was found in a series of oil and gas fields in north-east South Australia and the adjoining part of south-west Queensland. Most important of all were the discoveries at the Barracouta gas field, some 25 kilometres off the Gippsland coast, in Australia's first offshore well. Subsequent discoveries on the Gippsland Shelf in Bass Strait have made this area Australia's major source of oil and gas.

In the late 1970s, the rate of growth of the mining industry in Australia, which had been maintained for more than 15 years, began to slow. New mines had been developed overseas to meet a forecast demand for minerals, which resulted in a world oversupply. The Australian industry's costs had increased but mineral prices generally had not. The industry was largely dependent on exports and had to compete for sales with an increasing number of mines in other countries; some of which were less affected by cost increases, or were assisted in various ways by their governments. In Western Australia the gas discoveries on the North-West Shelf and the development of diamond mining at Argyle in the Kimberleys were important contributions to the state's resource base.

The 1980s was a time in which gold re-emerged as a significant contributor to mining income. Australian gold production increased from 18 tonnes in 1981 to 57 tonnes in 1985. In 1984 and 1985 alone, 24 new gold-mines were opened, and the re-treatment of old tailings began at several centres using new carbon-activated processing techniques.

The 1990s was a period of significant change for the mining industry. It was a period of consolidation, with considerable focus on further improving efficiency and safety of operations. Significant changes in mine safety legislation in a number of states were aimed at reducing the high injury and mortality rates associated with the industry. There was also a move by some of the nation's larger companies into overseas exploration and development. Companies such as Western Mining Corporation and BHP Billiton became major players on the world stage.

Exploration expenditure fluctuated through the 1990s before reaching new peaks of $981 million for petroleum in 1997–98, and $1149 million for minerals in 1996–97. When taking into account the changes in the value of the dollar due to inflation, this represented an expenditure of six times for minerals and three times for petroleum more than that of the 1960s. Gold continued to dominate exploration expenditure and this resulted in a dramatic growth in Australia's known economic reserves for this metal. Production increased from 2129 tonnes in 1990 to 4404 tonnes in 1998. Australia also retained its position as the world's largest exporter of black coal, with exports in 1999 exceeding 170 million tonnes and providing 10 per cent of total Australian exports.

By the end of the 1990s Australia's mining industry giants had grown beyond being a series of large national companies into global multinationals. Today the industry is diversified and integrated internationally through its exploration, mining and processing activities. The skills that have grown up along with the industry are now being exported, with the nation a world leader in the supply of information technology, engineering, construction and other mining services.

> **EXERCISES**
> 1 Outline the contributions that mining has made to the Australian economy and discuss its significance.
> 2 Discuss the influence of mining on settlement patterns and population distribution.
> 3 Construct a timeline on which you identify the major mineral developments in Australia since 1780.
> 4 Discuss the importance of mining in Western Australia compared to the rest of the nation.

## The cultural landscape

### Specialised mining settlements

Early mining towns such as Kalgoorlie, Norseman, Mt Isa and Broken Hill grew up at a time when the regions in which they are found were isolated from the main population centres. They needed to be self-sufficient to meet the needs of the population that was attracted to the employment and commercial opportunities that the regions offered. Over time they also acted as service centres to the newer mineral fields that were discovered in the surrounding areas. To a lesser extent, they also met the needs of the extensive pastoral industry that surrounded some of these early towns.

**Figure 5.1** Kalgoorlie town hall

Given the specialist function of these mining settlements, they tended to offer a narrow range of services with an emphasis on the requirements of the mining activities. Often they were poorly serviced by different state governments who believed, with some correctness, that these settlements would be of a temporary nature. Constructing new mining towns in remote locations is a very costly exercise and in Western Australia the last new mining settlements were Mt Newman and Goldsworthy. Both of these were jointly funded and serviced by the mining companies and the state government.

The cost of these projects and the increasing desire of workers to live close to the capital cities saw a radical change in the nature of mining settlements. There was an increasing trend to build quarters at the site of new mines and to fly the workers in and out on a rotational basis. By the early 1990s, Australia-wide, there were over 40 fly-in/fly-out mining operations, the majority being in Western Australia with others in Queensland and the Northern Territory. The importance of mining as a catalyst for population decentralisation and the growth of industry within the remote parts of the continent has declined and the founding of new towns within the arid and semi-arid areas in the future is relatively unlikely.

The nature of remote mining settlement is now one of sleeping quarters along with kitchen and dining facilities. Administrative offices, canteens or company stores, recreational facilities, medical facilities and in some cases a landing strip or heliport can also be seen. They provide no central place functions to the surrounding hinterland and are dismantled and completely removed once the mining operation ceases. These facilities can be seen at mines such as Mt Keith, Bronzewing and Granny Smith in Western Australia.

## The mine site

Mining operations fall into three categories. Underground tracked mining is the oldest form of mechanised tunnel mining in Australia. Compared to the other systems this is small scale and has the lowest output. It involves the sinking of shafts, which in turn have a series of interconnected tunnels and stopes that follow the ore body. Ore is moved towards the central tunnel by means of small carts, which are hauled along temporary tracks by electric trains. From here it is brought to the surface using ore cages. Once at the surface it is moved to stockpiles for transportation or initial processing.

Decline mining methods have largely replaced track mining operations in the areas where underground mining is carried out. While the preference is for open cut extraction, if the ore body is too deep and is rich enough to make mining feasible then the use of large rubber-wheeled earthmoving equipment has replaced the small carts of tracked mining. Large dump trucks follow a sloping tunnel underground until they reach the level at which the ore is extracted. Here, front-end loaders fill the trucks, which are then driven back to the surface or to a central shaft. Sometimes the tunnel is constructed as a corkscrew in order to remain within the mining lease. In these cases the trucks must be synchronised to prevent collisions as they move up and down the tunnel.

If possible the preferred system of mineral extraction is the open-cast or open-cut pit. In this system the top layers of rock and soil (the overburden) are removed to expose the underlying ore body. Very large machines can then be used to extract the ore and to transport it back to ground level. This is the most visible of the three mining landscapes with the cut often extending several kilometres in both length and width. Trucks within these sites can carry up to 240 tonnes at a time and may be filled by giant bucket shovels that grab 80 tonnes of ore in one scoop.

A variation on these mining operations is the re-processing of old waste or tailing dumps. This is more common with minerals such as gold, silver, lead, tin and zinc. In some instances the earlier mining activities were less efficient than current techniques and minerals were left behind, even after the ore was processed. These sites are now being re-treated and commercial quantities of minerals are being extracted.

The rehabilitation of mine sites, even in remote locations, has become much more important in recent times and a significant feature of many sites is the storage of soil and overburden for the eventual re-establishment of the natural ecosystems. This can be seen as large stockpiles near the mine site. Once mining is completed, mining equipment and structures are all removed (although in the past they were often left behind). The area is re-contoured and the topsoil and rock is replaced on the contoured site to provide a landform that blends with the surroundings and allows water to run off in a natural drainage pattern, or in some cases into a central lake created by the open-cut pit. Seed stock within the soil aids in the regeneration of the original species, while plants grown in nurseries are also transplanted. The eventual aim is to produce a landscape in which it is very difficult to detect that mining had occurred, and one where any toxic materials released during the operations have been removed or stabilised.

In 1990–91 the mining industry in Australia spent over $130 million on the rehabilitation and restoration of the environment. In Western Australia alone the mining industry plants in excess of 1 million trees annually, and in excess of 3 million Australia-wide. The mining landscape at Queenstown in western Tasmania is often cited as an example of the negative impact of mining and processing of minerals in Australia. In Figure 5.2 the effect on the vegetation of sulfur dioxide mixing with rainwater can be clearly seen. Ironically, this has now become a tourist attraction and the bare, colourful hills attract thousands of visitors each year.

**Figure 5.2** Queenstown, TAS

### EXERCISES

1 Outline the ways in which a mining town might differ from other urban settlements of a similar size and suggest reasons why.
2 Consider arguments for and against the proposition that mining has a limited impact on the environment.
3 List the main characteristics of underground and open-cut mining and then identify the reasons why open-cut mining is considered to be the most cost-effective.

## CASE STUDY 1

# Gold mining in the central study area

### Location

The location and concentration of gold within ore bodies is closely associated with the tectonic processes that have occurred with the development of the continents, or with the volcanic activity that is associated with plate margins. Located beneath the ocean, hydrothermal vents called 'black smokers' precipitate metals such as gold, silver, zinc and lead around their outlets. When this is sufficiently concentrated within receptive rocks such as greenstone it may be, economically, worth extracting. Deposits such as this are associated with Archaean greenstone belts and are common throughout the Yilgarn Block in central and southern Western Australia.

Often where rock is folded or faulted there has been an associated deposition of **quartz**, dolerite and telluride rocks containing gold. The Golden Mile at Kalgoorlie is a region of intense mineralisation associated with faulting and folding.

The margins of colliding or subducting tectonic plates are another location of recoverable gold deposits that are concentrated in volcanic magma. Their concentration is often too low to be mined. However, where the deposit is formed near the surface, its concentration increases enough to make gold extraction worthwhile. The discovery of this type of gold has encouraged mining companies to look at the sites of past collisions as potential mine sites.

Gold may also be found in locations away from its original formation. These secondary sites are mainly associated with fluvial erosion and deposition of alluvial soils and rock. Such gold deposits were often the first to be exploited by miners, who panned or sifted the soil to remove gold dust and nuggets. The alluvial deposits also indicated the potential location of primary gold sources within nearby lode-bearing rocks. The locations of gold deposits are illustrated in figure 5.3.

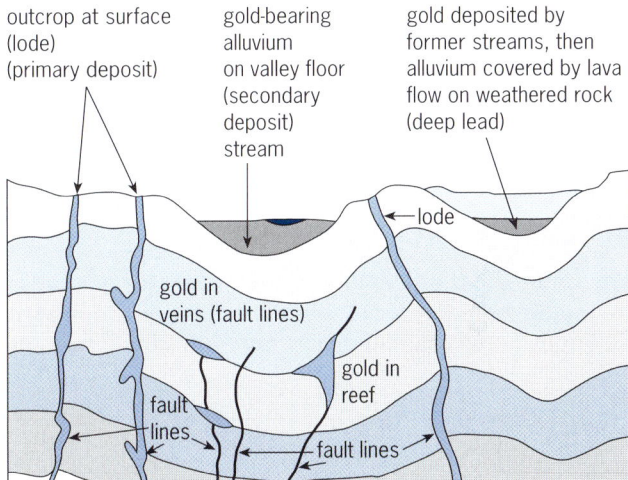

**Figure 5.3** The occurrence of gold

Within the Yilgarn Block there are over 20 separate sites that have each produced in excess of 100 000 ounces of gold, and many more have estimated reserves of more than 1 million ounces (there are approximately 32 ounces of gold to the kilo).

The Super Pit at Kalgoorlie is Australia's largest goldmine. It produces in excess of 500 000 ounces of gold per year and has known reserves of some 13.5 million ounces. When it is completed it will be an open cut pit around 4 kilometres long and 1.5 kilometres wide, extending to around 500 metres in depth.

Mt Charlotte, some 3 kilometres to the north of the Super Pit, is Australia's largest underground mine. Further to the north, near Leonora, the Sons of Gwalia deposits had produced almost 3 million ounces of gold up until 2000, and almost 450 000 ounces were recovered in 2001.

Kanowna Belle goldmine is located some 20 kilometres north-east of Kalgoorlie. The current underground reserves at the mine total 8.35 million tonnes of ore, containing 1.25 million ounces of gold, plus a further 2.66 million ounces in resources that the company is currently evaluating.

The St Ives leases, near Kambalda, cover an area of some 40 square kilometres and from 1980 to 1999 produced in excess of 4.6 million ounces of gold. Further exploratory drilling within this region has identified estimated reserves of 8 million ounces of gold.

Figure 5.4 indicates the location of some of the most significant mine sites in the central study area. It also shows the approximate location of the major goldfields. In all cases, these locations have been established goldfields for over a century and much of the current mining activity is associated with locations discovered during the early development of these fields.

> **EXERCISES**
>
> 1  Study the diagram of the occurrence of gold and describe its location and formation.
> 2  Using an atlas and the map of the goldfields of Western Australia, draw a map showing the location of the fields along with the main transport systems and towns.
> 3  Calculate the value of gold produced in each of the major mines, based on the current price of gold. Note that this is quoted in US dollars, so to complete this activity you will need to find the current value then the current exchange rate to convert your answer to Australian dollars.

## Mining and processing

The mining and processing of gold, while using similar methods to other types of mineral extraction, does have distinctive characteristics. The very low concentration

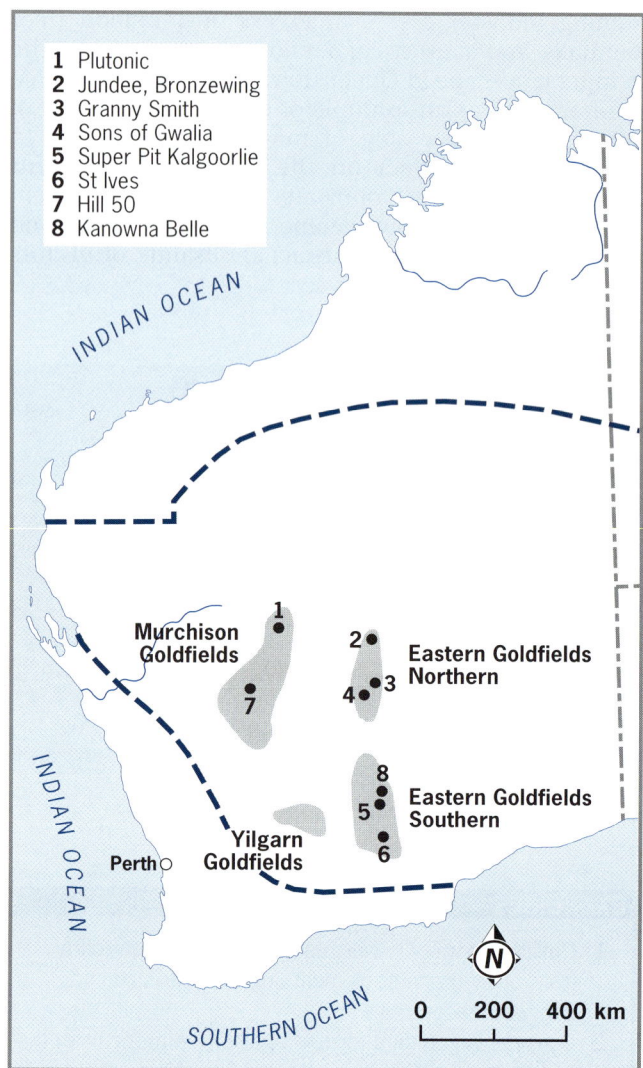

**Figure 5.4** Gold mining in the central study area

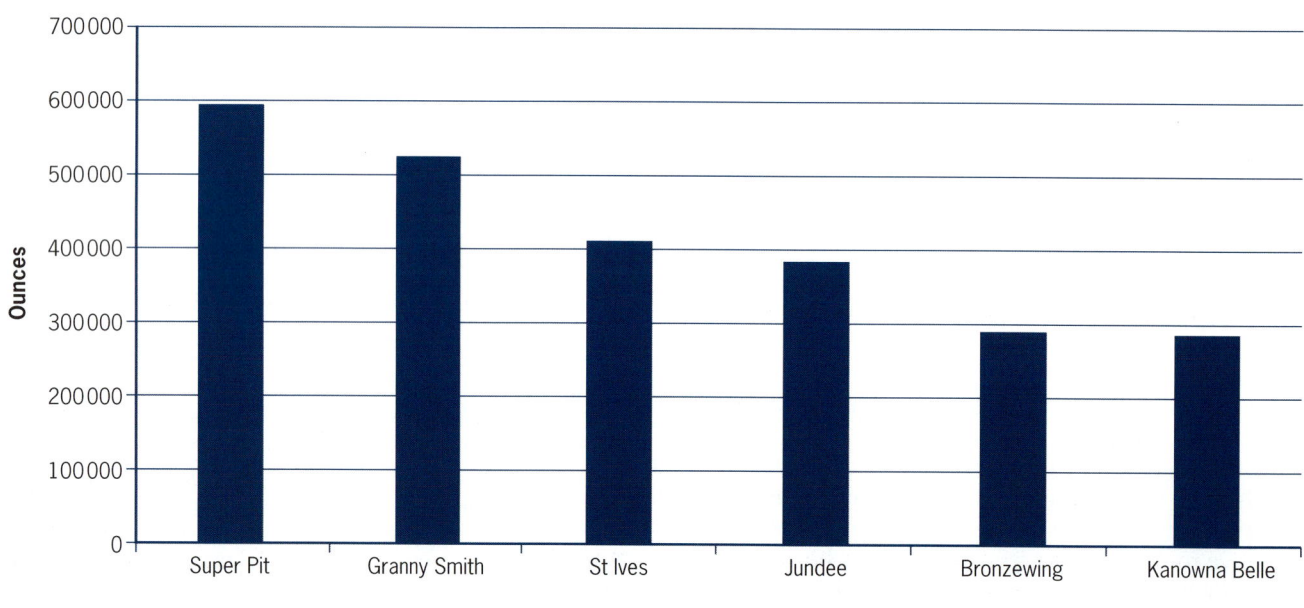

**Figure 5.5** Gold production (in ounces), 1999

of the precious metal means that large quantities of rock and ore must be removed in order to gain access to the gold. This poses special problems in storage of the waste material (**gangue**) during mining operations and in rehabilitation of the site when mining has finished. The low concentrations also make it necessary to carry out initial processing on site, as transport costs associated with moving the ore are extremely high. Some large open-cut mines in the central study area outlay around US$260 in order to recover gold worth approximately US$310 per ounce. In this section the different mining and processing methods will be discussed.

## Types of mining

### Alluvial mining

Historically, mining operations by individual small-scale operators involved the separation of gold from alluvial soil deposits, either by panning or sluicing. Sluicing involved placing soil in a sloping trough fitted with small rungs called baffles. Then the soil was washed down with water and any heavy particles were trapped behind the baffles—some of these included gold. Panning involved the use of a circular pan in which small quantities of soil were washed with water to eventually leave gold and other heavy metals behind. In a number of areas in the West Australian goldfields, the lack of water meant that other techniques had to be used. One of these was dry blowing, in which a small platform was constructed with a series of shelves. Soil was shovelled on to the top shelf and then the platform was shaken, letting the soil fall through small holes in the shelf. As it fell, the wind blew away the dirt and eventually the heavier materials, including the gold, would fall to the bottom shelf. With all of these methods a significant amount of the finer gold was not recovered.

The Yilgarn, Coolgardie, Kalgoorlie, Leonora and Murchison fields were opened up in the 1890s by thousands of miners using these processes. In doing so, they led the way for the larger operations that were to follow.

### Open-cut mining

Open-cut mines, such as the Super Pit in Kalgoorlie or the various pits at St Ives, allow the ore to be recovered directly from the surface. This commences with the removal of the overburden (rocks and soil covering the ore body), which is stockpiled for eventual rehabilitation of the site. Drilling rigs then drill a pattern of holes in which charges are set. After blasting, the ore and waste is removed using large shovel excavators and huge trucks (haulpaks). The pit is cut downward in a series of benches or steps, connected by ramped roadways along which ore and waste are transported to the surface. To prevent the edges collapsing, the sides of the pit are sloped inwards as the open cut descends (this will eventually limit the depth of the cut). If the ore body descends further then underground mining may be developed via a decline tunnel from the floor of the pit. The Bronzewing mine, north of Leonora, uses this technique.

As the pit deepens the watertable may be reached or subsurface water bodies may be breached, causing the

**Figure 5.6** The Super Pit at Kalgoorlie

pit to flood. A sump is then dug at one end of the pit and the accumulated water is pumped to the surface, where it is often used in the initial processing of the ore or for dust suppression and irrigation around the mine site.

The use of modern large-scale machinery together with more efficient techniques of separating gold from low grade ore bodies has stimulated open cut mining in the goldfields of the central study area. The majority of the sites indicated in the map of the goldfields (see figure 5.4) are open-cut pits.

### Underground mining

Access to deep ore bodies can only be gained by underground mining. Examples of this type of mining in the Eastern Goldfields may be seen at Kambalda, Fimiston (Boulder), Central Norseman and Mt Charlotte (Kalgoorlie). Prior to mining, the extent and quality of the ore body is determined by surface and underground drilling to extract samples of the rock. Once reserves are proven then the mining commences. A vertical shaft or a decline tunnel may be constructed, depending on the location and angle of the ore body. Where a shaft is dug then horizontal tunnels (levels) run outwards from this to the sections being mined. At the working face of the ore a large cavern, called a stope, may be developed as the ore is drilled, blasted and dropped on the floor for transportation to the surface or primary crushers. Often the ore is dropped down a sloping shaft, called an ore pass, towards the primary crushers, which are set up underground. Here it is broken down to make it easier to move. It may then be moved further towards the main shaft and stockpiled in tunnels from where it is lifted to the surface in ore cages or skips (see figure 5.8).

### Processing

#### Liberation

The preparation of the mined ore for refining or processing involves the crushing of the rock to release the gold from the waste material. Ore is initially broken up during the blasting of the rock. From the working face it is taken to the primary crusher, in the case of open-cut mining, by tip trucks. Here it is broken down using a cone or jaw crushers. In an underground mine the primary crusher is generally near the bottom of the main shaft and ore is gravity fed to it along a series of ore passes or tunnels. From this point it is then stockpiled for transportation to the surface. The crushed ore is further reduced in size using rod or ball mills, which grind the rock to a pulp and complete the liberation of the gold from the other minerals.

#### Refining

There is a variety of methods of refining the ore to release the gold, depending on its chemical complexity. Often it is simply a matter of using gold's high specific gravity to aid in the separation process. It can be spun in a centrifugal tank to allow the lighter material to spin

**Figure 5.7** Shovel and dump truck removing ore within an open-cut mine

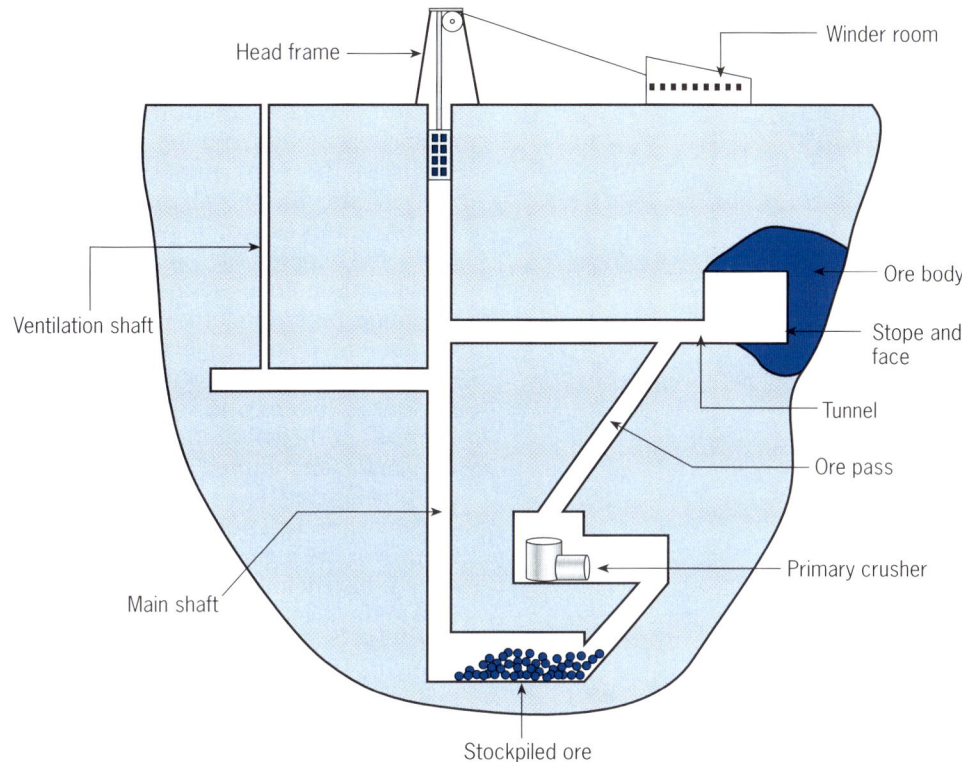

**Figure 5.8** Simplified underground mine operation

off and leave the gold concentrate behind. More complex ores (where the gold is mixed with other minerals) may require flotation techniques. Organic chemicals cause the gold and other minerals to float on the surface of water as a froth, which is then skimmed off. This is then roasted to turn base metal sulfides into sulfur dioxide and other metal oxides from which the gold is more easily recovered.

The slurry of gold-bearing material is now ready for further processing. This involves the mixing of the concentrate with sodium cyanide. Gold reacts with the cyanide and bonds. It is then fed into tanks containing carbon produced from burnt coconut husks, where it clings to the surface of the carbon pellets. After this has occurred the pellets are removed from the processing tanks and washed with a hot cyanide solution to remove the gold compound. Finally, electrolysis is used to extract the gold and make it ready for smelting.

*Primary smelting*

This step involves melting the collected gold and any other minerals that may still be mixed with it. As the gold is heated it melts and metals such as silver, copper and nickel are skimmed from the surface of the molten material. The gold is then cast in ingots for transportation to the **refinery**. Most gold is smelted at or near the mine site with the process achieving 75 per cent purity.

*Secondary refining*

In Western Australia the gold ingots are further processed at the Perth Mint, which is the largest gold refinery in Australia. Here the bars are remelted in a crucible and chlorine gas is bubbled through the molten metal. Chlorides of silver and other metals are removed for further reprocessing. The gold has now been refined to 99.9 per cent purity and is poured into moulds and allowed to cool as bullion. The ingots are then ready for marketing. From Perth gold is exported to the world with most of it going to South-East Asian countries. In 1999–2000 over 203 000 kilograms of gold, worth almost $3 billion, was mined in Western Australia, while the industry directly employed about 12 000 people.

### EXERCISES

1. Conduct research into mining techniques practised by the early prospectors and, using illustrations, describe the ways in which gold was extracted from alluvial deposits.
2. Outline the times when it would be more economic to develop an underground mine and not an open cut.
3. Study the illustration showing underground mining and describe the stages in extracting the ore and moving it to the surface.
4. Describe the distinctive features of an open-cut goldmine by referring to the illustration of the Kalgoorlie Super Pit.
5. Assuming that virtually all the gold mined in 2000 was exported, use the information provided in the pie graph as well as the information on total gold production to calculate the amounts of gold that went to the different countries.

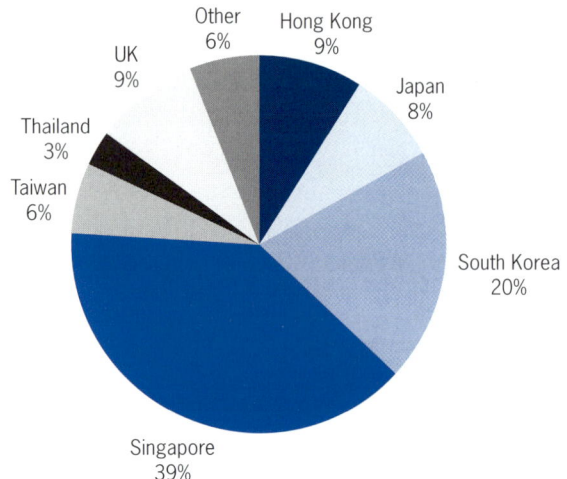

**Figure 5.9** Gold exports, 2000

## The cultural landscape

### The regional landscape

The cultural landscape associated with gold mining in the central study area is the product of over 100 years of exploration and development. It contains towns such as Kalgoorlie, Coolgardie, Mt Magnet, Cue, Norseman, Leonora and Wiluna. Along with a number of other settlements, these developed in the period from 1880 to 1900 in response to the Western Australian gold rushes. Located in isolated desert and semi-desert regions, they were the focal points for the surrounding goldfields. Around each settlement there is widespread evidence of the early gold mining activities. The small mine shafts of the early miners dot the landscape along with waste dumps, abandoned equipment and the tailing dumps associated with the processing of the ore. Mine and exploration roads and tracks radiate out from these settlements, often to isolated mines. Modern mining exploration and development groups have, in a number of instances, based their operations on the existing settlements, with the towns going through periods of boom and bust in line with the changing fortunes of the gold industry. Figure 5.11 illustrates the impact of abandoned mine sites as a feature of the cultural landscape.

Government-sponsored infrastructure is also a feature of the regional cultural landscape. Railway lines were constructed into the Murchison from Geraldton, and to the Eastern Goldfields from Perth. The goldfields water supply pipeline running from Mundaring in the Perth foothills to Kalgoorlie, a distance of some 600 kilometres, was a major undertaking when it was constructed in the 1900s. Today it stands as a reminder of the basic need for water in this arid region as well as forming an important part of the landscape.

The impact that early mining had on the natural vegetation remains to this day. Many of the desert trees were cut down in the areas surrounding the town sites

**Figure 5.10** Drilling ore

Figure 5.11 Disused open-cut mine near Ora Banda

and used to provide fuel for mine boilers, timber for shoring up underground shafts and tunnels and building materials for the temporary and roughly constructed housing that sprang up on each new field.

What is absent from the regional landscape is any significant agricultural development. While the gold-mining areas do have extensive pastoral leases, these have had limited influence on the cultural landscape when compared to the impact of mining. The presence of this form of agriculture may be seen in the sparsely dispersed windmills and small dams that provide watering points for the stock, and the fences that divide the leaseholds into very large paddocks.

The lack of a substantial agricultural population has produced settlements whose primary function is to serve the mining industry. A study of the comparative population statistics for the goldfields in figure 5.12 illustrates several significant differences. The higher percentage of males, as well as the higher percentage of people within the peak workforce age of 25–44 and the lower percentage of those in the older age groups, can all be related to the specialised function of the settlements and the emphasis on mining within the region. It is an industry that attracts more males than females. It is also an industry in which the potential working life span is limited, given the adverse conditions under which people generally work. In addition, it is a workforce which tends to be transient Along with the Pilbara, the goldfields has the lowest rate of people residing in the same location for more than five years (339 per 1000 compared with the state average of 470). Thus the nature of the population and the emphasis on mining will have significant impacts in determining the types of services or function offered by the urban settlements. Rental accommodation will be greater, as will entertainment and mining services. Emphasis on home improvements and furniture will be less. In general, the variety of services will be narrower within the mining towns.

## The mine site landscape

Operational mines are often temporary features on the landscape. The Chalice mine site is located 60 kilometres north-west of Norseman. The discovery of gold at this site was announced in 1993 and the deposits were proven in January 1995. By August of the same year, full production had commenced. Production at the mine site lasted for four years during which over 500 000 ounces of gold were produced from 4 million tonnes of ore. In January 2000, the company announced its decision to cease mining and operations came to an end. Since that time the company has completed all rehabilitation works associated with the 31 hectare Chalice mining area. This included the rehabilitation of all excavations associated with the exploration and mining activity, the closure of the Chalice–Higginsville haul road and the rehabilitation of the mill, office and workshop areas of the Chalice Gold Project.

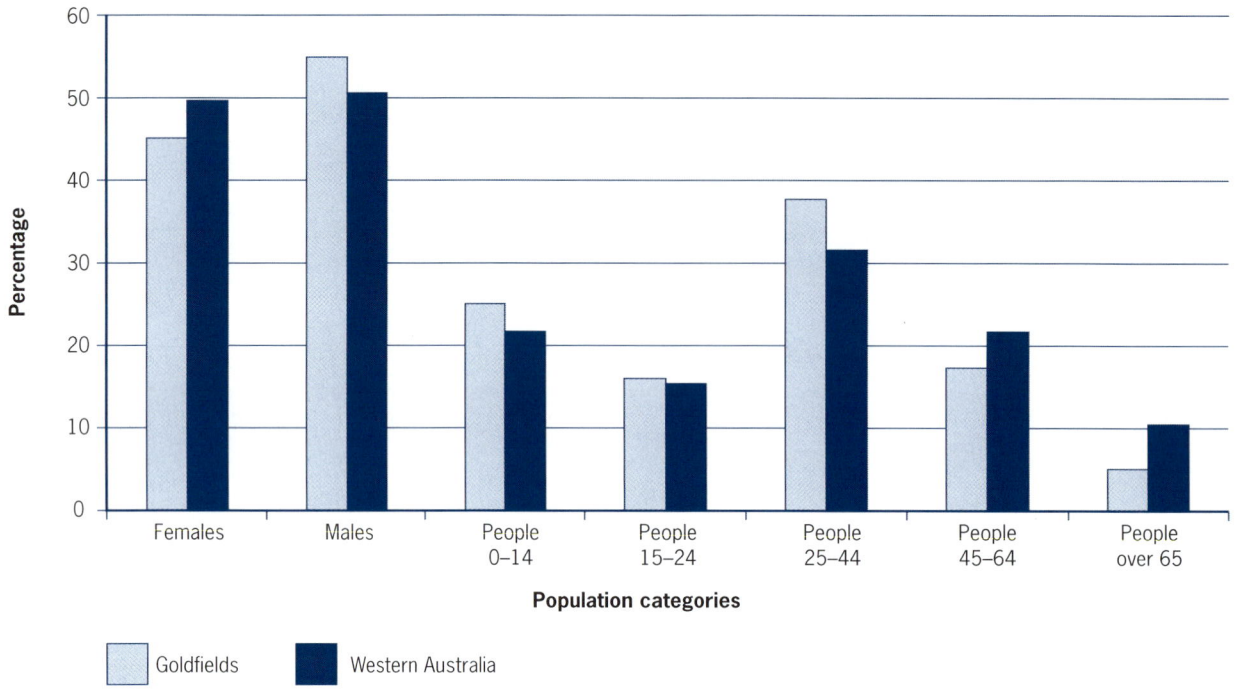

**Figure 5.12** A demographic comparison of goldfields population and Western Australian population, 2000

Mining projects generally follow the same sequence of events as illustrated by the Chalice goldmine: initial exploration is followed by development, production, processing and finally decommissioning. Each of these stages will produce different features of the cultural landscape.

### Exploration

The modern day geologist completes a great deal of the preliminary exploration using satellite images, geophysical surveys and historical data before any disturbance to the landscape occurs. If potential sites are found during this stage then exploratory drilling may be undertaken. If the site is remote then temporary roads may be constructed along which drilling rigs are transported to the site. The operators then set about extracting rock samples according to the sampling patterns calculated by the geologists. During this stage campsites for the field geologists as well as the drilling teams are established in the bush, bringing some disturbance to the immediate location. About one in every thousand exploration sites will be developed through to full production. Where the exploration does not lead to production, the company still has to rehabilitate the exploration site in line with environmental undertakings.

### Development

It may take up to 10 years from the initial exploration phase to the start of production. During this phase, development approvals, negotiations with traditional landowners, environmental management plans, financial backing and numerous other issues need to be resolved. The site is then prepared for the mining activity. This includes the construction of access and haulage roads, building of processing plants or mills, management offices and onsite accommodation if the location is remote.

After flora and fauna inventories have been conducted, the site is cleared and topsoil is removed and stockpiled for eventual rehabilitation. If the operation is open-cut mining then the overburden is also removed in order to expose the ore body. The construction phase will often require a larger workforce than that needed during the mining operation. The workers will be housed in transportable accommodation at the site. The development phase represents a significant change in the cultural landscape associated with the goldmine.

### Production

Once the mine site is in full production it operates 24 hours, seven days a week. In remote locations it is a self-contained operation. The pit or underground mine is connected to stockpiles of ore and waste by haulage roads or conveyors. Ore from the stockpile is transferred to primary or secondary crushers and then undergoes initial processing using centrifugal separators and the cyanide heap leach process. Settling ponds hold the leached gold solution until it is further refined in flotation and carbon in pulp chambers. The primary smelting process carried out in furnaces completes this phase. The gold is then held on site until ready for transport to the refinery. Many of these features can be seen at the Mt Charlotte and Fimiston treatment plants at Kalgoorlie Consolidated Gold Mines Pty Ltd, illustrated in figures 5.13–5.15.

The Plutonic mine occupies a 1 square kilometre site 500 kilometres east of Carnarvon, on the edge of the

**Figure 5.13** Mt Charlotte underground mine site and ore conveyor systems

Great Sandy Desert, the open cut appearing as a small, wavy oblong scar in the landscape. Mine buildings (administration, maintenance and mill) are situated close to the mine site and are linked to the open cut by well-constructed roads. A canteen and accommodation for workers ensure comfortable living conditions, and an airstrip enables the 300 person workforce to fly in and out on a rotational basis. Gold is also ferried to the Royal Mint by air for refining. The mine site is a hive of activity as the $3 million hydraulic excavator loads the 85 tonne trucks. About 50 000 tonnes of ore are transferred to the mill every 24 hours for processing.

*Decommissioning*

After production ceases, underground mines and open-cut pits are secured against intrusion by people and animals. Processing areas and other work sites are cleared, recontoured and revegetated. Waste dumps are recontoured to form low hills, which are covered with fertile topsoil and planted with local native vegetation. The industry has developed considerable expertise in establishing native ecosystems in the harsh, arid environment of many gold producing areas. Many decommissioned mines are retained by the companies for possible reworking should gold prices improve. The landscape will be substantially returned to something resembling the natural environment of the immediate area.

### EXERCISES

1. Compare and contrast the different impacts that mining has on the environment during the exploration, development and processing stages.
2. Describe the cultural landscape associated with a large-scale open-cut goldmine by referring to the layout and characteristics of the pit, processing plants and transport systems.
3. What factors would affect the length of time that a goldmine would remain operational?
4. Draw a sketch of the mine site gold recovery plant and label the ore and waste stockpiles, the heap leach dump, the cyanide recovery ponds and the gold recovery tanks and separators. Explain why the initial processing of gold ore occurs in close proximity to the mine site.

## Environmental issues

### Vegetation

A number of issues concerning the impact of goldmining upon the environment have emerged over the years. A major regional concern is the effect of the activity upon the natural vegetation of the area, while local concerns include air emissions from roasters and dust from waste dumps. Mining is an intensive form of land use and its immediate impact is confined to a relatively small area. In the early days of the industry, the demand for timber to fuel steam engines, to support tunnels and shafts, and to provide fuel for domestic uses was very high. As a result, the native cover of acacia and eucalypt woodland was extensively exploited. Clearing was widespread and unregulated and accounted for the removal of over 30 000 square kilometres of trees within the Eastern Goldfields. The clearing continued until the 1960s, when fossil fuels began to be used as a power source and Kalgoorlie was tied to the state power grid. Regeneration of species such as mulga is very slow, given the harsh environmental conditions that prevail; however, eucalypts are prolific seed bearers and areas where these trees were dominant are showing excellent signs of recovery.

Another threat to vegetation is the practice whereby small operators attach blades to tractors to skim away old alluvial workings so that fresh ground is exposed.

**Figure 5.14** Treatment plant at Kalgoorlie mine site

**Figure 5.15** Ore conveyor belt

Not only does this destroy areas that may have taken in excess of 90 years to revegetate, but it also exposes the thin topsoils to the strong winds which frequent the area, and to heavy sheet flow of water following thunderstorm activity. These methods are now considered to be environmentally unacceptable. A district inspector employed by the Department of Minerals and Energy must approve any use of mechanised clearing techniques by goldminers.

The implications of the removal of vegetation are serious, since the vegetation is vital for the maintenance of the stability of the entire ecosystem. Loss of plant cover has increased surface runoff and exposed the thin topsoil to serious gully and wind erosion. The benefits of vegetation for the suppression of salt movement and reduction of wind speed are absent in severely denuded areas.

## Water quality

Vegetation in the goldfields has also suffered from the effects of highly saline water, which is transferred from bore fields to the mine sites where initial processing occurs. The Department of Environmental Protection has expressed concern that leakage at pipe joints has led to the accumulation of large amounts of the saline water at widely dispersed locations. The high evaporation rates of the area have the potential to lead to the accumulation of salts at the surface, with detrimental effects upon vegetation. In addition, the saline water leaches into surrounding aquifers. Old tailings dams, which contain saline residues, can also create problems when seepage of dam contents pollutes the surrounding terrain. These old tailings dams also dry up. Their surfaces are contaminated with loose salt, which is blown by strong winds into surrounding bushland.

A further potential problem associated with some mining operations is the production of acidic water. This is particularly the case where water comes into contact with mine waste containing sulfides. This in turn may aid in the leaching of metals contained in the mine waste into streams or ground water, thus damaging local flora and fauna.

## Air quality

The sulfide ores of Kalgoorlie require 'roasting'. Unfortunately the location of roasters adjacent to the urban area of Kalgoorlie–Boulder has exposed the population to high levels of sulfur dioxide emissions, which pose a health threat. Sulfur dioxide can have a detrimental effect on the skin and can cause eye problems. Young children are particularly vulnerable, as are people who suffer from hay fever, bronchitis or asthma. The incidence of these ailments in Kalgoorlie is above the average for Western Australia. Sulfur dioxide dissolves in water to produce sulfuric acid. This is one of the causes of acid rain in Europe, and it can have a devastating effect upon vegetation. However, the low rainfall of the Kalgoorlie region minimises the risks of such an environmental problem developing. Mining companies in Kalgoorlie–Boulder are now required to monitor their sulfur dioxide emissions continually, in line with the *Environmental Protection Act 1986* (WA).

The Department of Environmental Protection also has three monitoring stations. If the average concentration of sulfur dioxide exceeds certain levels over particular time spans, the company can be subjected to heavy fines. Analysis in the mid-1980s indicated that sulfur dioxide levels exceeded acceptable levels established by the World Health Organization. For example, in 1987 it was reported that the WHO considered that the upper level of emissions (200 micrograms for a 24 hour period) should not be exceeded more than seven times per year. It was exceeded 61 times in the Kalgoorlie–Boulder area, resulting in considerable public outcry. In recent years the Gidgee roaster was established to the north of the city in an endeavour to reduce the severity of the problem; however, roasters continue to operate at three mine sites near the city.

Dust is also a problem in gold mining areas, and particularly in Kalgoorlie–Boulder. The many mullock heaps and tailings dumps have produced areas of hazardous waste pollution (cyanide from processing), a situation that is further complicated when the dangerous particles become airborne in high winds. Superficial mining in the area in 1989 involved the removal of large areas of topsoil, adding to the material available for wind dispersal. The new Super Pit development has resulted in surface disturbance for site development and roads, and has added to the waste rock material appearing in large dumps. On windy days the combined dust from all these sources substantially reduces visibility and spreads over the urban area. Future tailings dumps are to be located to the east of the city, not the south where they have previously been located, in order for the prevailing winds to blow any dust away from the urban area.

## Government environmental regulation

Regulation of environmental and rehabilitation standards is mainly the responsibility of state governments in Australia. In Western Australia the Department of Environmental Protection and the Department of Minerals and Energy oversee the legislative requirements relating to mining. Prior to the issuing of a mining lease a company has to develop an acceptable plan for how the site is to be rehabilitated, both during and after the cessation of mining activity. Consultation between the mining company and the relevant authorities takes place to determine the rehabilitation requirements for each project. The primary objective is to leave the site in a condition suitable for the agreed final land use, along with other goals relating to making the site safe, stable, non-polluting and maintenance free.

When Consolidated Rutile Australia (CRA) explored the Rudall River area, which lies on the north-western margin of the Central Australian study area, the company was required to lodge advance proposals for rehabilitation of the landscape before a license was issued. These proposals were lodged with the Department of Minerals and Energy and the Department of Conservation and Land Management (CALM). They included the routing and construction of tracks, the preparation of environmental surveys, the design of pits, and the rehabilitation of tracks and drill pads. Upon completion of the exploratory work the company was required to restore the surface and replace cleared vegetation, along with original topsoil. Any compacted areas were to be deep ripped, and the sites were left in a roughened state to reduce wind and water erosion and to assist the infiltration of rainfall. Sites were also to be left free of rubbish. On the Olympic Dam site (Roxby Downs) the company was required to monitor all mammals, reptiles, fish and birds in the area. There was to be immediate rehabilitation of disturbed dune sites, and any slopes were to be stabilised by 'hydro mulching', which involves spraying with a mixture of bitumen, hay and grass seeds. The aquifer of the region had to be monitored up to a depth of 200 metres. Monitoring was also required for gas and sulphur emissions, and tailings had to be stored in dams that were to be revegetated in due course.

In most cases in Australia, the lodging of a bond is required prior to the issue of a mining lease. Bonds can be required in one of two ways. The first type involves the company lodging a cash bond with the government, or with a bank in a trust account. The second type is in the form of a guarantee, lodged by a financial institution on behalf of the mining company. In both cases, the bond is returned to the company on successful completion of the rehabilitation requirements. In Western Australia in 1999–2000 high risk structures, tailing storage facilities and waste dumps with acid mine drainage incurred a bond of $15 000 per hectare, while plant sites and camp infrastructure sites were set at $5000. The Department of Minerals and Energy also encourages mining companies to adopt best practice methods in conserving and rehabilitating the environment. To this end they make an award to companies who have achieved outstanding results, in the form of certificates known as the Golden Gecko Award. This award symbolises company or individual commitment to go far beyond basic compliance with regulations, and provides public and industry recognition for their efforts. In 2001 the Plutonic mine site achieved one such award for their high standard of environmental management across the entire operation.

### EXERCISES

1. Outline and describe the steps taken by gold mining companies to rehabilitate a mine site.
2. List and discuss the ways in which state governments regulate the environmental impact of mining activities in the goldfields.
3. Describe the processes that a mining company must go through to gain approvals to develop a mine site.

## CASE STUDY 2

# Diamond mining in the northern study area

### Location

The Kimberley region in north-western Australia is the site of the continent's only commercial diamond producing mine. This is the Argyle diamond mine

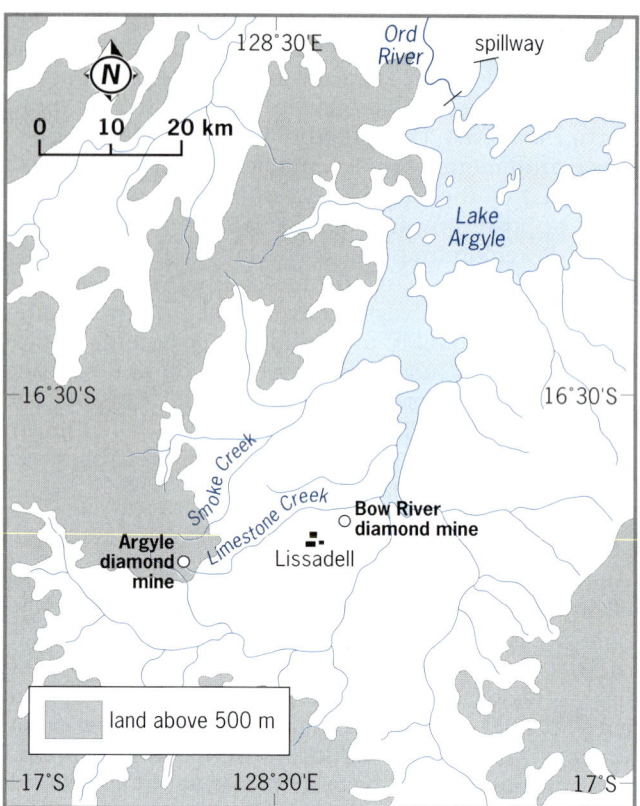

**Figure 5.16** Location of the Argyle diamond mine, WA

which is located about 160 kilometres south of Kununurra and 35 kilometres upstream from Lake Argyle, in the headwater region of Smoke and Limestone creeks (see figure 5.16). The mine occupies a valley in the Ragged Range and is named after the Argyle cattle lease, which was taken up in 1885 and managed by the Durack family for many years. Argyle Diamond Mines Joint Venture, which was purchased by Rio Tinto in 2000, recovers the greater majority of diamonds from an open-cut mine, although some diamonds are recovered from alluvial deposits at Smoke and Limestone creeks near the main mine site (see figure 5.17). Other known deposits of diamonds also exist at Bow River, to the north-east of the Argyle mine, and at Ellendale, in the west Kimberleys. Currently (2002), the latter is in the process of being investigated for economic viability by the Kimberley Diamond Company. Since 1994 they have investigated a number of potential sites over their 1000 kilometre Blina–Ellendale mining tenements. In 1994 they discovered an ancient buried river system (known as a paleo channel) in which diamonds up to 3.5 carats were found. By 1997 they had recovered 284 stones weighing over 70 carats in total. Since then the company has been searching for the source of these diamonds. When volcanic cones are eroded by running water the alluvial deposits that develop may contain diamonds which have been removed from the primary source.

The Argyle diamond pipe, known as AK-1, has economically recoverable reserves of ore in an area about 1600 metres long. It is an irregular shape, varying

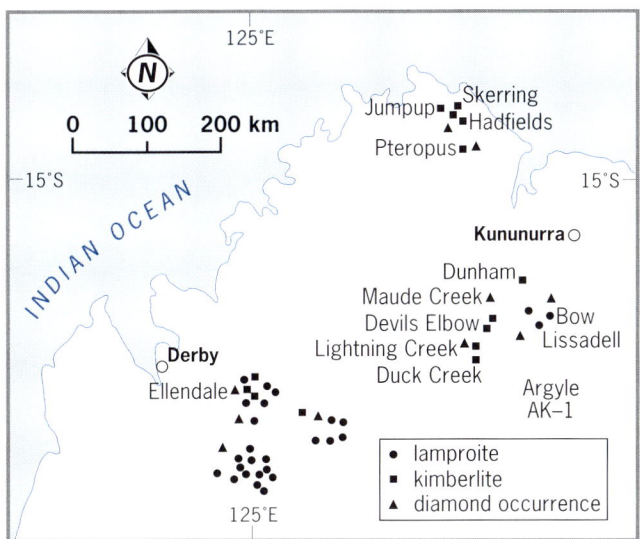

**Figure 5.17** Location of the Ellendale diamond deposits and other diamond occurrences

in width from 200 to 600 metres, with a north–south axis, and occupying an area of approximately 45 hectares. The top of the pipe, which was pushed sideways by faulting, occupies the entire valley floor. The Argyle pipe has a proven ore reserve of 61 million tonnes at an average grade of 6.8 carats per tonne. During 1988–89 the mine produced 34.4 million carats of diamonds. This output has remained relatively unchanged, with output in 2000 reaching 42.3 million carats and the amount of ore processed rising from 3.3 million tonnes to 10 million tonnes in the same time period. In 2000 the total value of Australian diamond production was $713 million.

The Bow River lease is held by Poseidon Bow River Diamond Mines Ltd. It is an alluvial diamond deposit and as of June 1996 was under care and maintenance. (In order to retain mining leases, companies must maintain a minimum level of operation. Some have one or two people on site who carry out security duties and basic maintenance of any equipment. If the company walks away or relinquishes the lease then the Department of Minerals and Energy can allocate it to any other interested group.)

## Geology and uses

Diamonds are made of pure carbon and are the hardest known substance, either natural or synthetic. They are marketed to meet a demand for their use in industry or in jewellery manufacture. Diamonds vary in size and quality. Their value may range from 50c per carat for small industrial stones to over $10 000 per carat for high-quality gems (a carat is 0.2 grams). Intensely pink Argyle diamonds command a price in the vicinity of $1 million per carat. Poor quality industrial gems are generally very dark in colour, with numerous flaws, and often have intrusions of other minerals within them. They are used for a variety of purposes including abrasive disks, drilling tips, machine bearings, glass cutting

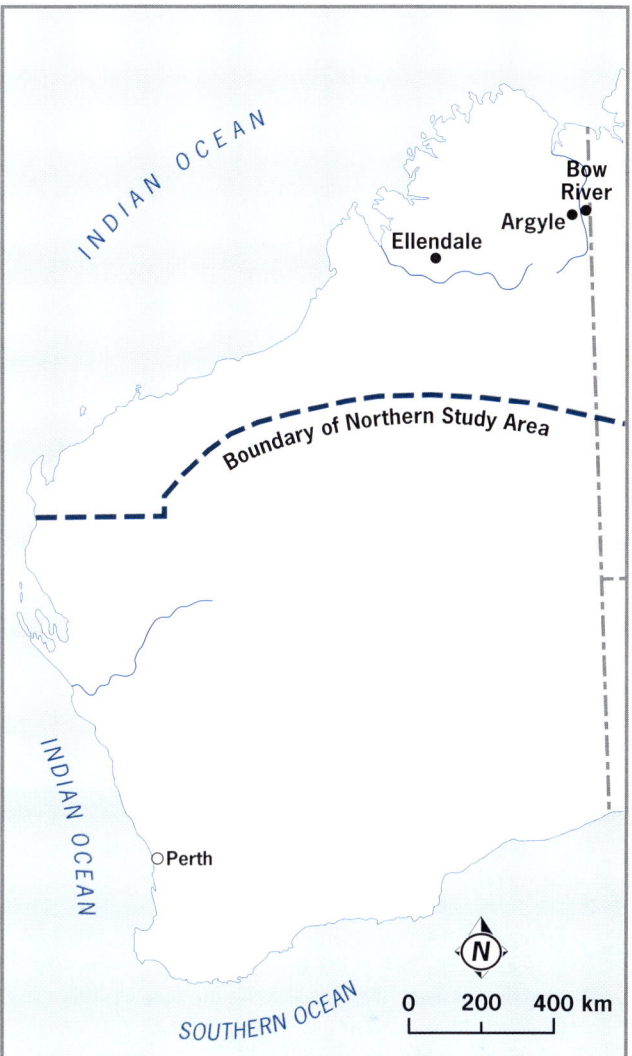

**Figure 5.18** Diamond mining area

and grinding and polishing. Gem quality diamonds are classed according to their size and degree of light reflectivity (refractive index). This gives the gem its brilliance—the 'fire' that is so highly valued.

Occurrences of diamond-bearing materials are rare, and these materials are usually found in an area of a few hectares within a zone of millions of square kilometres. Even then the diamonds may occur in concentrations as low as one part in a hundred million. An igneous rock called kimberlite is often the primary source of diamonds, although the primary source at Argyle is lamproite. The rocks are found in volcanic pipes and minor intrusions within the earth's crust. In its movement under pressure to the earth's surface, magma (molten rock) pushes into rock fissures within volcanic cones to form the pipes in which diamonds are sometimes found.

Diamond deposits in Western Australia vary in age. The AK-1 pipe at Argyle is about 1200 million years old, being formed during the Proterozoic era, while the Ellendale deposits in the West Kimberleys were formed about 20 million years ago.

## History of diamonds in Australia

Small quantities of diamonds from alluvial sources have been found in various locations throughout Australia for more than 100 years. Most were found more by accident than design, during gold and tin mining operations. Diamonds were first found in Western Australia in 1894 by goldminers working alluvial deposits at Nullagine in the north-east Pilbara region; however, the primary source of diamonds in the region has yet to be established. In 1972 the Kalumburu Joint Venture was established to search for diamonds in the Kimberley region. Alluvial diamonds had been found by other companies working in the area. With the added participation of CRA Ltd in 1976, the exploration partners became known as the Ashton Joint Venture, and activity by the joint venturers led to the discovery of diamonds at Ellendale in 1977. In 1979 alluvial diamonds were identified in samples collected at Smoke Creek to the north of Lake Argyle. Geologists followed Smoke Creek upstream and located the large AK-1 pipe. During 1980–81 two substantial alluvial deposits were located in upper Smoke Creek and Limestone Creek. Argyle Diamond Mines Pty Ltd was formed to manage the commercial production of diamonds from the AK-1 ore body and alluvial areas.

### EXERCISES

1. Describe how diamonds are formed and where they can be found.
2. On a map of the north-west of Australia show the location of the main diamond discoveries and the location of the only commercially operating mine site.
3. Identify the uses of diamonds and the factors that influence their quality.

## Mining and processing

### Open-cut mining

In most economic mines the amount of diamonds range from one part per million down to one part per 100 million. Once preliminary exploration has revealed alluvial deposits or volcanic pipes with sufficient concentrations of the gem then the site is pegged and the development of the mine commences.

Diamond mining is usually conducted using the open-cut method. The Argyle diamond mine is a conventional open-cut mine. The benches are about 15 metres high and present a vertical face for mining purposes. The drilling rig positioned on the mine bench drills at selected positions as a preliminary step to blasting. Once blasting is completed, 120 tonne dump trucks receive ore from hydraulic excavators. Dust suppression on the mine site is an important task within the mine site and this is done prior to the loading of the trucks by the use of water trucks and sprays. Open cut mines may extend to depths of 240 to 300 metres before removal of ore becomes difficult. If companies choose to mine underground, vertical shafts are dug adjacent to the pipe. Horizontal tunnels are then extended to the ore body so that mining can proceed. Open-cut mining at Argyle is estimated to continue until 2004, at which time underground mining may commence.

### Refining

Once loaded, the trucks negotiate the road system on the benches and, as indicated in figure 5.21, dump the ore into a 250 tonne hopper at the primary crusher. The crushing force is sufficient to break the host rock without damaging the diamonds contained in the rock. A conveyor belt then transfers the ore to a coarse-ore stockpile. Ore from the stockpile is subjected to a lengthy recovery process, which involves scrubbing and sizing, secondary and tertiary crushing and the concentration of diamonds in the heavy-media separation plant. The remaining material is then directed to huge tanks to undergo water recovery. In the heavy-media separation plant diamonds are separated out through gravity processes. This and all other operations are directed from the control room. The diamond-bearing material is dried and transferred to a recovery plant for X-ray separation of the diamonds. This recovery process is based on the principle that diamonds fluoresce when exposed to X-rays. Fluorescence is detected by photocells, which activate air-blast valves and eject the diamonds. The method is rapid, and retrieves more than 99 per cent of the diamonds. Acids are used to clean the diamonds a final time before they are transported to Perth for sorting. Value is added to the product when the rough diamonds are sorted, and top gem-quality diamonds are cut and polished at the company's Perth headquarters.

## The cultural landscape

### The regional cultural landscape

The relative isolation of the main diamond mining region in the Kimberleys, and the self-contained nature of the operations, has produced limited change in the regional cultural landscape. Historically, the area has been one of extensive cattle grazing and this is the dominant influence on the cultural landscape. Kununurra is the closest main settlement and its main purpose is to service the Ord River irrigation system, with its associated intensive farms. About 70 senior staff and support personnel who live in Kununurra fly in to the mining operation on a daily basis.

An important element of the essential infrastructure is the water main. The pipeline enters the Argyle complex from the north-west, after negotiating a 40 kilometre route from the southern end of Lake Argyle. Water is critical to the diamond separation process and to the residential complex. A reliable water supply is also required for successful landscaping and revegetation of some rehabilitation sites.

The mine site uses existing regional road systems including the Great Northern Highway, which runs from Perth to Kununurra. The mine is 15 kilometres to the east of the Great Northern Highway, and access is provided via the Lissadell Station Road (see figure 5.19).

## The mine site cultural landscape

### Exploration

The exploration phase involves similar processes to those of other forms of mining, such as goldmining. The study of satellite and geophysical data prior to field sampling occurs. Once potential sites are identified then a variety of search strategies is employed. These include the use of various instruments trailed beneath a helicopter to measure anomalies within the earth's gravitational and electro-magnetic fields, and drilling to extract core samples. One unusual technique developed by Kimberley Diamond Company is to sample the soil taken from termite mounds. Where these occur over a pipe the samples contain different elements. In a period from 1998 to 1999 the company collected over 4300 geochemical samples, which led to the discovery of 18 lamproite pipes. Once a potential site is found, drill samples are taken from the rocks and if these indicate promising results then bulk samples of around 200–300 tonnes are removed for processing. If these yield in excess of 5 carats per 100 tonnes then this would indicate a commercial mine site. During this phase of operations the impact on the landscape is comparatively minor, with signs of limited excavation.

### Development

Once a commercial site has been found the next stage is to develop the mine to the production phase. The total cost of developing the Argyle mine, regarded as one of the largest and most modern in the world, was about $430 million. The first stage of the project was based on mining the alluvial deposits in the region and had little effect on the cultural landscape. However, the second stage of the project, which began in November 1983, presented a tremendous challenge to the engineers and construction teams involved. People, equipment, machinery and building material (some of which was prefabricated in Perth) had to be transported 2200 kilometres from the south to the mine site. Construction moved very quickly and efficiently, despite the difficult environmental conditions characteristic of the region. The 22 million tonnes of accumulated waste rock that had to be removed before mining could proceed constituted the first major cultural imprint on the ancient landscape. Figure 5.19 shows the layout of the operation.

### Production

One characteristic of mining which generally sets it apart from other forms of economic activity is that production is concentrated in a relatively small area (see figure 5.19). For this reason, the cultural elements associated with mining are distinctive and easily observed. The most striking feature of the Argyle diamond-mining operation is the mine (see figures 5.20 and 5.21). The large open-cut pit, which is centred upon the ore-bearing pipe, is geometrically patterned with carefully engineered concentric benches. These benches allow the rock walls to be blasted with relative ease and the material to be loaded onto trucks for transport to the point of initial processing. A simple road network meets the requirements of the mining complex. The airstrip to the north-east of the mine (see figure 5.19) is linked to the mine site, processing works and residential area by an all-weather road. A comprehensive network of roads provides access for vehicles to the major dump areas that surround the mine and the alluvial plant located to the south of the airstrip. The large airstrip has been constructed so that jet aircraft can ferry employees to and from Perth on a '14 days on duty, 14 days off duty' basis. It also receives essential goods and services.

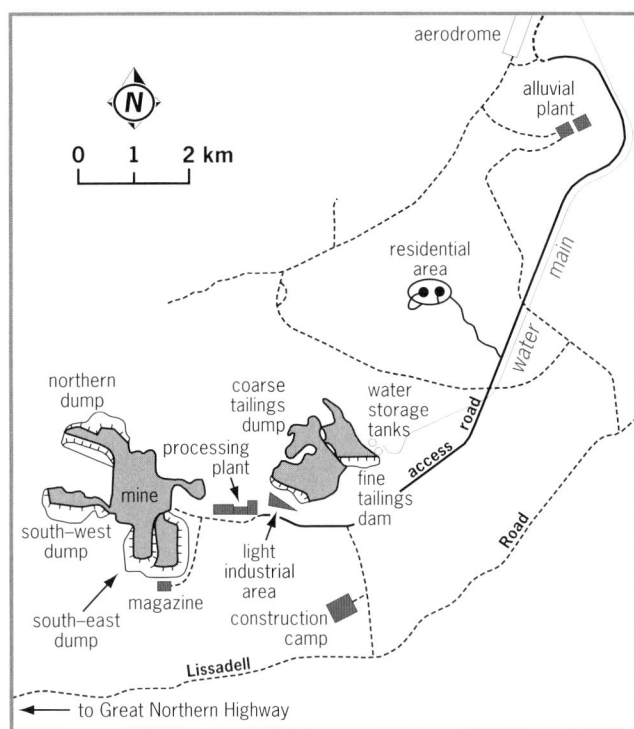

**Figure 5.19** Operational layout of the Argyle diamond project

As illustrated in figure 5.19, the processing plant is located close to the mine area, thus minimising the cost of transporting bulky ore for initial processing. The structures that are associated with the initial and subsequent crushing of the ore are common to the treatment of a wide range of minerals, with similar operations being found in the initial processing of gold or nickel. Perhaps the building that most clearly distinguishes a diamond operation from other forms of mining is the recovery plant in which X-ray separation of diamonds from waste material occurs. A number of other buildings are also part of the infrastructure. The first group has a service function and includes administration buildings, security offices, messes, a tourist building, change rooms and a first aid centre. Because of the remote location of the mining site, it is important that the operation be largely self-sufficient. To this end, the other group of buildings comprises warehouses, workshops, fuel storage sheds, a heavy machinery shop and a light industrial area (see figure 5.19). A $40 million company accommodation complex is set well apart

**Figure 5.20** Processing plant at the Argyle mine

from the mining, processing, service and maintenance operations. Employees live in high-standard, self-contained, single persons quarters. The attractively landscaped area includes facilities such as a swimming pool, a gymnasium, squash courts and a delicatessen.

> **EXERCISES**
>
> 1. Describe and account for the changing cultural landscape associated with the exploration and development of an open-cut diamond mine site.
> 2. Identify and discuss the extent to which diamond mining in the Kimberleys has influenced the cultural landscape of the region.
> 3. Using the illustrations and other information provided, describe the **site** and **situation** of the Argyle diamond mine. In doing this include information on distances from towns, major physical features, transport and other linkages, and location within Australia.

## Environmental issues

### Government controls

Before alluvial or open-cut mining could be commenced by Argyle Diamond Mines Pty Ltd, the company was required to submit an environmental review and management program (ERMP) to the Environmental Protection Authority (EPA), now the Department of Environmental Protection (DEP). An ERMP details the likely effect of a project upon the environment and is required when any mining company wishes to mine a mineral deposit in Western Australia. On the basis of the assessment of the plan the DEP granted approval to mine in the environmentally sensitive headwater region of Limestone and Smoke creeks. The decision by the mining company to use a commuting workforce has considerable environmental benefits, which were recognised by the company and the DEP. Because employees are not with their families at the mine site and are either housed in single units at the compact village or fly in on a daily basis from Kununurra, the need to allocate large areas of land for the development of a town site does not exist. This means that there is no need for extensive road systems, schools, hospitals, and large power and sewerage facilities. Pressure on the land is also reduced because outdoor family activities such as picnicking, bushwalking, fossicking and camping do not occur. Benefits also flow to Aboriginal communities and pastoral properties, which remain largely undisturbed by the mining activity. The original ERMP recognised that the project would lead to significant changes in the land surface and environment where mining had taken place and

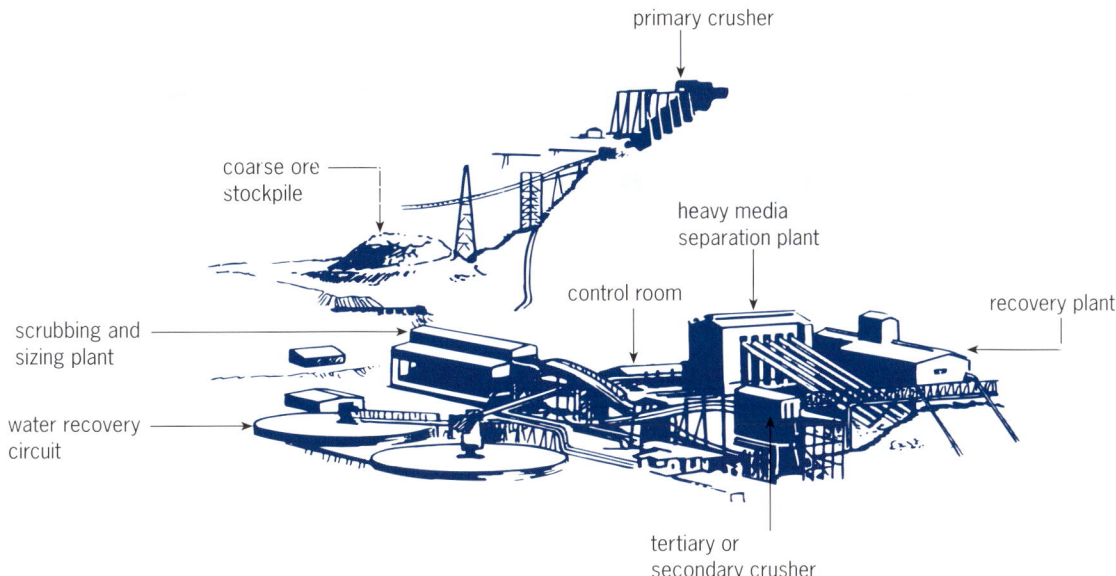

**Figure 5.21** Mining activity at the Argyle diamond mine

that strategies had to be put in place to minimise these impacts.

## Protecting ecosystems

The need to protect unique or rare ecosystems from mining activities was recognised by the company and the DEP. Where vegetation removal could not be avoided, seed was to be collected and propagated for use in rehabilitation work. The rehabilitation of riverine ecosystems, which would be unavoidably damaged by alluvial mining, was regarded as essential. The region has a major soil problem, which has a direct impact upon Lake Argyle. The problem is related to five factors: the composition and structure of the soils, which make them prone to erosion; steep terrain; torrential monsoonal rainfall, which leads to heavy surface runoff; degradation of natural vegetation due to overstocking and to grazing by feral animals such as horses and donkeys; and the poor land-management practices that occurred in the past.

In 1988 Argyle Diamond Mines Pty Ltd indicated that it intended to recommence mining the alluvial deposits on Smoke and Limestone creeks. The DEP expressed concern at the size of the area to be disturbed (about 155 hectares), the destruction of riverine vegetation, the potential for soil erosion, the likelihood of increased sediment deposits in Lake Argyle and the management of the dump used for oversized material. The company responded positively and engaged in a program that has minimised the effect of mining on the environment. For example, the project did not involve the introduction of new infrastructure (power, plant and roads). All alluvial tailings (wastes) were to be transferred to a special dam designed for this purpose. Topsoil was removed up to one month before mining began and replaced about one month after it has been completed in a certain area. The land is then reshaped by graders to blend in with the surrounding topography, and depressions incorporated to encourage the retention of water. Stony material from the alluvial treatment plant is used as a 'rock mulch' to assist in the rehabilitation of difficult clayey areas. Drainage ditches are constructed with a one degree fall to enable direct runoff into the main creek system. These drainage channels have silt traps constructed on them to reduce the amount of soil entering the downstream areas. The silt is periodically removed and spread over recontoured areas which are then revegetated with native plants grown in the company nursery. The area between the ditches is deep ripped to encourage the penetration of roots and moisture. As in all rehabilitated areas on the site, fertiliser is applied and the area sown with quick-growing natives and introduced plant species, followed by eucalypts. Introduced vegetation on the banks of Smoke Creek has been protected from flood damage by a deepening of the main creek channel. Waste material from the processing plant is used to protect the main channel from erosion. Fencing in the area has been upgraded to prevent damage from grazing animals. Regular inspections of areas under rehabilitation are made to ensure that the rehabilitation process is successful. Areas of poor regrowth undergo remedial treatment. An annual report is submitted to the DEP detailing the rehabilitation program and all other environmental works and investigations carried out.

### EXERCISES

1. Identify and discuss the main environmental issues that had to be considered by Argyle mines in developing a mine site within the Kimberleys.
2. Explain how the company overcomes the problems of soil erosion and rehabilitation of cleared areas.
3. Describe the ways in which the state government controls the environmental impact of diamond mining in the northern study area.

## CASE STUDY 3

# Coal mining in the south-western study area

## Location

While coal deposits have been found at Boyup Brook, Vasse near Busselton and Hill River and Irwin River to the south of Geraldton, the only current commercial field in the South-West is near Collie. The Collie field has been in continuous production since 1898. The coal deposits are found in 55 potentially economic seams and are broken into two basins covering an area of some 350 square kilometres (see figure 5.22). The Collie Coal Basin is small when compared to the coal deposits of the eastern study area, however it has been estimated that there are sufficient reserves to meet the energy needs of Western Australia for at least 100 years.

There are currently two companies involved in the mining of coal in the basin. Griffin Coal operates the Muja open-cut mine, about 12 kilometres south of Collie and the Ewington 2 mine to the north near the Collie power station, while Westfarmers Coal has developed the newer Premier open-cut mine several kilometres to the north of Muja. All mines are found in the Eastern Basin. While underground mining of coal occurred in both the western and eastern basins, the last operational mine ceased production in 1994. In 2001 the output from the Collie fields was about 7 million tonnes per year, with approximately 4 million tonnes being extracted from the Premier open cut and 3 million from the Muja and Ewington 2 open cuts. The primary consumer of the coal from the Collie Basin is Western Power. This is used to produce electricity at the Muja and Collie power stations. Both of these power stations are major contributors to the state's power grid. Coal is also used by Worsley Alumina Refinery, situated to the west of Collie. Lesser amounts of coal are also used in the production of synthetic rutile and cement and lime production.

## Geology and uses

Coal may be classified into three main groups. Loosely compacted fossil fuel with a high water content of more than 90 per cent is called peat. This is often found in swampy surface deposits and may contain recognisable vegetable matter. Lignite or brown coal is a relatively soft coal with a high water content and subject to moderate compression produced by overlying sediments. Black coal is a metamorphosed material, with a rock-like appearance. It varies in colour from grey to black and can be classified as either sub-bituminous or anthracite. Collie coal is sub-bituminous. It is a relatively clean burning coal with little ash and a low sulphur content, which makes it ideal for power generation.

The Collie Coal Basin resulted from faulting in the region during Permian times, about 250–280 million years ago. Figure 5.23 shows the main fault lines. The Muja Fault involves a displacement of about 1000 metres and it now forms the western edge of the Collie

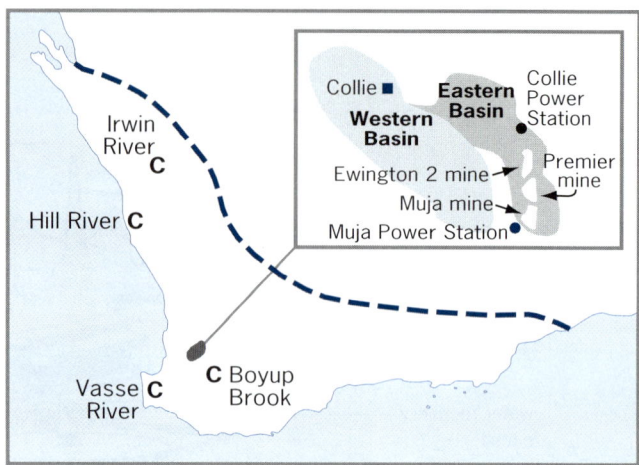

**Figure 5.22** Coal mining areas

Coal Basin. Occasional faulting continued and the graben (depression between two faults) began to accumulate successive layers of river-borne deposits of alluvium. These periods of deposition were followed by the growth and decomposition of vegetation, which led to the formation of peat. These layers of peat were buried by sands and clays, which were deposited in the basin. Brown coal and then black coal formed as the pressure of overlying material removed water and gas from the peat. The 350 square kilometre basin now comprises sandstones and shales in which the coal seams are found. It is surrounded by Precambrian granite of the Yilgarn Block. The section of the basin that bears coal is 1050 metres thick and consists of three coal-bearing measures (layers): the Ewington, Premier and Muja (or Cardiff) members (see figure 5.23). The Ewington and Premier measures are currently being mined.

Apart from power generation, coal has a variety of uses in the industrial and chemical industries. Coal may be processed to produce a variety of synthetic goods including nylon, tar, washing powders and detergents, paint, dyes, aspirin, perfume, explosives and fertilisers for agricultural use.

## History of the Collie Basin

Coal was first discovered beside the Collie River in 1883. In later years shafts were sunk beside the river and coal was found at about 2 metres depth. Premier John Forrest was keen to be independent of the eastern states for coal supplies and was convinced of the viability of the Collie coalfield. He ensured that a railway to the Collie area was included in the 1894 Loan Bill. The township was officially declared in 1897 and the 'tent city' image given to the area by early miners who lived near the mine sites soon disappeared as more permanent structures were built. Commercial coal mining began in earnest with the establishment of the Wallsend Colliery in 1898 (see figure 5.24). The infrastructure associated with this mining venture began to change the appearance of the landscape, as did the completion of the Brunswick–Collie railway line in 1898. The line linked the 600 inhabitants of Collie with major settlements on the coastal plain and ensured an

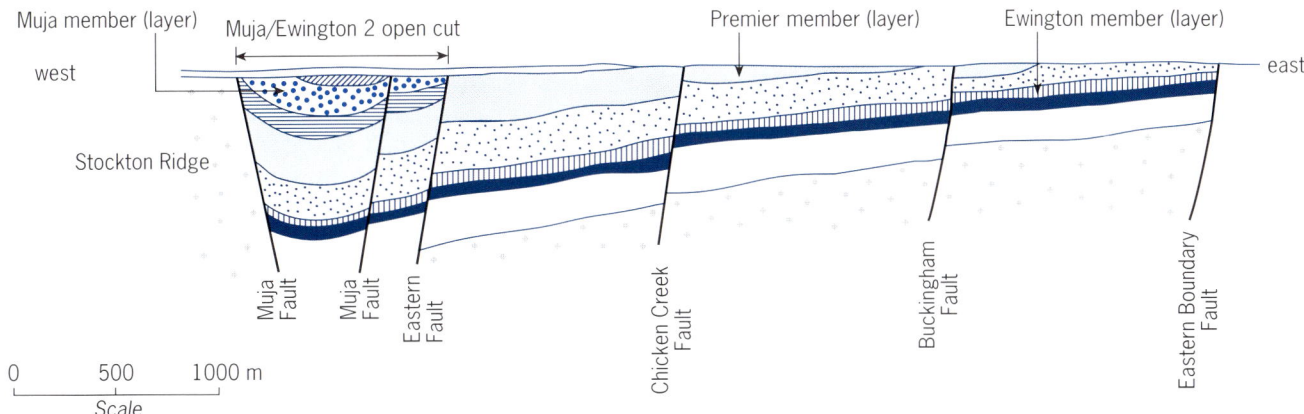

**Figure 5.23** Cross-section of the Collie Coal Basin

efficient means of transporting coal to the marketplace. This linkage was extended six years later with the completion of the Collie–Narrogin railway. By the turn of the century, Collie had produced 120 800 tonnes of coal and was entering a boom period in its history as the demand for coal grew in the developing state. The town boasted three large hotels, the Mechanic's Institute, a government school, a registrar's office, a courthouse, a hospital, a post and telegraph office and a police station. Small timber mills operated in the jarrah forest that surrounded the town, and mixed farming was practised on the fertile alluvial soils of the river floodplain.

The 1960s was a period in which the switch to diesel power generation and the use of diesel for rail transport resulted in a decline in the use of Collie coal. This brought about the closure of a number of underground mines. It was the world oil crisis of 1974 that produced a resurgence in the use of coal as an energy source. Power stations in Western Australia that had been converted to diesel were again modified to use coal to run the steam turbine generators. Open-cut mining at Muja was expanded, and in 1995 the Premier mine on the site of an older mining operation was commenced and brought into production in 1999.

### EXERCISES

1. Describe and account for the location of the Collie Coal Basin.
2. Outline the way in which coal is formed and explain the differences between peat, brown and black coal.
3. Identify some of the changes that have occurred to coal mining over the last 100 years. Include information on the changes in the type of mining as well as the effects of changing demand.
4. Why were the Collie coal fields developed?

## Mining and processing

### Production

Once coal seams are exposed, they are drilled and blasted in bench formations. Both overburden and coal are loaded separately by front-end loaders or hydraulic excavators, and transported by dump trucks. While the overburden is moved to dumps, any good quality sands encountered during mining operations are stockpiled for future drainage work in backfill areas. Sediments between coal seams are natural aquifers (water can pass through them). Excessive water in these sediments can destabilise slopes and create problems for the heavy equipment operating in the open cut. Underground mining at Collie was made difficult by the presence of water, and mining was made dangerous when the roofs of mine galleries were softened and weakened. About 11 million litres of water was pumped each day from open cuts and underground mines. The Muja Power Station receives about 55 000 kilolitres of this water daily for cooling purposes.

### Processing

Within the open-cut mines trucks transport coal intended for the power stations to a primary crusher, which reduces it to a maximum of 150 millimetres in diameter. Secondary crushers further reduce the coal to a thickness of 20 millimetres before it is stockpiled. The Westfarmers operation at the Premier mine site uses a

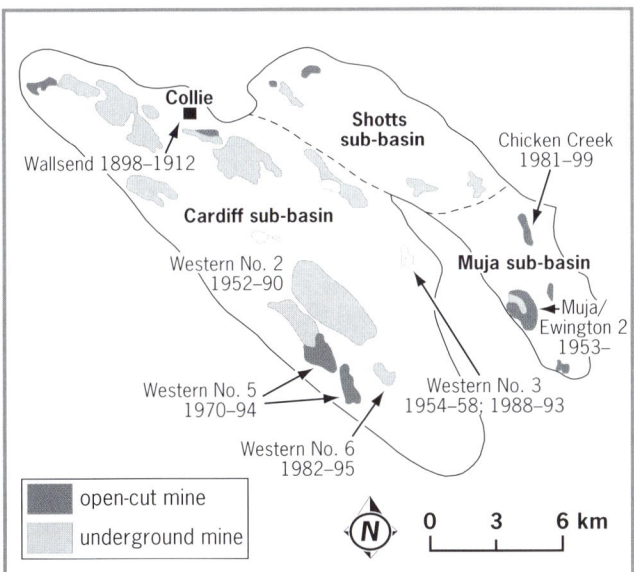

**Figure 5.24** The Collie Coal Basin, WA

rail-mounted stacker which travels up and down the stockpile to blend the coal to a consistent quality. The coal is then moved to either the Collie Power Station or the Muja Station via a conveyor belt system. Griffin coal also supplies the Muja Power Station, however they use trucks for this purpose. Coal is sometimes washed to remove pieces of rock or mineral that may be present. This involves immersing the coal in a liquid in which the coal can float and be recovered while the heavier mineral sinks to the bottom. Coal used for power generation is fed into mills, which pulverise it to the fineness of talcum powder. This pulverised fuel is blown by hot air directly into furnaces at the Muja and Collie Power Station boilers, where it is burnt to generate steam from water pumped in from the mines.

## The cultural landscape

### The regional cultural landscape

A distinctive cultural landscape has emerged following a long history of coal-mining activity in the region. Central to this landscape is the township of Collie, located on the Collie River (see figure 5.25). In 2000 the shire of Collie had a population of 8627, representing a decline of some 1000 people since 1989, which is a reflection of the closure of the last underground mine in 1994 as well as the increase in the open-cut mining system, which is much larger in the scale of operation and in the use of labour-saving machinery. In addition there is an increasing tendency for people to commute from areas outside the shire.

Early settlement of the town site occurred close to the river, but a series of devastating floods has clearly established the limits to which urban development can safely extend. Although housing in Collie is relatively modern, old homes are commonplace and many are occupied by the descendants of early miners. Old public buildings within the town are a reminder of the town's historical attachment to the coal-mining industry. Remnants of former mining settlements are distributed through the region and indicate the location of early collieries. The mine settlements were once thriving communities with populations of sufficient size to support a post office, store and primary school. Most of these, such as Cardiff, Collie Burn and Shotts, are still occupied and some, particularly in the case of Allanson, are attracting a growing number of residents. As indicated in figure 5.25, the Collie area is served by a comprehensive road and rail network. The main Bunbury–Narrogin railway links the Collie region with the port of Bunbury, Perth and an agricultural area to the east. This main railway allows coal to be transported in specially designed rolling stock to Western Power power stations at Bunbury, Kwinana and Muja. The railway also enables coal to be transported to Worsley for the production of alumina from Boddington bauxite, and to the Western Mining Corporation in Kalgoorlie. Feeder lines extend to existing open cuts and crushing and train-loading facilities. The imprints of former lines are clearly visible on the landscape. These lines once met the needs of mines and their associated settlements. With the decline of mining activity in these areas, the lines were eventually removed. The Coalfields Road is the principal road of the Collie Shire, and it negotiates a parallel route with the main railway. First-class roads that serve the mining areas and allow coal to be trucked to crushing facilities and to the Muja Power Station radiate to the south-east of the town site. Together with the unsurfaced roads that link up former mine sites, farms and timber-cutting areas, they form a complex transport network. Dumps of overburden from over 100 years of coal mining are common in the landscape. Some are made less obvious by natural revegetation, while others have been rehabilitated by the mining companies.

### The mine site cultural landscape

*Exploration*

As has already been noted, the discovery of most coal deposits in Western Australia occurred over a 100 years ago. These discoveries were based on the location of surface deposits. Further exploration from 1920 to 1966 in the Vasse River region, as well as near Boyup Brook and in the Perth Basin, has resulted in the identification of potentially economic deposits. The lack of significant tectonic movement necessary for the formation of coal seams has meant that the South-West is lacking in this energy source. In contrast the potential for natural gas deposits is much greater.

The initial identification of the extent of coal deposits in the Collie Basin is done in a similar way to other forms of mining. Preliminary drilling determines the quality, depth and size of coal seams as well as the type of overlying geology. This information is necessary in determining the way in which the deposit will be mined. While the initial exploration of the basin has a relatively limited impact on the landscape, the subsequent expansion or development of mine sites brings significant changes.

*Development*

The development of a coal mine is subject to a range of conditions imposed by government. Approvals must be obtained by the company from a number of state government agencies. These will be identified further under the environmental issues section of this case study. Once the planning has been done then the site can be prepared for mining.

While the objective is to minimise the amount of vegetation to be removed, the operations begin with the clearing of the area to be mined. In 1997–98 Westfarmers Coal initiated a program to maximise the use of any timber cleared in advance of mining. Jarrah removed from the Premier mine site was stockpiled for use as fencing poles, firewood, furniture timber and construction grade timber. Of the 156 hectares cleared, a significant percentage of timber that would have been previously left or burnt was redirected to these higher value uses.

To begin the mine after clearing, the topsoil is removed and either spread onto previously mined sites or stockpiled for future use. This operation is performed using bulldozers, excavators, front-end

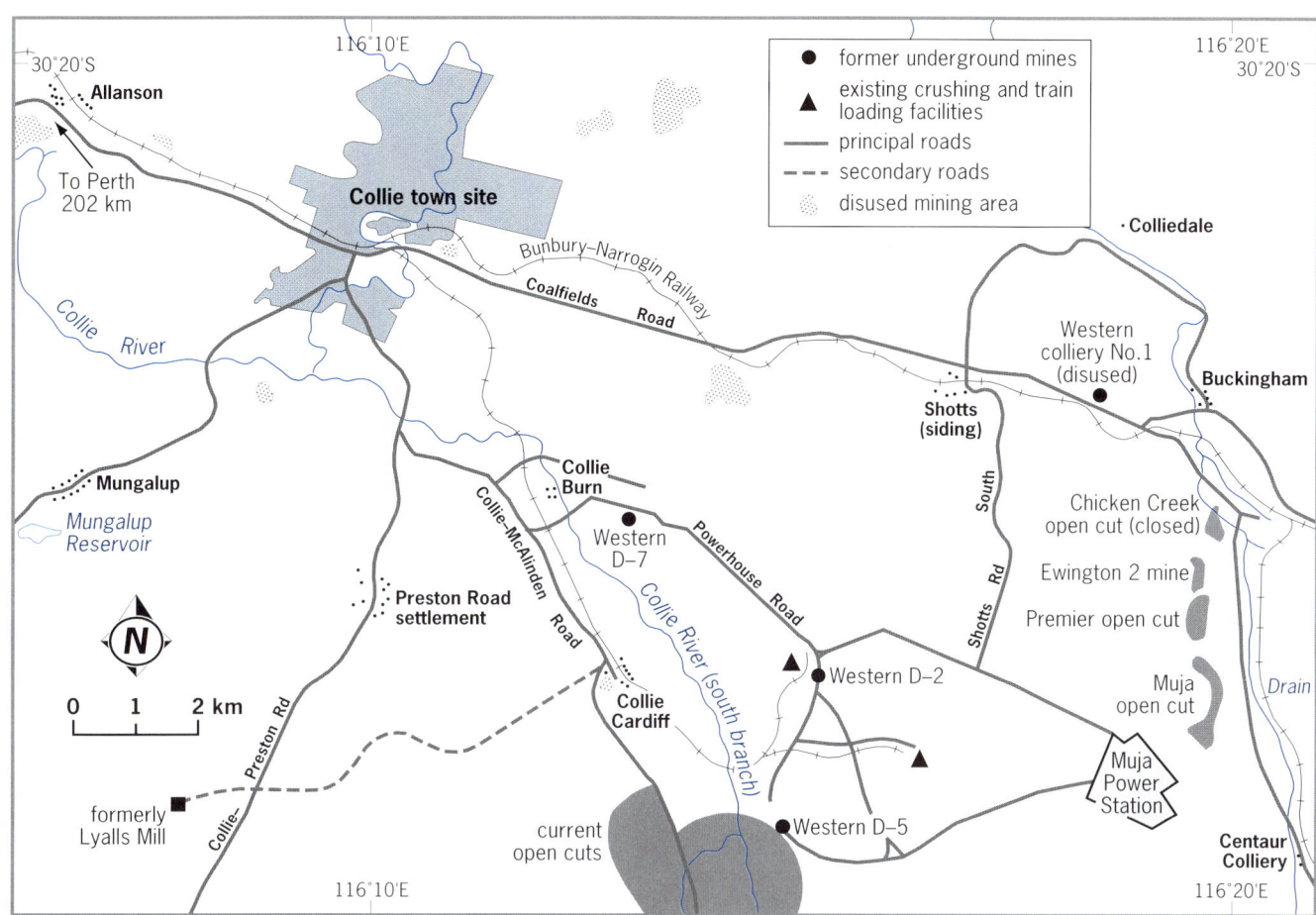

**Figure 5.25** The Collie coal-mining cultural landscape

loaders and dump trucks. Lateritic caprock is ripped by bulldozers or blasted and used as road base.

The remaining overburden is then prepared for blasting using rotary drill rigs and bulk explosives. Once broken up, it is removed by a hydraulic excavator operating with a dragline (this drags a large toothed bucket across the ground, picking up loose rock). The overburden material is then used to backfill older mine sites or shaped into new hills. Once the overburden is removed, then the top of the coal seam or roof is scraped clean using bulldozers. The development stage in open-cut coal mining has resulted in a number of changes to the natural landscape in the form of access roads, stockpiled soil, windrows and stockpiled vegetation, overburden and rock dumps.

*Production*

The Muja open cut, the largest in the Southern Hemisphere, is the centre of great activity. Huge trucks negotiate inclined ramps to reach the benches at the face from which coal is being blasted and loaded by equally huge Caterpillar front-end loaders. Infrastructure at the surface includes maintenance and welding workshops, an administration centre, fuel and lubrication facilities, electrical and water reticulation buildings, and warehouses. The Chicken Creek open cut was also mined by the Griffin Coal Mining Company Ltd and is located about 5 kilometres to the north-east of Muja. This is near the site of Ewington 2, Griffin's newest open-cut mine. The two open cuts operated by Westfarmers Coal are mined in similar fashion to Muja. Some of the evidence of past open-cut mining is being erased. One example is the site of the abandoned Collie Burn open cut, located 7 kilometres south-east of Collie (see figure 5.25), which was operated by Western Collieries Ltd. Mining in the open cut was seriously affected during winter, when heavy mining equipment frequently became bogged in the saturated sediment and it was common for trucks to break their axles. The mine relied heavily on pumps to clear excess surface water. Eventually the site was abandoned, but it has since been rehabilitated and landscaped.

The Muja Power Station, which is situated beside the Muja open cut, about 22 kilometres south-east of Collie, is an important element in the cultural landscape of the region (see figure 5.26). Four huge smokestacks rise above the large turbine hall of the station. The steam boilers are the equivalent height of a 16 storey building and each boiler consumes about 2300 tonnes of coal a day. To the east of this complex is the Western Power crushing plant, which subjects coal to primary and secondary crushing before it is fed into the power-generating system or stockpiled for future use. Pulverised coal is moved by conveyors for burning in furnaces to produce energy, which heats water in a

**Figure 5.26** The Muja open cut and Muja Power Station

boiler and produces pressurised steam to drive turbines. The electricity produced by Muja provides the 'base load' for Western Power's grid, which provides power for all of the south-west of Western Australia.

*Decommissioning*

The closure of a mine site is far more than rehabilitation. A range of tasks must be carried out. Buildings, concrete pads, power lines, buried cables, roads, rail lines, coal fines and waste dumps must all be removed. The ground must be monitored and made safe in the event of any possible subsidence where underground mining has occurred. Where an open cut mine is worked out or has become too deep to continue mining in this way, the pit is sometimes filled with rock, covered with topsoil and replanted with vegetation. Alternatively it may be allowed to fill with water to create an artificial lake. Several old pits in the Collie Basin have been allowed to fill with water by diverting small amounts from nearby rivers. When full, they will provide the possibilities for aquaculture or recreational activities. In 1999 Westfarmers Coal was planning to spend over $10 million on the rehabilitation and closure of two large pits, using this method. Decommissioned areas are designed to blend in with the surrounding forests of the region and from this point of view the objective is to minimise the imprint on the cultural landscape. However, the presence of artificial lakes and recreational park lands, with their associated roads, illustrates the evolutionary nature of the cultural landscape associated with coal mining in the Collie Basin.

### EXERCISES

1. List and describe the main characteristics of open-cut coal mining and suggest why this is a more cost-effective and safer method when compared to underground mining operations.
2. Outline the steps involved in the processing of coal in order to make it suitable for power generation. Include an explanation on how the energy of coal is converted to electrical energy.
3. What evidence is there to show that coal mining in the Collie region has had a significant effect on the cultural landscape over a long period of time.
4. Describe the cultural landscape associated with an open cut-coal mine.
5. Identify the ways in which coal mining has influenced the functions, characteristics and development of the town of Collie.

## The environmental impact

The environmental impact of coal mining in the area is localised but quite severe due to the intensive nature of the activity. Evidence of many years of coal mining takes the form of obsolete and current open-cut mines, fenced-off disused underground mines, dumps of overburden and coal stockpiles, old buildings and disused railway routes, and even 'hot ash' residue sites, which resist being extinguished. However, this environmental disturbance is largely confined to actual mine sites and the visual impact is minimised by substantial buffers of forested areas.

### Vegetation

The forests of the region supported a thriving timber industry before the establishment of the mining industry. Timber resources were greatly depleted once mining began. The onset of underground mining made heavy demands upon timber resources because large quantities of split jarrah were required for the roof-support systems of the mines. Leases held by various companies provided most of the mine timber needed; however, the influx of people into the region led to widespread cutting for home construction and extensive clearing for farmland and community facilities.

Of special concern in the Collie Basin is the potential effect of mining activities on the spread of jarrah dieback. This is most likely during clearing operations, removal of surface soils and rock and rehabilitation. The mining companies work closely with the Department of Conservation and Land Management (CALM) to map vegetation and dieback zones in order to plan for the separate removal of uncontaminated and contaminated soils and vegetation. (Dieback is caused by a fungal growth that attacks the feeding roots of trees, including jarrah, causing stress to the plant and resulting in it shedding leaves and branches from the canopy as it attempts to survive. If the plant is in fertile soil it often survives by growing new roots and branches. In less fertile and rocky areas the plants eventually die.)

Other procedures for controlling the transportation of the fungus into unaffected areas include washing vehicles, correct stockpiling of infected soils and controlling water movement within the mine site.

### Coal fines

Another issue of environmental concern which is linked to early mining is the once-common practice of dumping coal 'fines' in areas adjacent to mines following the removal of 'lump' coal, for which there was high market demand. Although much of the fines were recovered when a market was established, residues are prone to ignite during spells of hot weather or during forestry burns. Ignition occurs despite efforts by CALM to cover large areas of carboniferous materials (coal fines) and to revegetate the areas.

### Subsidence

Cave-ins sometimes occurred when mine workings were close to the surface. Although most of the subsidence was shallow, in 1947 workings from the Cardiff colliery caved in to a depth of about 23 metres over an area of one hectare. This area, which is on the Western Collieries Ltd lease, has since been rehabilitated. The Mines Department now insists that all excavations, underground shafts and tunnels are filled in and sealed.

### Open-cut mining and waste dumps

With the introduction of open cuts to the coal field in 1943, the potential for environmental degradation became considerable. Fortunately the scale of operations was held in check by wartime constraints on the availability of heavy equipment. The small bulldozers, trucks and loaders that were used limited the output of the open cuts and minimised the impact of mining on the landscape. The small scale of the early open cuts meant that overburden had to be dumped away from the mine, a legacy that still remains.

In today's large open cuts overburden is retained in the pit as backfill. Dumping of overburden had the unfortunate effect of inverting the soil profile, which led to the deeper strata being placed at the top of dumps. Heavy clays in some of this overburden created dumps that had little or no chance of developing covers of natural vegetation, and which were subjected to severe water erosion. As previously discussed, modern stockpile management strategies ensure that topsoil is kept separate from the rock overburden. Once the mining has finished these materials are used to recontour and rehabilitate the landscape in such a way as to make the area safe and maximise the chances of successfully creating a self-sustaining ecosystem.

### Air quality

Atmospheric emissions occur as a result of burning coal at the Muja and Collie power stations. Gaseous emissions comprise oxides of nitrogen, sulfur dioxide and carbon dioxide. Fortunately, the low sulfur content of the coal minimises sulfur dioxide emissions. The small volume of the total emission quantity and the large area over which the emissions are dispersed further reduce any risk of acid rain. However, carbon dioxide is a greenhouse gas and proposals to build another power station at Collie met with opposition because of the potential contribution to atmospheric warming. In a 1990 report in support of the power station, Western Power claimed that the contribution of carbon dioxide by Western Australia's power generation industry was very small in terms of worldwide total greenhouse gas emissions. The report acknowledges that a new coal-powered station would add to the warming process, but suggests ways in which carbon dioxide emissions could be reduced. These suggested safeguards included improvements in the energy-conversion efficiencies of boilers and turbines, an increase in overall plant efficiency leading to a reduction of about 36 per cent in emissions, the use of gas turbines during peak demand and the establishment of very large tree plantations to act as carbon sinks and to absorb some of the carbon dioxide.

Emission of particles is a characteristic of coal-powered stations. These emissions take the form of 'fly ash', the incombustible component of coal. A residue of about 85 kilograms of fly ash is generated from burning 1000 kilograms of coal at the Muja Power Station; however, electrostatic precipitators installed on smokestacks remove about 99.5 per cent of the emission, thus complying with the current National Health and Medical Research guidelines.

## Water quality

The issue of water quality is a most important environmental consideration because the Collie Coal Basin is within the catchment of the Wellington Dam. Salinity levels in the dam have been unsatisfactory due to extensive clearing in rural areas to the east. Large-scale tree planting has taken place in the catchment in an effort to reverse the situation. Western Power had planted about 750 000 trees in the catchment by 1990 to compensate for essential clearing of the site of the power station and transmission lines. Good quality water is stored in the aquifers of the region and controls required by the water authority are in place to minimise the impact of mining on the catchment area. Large quantities of water are pumped during mining operations and much of this is used for cooling purposes at the Muja and Collie power stations. The remainder is directed into the surface drainage system. This discharge is controlled by licence from the Department of Environmental Protection and administered by the Water Authority. Water from coal mines has a relatively high acid level, but studies have not indicated that the quantity released into the rivers has an adverse affect on the riverine ecosystems of the region. The input of mine water may in fact benefit native species because of the water's high oxygen level and low salinity.

## Rehabilitation

Both mining companies in Collie maintain an active rehabilitation program. Rehabilitation work normally begins several years after the commencement of open-cut mining. This enables a certain amount of pit space to be created to provide for the removal of overburden and the extraction of coal. Rehabilitation then aims to treat an area of land that is approximately the same size as the area being mined. Rehabilitation of mine sites was initiated by Western Collieries Ltd (Westfarmers Coal) in 1975 when topsoil was first stockpiled and trial plots of vegetation were established. These moves foreshadowed the Collie Coal Agreement Acts 1979 (WA) which details the requirements for environmental protection and management in the region. These requirements include monitoring, research and trial programs. Of particular importance is the requirement for companies to report on an annual basis, in addition to providing a comprehensive report every three years. At the time of the Collie Coal Agreement, Western Collieries Ltd had revegetated about 55 hectares of former mined land with native species, and the Griffin Coal Mining Company Ltd had introduced exotic pine trees, native species and pasture to land at its Muja mine site.

When mine dumps are created, they are first shaped to conform with the surrounding topography. Shaping involves the reduction of steep slopes in the dumps, which is done by bulldozers and scrapers. Topsoil is spread to a depth of about 30 centimetres and surface drains are constructed to assist with erosion control. The drainage pattern aims to direct water runoff from the reshaped surfaces to natural watercourses on undisturbed land adjacent to the site being rehabilitated. Soil samples are taken from the newly established surfaces and tests are conducted. This assists the mining company in determining the appropriate fertiliser to be used. If pasture is being established, lime and gypsum are usually applied to improve the condition of waste dump and topsoil, and the surface is then cultivated. Fertiliser (superphosphate and molybdenum) is applied before the onset of winter rains and the area is deep-ripped along the contours to a depth of one metre, at 1.25 metre intervals. This deep ripping traps the rainfall and encourages infiltration of moisture into the dump surface. In addition, ripping mixes the soil and assists root penetration. At the break of winter rains, areas intended for pasture are sown with one of five varieties of clover, together with oats or cereal rye, and tree species are planted in other areas. A wide range of seed from native species is collected from revegetated dumps or purchased from local suppliers. Both Westfarmers Coal Ltd and the Griffin Coal Mining Company Ltd are engaged in trials using different forms of mulch, agricultural crops, soil types and slopes. Today, environmental concerns stem mainly from open-cut mining. These concerns have lessened considerably because of successful rehabilitation programs, which have been in place since the mid-1970s. The alternative uses found for some open cuts have also allayed concern. For example, mines such as Wallsend near Collie have been used for municipal rubbish pits. This mine was later filled, levelled and compacted and has finally been transformed into Roche Park, a community recreational area.

### EXERCISES

1 List the main environmental issues associated with coal mining at Collie and then identify the ways in which each is being addressed.
2 Describe the different ways in which a mine site is used or rehabilitated after the operation has ceased.
3 Identify the ways in which mine site rehabilitation meets the needs of both the community and the environment.

# Chapter 6

# Australia's Population

The purpose of this chapter is to outline the main elements of demographic studies and to apply them to Australia, the South-West and the city of Perth.

## Introduction

Demography is the study of population. It includes the analysis of patterns of density and distribution, movements or migrations, growth rates, age and sex variations, and differences in ethnicity and socio-economic characteristics.

In accounting for demographic patterns, factors such as climate, landforms, soils, natural ecosystems, technology, economic development, political structures and policies, and cultural and historical influences can all be drawn upon.

The study of population has a number of significant implications. It influences government decisions at the national, state and regional level by providing information about how fast populations grow, where people are moving to and from, and the relative size of the workforce compared to the numbers of people in retirement. In addition, it reveals the rates at which people are being born or dying—all important considerations in planning for the future. Studies of population also raise fundamental questions as to what constitutes optimum population densities or total numbers, and how high or low population growth rates and the consumption levels of different societies will impact on finite resources or natural environments.

## Population density and distribution

### Density

Population density is expressed as the number of people within a given area of land. Often this is given as the number of people per square kilometre. If the population density is very low the rate may be expressed as a number per 100 square kilometres or, alternatively, the number of square kilometres per person. Thus, a density of 50 square kilometres per person is the same as 0.02 people per square kilometre.

In Australia, the **average** population density for the nation is 2.47 people per square kilometre. Half the area of the continent, however, contains only 0.3 per cent of the population, and the most densely populated 1 per cent of the continent contains 84 per cent of the population, which equates to 207 people per square kilometre.

### Distribution

Most of Australia's population is concentrated in two widely separated coastal regions. By far the largest of these, in terms of both area and population, lies in the south-east and east. The smaller of the two regions is in the south-west of the continent. In both coastal regions the population is largely concentrated in **urban** centres, in particular the state and territory capital cities. Figure 6.1 demonstrates this distribution pattern.

The eastern region of the continent contains the bulk of the nation's population, with most people living within 100 kilometres of the coast. Australia's three largest cities are found within this area—Sydney, with a total population of 4 085 600 in 2000; Melbourne with some 3 466 000; and Brisbane with 1 626 900. These three cities alone contained 47 per cent of the nation's population. When the populations of the other state and national capital cities are added to these three their combined total represents approximately 64 per cent of the nation's population. The dominance of the capital cities in Australia's population distribution, known as **urban primacy**, is significant and far outranks that in countries such as the United Kingdom and the United States.

Of the next 10 largest cities or population centres in Australia (after the capital cities), eight are located on the east coast. These include Newcastle, the Gold Coast, Wollongong, the Sunshine Coast, Townsville and Cairns (see figure 6.2). It should be noted that regions such as the Gold Coast and the Sunshine Coast represent population centres where smaller settlements have

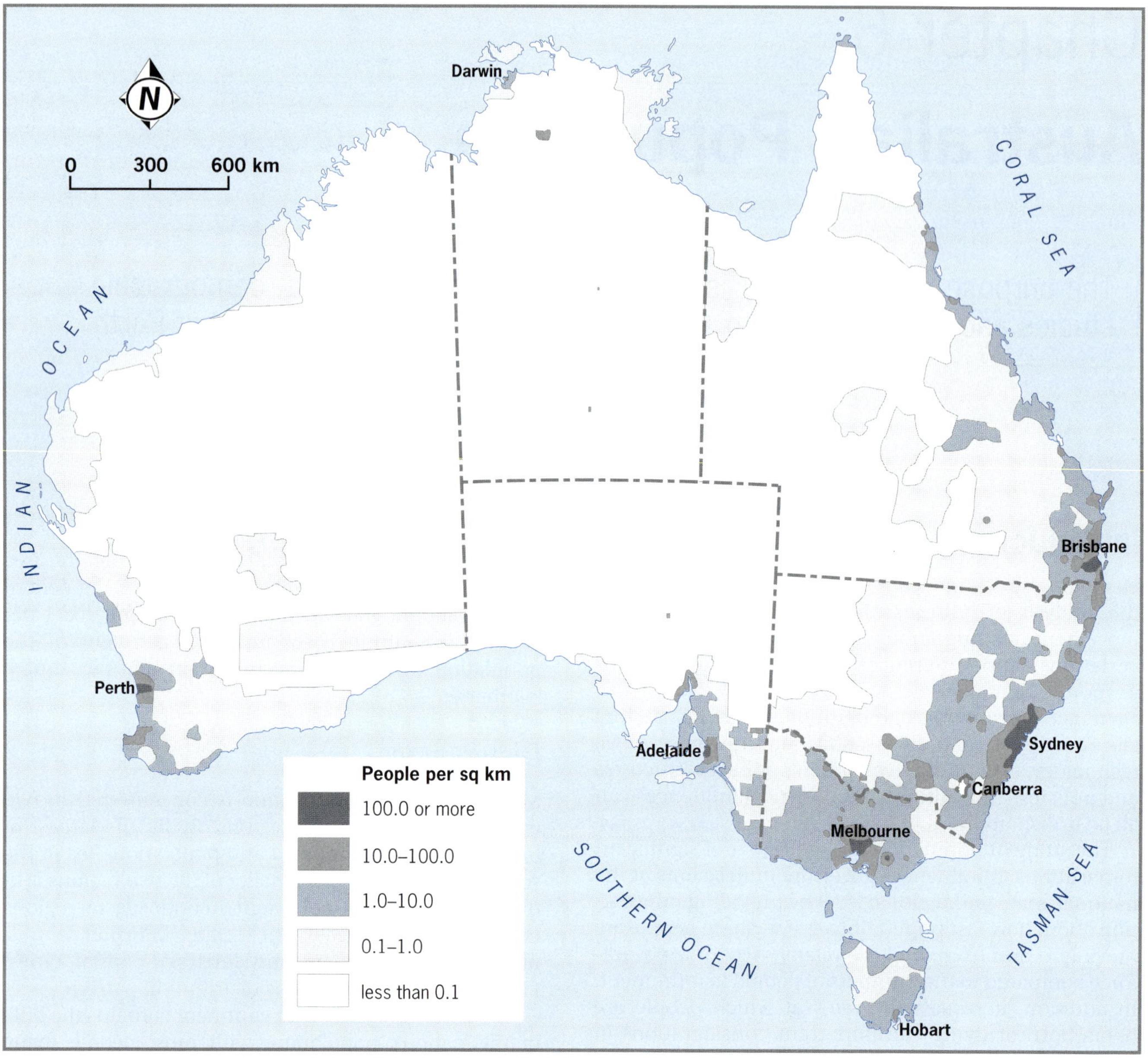

## 122 LANDSCAPES AND LAND USES

**Figure 6.1** Population density and distribution, Australia, 1996

People per sq km
- 100.0 or more
- 10.0–100.0
- 1.0–10.0
- 0.1–1.0
- less than 0.1

**Figure 6.2** High density settlement on the Gold Coast, Surfers Paradise to Burleigh Heads

**Figure 6.3** The settlement of Alice Springs in the McDonnell Ranges of Central Australia

amalgamated to form urban **conurbations**. The Australian Bureau of Statistics classifies both localities as population nodes or regions.

In contrast to the inhabited regions of the continent, vast areas of the central, west and north are virtually uninhabited. These regions contain sparsely scattered rural populations associated with extensive pastoralism, or small concentrations of population within remote mining and urban settlements. Alice Springs, with approximately 25 000 people (June 2000 figures), contains much of the population of Central Australia (see figure 6.3). Central Australia is mainly a region of arid and semi-arid lands, national parks and Aboriginal lands.

## Factors influencing Australia's population distribution

Whether at the local, regional or national level the factors that influence the density and distribution of population are essentially the same. There are factors that encourage high concentrations of population and those that act against it. They include environmental or biophysical factors, cultural and historical factors and economic, political and technological factors.

### *Biophysical factors*

#### Climate

Climate is one of the most significant natural influences on population distribution. Areas with adequate rainfall and moderate temperatures provide good agricultural possibilities and living conditions. Areas of climatic extremes or aridity are more difficult to settle and generally have low population densities. In Australia, the majority of the population is found within the temperate humid climates, while the arid and semi-arid areas have very low population densities. A comparison of the climatic regions discussed in chapter 1 with a map of population distribution will clearly illustrate this relationship. It can be seen that most of Australia's population can be found within the tropical and temperate maritime and Mediterranean climatic zones.

#### Landforms

In addition to climate, landforms also help to shape the patterns of population distribution. The concentration of population on coastal plains and along river valleys or on floodplains can be seen clearly in many parts of Australia. On the eastern seaboard the population concentrates on the narrow discontinuous coastal strip, with its westward expansion limited by the Eastern Highlands. The situation is similar on the west coast, where the population concentrates on the Swan coastal plain and decreases sharply once the Darling Escarpment is reached. In figure 6.4 the impact of landforms on the rural population is clearly evident. Intensive farming occupies the valley floor, with the higher and steeper areas being left under forest.

**Figure 6.4** Farming in the Mt Lofty Ranges, SA

#### Soils

While of lesser importance than the previous factors, soils have influenced the location of rural populations, especially in combination with climate and landforms. The alluvial soils of the Murray–Darling river system are important regions of intensive irrigated farming and this partially accounts for the higher population densities found within the inland areas of Victoria and New South Wales. Areas of sandy or infertile soils have generally supported only low rural population densities.

### *Human factors*

#### Historical factors

Prior to European settlement of Australia, the size of the Aboriginal and Torres Strait Islander population was estimated to range from around 300 000 to over one million. In the years following colonisation this population declined dramatically under the impact of new diseases, repressive and often brutal treatment, dispossession, and social and cultural disruption and disintegration. The available data suggest that the Aboriginal and Torres Strait Islander population had declined to 60 000 by the 1920s. There is some difficulty in determining the population size of the group, given that those people who lived outside the major cities and towns were not counted in the nation's census until 1971. In recent times, the change in the determination of who is Aboriginal or Torres Strait Islander has had a significant impact on the numbers being recorded in census data. Partly due to changes in definition, and partly to an increased willingness by people to identify themselves as Aboriginal or Torres Strait Islander, this group is now one of Australia's fastest growing, with an estimated total of some 310 000 in 2001.

European settlement commenced in 1778, when about 1000 people landed at Sydney Cove in New South Wales. Following the establishment of this coastal settlement the population gradually expanded into the hinterland—a pattern that was to be repeated at a number of locations around the Australian coast. Settlement frontiers based on the original colonial capitals of Sydney, Melbourne, Adelaide, Brisbane and Perth expanded outwards to provide a pattern of population that is still apparent. Thus, the capital cities today contain over 60 per cent of the total national population, with another 20 per cent found within close proximity to these major nodes.

Where variations to this general pattern occur they can be explained by the discovery of minerals such as gold in the 1850s and 1890s or the development of intensive farming around valuable water and soil resources, such as along the Murray River in New South Wales and Victoria.

Significant social and economic events such as World Wars I and II, changes to immigration policies, and periods of economic growth and decline can all be linked to the ways in which the national population expanded and settled the land. Changes in migration policies, as well as changes in the rate of natural population increase following World War II, brought significant changes to Australia's population densities and distribution and this will be expanded on further in the section in this chapter on population growth.

Centralisation in the large urban regions of Australia of the wealth gained from the nation's mineral and agricultural resources has been a consequence of political, social and economic forces. A study of road, rail, air and sea links in Australia clearly shows how they radiate from the state capitals into their hinterlands. The effect of this pattern of transport systems has been to focus the resources of these areas on the capitals. In turn, this has encouraged the concentration of commercial and industrial activity in these favoured locations. Over time the state capitals have grown both in size and influence, outstripping all other regional centres within their sphere of influence. The map in figure 6.5 illustrates the way in which Perth has influenced the movement of people and goods within the hinterland and beyond.

## Agricultural land uses

The density and distribution of rural populations is closely related to the type and intensity of agricultural land uses. A comparison of figure 3.5, which shows the variations of rural land use intensity, and figure 6.1, Australia's population density and distribution, illustrates this relationship. The regions with the smallest farms are also the areas with the highest rural population densities, while regions with very large rural properties have correspondingly low rural population densities.

### Densely populated rural areas

Where agricultural land is worked intensively, farm properties are small and farmsteads are closely spaced. This results in a close rural settlement pattern and a high rural population density. The south-eastern coastal areas are relatively well watered and contain many major markets. These areas are therefore suited to intensive forms of agriculture, which result in high population densities. High rural population densities give rise to closely spaced networks of towns and regional centres. A population of 1000 rural workers requires about the same number of people working in local towns and regional centres. Twice as many people are needed in major cities to provide manufactured goods and a wide range of specialised services. As a result, large urban centres with high population densities have been established throughout the closely settled agricultural areas of Australia. The largest are the regional service centres, many of which have been established on the coast to facilitate the export of agricultural produce. For example, the large coastal centre of Townsville was established on the coast of the sugar-producing area of Queensland.

The most densely populated rural areas in Australia (with averages of 5–10 people per square kilometre) result from the intensive agricultural forms of specialised crop farming, including intensive sugar farming, market gardening and irrigated horticulture. Farms in areas around the capital cities, in various coastal pockets and in irrigated lands along the Murray River average less than 50 hectares in size, and the towns that support them are relatively close together. The coastal areas of south-central Queensland, New South Wales, Victoria, South Australia's Gulflands, northern Tasmania and the South-West are relatively well watered and have moderate temperatures and relatively fertile soils (particularly those alluvial soils in coastal river valleys and flats). These were among the first regions settled by Europeans. Intensive cropping came to be the dominant form of agriculture close to the state capitals, where a frequent supply of relatively perishable products is required by the large population of the cities. The small areas of land needed to produce high yields means that intensive farms can 'outbid' other more extensive land uses for land adjacent to the cities. Figure 6.6 shows a typical landscape associated with high density rural populations. Farm sizes are small, with the farmhouses often within sight of each other.

The better watered coastal locations are also primarily used for intensive forms of pastoralism (such as cattle and lamb fattening, and dairying) as well as for

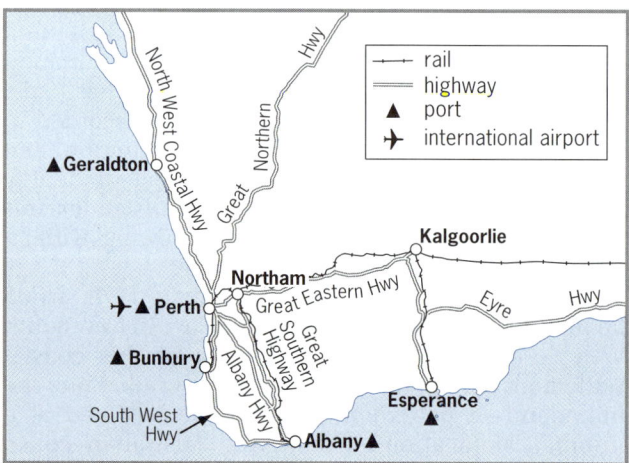

**Figure 6.5** Major transport systems in the South-West

**Figure 6.6** High density rural population near Mt Gambier

intensive forms of mixed farming. Population densities associated with this farming activity are lower than for intensive cropping because farm sizes of between 50 and 150 hectares are common. Population densities of 2–8 people per square kilometre are usual, and the towns that service the areas are more widely distributed than in the areas of intensive cropping activity.

Many of the factors that attracted intensive cropping to various coastal locations also attracted the more intensive forms of pastoralism. The relatively wet and more reliable rainfall conditions of coastal plain and valley areas were conducive to the growth of introduced pastures in areas of relatively fertile soil. Areas of forest that originally covered the coastal plain and inland ranges were cleared to make way for the introduced pastures that made intensive grazing possible. Also, as explained in chapter 3, these intensive forms of agriculture can afford near-city coastal locations, as only small parcels of land are required for grazing in areas where environmental conditions are satisfactory.

*Moderately populated rural areas*

The mixed wheat–sheep farming areas of the interior slopes and plains in central New South Wales, the Wimmera area of north-western Victoria and much of the South-West contain much larger farms (over 500 hectares) than those found in coastal areas. These areas are moderately populated, with population densities ranging from less than 1 person per square kilometre to 8 people per square kilometre. There are far fewer service towns than in the coastal region, and they are more scattered.

The development of the wheat–sheep belt and its settlement patterns occurred in response to the efforts of early explorers and settlers, the development of overseas markets and ideal physical conditions. Early European settlement in the south-east was restricted to the coastal plain and the valleys extending inland from it until 1813, when a passage was found through the agriculturally unproductive escarpments and ranges (the Blue Mountains) that mark the inland boundaries of the coastal plain. This discovery allowed the land beyond to be opened up for farming. The interior plains were initially given over to large pastoral (sheep for wool) farming enterprises run, at first illegally, by squatters. Later, government intervention saw the most easterly properties divided up to facilitate the establishment of wheat farming. This occurred in response to the development of overseas and local markets for wheat. The drier climates of these inland areas proved ideal for the establishment of diversified crop and livestock farming. Except for the irrigated lands of the Murray and Murrumbidgee rivers, most of this inland area is totally unsuitable for small-scale intensive farming.

*Sparsely populated rural areas*

In most of Western Australia and South Australia, the whole of the Northern Territory, Far North Queensland,

and western New South Wales, only populations of less than 0.3 people per square kilometre can be supported. Smaller areas in the highest parts of the Eastern Highlands and in the mountainous country of south-west Tasmania also have extremely low population densities. Some areas, such as the Simpson and Gibson deserts, are virtually devoid of any permanent settlement. Aridity contributes to these low population densities, particularly in the centre and north of the continent. As discussed in chapter 1, the amount and the reliability of rainfall decreases as you travel inland from the north-east, south-east and south-west coasts, so that most forms of intensive agriculture are not viable in central and northern Australia (except for intensive horticulture in irrigated areas such as Carnarvon and the Ord River in Western Australia). Surface runoff is negligible because of the lack of effective rainfall. Even in the extreme north of Australia, where heavy summer rainfall occurs, evaporation rates are also very high. Generally, the interior areas have limited surface water and are lacking in long perennial rivers. Due to these limiting factors, rural land use has been restricted to extensive pastoralism. Other physical factors, such as extreme temperatures, combine with the low and unreliable rainfall to create inhospitable environmental conditions. The poor grassland sustained by the arid climate can support only low stock numbers. Many areas have very poor, thin soils derived from the ancient rocks of the interior plateau. Limestone regions such as the Nullarbor Plain and the Barkly Tableland cannot retain surface water and consequently have only stunted vegetation. This also results in low carrying capacities. The combination of high temperatures and high rainfall in the north produces high levels of humidity, and therefore uncomfortable living conditions. Apart from extensive pastoralism, economic activity in these regions has been restricted to mining, exploration, and hunting and gathering by some Aboriginal groups. The result has been large, very sparsely populated regions with population densities of more than 65 square kilometres per person. The landscape is often viewed as wide open spaces devoid of people, similar to that illustrated by figure 6.7.

**Figure 6.7** Extensive grazing region in Central Australia

## Mining and settlement

The location of commercial mineral deposits has had a strong influence on the decentralisation of population in Australia. Most of the nation's large inland towns began as mining centres and where the mineral deposits have been extensive, the towns have continued to prosper and grow. A study of the map of Australia's population density will show the relationship between higher concentrations of population in the arid and semi-arid regions and mining settlements such as Kalgoorlie, Broken Hill and Mt Isa. It should be noted, however, that when mineral deposits are exhausted many mining settlements cannot be sustained and their populations decline rapidly. In addition, there is an increasing emphasis on fly-in, fly-out mining operations. These populations are not recognised as forming permanent settlements, and so do not influence the nation's distribution patterns in the same way as towns with resident populations.

## Economic development

In addition to rural and mining activities, the many other commercial enterprises that contribute to the national economy also influence population density and distribution. The tourism industry in the northern and central regions of Australia has brought permanent as well as temporary populations into areas of low population density. Alice Springs and the surrounding areas receive some 500 000 domestic and international visitors per year and this has contributed to the dramatic growth of the town and the region. A number of tourist-based services have been developed and staffed to cater for this industry. These include tour group services, souvenir shops, motels, hotels and backpacker accommodation, theme parks and restaurants. Tourism has also had a significant effect on population along the eastern seaboard, with centres such as the Gold and Sunshine coasts and Cairns growing as a result of this industry.

Industrial and commercial centralisation in the larger towns and cities has produced centres with the highest population densities and has contributed to their continual growth (often at the expense of smaller settlements as the population drifts towards the larger centres). Most of the nation's commercial and industrial economic activity has gradually been concentrated in the Sydney and Melbourne metropolitan areas and this in turn has encouraged further population growth.

### EXERCISES

1. Study the map of Australia's population density and distribution (figure 6.1) and then answer the following questions.

    (a) Compare and contrast the density and distribution of Australia's urban population on the east and west coasts. Account for the differences observed.

    (b) Identify the main characteristics of Australia's population density and distribution within the central study area (arid and semi-arid regions) and then account for the pattern.

(c) On a map of Australia, show the areas of high, medium and low rural population density (do not include urban centres). Compare this with the land use map found in chapter 4 and identify the associated land uses. Explain the variations in rural population density.

2 Design a model (diagram) that illustrates the factors affecting population distribution and density. This should include the identification of factors that attract or repel settlement (i.e. **push–pull factors**) as well as distinguishing between biophysical and cultural factors. The diagram could be two concentric circles with the low density forces on the outside and the high density ones on the inside.

3 Identify the three or four most significant characteristics of Australia's population distribution and then justify your selection.

## Changes in Australia's population

The rate at which population grows is determined by the rate of natural increase combined with the level of migration. This can be expressed as the formula:

**Population growth rate =**

$$\frac{(\text{Immigration} - \text{Emigration}) + (\text{Births} - \text{Deaths})}{\text{Total population at commencement}} \times \frac{100}{1}$$

In 1999, the natural increase (births minus deaths) in Australia's population was some 120 000. In the same period, net immigration was also around 120 000. When compared to the total population in 1998 of 18.7 million, this represented a growth of 1.2 per cent. The data in table 6.1 indicate how this compares with selected locations and with the world in general. Compared to other developed nations, Australia has a significantly higher growth rate. This growth rate relies strongly on the level of net migration. If this is removed from the calculation then the pattern is more in keeping with other developed nations. The natural increase in Australia's population has been in general decline over the last 50 years, due largely to falling birth rates. In 1954, the average number of children per woman in the child-bearing age range was 3.8. In 1998, this had fallen to 1.7. This figure is below population replacement and would result in the Australian population declining if there were no immigration.

**Table 6.1** Average annual growth rates for selected regions

| Region | Period | Rate |
|---|---|---|
| World | 1990–2000 | 1.4 |
| Less developed countries | 1990–2000 | 1.7 |
| More developed countries | 1990–2000 | 0.3 |
| North America | 1990–2000 | 1 |
| Europe | 1990–2000 | 0.1 |
| Indonesia | 1998–99 | 1.4 |
| Australia | 1998–99 | 1.2 |

The changes that occur in levels of births and deaths progress through stages as a country moves from low to high levels of development and living standards. This progression is termed the **demographic transition** and is illustrated in figure 6.8.

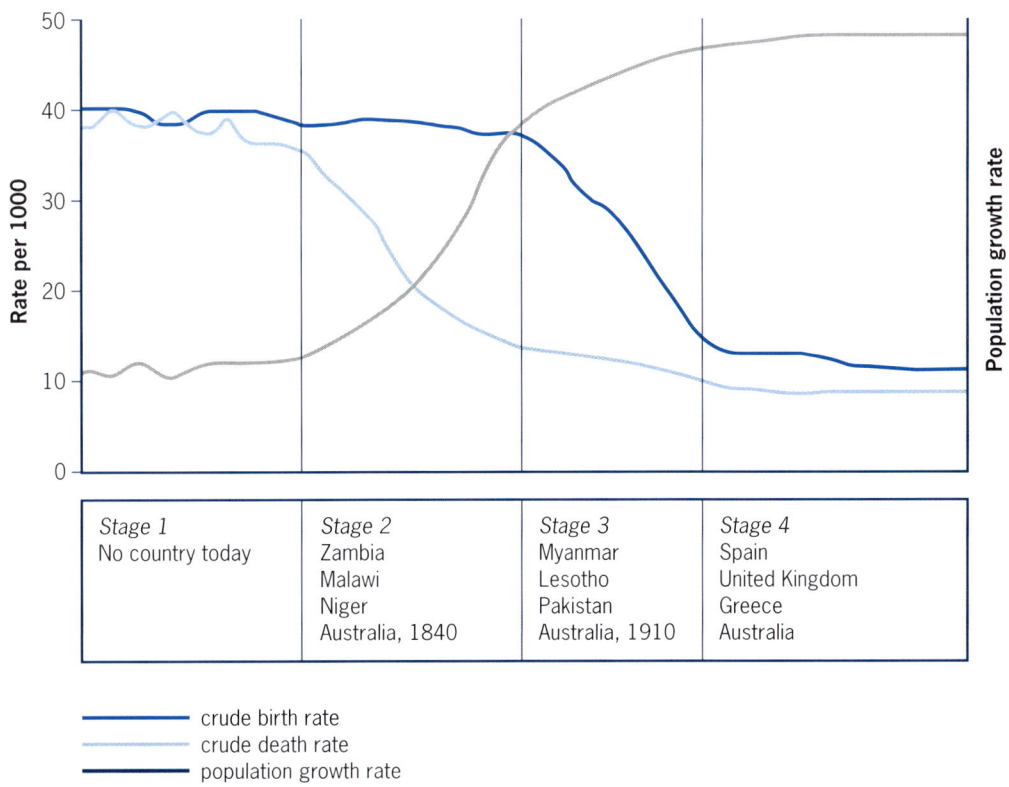

**Figure 6.8** Demographic transition

## The demographic transition

*Stage 1* of the transition is a period of high birth and death rates. Families are large, however low life expectancies and high infant mortality rates result in little addition to the total population. Periods of famine or disease often push death rates above birth rates, with subsequent decreases in total population. This stage reflects the situation throughout most of human history, with world population growing very slowly.

*Stage 2* commences when a sustained decline in death rates occurs. This period is initiated by improvements in food supplies or in general community hygiene and health standards. During this time birth rates remain high, accompanied by a decrease in infant mortality rates. Most European countries entered this phase in the eighteenth and nineteenth centuries. It appears to be unlikely that Australia progressed through stage 1, given its settlement in the eighteenth century. Rather, the population transplanted from Europe was already at stage 2. It is a period of increasing population growth rates.

*Stage 3* begins with a sustained decline in birth rates. Factors such as low infant mortality rates reduce the need to have larger families. Economic development and shifts from rural- to urban-based occupations, along with technological improvements, reduce the need for large families to meet the rural labour needs. Continued improvements in community health and general medical advances maintain low death rates. Countries in this stage enter a period of declining population growth rates. In 1901, Australia infant mortality rates were around 103 per 1000 and the **total fertility rate** stood at 4. By 1999, the fertility rate had declined to 1.7, while the infant mortality rate had dramatically decreased to around 5.3 per 1000.

*Stage 4* is a period of low birth and death rates. Population growth rates are low, with some countries approaching zero population growth (ZPG) or even declining in population. In Australia, the period from 1901 to 1999 saw the nation's demographic transition from stage 3 to stage 4. This is shown in figure 6.9. The trend lines for crude birth rate and crude death rate indicate the narrowing of the gap between the two measurements. This corresponds to a slowing of the rate of natural increase in the total population.

## Age–sex distribution

At the beginning of the twentieth century, more than one-third of the Australian population was less than 15 years of age. At the end of the century, this had declined to about 20 per cent, or one-fifth of the population. In contrast, the proportion over 65 years of age had increased from 4 per cent to over 12 per cent and the median age had moved from 22.6 to 34.9 years. The four 'population pyramids' in figures 6.10 to 6.13 illustrate these trends.

This ageing of the population was not continuous, with the post-war baby boom (seen in the 1954 population pyramid) producing a temporary rejuvenation of the population. However, since that time there has been a return to the situation of declining birth rates and increasing proportions of the population within the older age groups. The baby boom has, in fact, made this ageing pattern more obvious as the 'baby boomers' move towards retirement. This is clearly demonstrated in the 1999 population pyramid.

Changes in the shape of the population pyramids can be linked to the demographic transition, with the fall in the proportion of younger people, the increase in **life expectancy** and the growth in the older age groups

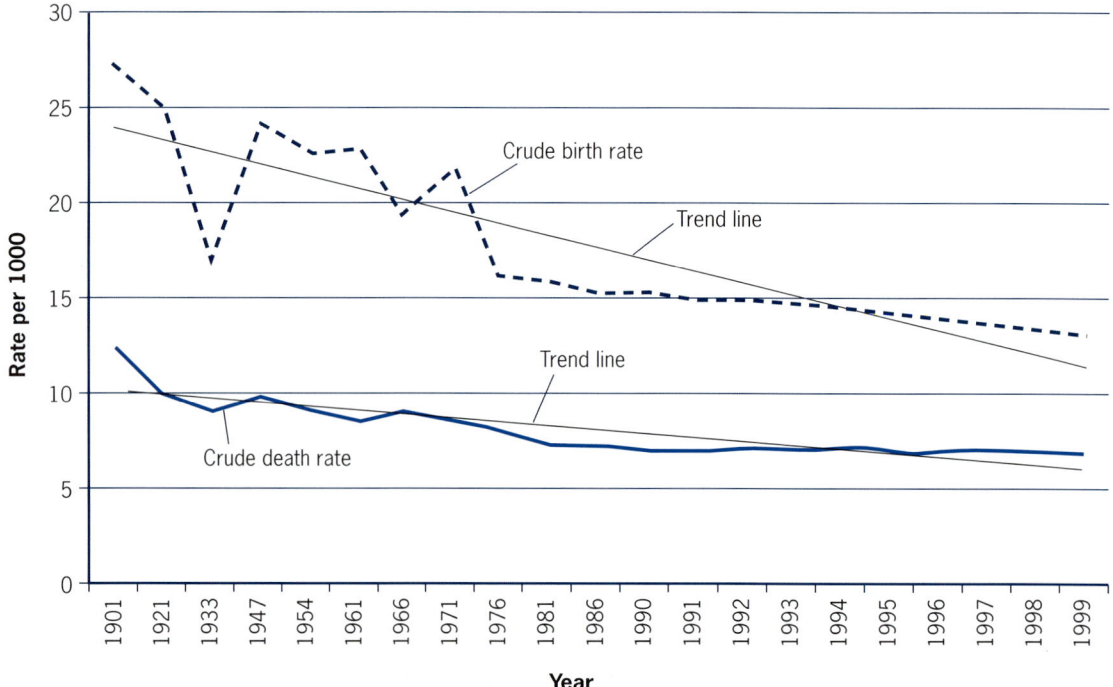

**Figure 6.9** Australia's demographic transition

corresponding with the third and fourth stages of the transition. (The first and second stages would be illustrated by a pyramid with a large base and a rapid tapering off within the older age groups.)

The age and sex distribution of a population will have significant effects on the economic and social characteristics of a nation. The demands for age-specific activities, educational services, housing and health-related services are just some of the areas that can be closely linked to the distribution pattern. Age and sex distributions will also influence the size and composition of the workforce, which mainly corresponds to the 15–65 age group.

An additional feature of the changes in the twentieth century is the proportion of males to females. The gradual shift towards greater numbers of females in nearly all age groups and, more significantly, in the older age groups is clearly evident in the population pyramids seen in figures 6.10 to 6.13.

For the first two centuries of European settlement in Australia, males outnumbered females, with males making up the larger proportion of immigrants in that period. By 1901, there were 110 males for every 100 females in Australia. This gap progressively closed as life expectancy for females continually outstripped males until the 1970s, when both sexes were approximately equal in number. In addition, a changing pattern of immigration in the last 30 years has seen female settlers consistently outnumber male settlers. By the 1980s, females outnumbered males and by 1999 there were 99 males for every 100 females, with this proportion being significantly greater within the older age groups due to the difference in life expectancy. In 1999, Australian males could expect to live to 74 and females to 80 years.

## Population migrations

### International migrations

Over the last 100 years the impact of immigration on the size and composition of Australia's population has been significant. During this period, World War II produced the biggest change in the nation's policies and attitudes towards migration. One of the first acts of the Commonwealth Government after federation was the introduction of the White Australia Policy. This was to ensure that the nation's immigrants were of European origin and, in particular, **Anglo-Celtic** origin. Following World War II, however, it became increasingly more difficult to attract migrants from the United Kingdom and the existence of several million displaced persons in Eastern Europe encouraged a change in immigration policies. Australia accepted around 300 000 displaced persons and, despite concerns that such large numbers of people from non-Anglo-Celtic backgrounds would not be able to adjust, they did so with great success. This then led to settlers being recruited from elsewhere in Western and Northern Europe. In the 1950s, migrants were attracted from Southern Europe and, following that, in the 1960s from parts of Eastern Europe and the Middle East. Over these decades the White Australia Policy was gradually dismantled until, by the mid-1970s, it was totally abolished. From this point on Australia's immigration policies have been based on factors other than racial or ethnic background.

The effect of migration on the growth in Australia's population following World War II was dramatic. From 1947 to 1999 the nation's population grew by some 11 million people, with migration estimated to be directly responsible for approximately 7 million of this increase.

### Interstate migrations

A major factor changing the distribution of Australia's population is internal migration. During 1999–2000, 367 390 people moved from one state or territory to another. This movement was similar to that experienced in 1998–99. In 1999–2000, only Victoria and Queensland recorded net interstate migration gains. Tasmania's population declined by about 430 people, as the natural increase in the state was offset by a

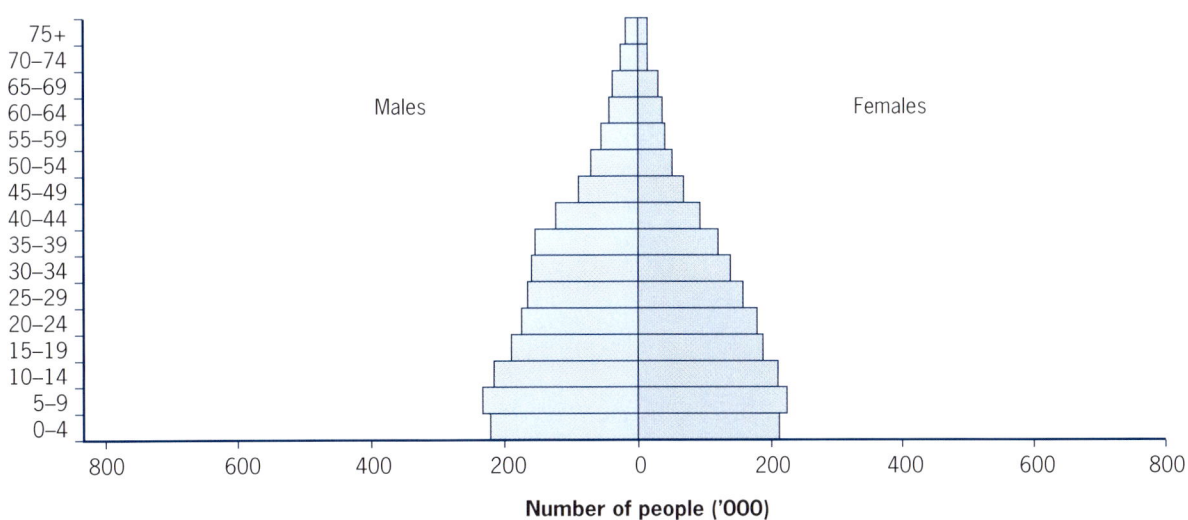

**Figure 6.10** Australia's population pyramid, 1901

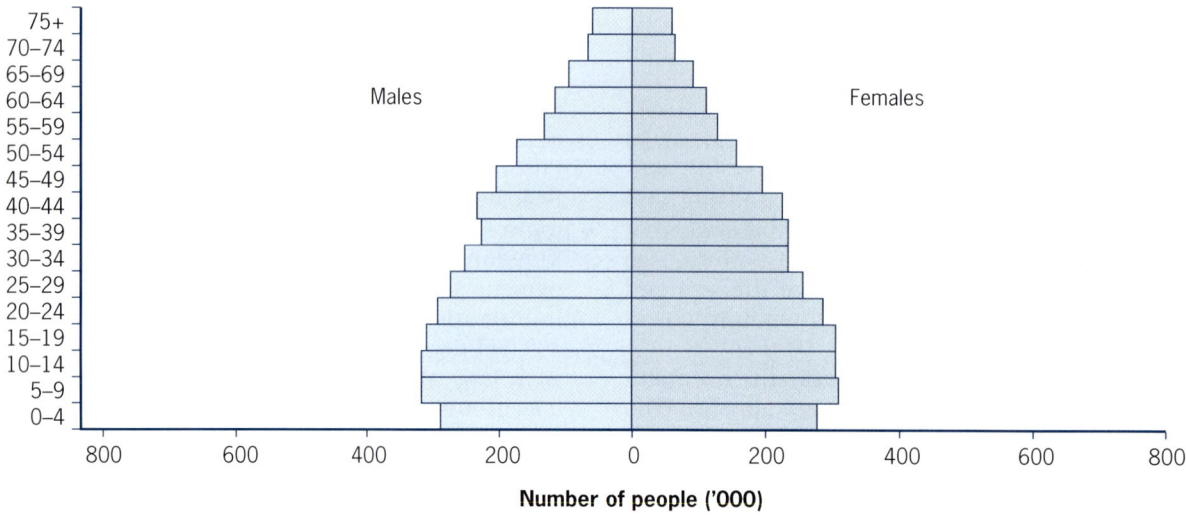

**Figure 6.11** Australia's population pyramid, 1933

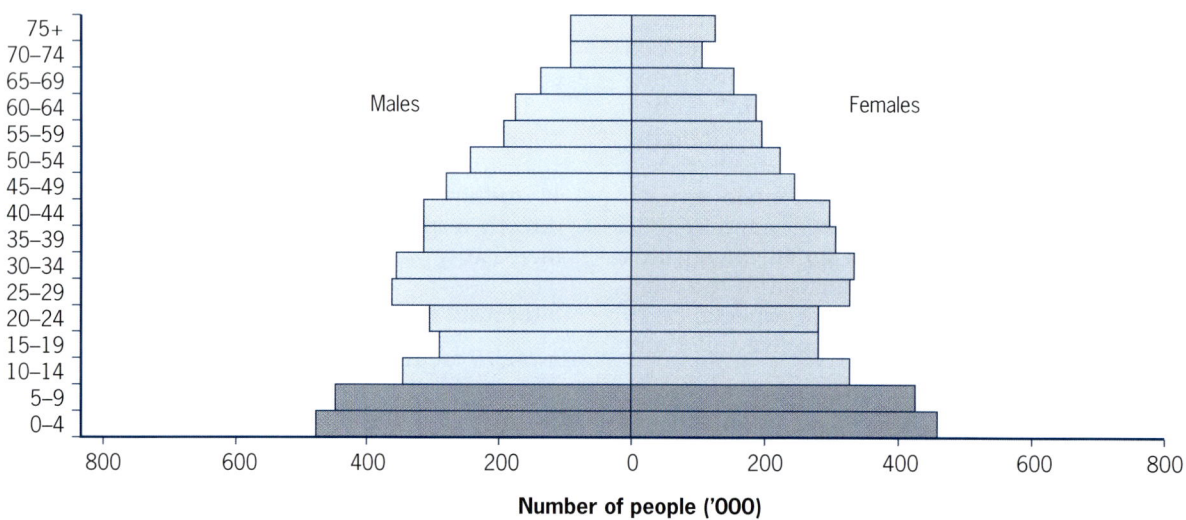

**Figure 6.12** Australia's population pyramid, 1954

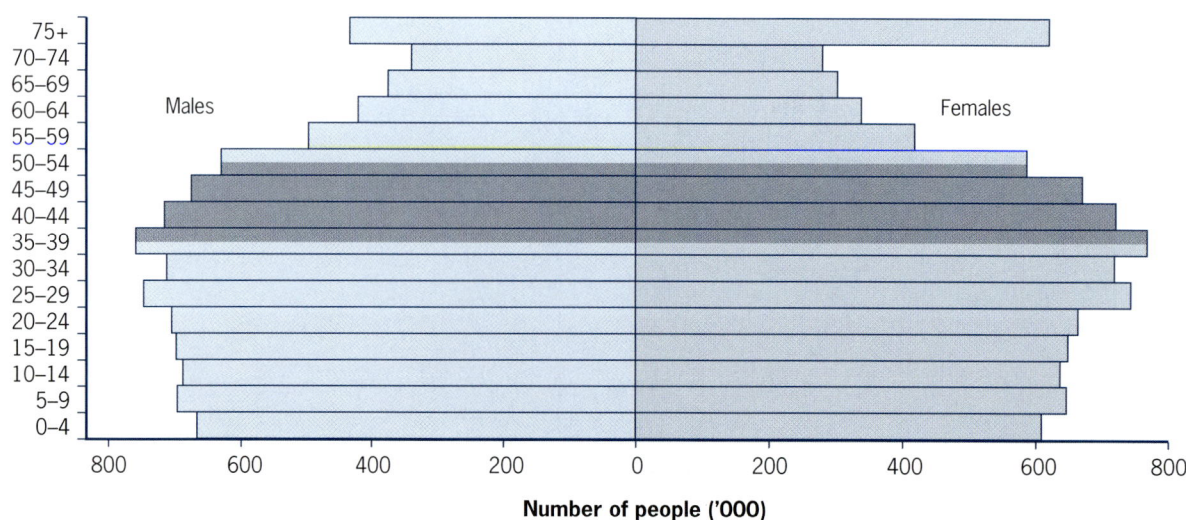

**Figure 6.13** Australia's population pyramid, 1999

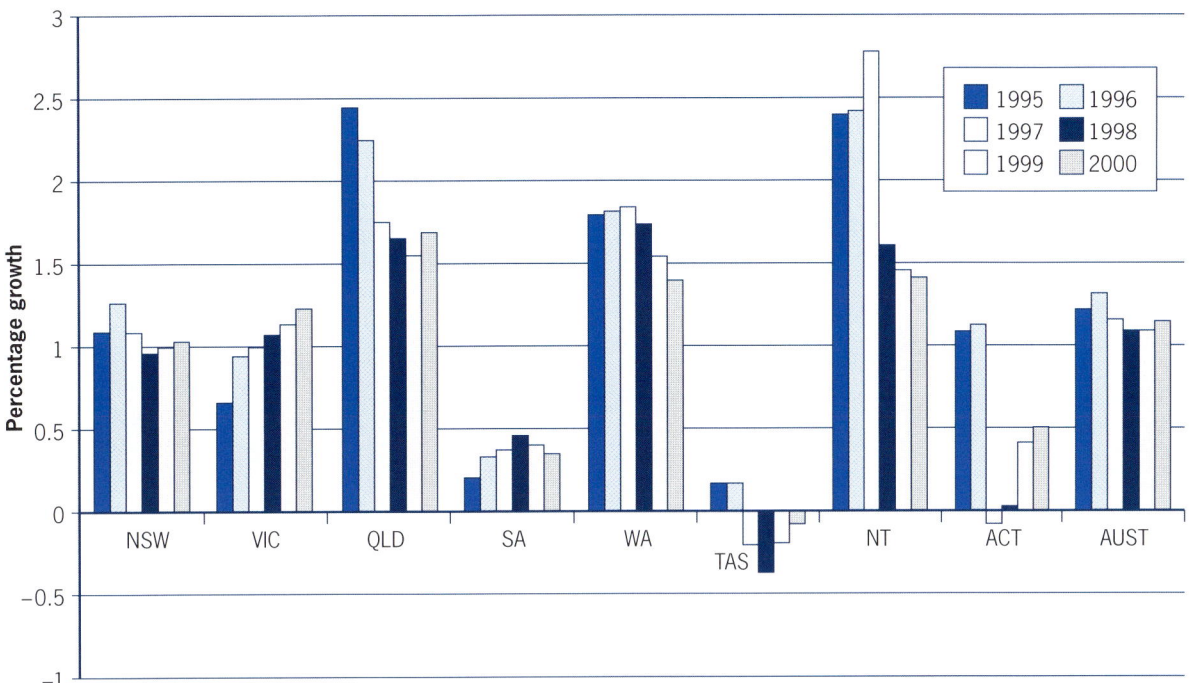

**Figure 6.14** Population growth rates, 1995–2000

continued exit of people from the state. The growth in Queensland's population due to interstate migration has been a feature of that state's population growth for over 30 years. Retirees from the southern states of New South Wales and Victoria have moved northwards into the warmer climates along the Sunshine and Gold coasts. Figure 6.14 illustrates the relative growth rates of the states and territories as well as the nation. This graph includes all aspects of population change, not just interstate migration. However, as the growth rates are shown as percentages, it must be realised that where the starting population is low then a small number of new arrivals can produce a large percentage increase. This is the case in the Northern Territory and the Australian Capital Territory.

As a consequence of variations in population growth, over time the distribution patterns of Australia's population will undergo change. Queensland is projected to replace Victoria as the second most populous state between 2026 and 2038, while the population of the Australian Capital Territory could overtake that of Tasmania between 2041 and 2047. The Northern Territory could overtake the populations of both Tasmania and the Australian Capital Territory between 2044 and 2048. Tasmania is predicted to be the only state that will experience a sustained decline in population over the next 50 years.

## Urban growth and urbanisation

Urbanisation and urban growth are processes that account for the way in which the population of a region grows in urban centres compared to the rural areas. The term 'urbanisation' is used in two distinct ways. The first is as a statement of the proportion of urban dwellers compared to rural dwellers. In this context a region or nation may be described as highly, moderately or lowly urbanised. Thus, in 1999 in Australia, the degree of urbanisation of the population stood at 85 per cent, which represents a highly urbanised society. The second way in which the term may be applied is in describing the changing proportions of urban and rural dwellers within a region over a given time period.

In 1911, 43 per cent of Australians lived in rural areas. This proportion steadily declined and, by 1976, only 14 per cent of the population lived in rural areas. During the period from 1976 to 1991 the urbanisation of the population halted. However, by 1999, the process had once again shifted towards increasing proportions of urban dwellers (see figure 6.15).

Some of the factors that account for this high degree of urbanisation include:

- the inability of large areas of Australia to support close rural settlement;
- the early emphasis on commercial agriculture such as wool production, which was both extensive and had limited permanent labour needs;
- the absence of a long-established subsistence-type rural society;
- the dominance of the colonial and later the state capital cities;
- the long periods of high economic growth that favoured the coastal towns and cities;
- the nature of migration, whereby urban dwellers from other countries settled in the larger urban centres in Australia rather than in the rural areas;
- the declining need for labour in rural areas associated with increased mechanisation of farming processes and the amalgamation of rural land holdings into larger units;
- the job opportunities created by the growth of manufacturing and service sectors of the nation's economy within the larger urban centres.

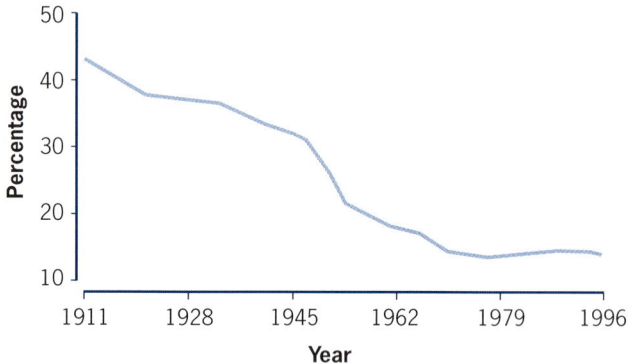

**Figure 6.15** Rural population as a percentage of total population, Australia

A concept related to urbanisation is that of urban growth. This refers to the increase in the population of an urban centre. It may or may not occur along with urbanisation, as indicated by figure 6.16. In Country 1, the nation's population has grown over the time period from 2.5 million to 20 million. This growth has occurred evenly in both urban and rural areas. Thus, the proportion of urban dwellers has not altered even though the number of people in the cities has increased from 625 000 to 5 million. It has remained constant at 25 per cent of the total. Country 1 has experienced urban growth, but not urbanisation.

In Country 2, there has been urbanisation as well as urban growth. The percentage of urban dwellers has increased from 25 per cent in 1900 to 70 per cent in 1990, thus indicating urbanisation has occurred. In the same period, the urban population grew from 625 000 to 14 million, so the country also experienced urban growth.

About 70 per cent of Australia's population growth between 1995 and 2000 occurred in the capital cities, the most significant increases being on the outskirts of these metropolitan regions. Of all the capital cities, Sydney and Melbourne had the largest growth in the five years to 2000, with increases of 264 000 and 222 000 people respectively. The fastest growth in capital city population over the 1995–2000 period occurred in Darwin, which experienced an average of 2.3 per cent growth per year. Brisbane and Perth had the next fastest growth rates, with average annual growth rates of 1.8 per cent and 1.7 per cent respectively. Other major population centres experiencing significant population increases between 1995 and 2000 were the Gold Coast–Tweed and the Sunshine Coast, which grew by 3.5 per cent and 3.4 per cent per year, respectively. Cairns and Kalgoorlie–Boulder were slightly slower at 2.4 per cent and 2.2 per cent per year, respectively. When compared to an average annual population growth rate for the nation of 1.2 per cent during this time period, it can be seen that the larger cities are growing at a faster rate than the average. This indicates that the majority of the natural population increase, as well as the net migration, is being concentrated within the urban rather than the rural areas, resulting in an ongoing process of urbanisation.

Urbanisation is not necessarily constant across the full spectrum of urban centres. It can vary markedly from state to state and from one local area to the next. Smaller urban centres within rural areas often undergo a decline as population relocates to larger urban centres. This is accompanied by a corresponding loss of rural population within the surrounding hinterlands. The rural urban drift and small town decline is associated with factors such as the rationalisation of government services, depressed rural economies, technological changes, declining rural employment and the strong attractions of the larger urban centres.

## Population projections

By June 2001, Australia's population reached 19 387 000 —five times the population in 1901 (3.8 million). From European settlement in 1788 it took 70 years for the population of Australia to reach 1 million and 19 years

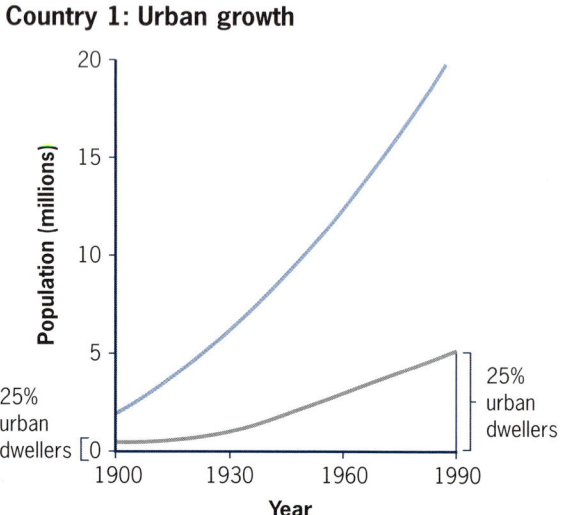
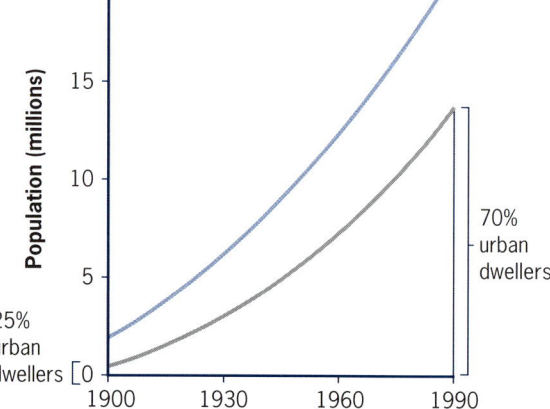

**Figure 6.16** The processes of urban growth and urbanisation

to achieve the second million. Subsequent millions were added in progressively shorter time intervals, with the 6 million mark reached in 1925. The growth in population slowed during the Great Depression and World War II due to declining fertility and lower immigration. This increased the time that it took for the nation to reach its seventh and eighth million population mark. Following the war, the baby boom and post-war migration programs increased population growth and, since 1949, when the population reached 8 million, each successive million has been added in around four to five years. The seventeenth million, reached in 1990, was the fastest million added, achieved in just three years and nine months.

In mid-2004 the population is projected to reach the 20 million mark—four years and nine months after the 19 million mark in 1999. The time interval between each subsequent million is projected to increase in the twenty-first century as population growth slows due to the ageing population and lower fertility rates.

The ageing of Australia's population is one of the most significant factors in making predictions about future population characteristics. It becomes even more important when linked to the declining levels of childbirth or natural increase. The combination of these two factors will see the median age increase from 35 years in 1999 to around 45 years by 2051. The population pyramid shown in figure 6.17 illustrates the differences between the population of 1999 and that expected for 2051. Despite the larger population, fewer children will be born in 2051 compared to 1999 and the numbers of people within the older age groups will be much larger. By 2051, the population aged 65 years and over is projected to be at least double its present size, increasing from 12 per cent of the population in 1999 to 25 per cent in 2051.

### EXERCISES

1 Using the formula for population growth rates, calculate the rate of increase or decrease for the following three examples. Compare and contrast the differences illustrated by your calculations.

| Region | Immigration | Emigration | Births | Deaths | Commencement population |
|---|---|---|---|---|---|
| Country A | 1 100 000 | 35 000 | 375 000 | 120 000 | 15 000 000 |
| Country B | 25 000 | 40 000 | 39 000 | 33 000 | 21 000 000 |
| Country C | 250 000 | 8 000 | 150 000 | 85 000 | 110 000 000 |

2 Describe the changes in Australia's population growth rates over the last 50 or so years and, with reference to migration rates and rates of natural increase, account for this change.

3 Draw the diagram for the typical demographic transition and for each of the four stages summarise the main characteristics.

4 Study Australia's population pyramids for 1901, 1933, 1954 and 1999 (provided in this chapter) and then match them to the different stages in the typical demographic transition. Discuss the demographic characteristics that would be expected for these particular years and explain why you matched the pyramids to the particular stages that you chose.

5 Study the population pyramids shown for 1901, 1933, 1954 and 1999. Describe the changes that have taken place in the age and sex distribution over this time period. Consider the effects of changes in birth and death rates as well as migration.

6 Study the population pyramid showing Australia's population in 1999 and 2051. Identify the patterns or trends revealed and suggest reasons for these.

7 In predicting future population growth, what factors might increase or decrease growth rates, change the patterns of age distribution and the proportions of males to females?

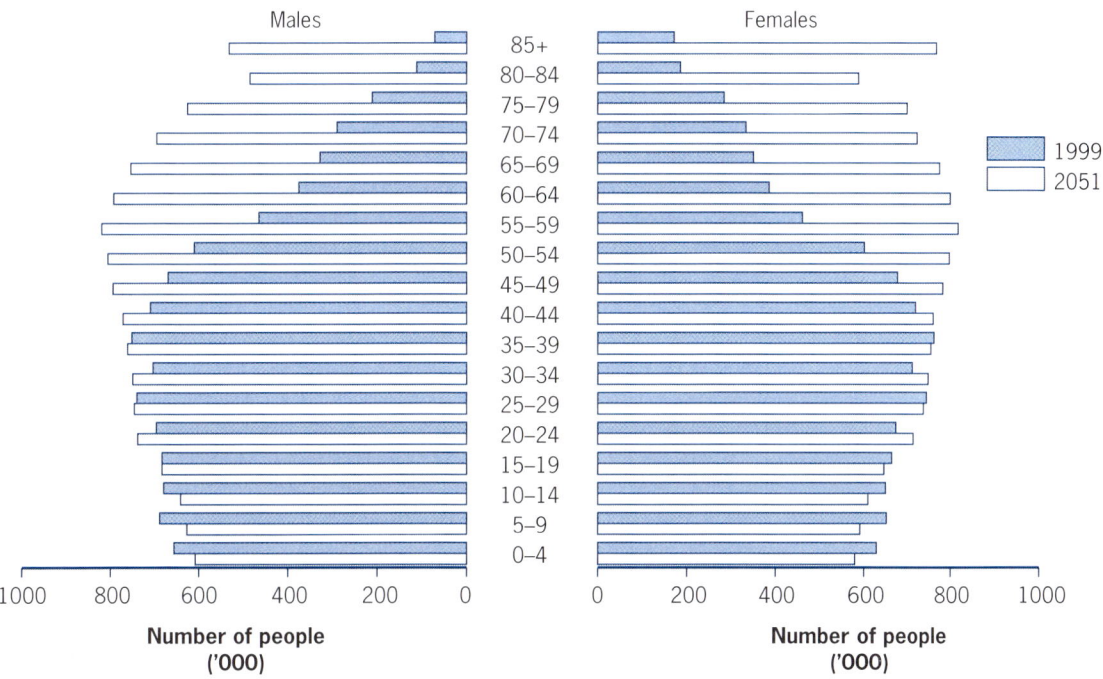

**Figure 6.17** Australia's projected population age structures

# Population patterns in the South-West

In 2000, the south-west of Western Australia contained 1.693 million people, representing 89 per cent of the total state population and had an average population density of 3.62 people per square kilometre. The average density for the whole state was approximately 1.36 people per square kilometre.

The growth and distribution of the population in the South-West is illustrated by figure 6.18. Medium to high population densities occur along the west coast, from Perth in the north to Vasse in the south. This has also been the region of greatest growth. Significantly, the population increase in the Vasse region indicates possible longer term shifts in the South-West's distribution pattern. This area contains settlements such as Busselton, Margaret River and Augusta. Closer to Perth, the coastal regions of Moore and Dale are significant in that their growth rate has exceeded that of Perth from 1996 to 2000. This is partly accounted for by the expansion of Perth's population beyond its existing statistical boundaries into these relatively unpopulated coastal areas. In addition, the region of Dale contains Mandurah, one of Australia's fastest growing urban areas.

Traditionally, the Guildford and Pinjarra loams found at the foot of the Darling Scarp have encouraged intensive forms of agriculture and higher rural populations along with associated urban settlements in the Dale and Preston regions. Towns such as Bunbury, Pinjarra, Harvey and Waroona continue to show steady growth in this area. In the period 1996 to 2000, Bunbury grew by 8 per cent with the Harvey shire exceeding 27 per cent and the Murray shire (Pinjarra) growing at 12 per cent. Industrial and commercial development, as well as the growth in hobby farms in these areas, accounts for this growth.

Inland from the coast the growth has been less spectacular, with some wheat belt regions still experiencing small declines in numbers. Factors such as ongoing farm amalgamation, declining demands for rural labour and uncertain economic conditions help to account for this population loss. Significant declines in population in the regions on the north-eastern fringes of the South-West are related to changing fortunes in the mining industry. These areas have very low population densities and any small movement in the numbers will translate into large percentage changes.

Along the south coast, the growth in the statistical divisions of King and Johnston relate mainly to the growth of the towns of Albany and Esperance. These coastal settlements are attracting small numbers of retirees from the surrounding regions, with Esperance

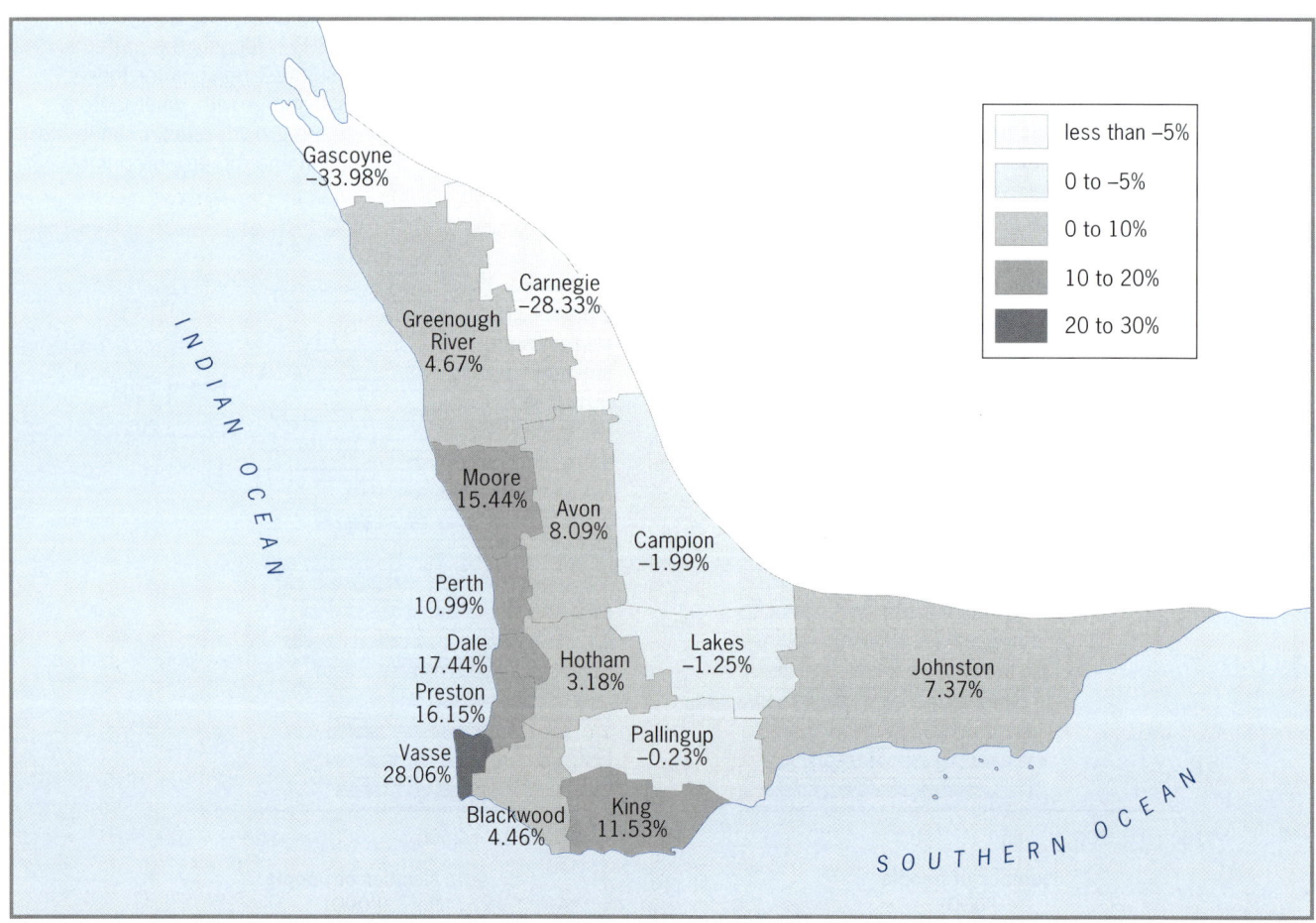

**Figure 6.18** Percentage change in population in the South-West, 1996–2000

long seen as an attractive retirement centre for people from Kalgoorlie and surrounding mining areas.

One feature that areas outside the Perth region have in common is the sex ratio of males to females. While Perth has slightly more females than males, this is not the case in the rural areas of the South-West. Virtually all of these areas show a dominance of males. The shire of Narrogin within the Hotham district has 137 males for every 100 females, while the shire of Albany in the King district has 103. Within the urban centres of the South-West the ratios are more mixed. The town of Albany has 89 males per 100 females; the town of Bunbury has 100 males per 100 females; Collie has 108 and Merredin 113.

The decline in the female population in the rural areas can be attributed to several factors:

- the greater tendency for young women to move to Perth for educational and employment opportunities
- the tradition of male inheritance of rural landholdings, encouraging the retention of males and the movement of older females to coastal towns or Perth in retirement.

### EXERCISES

1. Construct a column graph to show the variations in population growth for the period 1996 to 2000 for the statistical areas in figure 6.18.
2. Suggest reasons for the higher population growth rates in the statistical regions along the west coast of the South-West.
3. Using the figures provided in the following table, construct a graph for the ratio of males to females and then describe the pattern.

**Table 6.2** Town population, 1996

| Region | Males | Females | Ratio |
|---|---|---|---|
| Perth | 687 313 | 693 814 | 99 |
| Dale | 30 511 | 30 462 | 100.1 |
| Preston | 38 844 | 37 844 | 102 |
| Vasse | 5 191 | 4 762 | 109 |
| Blackwood | 8 940 | 8 557 | 104 |
| Pallingup | 6 231 | 5 737 | 108.6 |
| King | 26 267 | 25 861 | 101 |
| Hotham | 7 537 | 6 963 | 108 |
| Lakes | 2 749 | 2 361 | 116 |
| Moore | 7 084 | 6 296 | 112.6 |
| Avon | 14 327 | 13 345 | 107 |
| Campion | 6 411 | 5 523 | 116 |
| Gascoyne | 5 036 | 4 719 | 106 |
| Carnegie | 3 334 | 1 801 | 185 |
| Johnston | 8 461 | 7 762 | 109 |

4. Trace a map of the statistical regions of the South-West and, using the ratio of males to females in the table provided in question 3, show the pattern of distribution for this data. You will need to make up an appropriate scale, using around four to five categories. Suggest possible reasons for the variations illustrated by the map.

## Population characteristics of Perth

### Development and distribution

Within the Perth metropolitan area in 2000, population densities varied from 11 people per square kilometre for the rural urban local government area (LGA) of Serpentine–Jarrahdale through to 2480 people per square kilometre for the inner city LGA of Vincent. Factors such as the percentage of undeveloped land, industrial land use and parkland or conservation reserves, as well as the density of residential development, all account for the differences in the population densities (see figure 6.19).

This map shows a gradual decrease in population density as you move away from the areas along the Swan and Canning rivers. This decline is most noticeable in the inland regions, as the trend is for people to concentrate in near-river and coastal locations. Other site factors such as the coastal plain and the Darling Escarpment have also played a part in the population distribution of the city. The sandy coastal soils, while of limited agricultural value, have proved to be cost-effective sites for constructing residential and industrial properties. They allow for generally low site costs associated with the laying of concrete slabs and footings. In contrast, the higher and rugged scarp and hills to the east of the city have higher construction costs due to the constraints placed on development by this environment. In addition the location of major water catchments, forest reserves and national parks within this region has meant that the land was not suitable or available for urban development.

The history of the initial settlement and subsequent growth of Perth has also helped to shape the present pattern of population distribution. Early development occurred along the Swan River, with small villages located at Fremantle, Perth and Guildford. Other rural villages also developed at Midland and Armadale as the colony progressed. However, most of the growth in the first 75 years was around the original nodes and along the river. Events such as the discovery of gold at Kalgoorlie in the 1890s and the expansion of the agricultural regions in this period boosted the economy, supported the growth of the city and paid for the development of transport, water and power infrastructure. From the 1900s up to the 1950s the city grew outwards along suburban rail links, giving it a distinctive radial pattern of settlement. The city contained approximately 450 000 people by the end of the 1950s, with most of the population concentrated along the corridors westwards from Perth City to Fremantle and eastwards to Midland. Expansion southward along the Armadale railway link was also occurring.

The formalisation of the Stevenson–Hepburn metropolitan planning scheme into the Corridor Plan in the 1960s focused the metropolitan population into a series of corridors based on transport arteries extending outwards from the original site of the settlement of Perth. In 2000, the population reached 1 381 000 residents with the majority of these found along the coastline from Joondalup in the north through to Rockingham in the south and extending inland to the

136  LANDSCAPES AND LAND USES

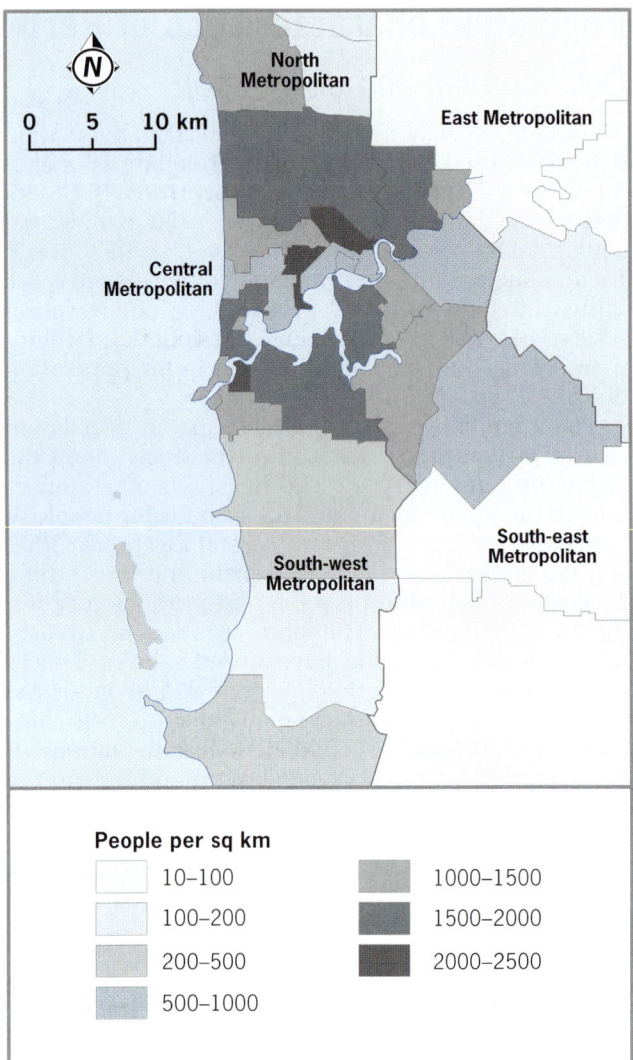

**Figure 6.19** Population densities, Perth metropolitan area, 2000

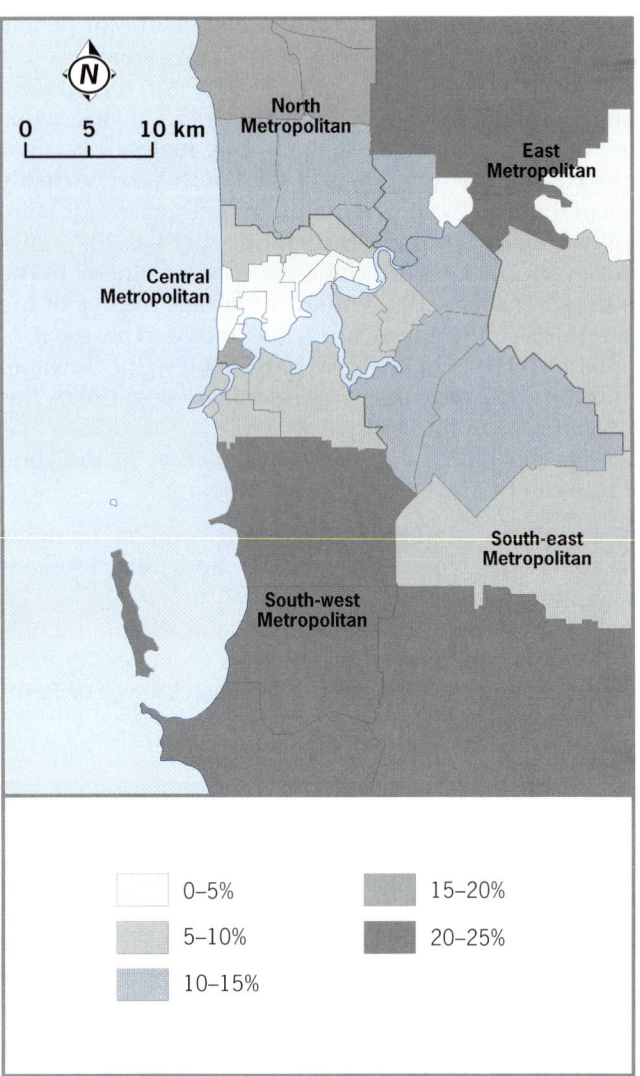

**Figure 6.20** Perth metropolitan growth rates, 1996–2001

base of the Darling Scarp. The pattern of Perth's population growth is illustrated in figure 6.20 and indicates the generally higher rates of population increase within the fringes of the city.

It is expected that the future growth of the Perth region will be focused on the coastal strip, with the city extending both north and south in ever-increasing distances. The proposed development of the southern rail link to Mandurah, some 85 kilometres south of the **central business district** of Perth, will encourage further growth within this satellite city. Future growth to the north of Joondalup will also be significant in influencing the population distribution of the metropolitan region. One trend that is reducing the pressures for outward growth of the urban population is the increasing trend towards higher density residential development and redevelopment and decreasing family size. In 2002 in Perth, approximately 50 per cent of all housing construction occurred as redevelopment within the established areas. This is perhaps best illustrated by the shift towards apartment living that is taking place within the inner city precincts (see figure 6.21). In this regard, Perth is following a pattern that is being repeated in all of Australia's capital cities. Associated with this re-use of the existing urban land stock has been an increasing desire to protect remnant ecosystems in urban and near-urban locations from **urban sprawl**. This has made it more difficult for property developers or public authorities to initiate new subdivisions on what are termed 'green fields' sites.

## Age structures

The variations in age groups within the Perth metropolitan area is a reflection of the way in which the city has developed, the changes in birth and death rates and the patterns of international and internal migration. Perth has an ageing population, which is in line with the national trend. It has been ageing steadily over the last 20 years, with the median age rising from 28.2 years in 1980 to 34.2 in 2000.

### The 0–14 age group

In 2000, this age group represented 20 per cent of Perth's population. However, its distribution

**Figure 6.21** High density residential redevelopment in East Perth

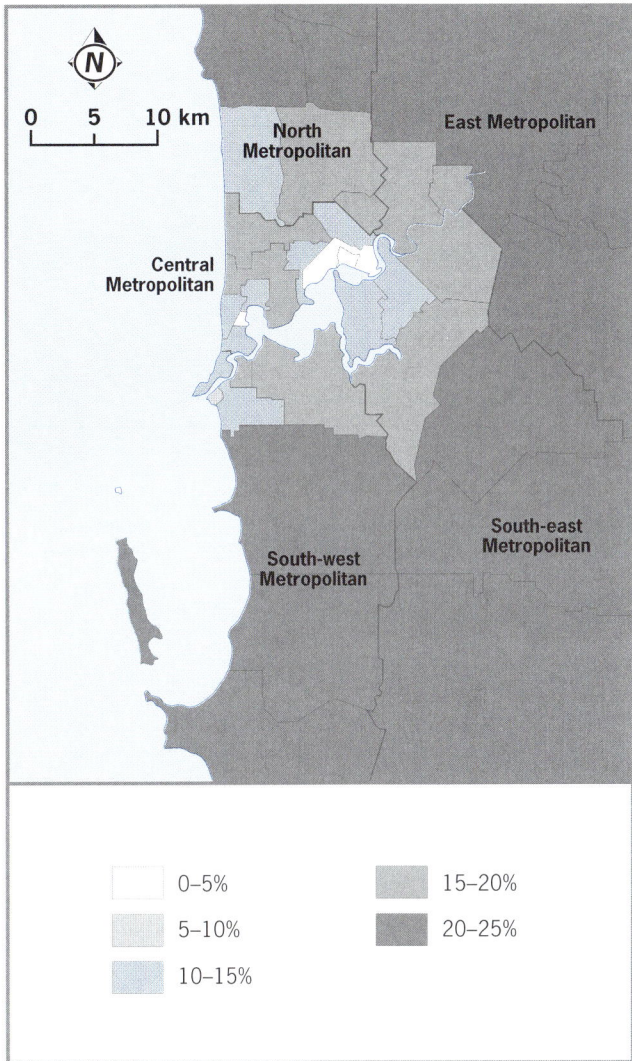

**Figure 6.22** Percentage of Perth's population under 15 years, 2000

throughout the metropolitan area varied significantly. Figure 6.22 illustrates this variation. Very few young people are found in the inner city areas—the central Perth area having around 5 per cent of its resident population in this age group. One exception to the general pattern is the shire of Peppermint Grove. Here, the resistance to higher density dwellings has allowed the area to retain its family-style housing and this has attracted families with young children, resulting in a **rejuvenation** of the population. The tendency for dual income working couples without children to live in the near-city areas has contributed to the lower numbers within the younger age groups. In addition, the inner city areas are long established and so tend to have a higher proportion of older people. In contrast, the outer suburban areas contain the highest percentages of people under 15 years of age. Here, the development of new housing estates has attracted couples wishing to start a family and looking for relatively cheap residential properties. Often these housing estates are promoted as ideal environments in which to raise a family and facilities and services are aimed at populations that fit this particular demographic profile.

## The 15–64 age group

This age group is distinguished from the under 15 and over 65 components of the population because it corresponds to the section making up the workforce. The highest proportion of this age group occurs within the central Perth, Fremantle and Subiaco regions of the metropolitan area. Statistically, this group represents some 69 per cent of the total Perth population. This group is relatively mobile, shifting within the metropolitan area in response to changing economic and social conditions. Its members may have migrated from the outer regions to the inner city areas on entering the workforce, and then on starting a family will look to the edge of the city to buy or build the family home.

## The 65 and over age group

This age group is mainly made up of retired persons. In 2000 in Perth, it represented 11 per cent of the metropolitan population. Claremont and Victoria Park contained the highest percentage of people aged 65 and over, with approximately 18 per cent of people within this age group. In the inner city areas, much of the population has aged along with the suburbs and this partly explains the concentrations of this age group here. In addition, these are areas where the ageing population has been relatively well catered for with retirement unit development that has allowed people to sell their family homes and still remain within the suburb.

In general, the lowest percentages of this age group were found in the outer metropolitan areas, where regions such as Joondalup, Swan and Wanneroo had less than 7 per cent of people aged 65 and over. One exception to this pattern was Rockingham, where almost 12 per cent of the population fell into this category. Further south in Mandurah, the percentage was even higher at 17 per cent. These two localities have traditionally been seen as desirable places in which to retire and this is reflected in the demographic profile. The distribution of people within the 65 and over age group is illustrated in figure 6.23.

# 138 LANDSCAPES AND LAND USES

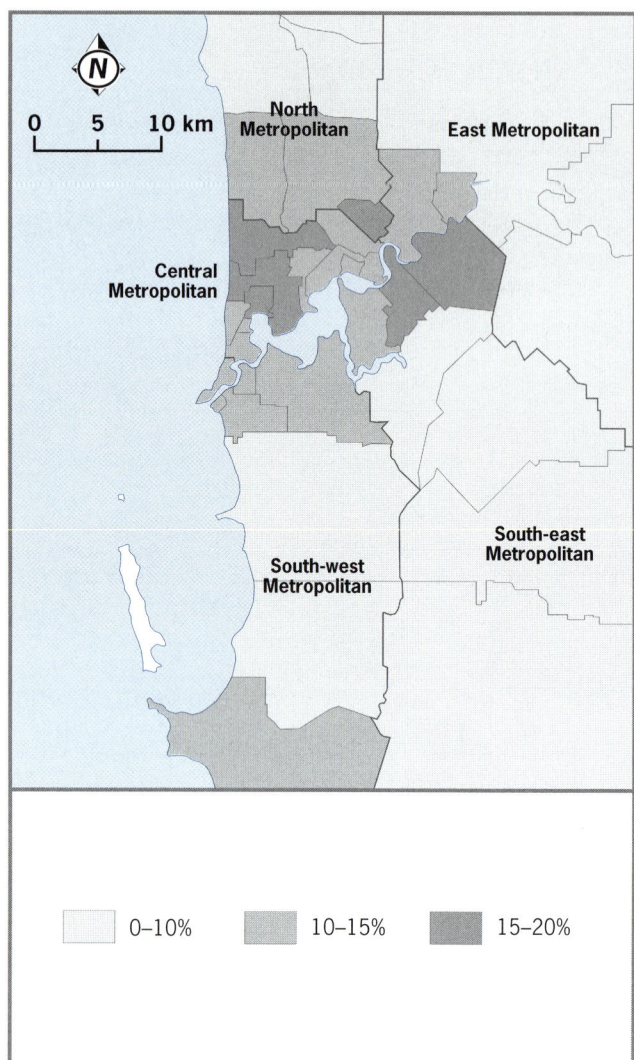

**Figure 6.23** Percentage of Perth's population 65 and over, 2000

Locations with high concentrations of females include the adjoining suburbs of Claremont, Peppermint Grove and Mosman Park. Here, several patterns emerge. There is a dominance of females in the 10–19 age group. The presence of several large girls colleges in this area, with significant numbers of boarders, can account for this feature. These suburbs also have a dominance of woman in the 30–49 age group, as well as the 65–85 age group. The attractiveness of these suburbs in terms of location and facilities has encouraged single professional females to live here, while the longer life expectancy of females can account for the larger numbers of women over 65.

The majority of the statistical areas fall close to the ratio of 100, with those in the 90 to 100 group having slightly more females than males, while those in the 100 to 110 group have higher numbers of males. The latter generally corresponds to those regions with industrial or rural concentrations of employment, while the former is mostly accounted for by the longer life expectancies of females in the older age groups.

## Sex ratios

The sex composition of a population can be expressed as the ratio of males to females. The Australian Bureau of Statistics calculates this ratio as the number of males per 100 females. Where the sex ratio is less than 100, females outnumber males within the population; however, if the ratio exceeds 100 then there are more males than females. While the ratio for Western Australia was 101.4 in 2000, the Perth metropolitan area had slightly more females than males with a ratio of 99. However, within the metropolitan area this varies significantly, as shown in figure 6.24.

Males significantly outnumber females in the central Perth area, with ratios of 130 to 140. This dominance is most pronounced in the 24–74 age groupings and can be partly accounted for by the greater tendency for single males to locate closer to the central business district, both for work and entertainment. This pattern is also repeated in the central Fremantle area, where there are concentrations of males within similar age groups.

### EXERCISES

1. Describe the variations in population density found within the metropolitan area of Perth and suggest possible reasons for the differences.
2. Compare the two maps showing distributions of persons under 15 with persons aged 65 and over. Identify and describe any apparent relationships between the two sets of data.
3. Using the information provided in all the demographic maps of the Perth region, compare and contrast the characteristics of the suburbs located along the Swan River, between central Perth and Fremantle, with those suburbs found on the eastern margin of the metropolitan area.
4. Construct a series of population pyramids using the data provided in table 6.3 and then answer the questions that follow. Note that the suburb names have been omitted so you can use the demographic information within this chapter to find their identity.

    (a) Compare and contrast each of the locations by analysing age and sex distributions. Suggest possible reasons for the similarities and differences observed.

    (b) The four locations represented by the data are Victoria Park, an older inner city suburb; Peppermint Grove, a suburb found on the Swan River half way between central Perth and Fremantle; Serpentine–Jarrahdale, an outer metropolitan locality; and central Perth. Based on the maps provided, as well as the text information on age and sex distributions within the Perth metropolitan area, suggest which sets of data relate to each location.

    (c) Compare the pattern of each pyramid to the national pattern shown for 1999 in this chapter and describe the similarities and differences.

## Table 6.3 Selected Perth suburbs—age and sex data

|  | 0–4 | 5–9 | 10–14 | 15–19 | 20–24 | 25–29 | 30–34 | 35–39 | 40–44 | 45–49 | 50–54 | 55–59 | 60–64 | 65–69 | 70–74 | 75–79 | 80–84 | 85+ |
|---|---|---|---|---|---|---|---|---|---|---|---|---|---|---|---|---|---|---|
| Male | 42 | 65 | 60 | 64 | 46 | 44 | 42 | 15 | 53 | 81 | 56 | 30 | 41 | 27 | 16 | 15 | 24 | 10 |
| Female | 38 | 60 | 157 | 205 | 59 | 28 | 21 | 44 | 80 | 73 | 65 | 34 | 39 | 32 | 20 | 21 | 23 | 18 |
| Male | 62 | 43 | 75 | 353 | 553 | 357 | 264 | 258 | 269 | 274 | 245 | 211 | 167 | 130 | 103 | 80 | 40 | 42 |
| Female | 51 | 47 | 47 | 315 | 555 | 347 | 189 | 145 | 150 | 135 | 136 | 139 | 105 | 84 | 74 | 72 | 47 | 42 |
| Male | 422 | 545 | 487 | 411 | 358 | 320 | 357 | 439 | 543 | 519 | 457 | 359 | 260 | 177 | 143 | 88 | 32 | 45 |
| Female | 397 | 444 | 470 | 452 | 258 | 298 | 378 | 505 | 499 | 430 | 394 | 314 | 220 | 141 | 134 | 68 | 53 | 65 |
| Male | 569 | 520 | 484 | 762 | 1661 | 1644 | 1288 | 1200 | 962 | 850 | 773 | 506 | 401 | 333 | 450 | 430 | 285 | 252 |
| Female | 566 | 438 | 424 | 878 | 1814 | 1561 | 1115 | 996 | 827 | 712 | 700 | 428 | 470 | 565 | 667 | 650 | 558 | 762 |

## Socioeconomic characteristics of Perth

The analysis of socioeconomic data helps to identify apparent differences in a city's residential population. Census data is collected with this purpose in mind. The data can then be used to formulate policies that address the different needs of distinctive metropolitan communities. Factors such as percentages of home ownership compared to rental accommodation, household income levels, property valuations, educational achievement, employment levels and variations in the ethnic composition of different suburbs all represent socioeconomic elements.

Any data relating to social characteristics of different statistical areas should be treated with some caution, as there is often considerable variation within the statistical regions. In addition, specific areas often undergo significant change in relatively short periods of time, which can make previous analysis out of date or misleading. The study of **spatial** variation in socioeconomic factors in this text has been limited to the analysis of household income levels, home ownership, educational achievement, employment levels and the distribution of major migrant or ethnic groups within the Perth metropolitan area.

### Spatial variations in household income

Household income levels can be expressed in several ways. These include total incomes for all members of the household, average weekly income per person and median weekly income per person. Each type of measurement will reveal different characteristics at both the household and the regional level. Figure 6.25 indicates several significant spatial variations in median income within the metropolitan area.

The variation in weekly median income per person is based on those people within the statistical region receiving an income from employment, investments or from social benefits. It is the central or fifth decile income within the sample group and, unlike an average, is not affected by an uneven spread of incomes. For example, 10 per cent of the group earning 90 per cent of the income would give a misleading impression if expressed as an average.

Low-income areas may be accounted for by such factors as the level of unemployment or the numbers of pensioners. The presence of these low-income areas also reflects variations in the general levels of employable skills or qualifications possessed by the population in different parts of the city. The concentration of low-income earners within specific regions may also be accounted for by the percentage of public housing within an area. This may be available for rent or purchase and is supplied by government agencies at lower rents or repayments. The area may also be in an industrial zone and attract unskilled and semi-skilled

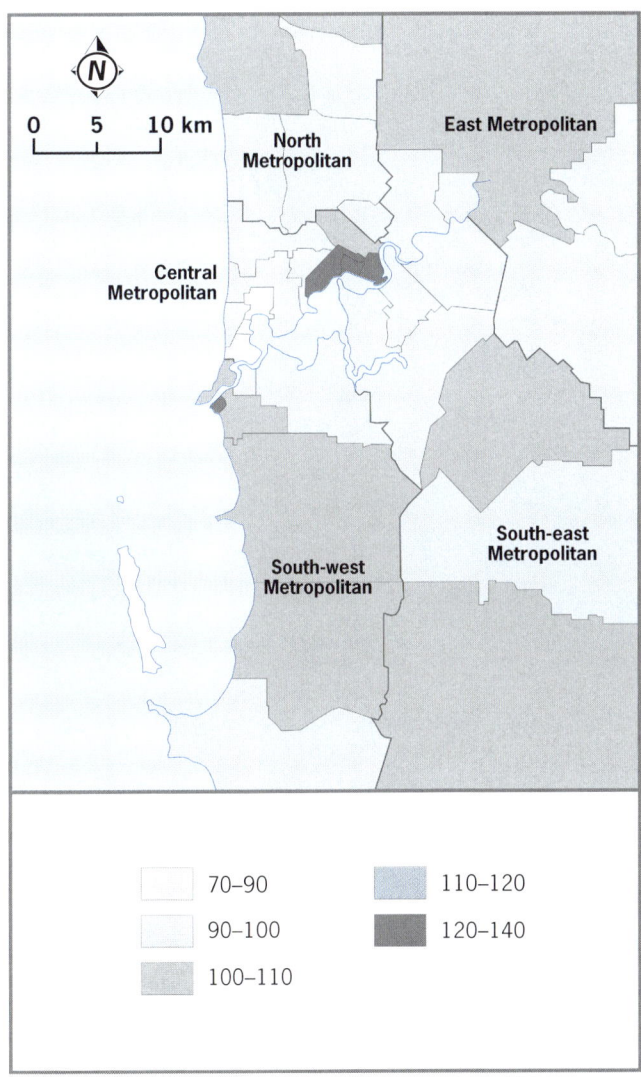

**Figure 6.24** Perth's population sex ratios, 2000

# 140 LANDSCAPES AND LAND USES

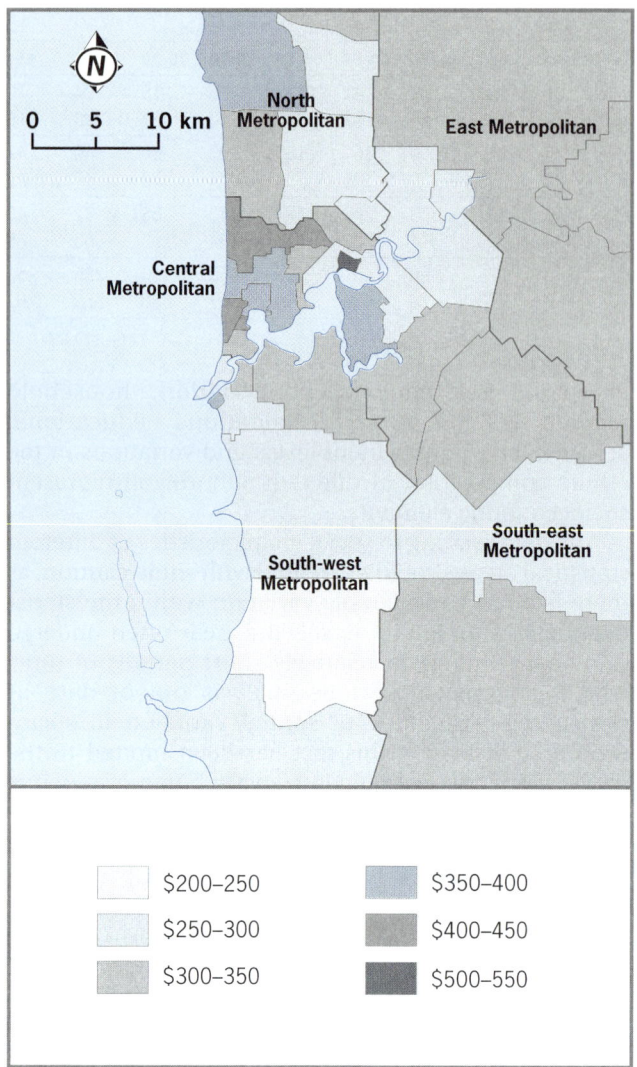

Figure 6.25 Median weekly income, Perth, 1996

## Spatial variations in home ownership

Whether people are buying the home that they live in, own it outright or are renting public or private accommodation reflects differences in the socioeconomic characteristics of an area. Within the metropolitan area, there are locations where home ownership is low and rental properties are common. Alternatively, other areas will have very high percentages of home ownership. These differences can be accounted for by factors such as the cost of properties, the characteristics of the population and the age of the suburb. Figure 6.26 illustrates significant variations within the Perth metropolitan area.

The level of home ownership forms a concentric pattern around the Perth CBD. Within the inner city areas the percentage of homes owned or being purchased by the occupier is very low. It is an area with a high percentage of rental properties. These properties offer an attractive return to investors, with tenants willing to pay higher rents due to the proximity to good employment opportunities and entertainment facilities. The area tends to be less attractive to

workers to that location. The Kwinana region in the south of the metropolitan area is one such location.

High income areas are those where the skilled and qualified sectors of the workforce tend to concentrate. These areas have a high proportion of professionals or owners of medium to large private enterprises. They are also regions of high home ownership and high real estate values. They often have attractive environmental or locational features, such as views, closeness to major services, high accessibility, good recreational or cultural facilities and desirable climatic characteristics. Proximity to the river, the coast and the central business districts of Perth and Fremantle can be seen in the location and pattern of regions with high median incomes. Local government areas such as Cottesloe, Cambridge and Peppermint Grove illustrate this pattern. One distinctive feature revealed by the map is the median income level associated with the Perth CBD or inner zone. Here, a recent trend towards luxury apartments with very close proximity to the commercial heart of the city has attracted the highest income classification of all of the statistical areas.

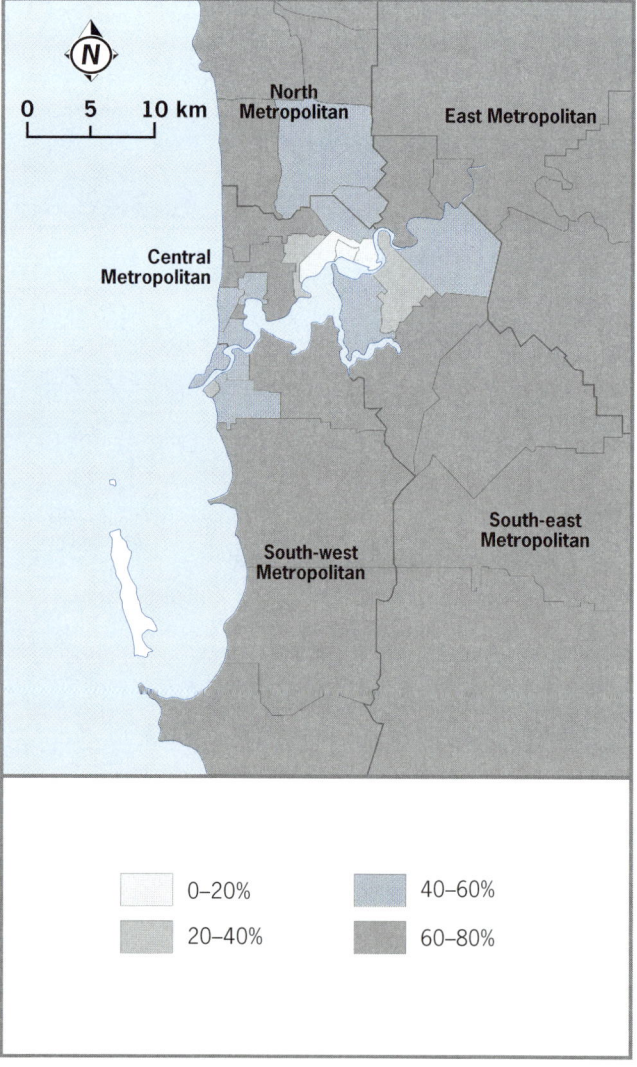

Figure 6.26 Level of home ownership, Perth, 1996

owner–occupiers, who often have families and are more attracted to the middle and outer suburban areas. The central Perth region has around 17 per cent home ownership and 56 per cent rented accommodation—far higher than in any other area. It is an area that also has significant numbers of unoccupied residential properties, with some 18 per cent of the available houses falling into this category. These are often lower quality properties and are held by investors who do not wish to supply and maintain rental accommodation, but are looking towards future redevelopment.

The highest levels of home ownership generally occur on the fringes of the city and in the more desirable locations. Some of the highest percentages of ownership can be found in Joondalup and Wanneroo, on the northern fringes of the metropolitan area. Here, approximately 75 per cent of all residential properties are owner-occupied. Being relatively new suburban areas, they are also regions where the proportion of houses being purchased compared to those fully owned is also high. There are approximately 150 mortgaged houses per 100 owned houses. In contrast, while the regions of Cambridge and Nedlands within the central metropolitan area have high ownership, there are lower levels of properties being purchased. Here, for every 44 houses being purchased 100 are fully owned. This reflects the greater age of these suburbs and the longer periods of home ownership.

## Spatial variations in educational attainment

Educational attainment is often used as an indicator of variations in socioeconomic characteristics of a region. In figure 6.27, the data for the main areas of Perth have been taken from the 1996 census to show spatial variations in tertiary qualifications, including degree and postgraduate qualifications gained from a recognised tertiary institution. When compared to other socioeconomic data, such as median income and unemployment levels, a relationship can often be observed. Regions with high levels of tertiary qualifications also tend to have higher median incomes and lower rates of unemployment. People holding tertiary qualifications will tend to be attracted to areas where skilled or professional employment opportunities are greatest. The map reveals a general pattern of declining percentages of people with tertiary qualifications as you move away from the central Perth area.

The pattern over time has been for the percentage of the population with post-secondary qualifications to increase. Along with the rising proportion of people holding degrees, there has also been an increase in the number of people with trade and diploma qualifications. Emphasis by both state and federal governments is on encouraging people to increase their standard and level of education and this is reflected in the retention rates in secondary schools as well as the increase in post-secondary enrolment rates. A further factor influencing this trend is the growing requirement for qualifications within many areas of employment, and the emergence of new fields of specialist employment.

## Spatial variations in unemployment

A person's ability to find a match for their skills and qualifications amongst the available job vacancies often affects their employment status. People with specialist qualifications that are not required by the labour market will often have to look outside their field of expertise and can experience periods of unemployment. Unskilled or inexperienced workers may also find the search for full-time employment difficult. Sometimes, people in this category will move to locations where the living costs are lower or where they believe that employment opportunities are better. The pattern of unemployment illustrated in figure 6.28 indicates several areas of higher unemployment. These include the central Perth area, Fremantle and Kwinana. All of these areas are in close proximity to large commercial or industrial precincts where there is an increased potential to eventually find employment. They are also areas where low-cost rental housing can be found. In the central Perth region, flats and medium density dwellings account for over half of the housing

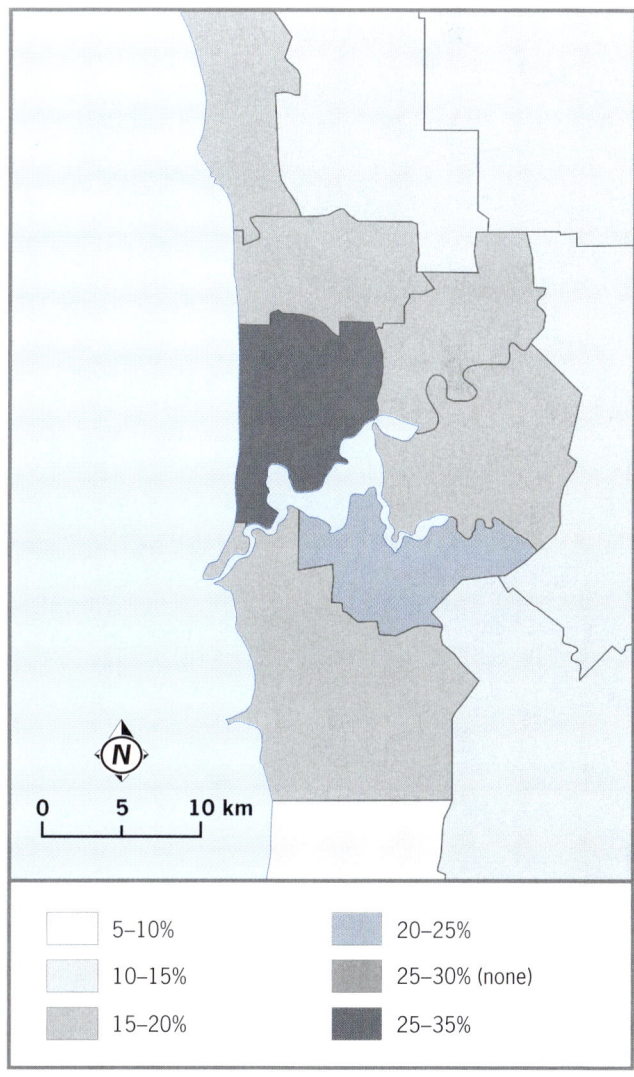

**Figure 6.27** Percentage of the population with tertiary qualifications, Perth, 1996

stock. Here, a relatively high proportion of flat dwellers are lower income earners or people living on social security, who may share facilities to further reduce costs.

## Spatial variations in ethnic distributions

Figure 6.29 shows the distribution of the Australian-born population in 1996. In all statistical areas, it exceeds 40 per cent. Compared to the distributions in 1986, there have been small decreases in the percentages in some of the western suburbs such as Nedlands and Claremont, while the outer suburbs of Mundaring and Serpentine–Jarrahdale have also shown an increase. Between 1986 and 2000 the Australian Bureau of Statistics made several changes in the statistical regions, so it becomes difficult to determine trends in the percentages with accuracy. However, given that the number of migrants settling in Perth is roughly equal to the natural increase in population, significant changes in proportions in the medium term are unlikely.

Areas with traditionally high numbers of migrants and fewer Australian-born residents include the central Perth area around suburbs such as Highgate, Northbridge and North Perth, and the inner Fremantle district. Migrants from Italy, Greece and, in more recent times, Vietnam and other South-East Asian countries have settled in these areas. In 1986, there were approximately 24 000 Italian-born migrants in the metropolitan area, with this group accounting for about 10 per cent of the population in the North Perth area. Compared to these ethnic groups, migrants from the United Kingdom (UK) and Ireland have a far more dispersed pattern of settlement. In figure 6.30 the number of UK and Irish-born migrants per 100 migrants from all other countries has been calculated. Where the number is less than 100, migrants from the UK and Ireland represent less than 50 per cent of the total number of migrants within that statistical area. In 17 out of the 36 statistical regions, Anglo-Celtic migrants were in the majority. It can be seen that this migrant group is predominantly concentrated in the

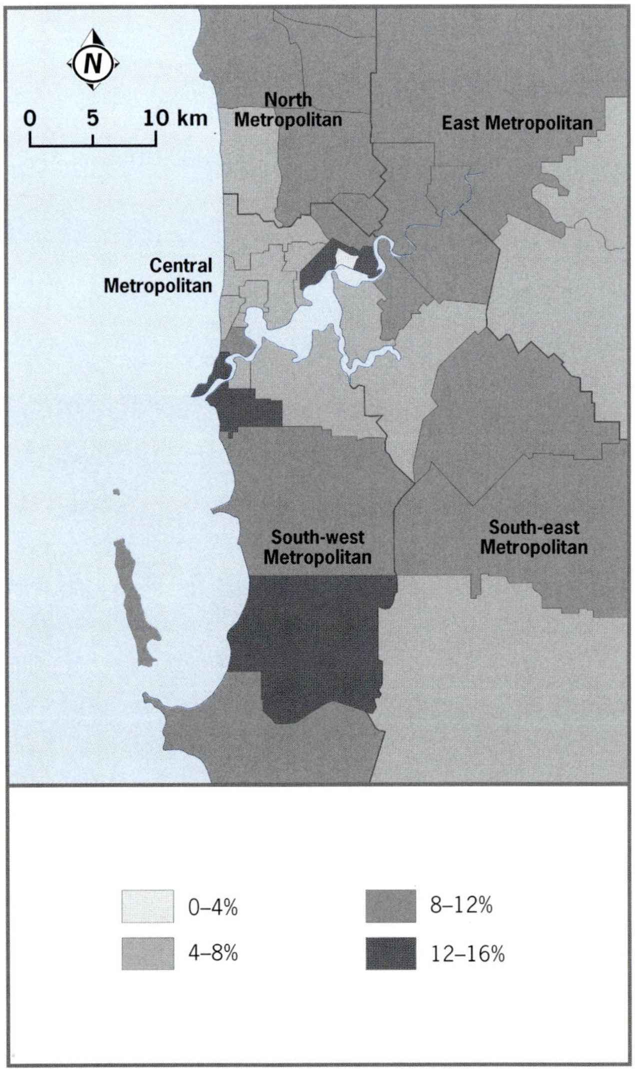

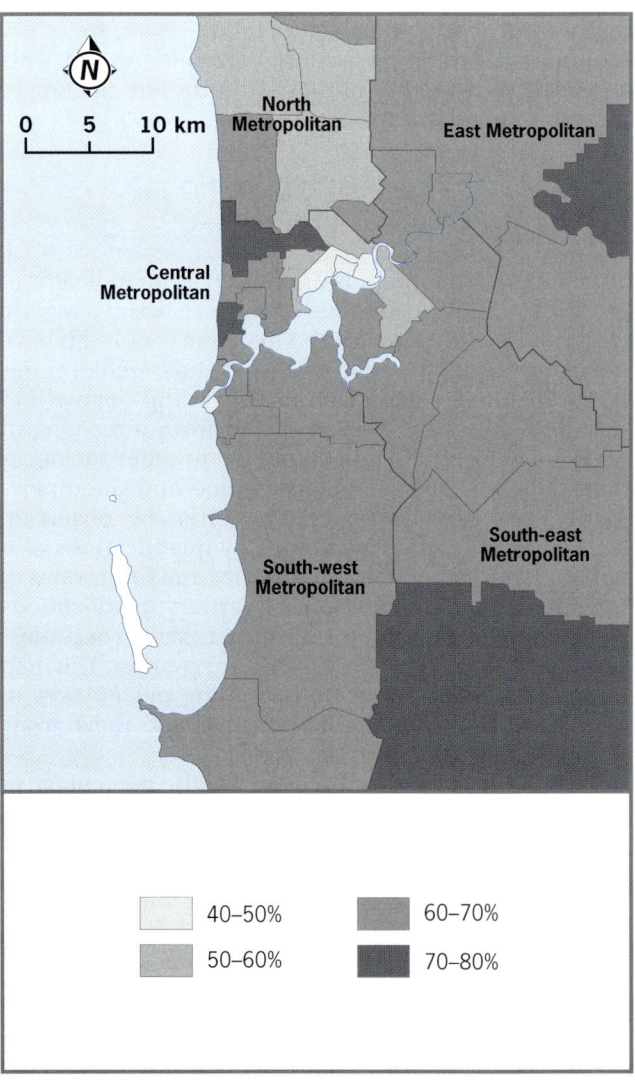

**Figure 6.28** Percentage of the labour force unemployed, Perth, 1996

**Figure 6.29** Percentage of Australian-born in the population, Perth, 1996

AUSTRALIA'S POPULATION  143

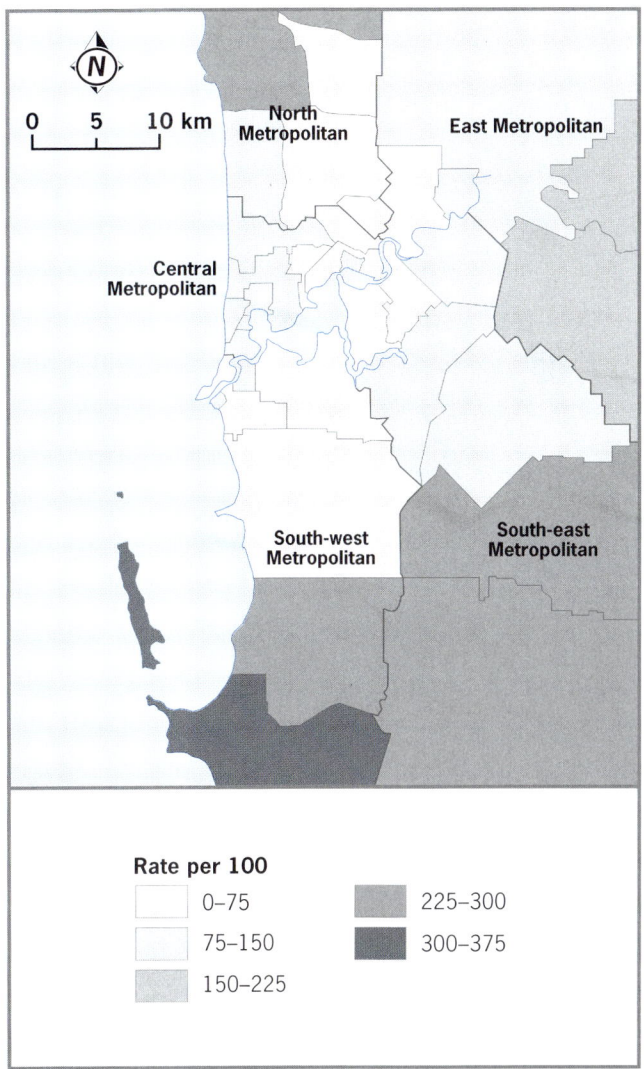

**Figure 6.30** Ratio of Anglo-Celtic migrants to other migrant groups, Perth, 1996

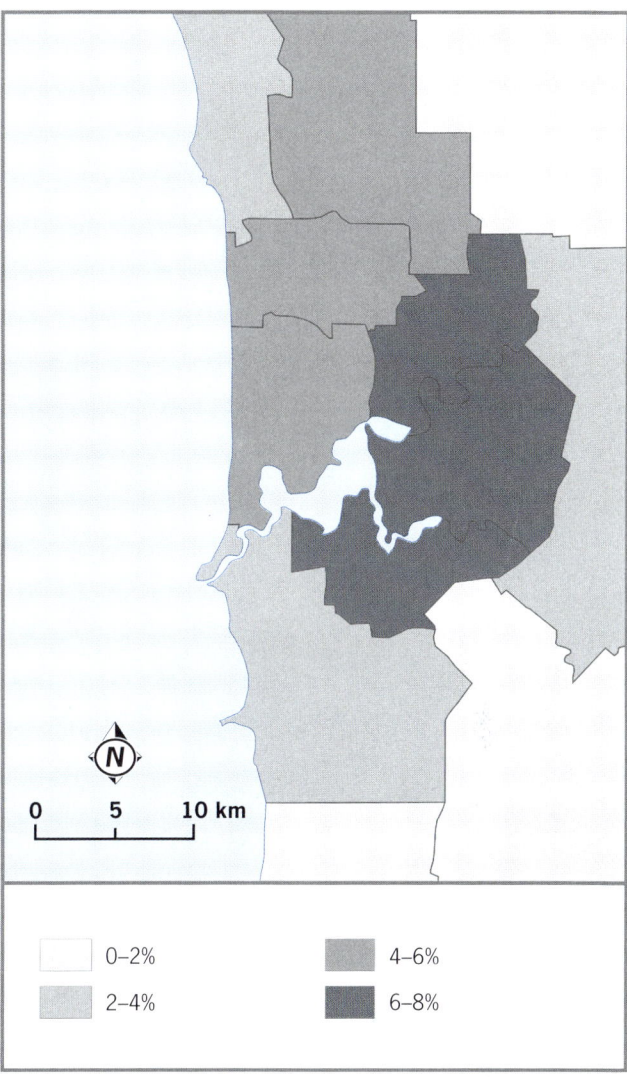

**Figure 6.31** South-East Asian migrants as a percentage of the total population, Perth, 1996

outer suburban areas, with the Rockingham area having the highest ratio. In this region there are 367 people from the UK and Ireland for every 100 from all other countries. The more dispersed pattern of this migrant group can be partly accounted for by its relatively large numbers, which in 1996 amounted to over 50 per cent of all migrants. In addition, the capacity to speak English allowed them to enter more readily into the general population. In contrast, people from non-English-speaking backgrounds are composed of different ethnic groups that often cluster together in inner city locations in order to provide mutual support and assistance.

### EXERCISES

1. Using the two maps (figures 6.31 and 6.32) showing the distribution of South-East Asian and Southern European migrants in the Perth region, compare and contrast the patterns revealed and identify possible locational influences.

2. Study the map of home ownership in Perth and describe the variations that occur as you move away from the central areas to the outer suburban areas. Suggest reasons for these variations and how the pattern might influence the distribution of rental accommodation.

3. Carry out an analysis of selected locations within the Perth metropolitan area by completing the following activities.
   (a) Complete the table below by finding the information identified for each of the suburbs. To locate these suburbs, use a road directory or similar resource.

| Suburb | Home ownership | Unemployment rate | Level of tertiary qualifications | Median weekly income |
|---|---|---|---|---|
| East Victoria Park | | | | |
| Safety Bay | | | | |
| Nedlands | | | | |
| Ocean Reef | | | | |
| Roleystone | | | | |

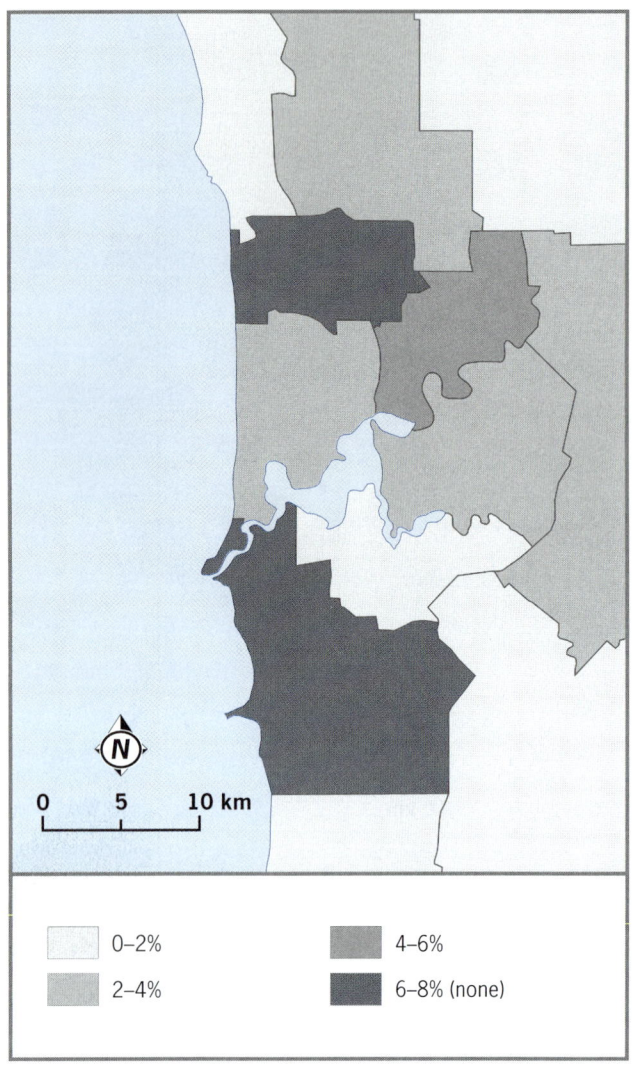

**Figure 6.32** Southern European migrants as a percentage of the total population, Perth, 1996

> (b) Compare and contrast the different locations and suggest reasons for the variations in each of the factors.
>
> **4** Compare figure 6.30 with figures 6.31 and 6.32. Identify, and suggest reasons for, any similarities and differences.

# Chapter 7

# Settlement Patterns and Urban Networks

The purpose of this chapter is to introduce the concepts of urban networks and hierarchies, with particular emphasis on the South-West. A case study of the Avon statistical area applies the various concepts to the urban network of this region.

## Settlement patterns

Settlement may be defined as any form of human habitation that is of a relatively permanent nature. Settlement structures include the houses and buildings and the infrastructure that has been developed to support the settlement.

Settlements occur as a consequence of people's occupation and use of an area of land and can vary in size from small farmhouses and farm buildings through to large cities. Temporary structures such as caravans and tents are generally associated with nomadic activities such as mining exploration or tourism and for the purpose of the discussion in this chapter have not been included in the study of the functions and structures of settlements.

### Rural and urban settlements

A rural settlement is a centre in which the majority of people are engaged in activities associated with the primary industries of agriculture, fishing, forestry and mining. Rural settlements are located in country areas (the countryside) and consist of isolated dwellings (homesteads) or clusters of dwellings. They have a dormitory function, with the bulk of the population living and sleeping in that settlement for at least part of the year, but working on the surrounding land or sea.

The most common rural settlement in Australia is the farmstead, although isolated mining camps or fishing settlements are also considered to be 'rural'. These settlements are usually quite small in size. If they were any larger they would probably attract non-rural functions (secondary and tertiary) and would therefore be classified as urban settlements. Urban settlements are those in which the majority of the workforce is engaged in secondary or tertiary industry, ranging from manufacturing, processing and fabricating through to retail, professional, financial and social or community services. Urban settlements have a mixture of secondary and tertiary functions, which often form distinctive distribution patterns within the town or city. In addition, they have relatively large residential precincts providing accommodation to the people engaged in these non-rural activities.

Apart from the difference in functions of rural and urban settlements in Australia, the other main method of distinguishing between the two is the pattern of distribution. As discussed in previous chapters, rural settlements tend to have a dispersed distribution pattern, while urban centres are nucleated in nature.

A further method of classification between the two types of settlements is on the basis of population. This can vary widely from country to country and has limited usefulness in the study of settlement geography. For example, in Australia the Australian Bureau of Statistics (ABS) classifies rural settlements as any town or locality with less than 1000 people, while those with more than 1000 people are classified as urban. This system takes no account of the functional activities within the settlement, with many small towns in Australia of less than 1000 being predominantly urban in structure and function.

### Classes of settlements

Rural and urban settlements can be classified both on the basis of their dominant function and on their size. Settlements can range in size from small homesteads through to large cities. They can form distribution patterns in which smaller settlements are more common and more closely spaced, and where larger settlements are fewer in number and set further apart.

**Farmstead**

This is the smallest type of settlement in Australia. It is rural in function and usually consists of one or two houses and the buildings and structures associated with the type of farming being practised. In intensive farming regions they occur in close proximity to each other, while in the more extensive farming areas they may be located many kilometres apart. In figure 7.1 the

**Figure 7.1** Rural homestead in the wheat–sheep region of Western Australia

homestead is one associated with the wheat–sheep belt of Western Australia. It has a shearing shed on the right and seed bins and machinery sheds behind the main house.

## Hamlet or village

This is the smallest of the urban centres in Australia and in the last 30 years they have declined significantly in number as rural population changes have occurred. They often have one or two multi-function services and a small number of houses. These may house rural workers or service personnel. Their appearance lacks the separation of functions associated with the structure of larger urban settlements.

## Town

The distinction between the town and the hamlet is based both on size and functional complexity. The town offers an expanded range of functions to the residents as well as the surrounding rural populations. There is some spatial differentiation of functions, with commercial, residential and industrial areas apparent. The town may function primarily as a centre for services used by the surrounding rural populations, or it may have developed in response to some specialist purpose. The latter might include tourist centres, religious settlements, fishing villages, mining settlements or timber towns. The settlement of York in the South-West began as a central place, but in recent years has grown as a special purpose settlement with its success as a tourist centre (see figure 7.2).

## Regional city

Groups of villages and towns are often linked by transport networks to larger settlements, where functions that require a greater number of consumers or customers are located. These regional cities have a variety of retail, financial and social services as well as regional offices associated with government departments and private companies. They will often have light industrial activities, with fabrication and manufacture of building and farming products taking place. In Australia, the classification of 'city' is generally given to settlements with a resident population of 20 000 or more.

## Metropolis

The main city of a large region or state is usually called a metropolis. It functions as the commercial and political centre of the area, with a large and diverse population. It is functionally very complex, containing a wide variety of services. These can range from everyday services through to highly specialised functions, such as those associated with health, finance and technology. In addition, it will have a range of industrial activities including both heavy and light manufacturing functions. The metropolis occupies large areas of land and shows complex distributional patterns of the different functions. A significant proportion of the total urban area is occupied by residential areas and services. The capital cities of Australia fall into this category, given their size and functional complexity (see figure 7.3).

# SETTLEMENT PATTERNS AND URBAN NETWORKS 147

**Figure 7.2** The town of York, WA

## Conurbation

A conurbation is a very large urban centre, in terms of both its area and its population. It is formed when two separate centres expand and coalesce. This can happen when the centres grow outwards along transport links and meet, with the intervening rural land use replaced by urban functions. In some cases, the outward growth of a major city will absorb smaller settlements into its metropolitan area. All Australian capital cities indicate some degree of coalescing of separate urban centres associated with their urban expansion. The joining of Botany, Sydney and Parramatta is one such example.

In some of the larger population regions of the world, large cities have expanded outwards to such an extent that they form an unbroken belt of urban land use that stretches over hundreds of kilometres. The term **megalopolis** has been applied to these settlements. Along the north-eastern coastline of the United States, the cities of New York, Boston, Baltimore, Philadelphia and Washington have coalesced to create an almost unbroken belt 650 kilometres long on the Atlantic coast. With the rate of urbanisation occurring in Asia and Europe, this process will continue to happen in these regions as well.

> **EXERCISES**
>
> 1 Describe the different types of settlements and suggest reasons for their existence.
> 2 Find illustrations of different types of settlements in magazines or on the Internet and produce a collage or poster to illustrate their variety.
> 3 On a map of Australia show examples of regional towns, cities and conurbations. Research several of these and write brief descriptions. Note that there are many web sites for towns on the Internet.

## Location of settlements

A variety of factors account for the location of settlements. These include environmental, historical, economic, social, political and technological factors and they will be discussed further in the sections on urban networks of the South-West, as well as in

**Figure 7.3** Sydney, Australia's largest city

chapter 8. Two concepts used to describe the location of settlements are site and situation. The site of a settlement is the biophysical environment on which the buildings and associated infrastructure have been constructed, while the situation is its relative position or location in its regional setting.

### Site

There are a number of features to observe when analysing the site of a settlement. These include:

- the type of soils or geologic structures
- slope or gradient of the land and landform types
- drainage patterns and features
- natural vegetation formations
- **microclimatic** characteristics.

In selecting and developing a settlement site, planning considerations can include factors such as defence, accessibility, ease of construction, and the impact of events such as flooding, earthquakes, storms and volcanic activity. In addition, climatic conditions such as temperature, rainfall, humidity and wind patterns often influence the location of different functions within the settlement, or where the settlement will be built. Where a settlement has been carefully planned, the selection of the site can also take into account the potential for future growth or expansion.

### Situation

In analysing the situation of a settlement, factors that identify its relationship to the wider region are considered. These include:

- its position within the urban network or hierarchy
- its setting within the biophysical environment
- strategic considerations, such as its location on trade routes, within productive **hinterlands**, at transhipment points or nodes
- its global position as identified by its regional, national or international position, and by its latitude and longitude.

The analysis of situation of a settlement focuses therefore on its distance to other settlements of smaller, larger or equal size, as well as some explanation for its position with the urban network. It also looks at the settlement's position within the landscape, taking into account such factors as vegetation, climate, and landform regions and features. A settlement that functions as a seaport or railhead has a situation that is different to one which is located within an isolated mining region.

Finally, the analysis of the global situation identifies its relationship to the rest of the world. Is it simply a dot on the map or does it exhibit a strong global relationship? National capitals and world financial centres such as New York, London and Paris generate a strong global influence through the effects that their financial institutions have on companies in all parts of the world.

## Urban networks and hierarchies

A study of the ways in which towns have developed within a region will reveal patterns of different size settlements linked together by a variety of transport and communication systems. In turn, these settlements will show links or relationships with their surrounding rural hinterlands. In analysing these features we can in many cases see causes or explanations for the patterns of the network and, in doing so, develop geographic theories to test these causes in each new situation. Two such explanations of urban networks and hierarchies are urban primacy and **central place theory**.

### Urban primacy

A study of Australian networks will often reveal the existence of a dominant settlement within a state or region. It assumes primacy or high-order position and exerts considerable influence over the surrounding settlements. It is much larger than the next largest centre within the network and, in addition to the functions that are common to other centres, it has functions or services that are unique. In other words, it has a much higher level of functional complexity. Figure 7.4 illustrates the functional and population relationship between the primate city and the other urban centres within the hierarchy. There is a clear distinction between the first and second centres in the rank.

The measure of urban primacy can be determined by the primacy index. This compares the population of the largest centre to the region's population. Where a city contains at least 40 per cent of the total population it can be classed as a primate city. From a national point of view, there is no city in Australia containing 40 per cent of the nation's population. However, within most states the capital cities often achieve this figure. In figure 7.5 the population of each state capital is compared to the state population by calculating the percentage of people located in each of these cities. The graph, which is based on the year 2000 population figures, shows clear levels of primacy in four of the seven states and territories, with Hobart in Tasmania having the lowest percentage of the total state popula-

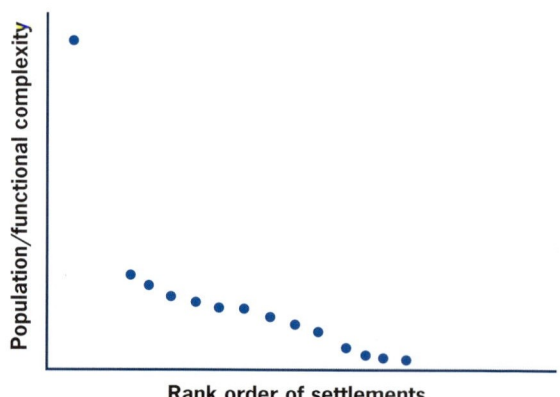

**Figure 7.4** Primacy hierarchy

tion. Even so, at 41 per cent, it achieves the primacy index cut-off point.

A second method of determining primacy is to compare the size and functional complexity of the first- and second-ranked urban centres. In figure 7.6, the next largest city within the state or territory is compared to the capital city to determine the relative size. This is expressed as a percentage, and the lower the percentage the higher the level of capital city primacy. Some researchers have suggested that the primate city should be at least eight to ten times the size of the next largest city. In other words, the second-ranked settlement should have about 12 per cent or less of the primate city's population. By applying this definition to 1996 population data, regional primacy occurs in New South Wales, Victoria, South Australia and Western Australia.

In conclusion, urban primacy occurs when one city within a region, whether the region is a state or a nation, achieves a level of political, economic and population dominance that significantly outranks that of all other towns within its sphere of influence. The two methods used to measure this dominance are comparison with the total population of the region and comparison with the next largest urban centre within the region. To some extent, the determination of primacy is arbitrary—it depends on where the index of primacy is set and the way in which a region is identified or defined. It might be argued that within a small region, such as a farming region where the largest town has 10 000 people and the next largest has 1000, then there is a form of regional primacy.

## Factors influencing regional primacy

A number of historical, economic, demographic and political factors are significant in explaining the high degree of regional (state/territory) primacy displayed by Australian capital cities.

### Colonial settlement

From their original function as colonial outposts, cities developed commercially as ports, transport nodes, major centres of manufacturing and tertiary industry, and administrative centres. The separate origins of the states as sovereign colonies initiated the development of one main centre within each state. These colonial centres became port cities and acted as transhipment points for exporting primary produce from the surrounding areas, and for receiving imports for distribution to the rural hinterland. Trade and administration therefore became centralised in one coastal location. In this way each colony had an administrative centre, assuring it a degree of independence. This independence, and the vast distances between colonies, saw each state develop a settlement pattern centred on its port city. Separate development of the administrative centre was enhanced by the establishment of customs posts between colonies and the introduction of different gauge railways. The railway system in each colony or state was constructed with different widths between the rail lines and, as a result, trains could not run from one state to another. This restricted the transportation of goods between the states and 'protected' each local economy from competing interstate products.

### Agricultural development

Even from early times the rural hinterlands of each colonial settlement did not develop as large areas of closely settled rural communities focused on subsistence farming. The demands of overseas markets

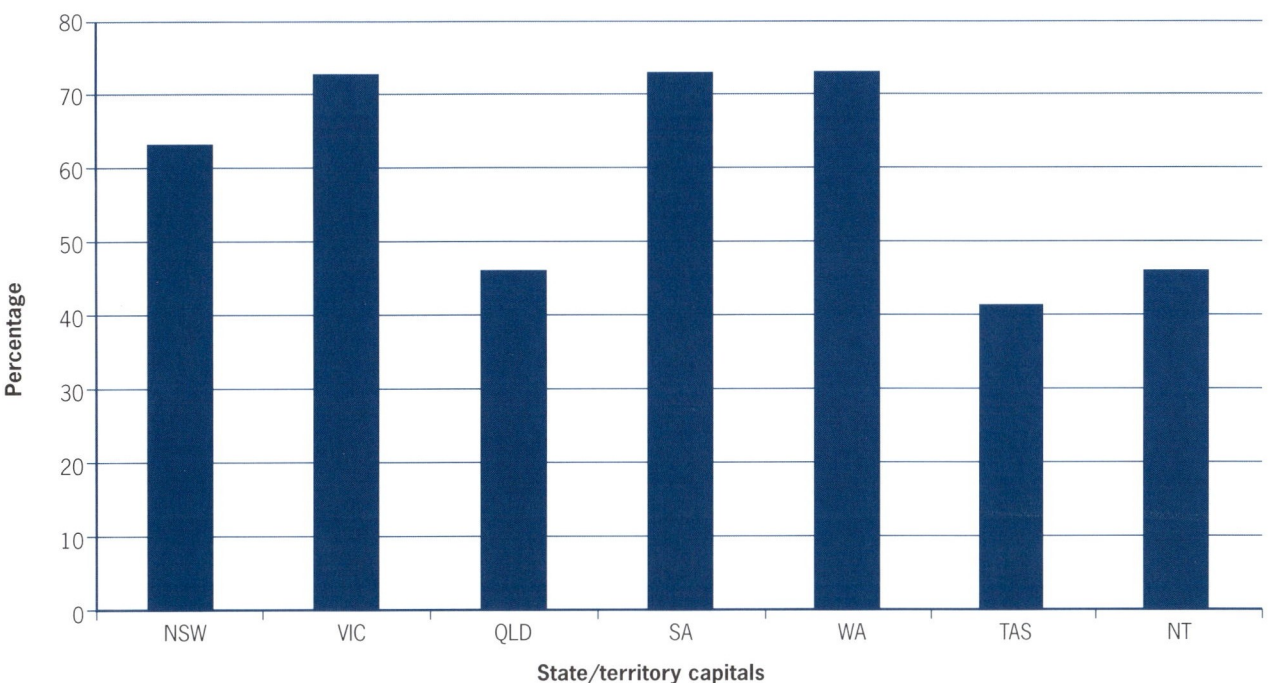

**Figure 7.5** Capital city primacy index

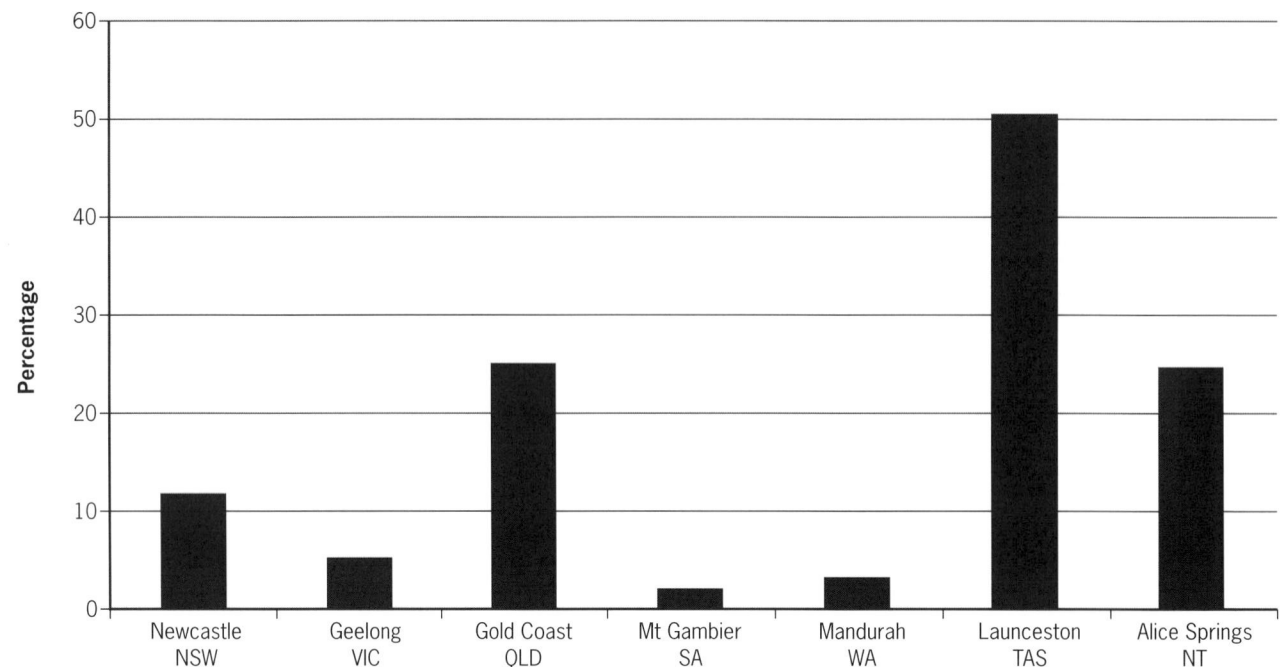

**Figure 7.6** Second-rank regional city populations as a percentage of the state or territory capital city population

instead encouraged the development of commercially-based rural produce such as wool, which was highly sought after by the woollen mills of the United Kingdom. As the highly efficient, commercially-oriented agricultural system required few workers, the majority of the population remained in the coastal cities. This situation, together with the lack of alternative sites for major settlements, increased the commercial and administrative dominance of the port cities. The increasing sophistication of the transportation network allowed for the further growth and dominance of state capitals. In the late nineteenth century, a radial network of railway lines developed, linking interior agricultural lands with the coastal capitals. As a result, the trade area of these cities expanded considerably, and they became even more important as transport nodes.

### Industrial agglomeration

The concentration of manufacturing in the colonial capitals further boosted their trading role and the rate of growth of these coastal centres increased. Any chance of one capital city becoming more economically dominant than the others was prevented, however, by the use of different gauges by the state railway systems. The lack of good harbours for overseas shipping also prevented the development of competitive regional centres outside the capital. The concentration of population in the state capitals, particularly Melbourne, Sydney and Adelaide, created a large domestic market and a skilled labour force that was attracted to the development of secondary and tertiary industry.

The peak of metropolitan dominance of these cities actually coincided with the period of maximum industrialisation in the 1950s and 1960s. Other state capitals also sought to increase their dominance by encouraging the development of industry. Perth, for example, grew rapidly in the period between 1966 and 1971 as a result of mineral exploitation and associated tertiary activities. The increased secondary and tertiary activity in the capital cities had a cumulative effect, generating further growth of the population and economic activity and therefore creating an even greater imbalance in importance between the capitals and other smaller urban centres in their states.

### Migration and natural population increase

Both rural–urban migration and natural increase have contributed greatly to the increasing concentration of the populations of most states in their capital cities. Increasing farm technology, mechanisation and the amalgamation of farms have led to rural depopulation. Many farmers have been attracted to the capitals by employment opportunities, though some 'city farmers' today actually reside in the primate city and commute to their farms. European and other migrants who have entered Australia through the capital cities have tended to stay there, attracted by the employment opportunities and the urban lifestyle that they were used to in their homelands. Since most of these people have migrated at a young age, they have contributed to the growth of the population in these cities by accelerating the birth rate (the growth of Melbourne and Sydney during the 1970s was almost entirely the result of migrant intake). Natural population increase has also contributed to the growth of the capital cities, since they have the largest youthful populations and, therefore, the greatest rate of natural population increase.

### Political factors

The dominance of the primate cities can also be attributed to a number of important political factors. In

Australia, the first point of settlement in each state became the administrative centre responsible for the government of the colony, until federation in 1901. Even after federation, the states maintained responsibility for most aspects of government. In this way the independence of the states was established. The administrative function of the state capitals exerted a centripetal (attracting) **force** on other activities, enabling the cities to increase their regional dominance. The limited role and late development of a system of local government restricted its ability to compete against the influence of the state government. In all cases, local government was initially introduced in the state capital city and other major cities. Only later was it established in country towns and rural areas, and even then it had only limited powers. Local residents tended to rely on the state government for a range of services, which enabled centralised political dominance to be maintained.

## Issues of primacy

The impact of capital city primacy on urban networks in Australia has a number of consequences. The centralisation of functions and development within several large cities has allowed the country to make use of economies of scale. Large concentrations of industrial and commercial functions within a relatively small area encourage collaborative forms of production and enterprise. One company provides the necessary materials for another, while ancillary services such as business advice, research and development, and venture capital suppliers are located within the same urban centre. In addition, the number and variety of employment opportunities encourages skilled and qualified staff to settle in these centres. Further, the large metropolitan centres provide a ready market for many manufacturing enterprises. In states such as Western Australia, where over 70 per cent of the population resides in Perth, local and state governments can provide services to a concentrated region of population more cost-effectively than if the population were to be found in widely dispersed settlements.

The high degree of regional primacy in Australia has, on the other hand, given rise to a number of problems. One area of significance is the impact on the other settlements within the region or state. The primate city draws population away from these centres, thus reducing their capacity to grow and develop. This is especially apparent among the young and educated sector of the workforce, where the rural–urban drift is most pronounced. In a number of areas within the South-West the trend is even more noticeable among females in the 19–45 age group and this has resulted in many rural regions having a higher number of males (as discussed in chapter 6). Over the last 20 or so years, a process of rationalisation of government and business services in the smaller urban centres has hastened this population decline and has had the effect of reducing the quality and number of services offered to the remaining population. In the primate city, sustained and often rapid growth has produced pressure on the city margins, with urban growth extending into productive rural lands. In addition to this urban sprawl, there has also been pressure for redevelopment within established areas, resulting in higher density residential construction.

### EXERCISES

1. With reference to Australia's capital cities, explain the two methods used to calculate urban primacy.
2. List the factors that would favour the development of an urban primacy hierarchy.
3. Identify the advantages and disadvantages of urban primacy.
4. Explain how Australia's economic and population growth has encouraged the development of capital city primacy.
5. Suggest reasons why capital city primacy is greater in Western Australia than in Queensland.

## The central place model

A German geographer, Walter Christaller, developed this model or explanation of urban networks and central places in the 1930s. He put forward the view that urban networks were orderly systems that could be explained by the way in which towns developed in order to meet the needs of the rural populations. The town, or central place, represented the most effective way of supplying goods and services to surrounding rural inhabitants, and so it acted as a nodal point for this population. Goods and services flowed from the town into the surrounding region or hinterland, while raw materials from the hinterland were brought to be marketed or processed within the town. In this way, the town or central place became the focal point for movements of people, goods and services within the hinterland. This arrangement may be defined as a **functional region** (functional and formal regions are discussed further in chapter 9). Figure 7.7 illustrates the association between central places and hinterlands.

Christaller also noted that the size and spacing of central places and their hinterlands appeared to form an orderly hierarchy, with different discrete classes or orders (see figure 7.8). Within the different towns, services called central place functions varied in type and number. He further suggested that there was a set ratio between the number of central places in each order and the size or area of each central place hinterland. In a theoretical four-order central place hierarchy there would be one fourth-order centre, two third-order, six second-order and 18 first-order central places. These would occur in discrete classes. In explaining this pattern, Christaller identified two factors that appeared to cause this organisation. These were market or population threshold and range.

**Market threshold**

Market threshold has been defined as the minimum population needed to support a given service or supplier of particular goods. This definition suggests that there is a level of income that the owner or supplier of a good or service needs to achieve to be commercially viable. Should the population or number of

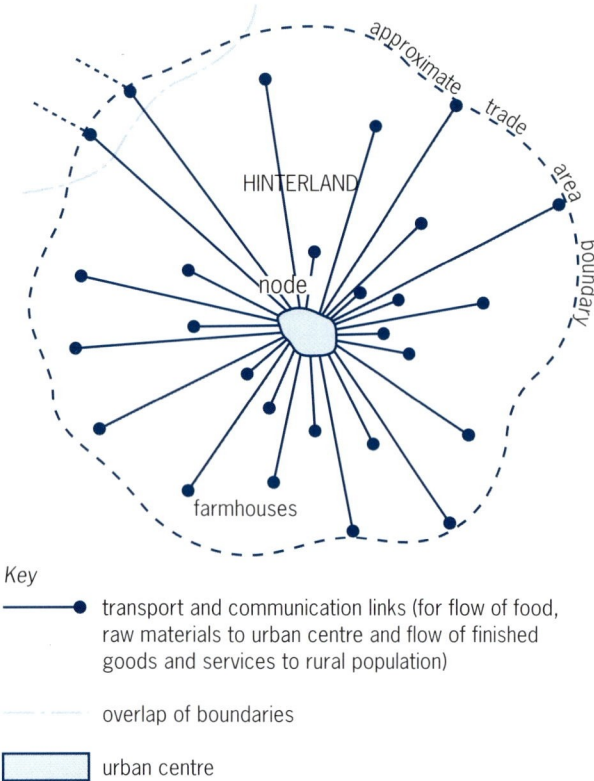

**Figure 7.7** A functional region

usage is infrequent. Similarly, if the percentage of the population that uses the service is high then this will also produce a low threshold. Factors that influence the frequency and degree of use are necessity, cost and durability. Essential, low-cost temporary services or perishable products tend to be consumed more frequently by a greater percentage of the population, while non-essential, high-cost durable goods or services are used less frequently by a smaller percentage of the population.

A legal service supplied by a solicitor would be used very infrequently by the average consumer and only by a small percentage of the total population. The solicitor would need a large population base in order to attract sufficient consumers to make this function commercially viable. Figure 7.9 illustrates the variables influencing threshold populations for various central place functions by comparing the newsagency and the solicitor. The solicitor provides a high cost service, has a low frequency of use, may produce a lasting solution and tends to be used by a low percentage of the population. The four graphs thus show this service as having a high population threshold. In contrast, the newsagency selling newspapers and various stationery items has products with a low cost and high frequency of use that are non-durable, and is used by a larger percentage of the population. Thus, it has a low population threshold and would be a more common service within the central place.

Other factors that may impact on the population threshold include changes in the nature of consumer demand, income levels, technology and population demographics. Along with the other variables influencing threshold population, these considerations will be investigated more fully in the case study of the Avon region of south-western Australia.

consumers fall below this level of demand, the central place function will cease to operate. Market thresholds can be estimated by dividing the population of the central place and the hinterland by the number of that type of function. Thus, if there are two newsagencies in a town and the total rural and urban population is 500, then the threshold is 250. This calculation assumes that 250 is the minimum population needed to support one newsagency. Should the total population fall below 500, one agency will disappear and if it falls below 250, the second agency will disappear.

One variable in determining the threshold is the frequency of usage. If the function is used frequently then the threshold is likely to be lower than where the

### Range or draw area

The range of a good or service is the maximum distance that people are prepared to travel in order to obtain or use the particular central place function. Like threshold, the consumer's willingness to travel a given distance will depend on factors such as the cost of the service and the frequency of their use. High-frequency, low-cost services are termed 'convenience goods' and have a limited draw area or range. Low-frequency, high-cost services are called 'comparison goods' and have a large **draw area** or **range**. In figure 7.10 the difference between the range or draw area of a solicitor and a newsagency is compared.

The purchase of convenience goods or services, as the name suggests, is influenced by the ease of accessibility. The consumer looks for the closest available supplier. Cost is not an overriding consideration as it represents a small proportion of total expenditures. In contrast comparison goods, with their higher price and lower frequency of purchase, often involve greater expenditure of time and money in seeking out the best product or service at the most competitive price. The purchaser is willing to travel greater distances in order to satisfy their needs when seeking comparison goods or services.

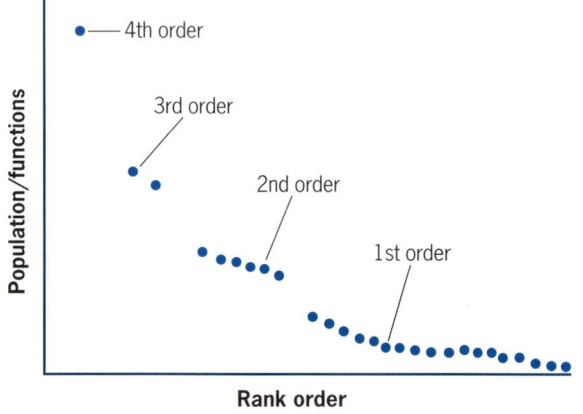

**Figure 7.8** Central place hierarchy

## Central place hinterlands

Each central place exists to supply goods and services to the people who live within the centre as well as those who live within its hinterland or draw area. The concepts of range and threshold are closely linked to the size and shape of the hinterland. Low-order goods and services (convenience functions) generate small hinterlands. This in turn means that these functions will be found in all central places, both small and large in an urban network or hierarchy. Low-order functions thus have a small threshold and a limited range. Where

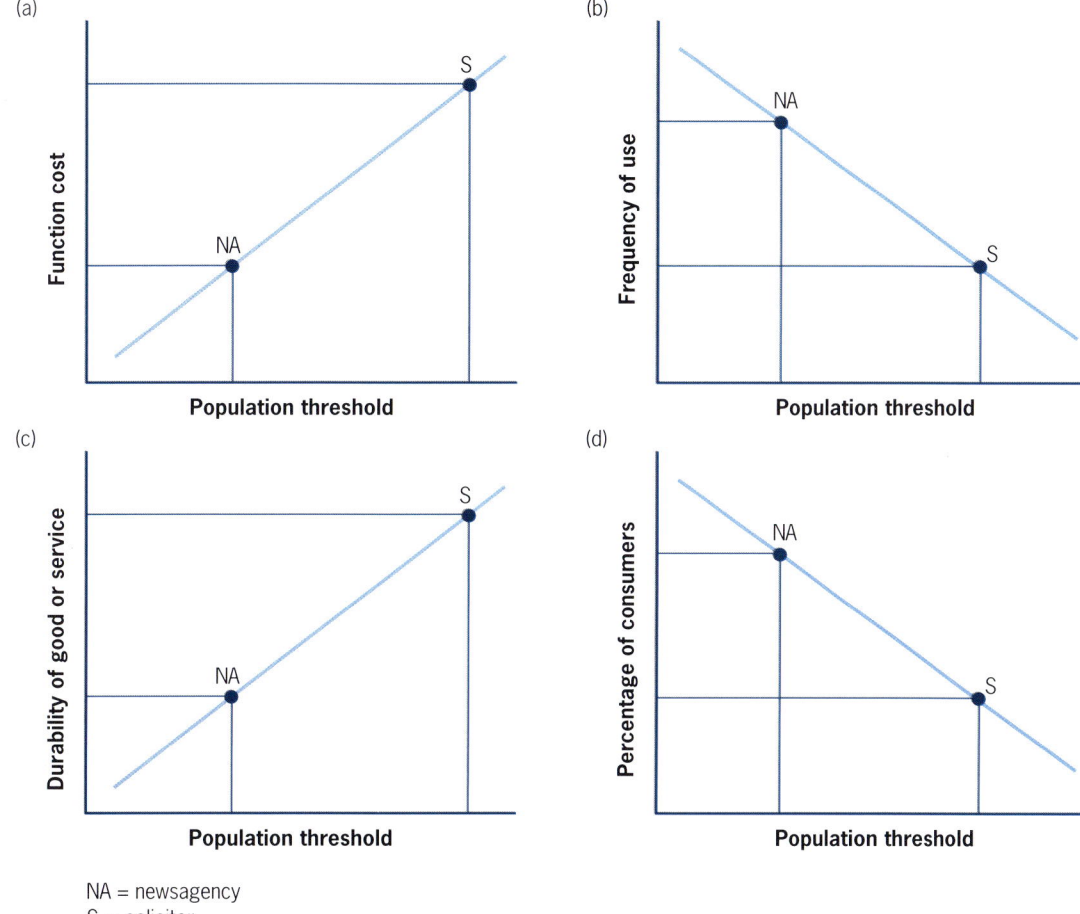

NA = newsagency
S = solicitor

**Figure 7.9** Factors influencing threshold population

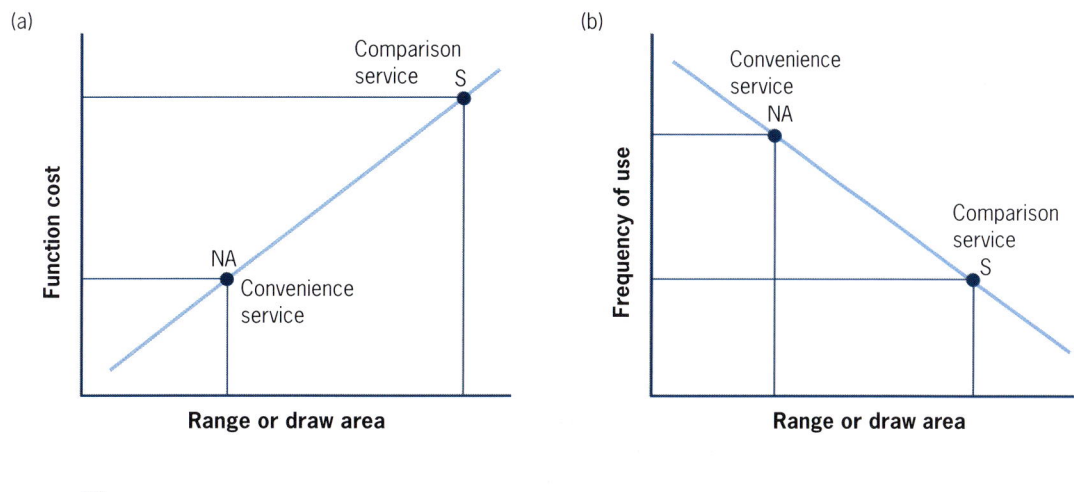

NA = newsagency
S = solicitor

**Figure 7.10** Factors influencing range

the function is a higher order comparison good or service, then the hinterland is larger and the threshold population is greater. Comparison goods and services tend to be located in larger central places within an urban hierarchy.

In the central place theory, Christaller suggested that if the landscape on which the urban network had been placed were entirely uniform then the towns at each level would be evenly spaced and their hinterlands would be the same size and shape. He identified the hinterland shape as being hexagonal. This he reasoned would allow for all rural dwellers to fall within the draw area of a service, while allowing for a similar distance to be travelled by the furthest consumer. He also stated that smaller hinterlands would fall within the range of larger hinterlands, thus every person would have access to functions that were both low order and high order. His theoretical urban hierarchy would provide the optimum number of central places, along with the most efficient pattern of hinterlands. Figure 7.11 illustrates the spatial pattern of hinterlands and central places in Christaller's central place theory.

Geographers have debated the value of the central place model in more recent times as more and more studies of urban hierarchies failed to repeat the patterns identified by Christaller. In one sense this gives value to the model. It can be used as a fixed reference point against which other studies can be compared. It also allows researchers to find factors that account for the differences between the model and the urban network being studied. Table 7.1 summarises a number of factors that account for variations in range and threshold and therefore in the size and spacing of central places and the shapes of their hinterlands.

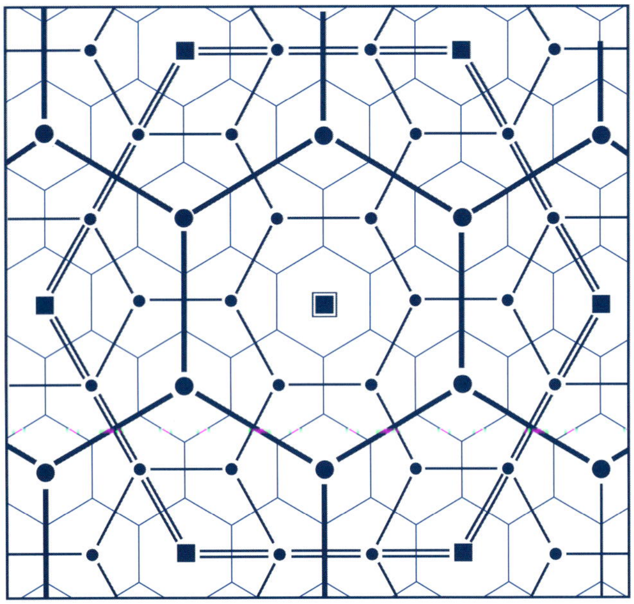

**Table 7.1** Factors responsible for the variations in ranges and thresholds

| | |
|---|---|
| Physical characteristics of hinterland | • Fertility of the soil<br>• Nature of topography<br>• Climatic conditions<br>• Natural vegetation<br>• Drainage conditions |
| Economic characteristics | • Agricultural type and intensity<br>• Rural population density<br>• Transport networks<br>• Other functions of the central places (e.g. industrial activities) |
| Political/historical aspects | • Government policies (e.g. planning)<br>• State borders<br>• Historical development of the area |
| Consumer characteristics | • Income levels (spending power)<br>• Consumer mobility<br>• Age<br>• Sex<br>• Tastes/values<br>• Cultural characteristics |
| Marketing characteristics | • Advertising<br>• Shop layout and attractiveness<br>• Availability of credit<br>• Service type and quality |
| Changing technology (as it affects other characteristics) | • Transport technology<br>• Refrigeration technology |

**Figure 7.11** Walter Christaller's 'central place theory': the arrangement of central places given uniform conditions

Central places:
- ■ city......... fourth order
- ■ town....... third order
- ● village..... second order
- • hamlet.... first order

Hinterland boundaries:
- ═══ city
- ━━━ town
- ─── village
- ─── hamlet

### EXERCISES

1. Define the following terms:
   - urban network
   - function
   - population threshold
   - hierarchy
   - hinterland or draw area
   - central place
   - range of good or service
   - central place order
   - functional complexity

2. With reference to figure 7.11:
   (a) Draw a central place model graph to show the number of central places in each of the four orders illustrated in figure 7.11. This should be similar in arrangement to figure 7.8. How many towns occur at the first, second, third and fourth order levels and do they illustate a ratio? Suggest reasons for any patterns that you have observed.
   (b) Suggest the possible pre-conditions that Christaller might have set in developing his theoretical model. In answering this question, consider the need for uniformity in all variables.
   (c) Suggest reasons why the model uses hexagonal rather than circular or square hinterlands.
   (d) Suggest at least four reasons why it is impossible to find urban networks that are the same as the theoretical pattern.
   (e) What do you consider is the value of Christaller's central place model or theory?

3 Study the following illustrations of range and threshold and then explain the relationship between these two concepts in each instance. It is important to have a good understanding of the definitions for these two concepts when explaining the relationships. Assume that the same number of people make up the threshold population in each instance and that the circle represents their distribution.

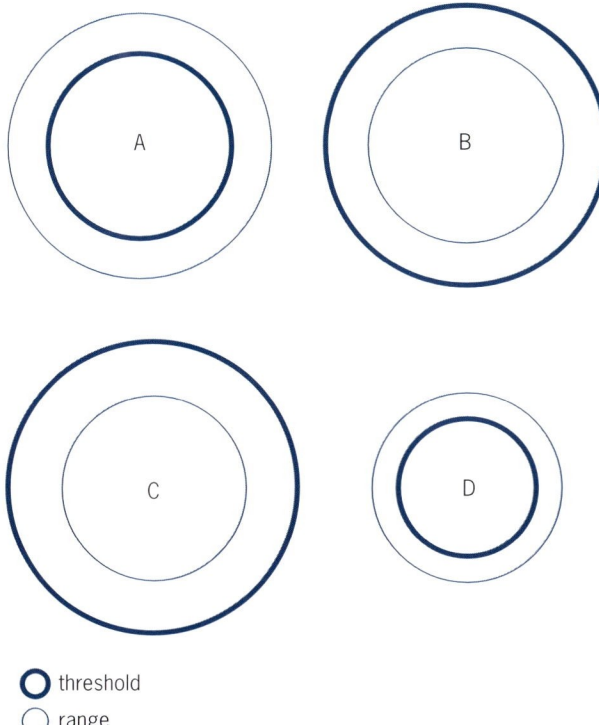

○ threshold
○ range

**Figure 7.12**

## The urban network of the South-West

Within the South-West it is possible to study the size and distribution of central places from a number of different perspectives. An analysis of populations in different centres will reveal a primate urban hierarchy centred on Perth. In table 7.2 this is clearly revealed. The growth of Perth at the expense of other towns in the region has been an ongoing process. In 1911, 42 per cent of the population lived in the city. This had increased to 60 per cent following World War II and in the 1980s had reached 70 per cent. In 2000, 73 per cent of the population lived in the city. The long-term dominance of Perth has focused the greater proportion of high-order functions within the city and has limited the growth of the other regional towns.

Below Perth, a cluster of five centres forms the next level of the hierarchy. These include Kalgoorlie, Geraldton, Mandurah, Bunbury and Albany. As figure 7.13 shows, there are discrete orders within the South-West and, depending on the selection process, it is possible to argue that the urban hierarchy has five or six orders. As the graph reveals, for all centres over 200 people there about four steps as the population jumps from one order to the next. These, however, are not particularly strong. The conclusion that may be drawn from this is that, although the urban network of the South-West shows elements of the central place model with its discrete classes, it more closely resembles the primacy hierarchy.

In figure 7.14, Perth has been included and the pattern of urban population distribution shows the strong degree of primacy exhibited by the urban network of the South-West. In order to show Perth and the other centres on the same graph, a logarithmic scale has been used. If drawn using a set scale then the gap would be much greater between Perth and Mandurah, which is the next largest centre.

Table 7.2 shows the settlements of the South-West ranked in order of population size and this data was used in the construction of the graphs in figures 7.13 and 7.14.

## Factors influencing the urban network

There are a number of processes and factors that have acted to shape the urban network of the South-West. These include variations in the biophysical environment, variations in land uses (both rural and industrial) and variations in transport patterns and levels of accessibility. In addition social, political, economic and technological factors have over time interacted in various ways to shape the network.

### The influence of rural population distribution

The pattern of urban settlements in the South-West can largely be explained by the rural population densities of various agricultural regions within the study area. These population densities in turn can be attributed to the biophysical conditions of climate, relief and drainage (and to the economic considerations of the relative profitability of agricultural enterprises in a given area). The restrictions to agricultural development in the area are the lower, very marshy lands and infertile soils of some coastal regions; the steeper, dissected slopes of the Darling Scarp (and its river catchment region); the slopes of the Porongorups and the Stirling Range; and the salt lake areas to the south-east and north-east of the study area. With the exception of these areas and other regions that have been left forested, agriculture developed unhindered over most of the coastal plain and plateau. Along the wetter coastal and river plain areas to the south of Perth, intensive agricultural development has occurred (market gardening, orcharding and intensive pastoralism), resulting in relatively high rural population densities. This in turn has encouraged the establishment of relatively large numbers of urban centres of various orders to service the farming population.

While the high rural population densities of the coastal plain can largely be attributed to the more favourable physical conditions for agriculture, the 'thinning' of the rural population towards the east results from drier climatic conditions and the development of more extensive agricultural land uses. The wheat–sheep lands have lower population densities as a result of the larger farm sizes required. Drier areas further inland (beyond Merredin and Lake Grace) have even lower population densities, because rainfall is lower. The number of urban settlements diminishes with increasing distance inland as there are insufficient people in the area to support the establishment of a large number of central places. Towns are spaced more widely and there are fewer lower-order centres. The towns that do exist in the wheatbelt area have hinterlands more than double the size of the territory of a similar-sized town on the coastal plain. Figure 7.15

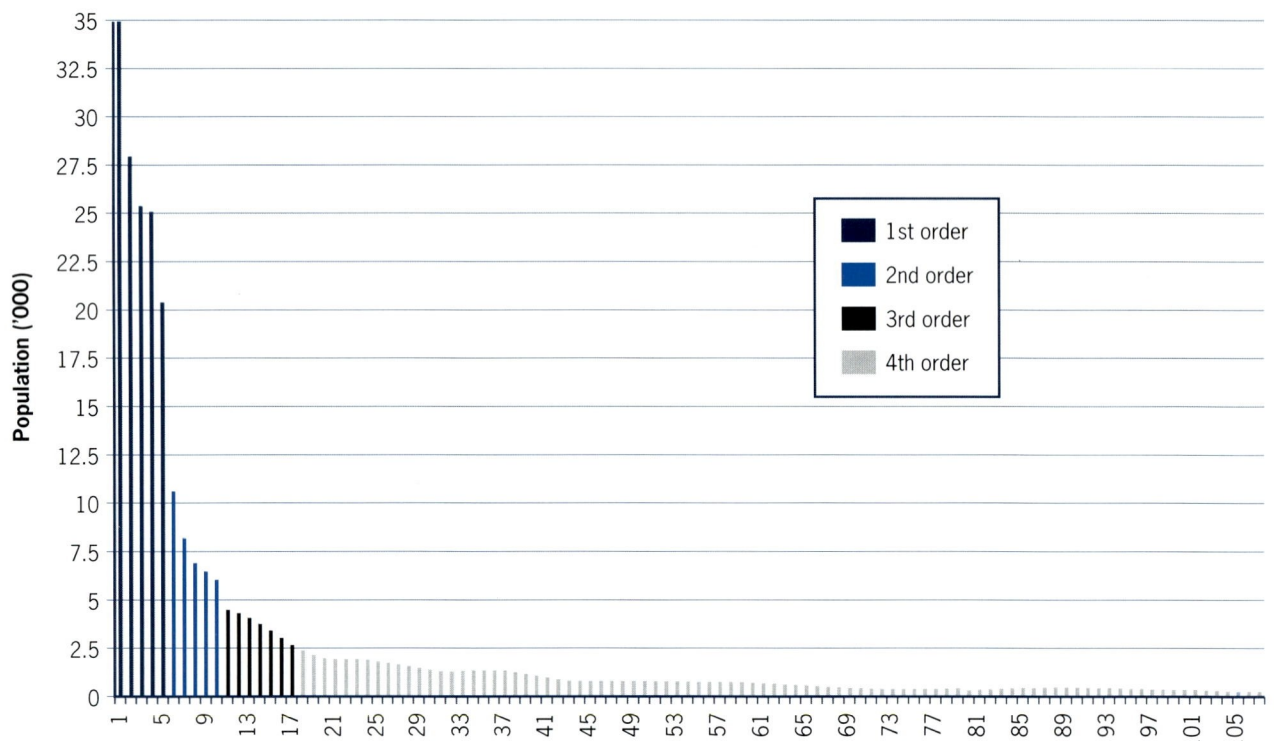

*Note:* Settlements under 200 people are not identified by the ABS—their populations are counted in with the closest centre over 200 people

**Figure 7.13** Regional towns of the South-West (excluding Perth)

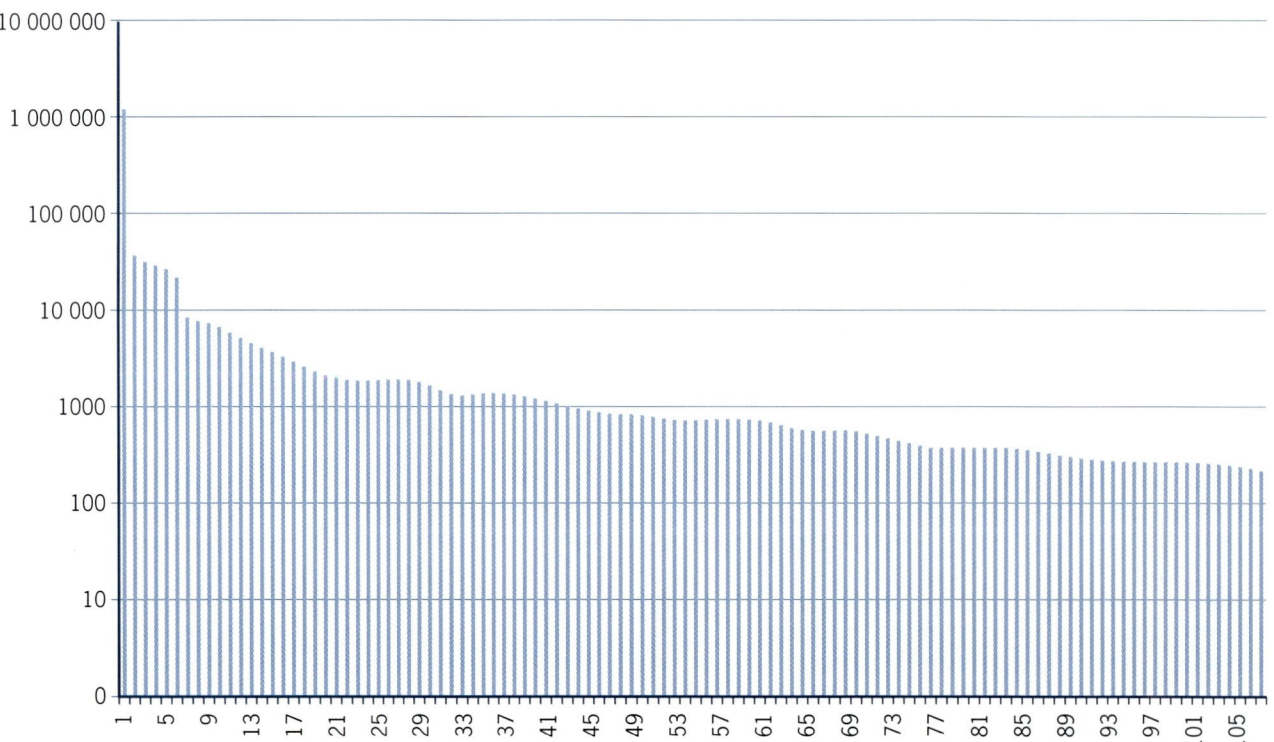

**Figure 7.14** Primacy network of the South-West (including Perth)

shows the pattern of urban centres in the lower South-West as they relate to the population densities of the various shires. In general, population densities decrease from western coastal areas to the east, and the largest concentration of settlement occurs on the coast (especially at river mouths) and along the coastal plain. The location of concentrated settlement corresponds generally to intensive agricultural production in the area. An absence of settlements in the scarp area can be attributed to the lack of agricultural development in this forested area. Forested areas also occur around Nannup and between Manjimup and Walpole. The wide spacing of the relatively few urban centres in the marginal wheat–sheep belt areas is the result of the large farm sizes and low rural population densities.

## Transport patterns

The distribution of urban settlements in the South-West has been strongly influenced by transport links, creating linear patterns of urban settlement along major highways and the railway lines that run often parallel to the highways (see figure 7.16). The reason for the particularly pronounced alignment of centres along these routes (and at their junctions) is the high degree of accessibility provided by the transport links. The number of settlements located along a major road or railway relates directly to the intensity of the land use and rural population densities. The South Western Highway (and railway) between Perth and Bunbury is only 160 kilometres long, but has 21 settlements located along it, many of which are small although others such as Pinjarra, Waroona and Harvey are slightly larger.

Along the Great Southern Highway and railway between Northam and Albany, settlements are more widely scattered (29 over a distance of approximately 450 kilometres), reflecting the lower rural population densities of these regions. York, Beverley, Narrogin, Katanning and Mount Barker are relatively high-order centres along this route. This strong linear pattern of widely spaced central places is repeated along the Great Eastern Highway and transcontinental railway. The pronounced alignment of settlements along the railway line/road systems in the wheat belt areas occurred largely because of the need for 'collection and

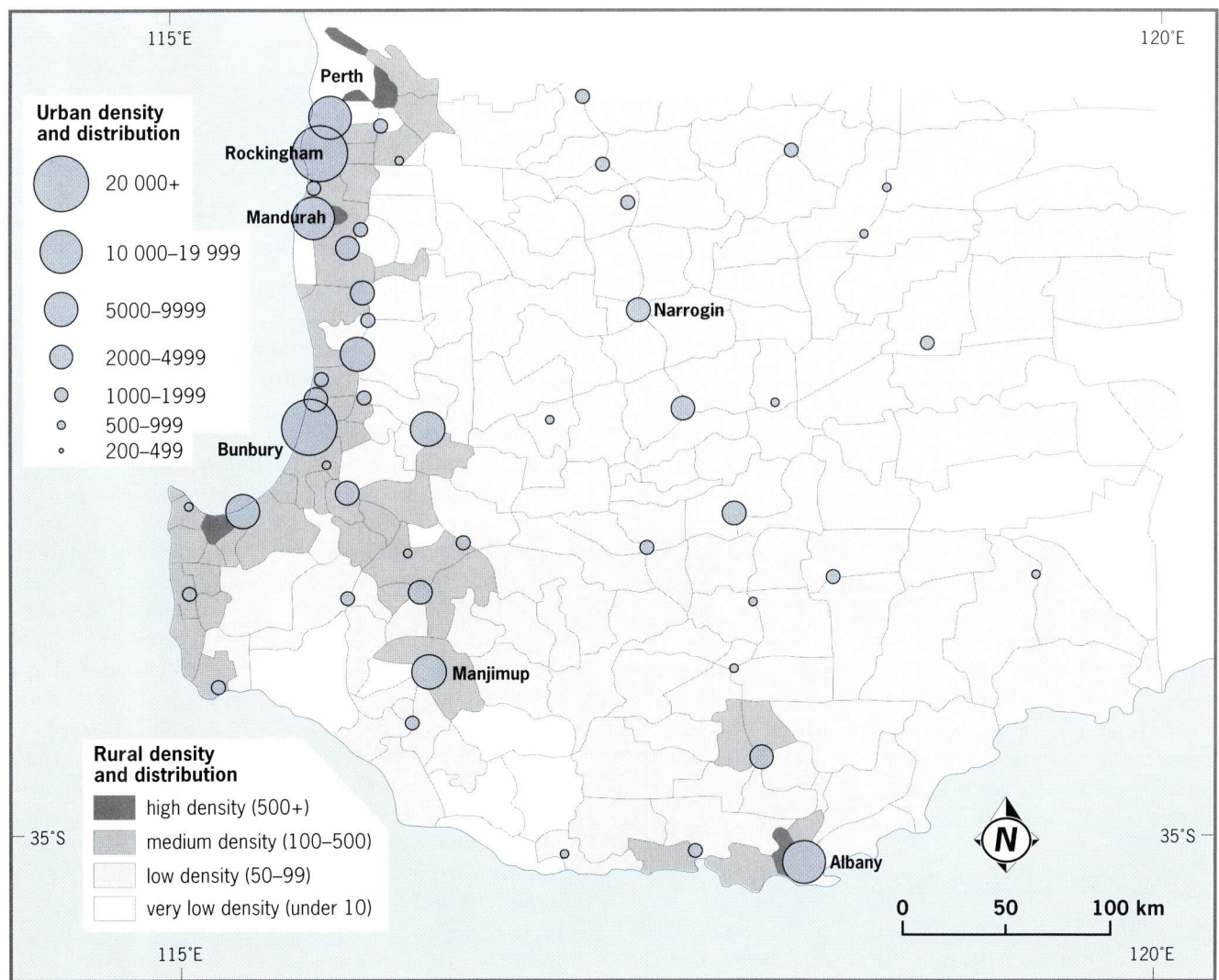

**Figure 7.15** Relationships between urban centres and population densities of shires in the lower South-West

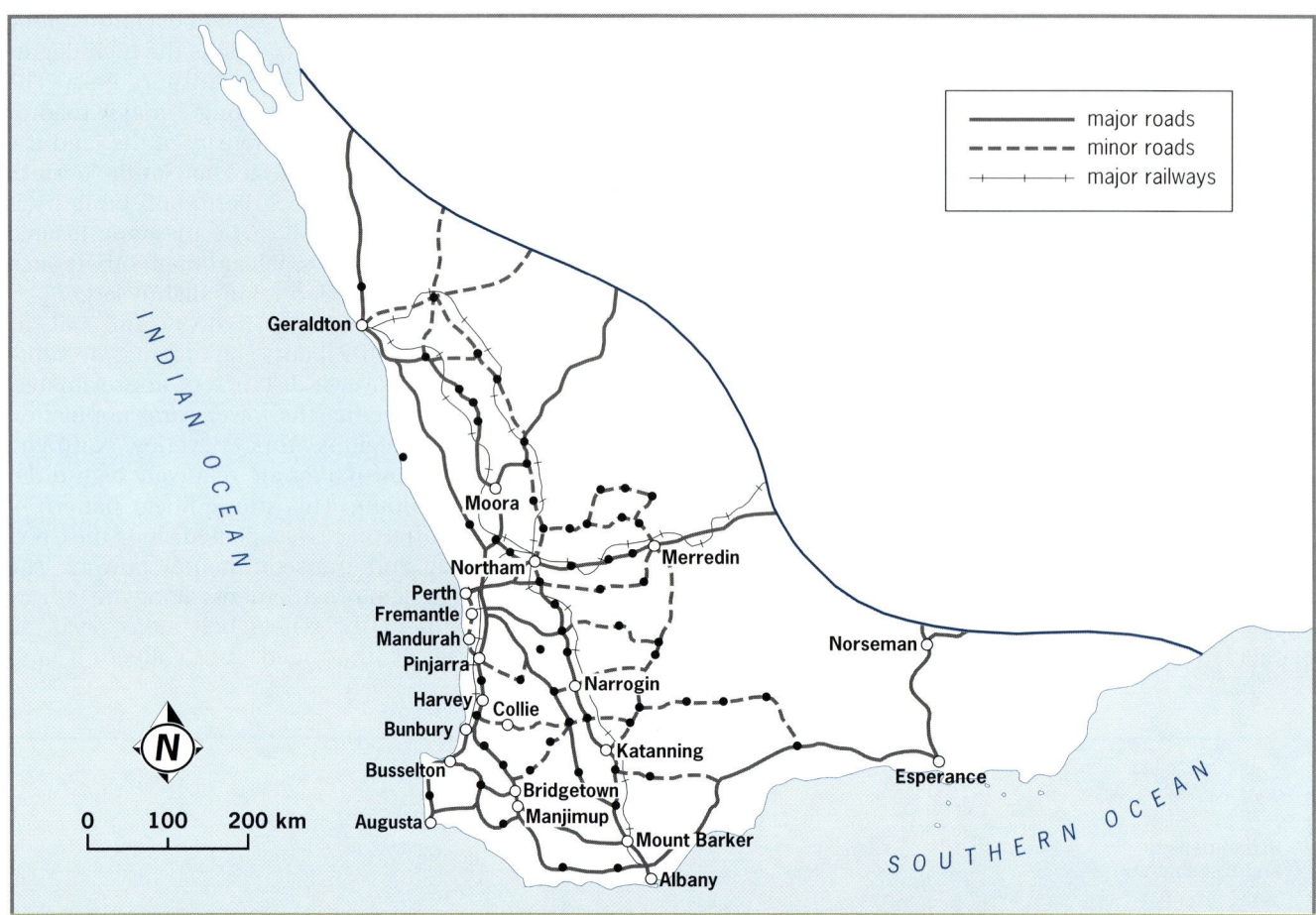

**Figure 7.16** Relationship between location of urban centres and major transport links in the South-West

distribution' points as the area under wheat production expanded in the 1920s and the railways pushed into the interior. The role of many of these centres was to collect wheat from the hinterland farms. Railways provided transportation to the port of Fremantle, from where the wheat was shipped to overseas markets. Goods that were required by farmers were transported back from Perth along the same route and were distributed to the farms through the towns. These centres also provided the central place functions that were required by the rural population of the hinterland. The effect of the linear development of transport links has been to increase the distance that people are prepared to travel to buy goods and services offered by central places. As a result, the boundaries of trade territories tend to overlap along the highways. The linear pattern of settlement has also resulted in the demise of many smaller central places (as explained more fully in the case study of the Avon).

## Special function centres

Not all the urban centres are central places. Some exist as special function (special purpose) centres. Unlike central places, they are not wholly dependent on a surrounding rural hinterland. Rather, special function centres are established and grow in a specific location, mainly in response to the presence of some resource or as a result of some special influence. Types of specialist centres include mining, fishing, sawmilling, retirement and tourist settlements.

Although special purpose towns in the South-West differ in terms of their general function, they have many similar characteristics. The goods and services they provide tend to reflect their dominant function. Some specialist centres offer limited functions relative to their large populations (for example mining centres), while others such as tourist centres have relatively small populations but offer a disproportionately large number of functions. The age–sex structure of the population of a specialist town also reflects its primary function. An example of this is the town of Boddington (bauxite and gold) which is a predominantly male community with a few retirees. Many of the towns in the South-West are multi-functional, as reflected in the range of goods and services they offer. The town of Manjimup, for example, is the centre of a large forestry and agricultural region. Yarloop is also a 'timber town' and it services a dairying hinterland. Bridgetown's hinterland contains areas of millable timber and softwood plantations. In addition, sheep and cattle grazing and fruit growing are important economic activities.

## The evolution of the network

The sequence of occupation of the land over time in response to the growth of agriculture, mining, forestry and fishing in the region illustrates the importance of historical influences in shaping the urban network. From the centres of Perth and Albany, which acted as 'springboards', settlement pushed into the interior in reasonably distinct stages from 1829 onwards.

### 1829 to 1880

The period 1829 to 1880 saw the establishment of the initial settlements of Perth, Albany, Bunbury, Busselton, Australind and Augusta. These years also saw the limited expansion of agriculture and the consequent establishment of inland towns in the South-West. The first towns to be established away from the coast were Pinjarra, to the south, and Toodyay, Beverley, York and Northam on the Avon River (details regarding the establishment and growth of Northam are provided in chapter 8). Between the 1840s and the 1880s, limited settlement of the Swan Coastal Plain occurred. Geraldton and other northern towns were established at this time and a few towns (for example, Kojonup) appeared in today's western wheat belt area. In general, the number of towns established before the 1890s remained low as a result of the sparse rural population.

### 1880 to 1920

Most of the town sites in the South-West were established between 1880 and 1920, largely in response to the gold rushes of the 1890s, the influx of population and the development of the rail network. The development of the Eastern Goldfields centres and their linkage by rail to Perth, as well as the wealth that gold brought, helped in the expansion of agriculture to the east of the broad coastal area. Development of the railway network led to a rapid expansion in the number of urban settlements along the lines, creating a linear pattern of development.

### 1920 to 1950

The period to 1950 saw the extension of the wheat belt to the east and north-east and the establishment of more towns as collection and distribution centres along the railway. Schemes that catered for returned soldiers (for example, the Group Settlement Scheme) resulted in the opening up of the deep South-West for intensive pastoralism, and in the establishment of towns such as Northcliffe, Witchcliffe and Walpole. Settlement of the coastal plain was also intensified because of these schemes. The area just north of Esperance was opened up at this time.

### 1950 to the present

After 1950, three broad types of settlement were established. Those such as Jerramungup and Fitzgerald serviced the newly opened agricultural lands west of Esperance. A number of service towns were also founded on the Old Coast Road between Perth and Bunbury including Myalup, Binningup and Eaton. To the north of Perth, the coastal cray fishing centres of Jurien, Two Rocks and Leeman were established.

## Changes in the urban network

The urban network of the South-West has undergone changes in the past and will continue to change as the central places and hinterlands re-adjust to new influences or forces. Changes affect both individual centres within the network and the network hierarchy as a whole. The change in the functional complexity of any node is directly related to changes in its hinterland. The most important changes that are occurring in the South-West regional network are the increasing dominance of the primate centre (Perth) and the stagnation and even, in some cases, disappearance of some of the smaller towns in the hierarchy.

### Urban primacy and changing transport technology

As discussed earlier in this chapter, Perth has increased its dominance (functional complexity) at the expense of other centres within the urban network. Most of the smaller centres, especially hamlets and small country towns, are either stagnating or declining. Only those fortunate enough to have a strategic location within the network, or centres that have been successful in expanding their economic base, are still developing. Technological change has been the primary force behind the changing face of the urban network of the South-West. Better roads, faster cars, a more extensive rail network (and even faster, cheaper air travel) have overcome the problem of distance. As a result, the range of goods and services offered by urban centres has increased since the days of 'horse and buggy' transport, and shoppers may tend to bypass smaller towns in favour of others. This has occurred especially since the advent of 'one-stop', 'once per week' shopping in large centralised shopping centres for low-order goods, which can now be refrigerated. The population thresholds for many low-order goods and services have dropped in many centres, with the result that the function can no longer be offered there. This situation has resulted in the demise of many small centres in the South-West. Typical of such centres are Roelands on the South Western Highway, Konnongorring near Wongan Hills and Gutha near Morowa. Other northern centres such as Mingenew, Three Springs and Carnamah have also experienced substantial population loss, which can be attributed to the diversion of the main north-bound traffic from Perth away from Midland Road to the Brand Highway. The latter negotiates a shorter coastal route to Geraldton.

### Rural amalgamation

Another reason for the demise of small centres is the rural depopulation that occurs as a result of the mechanisation of agriculture and the subsequent amalgamation of farms. The aim of this process is to consolidate farming operations into more economically viable units. This replacement of some farm labour by machines and the absorption of one farm by another (to enable greater value and efficiency to be gained from the use of the machinery) leads to a decline in rural population, thus affecting a town's market threshold. This situation has a snowballing effect, as people employed in town businesses are forced to move to larger centres to find employment.

## Economic factors

Downturns in commodity prices associated with economic recessions also have a significant effect in reducing rural population. While the demise of small country towns can occur gradually over time, that of more specialised centres can occur very suddenly. The creation of 'ghost towns' from once-thriving mining centres in the South-West has been well documented. Forestry towns can also have a short lifespan. Grimwade was a small forestry town located about 80 kilometres south-east of Bunbury. It was established in 1910 by Millar's Trading Company for employees working at a large jarrah sawmill at East Kirup. The town depended on this forestry function. In the late 1980s the town had only 40 residents, but in 1990 it was closed because many functions relocated to nearby Kirup, 17 kilometres away. The effect of the state government's recent decision to stop logging of old growth forests has had a similar impact on the timber towns of Pemberton and Manjimup, where there have been significant job losses in the industry.

The impending decline of some country towns has been halted by a broadening of their economic base. Many towns have been able, because of their locational advantage and their historical interest, to tap a growing leisure, tourist and 'hobby farmer' market. York and Toodyay illustrate this change and these will be discussed further in the Avon region case study.

### EXERCISES

1. List the factors that have influenced the number and type of urban centres in the South-West.
2. Describe the distribution pattern of urban centres in the South-West.
3. Study the graphs of central places in the South-West and discuss the extent to which they illustrate primacy and central place hierarchies.
4. What factors have contributed to the decline in the number of low-order centres in the South-West?
5. Using examples, explain the difference between a specialist and a central place settlement.
6. Explain why the range of goods and services and the size of hinterlands varies from west to east in the South-West.
7. Using the data provided in Table 7.2, group the urban centres into five or six different orders and then plot these on a map of the South-West. Identify and account for any apparent patterns of distribution of the different orders.

**Table 7.2** Population of urban centres of the South-West

| Urban centre | Population, 1996 | Urban centre | Population, 1996 |
|---|---|---|---|
| Perth | 1 192 820 | Corrigin | 703 |
| Mandurah | 35 945 | Dalwallinu | 697 |
| Kalgoorlie | 28 087 | Morawa | 692 |
| Geraldton | 25 244 | Toodyay | 674 |
| Bunbury | 24 945 | Chidlow | 672 |
| Albany | 20 493 | Jurien | 636 |
| Busselton | 10 642 | Yarloop | 619 |
| Esperance | 8 647 | Wundowie | 613 |
| Collie | 7 194 | Lancelin | 597 |
| Northam | 6 300 | Mullewa | 591 |
| Australind | 5 694 | Bruce Rock | 579 |
| Narrogin | 4 491 | Lake Grace | 575 |
| Manjimup | 4 390 | Boyanup | 575 |
| Katanning | 4 035 | Allanson | 572 |
| Golden Bay | 3 671 | Boyup Brook | 553 |
| Merredin | 2 911 | Gingin | 549 |
| Margaret River | 2 846 | Leeman | 531 |
| Harvey | 2 570 | Brookton | 526 |
| Bridgetown | 2 123 | Nannup | 521 |
| Byford | 2 102 | Goomalling | 482 |
| Denmark | 1 978 | Cervantes | 480 |
| York | 1 923 | Narembeen | 459 |
| Pinjarra | 1 892 | Rottnest Island | 412 |
| Dongara | 1 874 | Three Springs | 411 |
| Waroona | 1 883 | Greenbushes | 403 |
| Kalbarri | 1 788 | Dwellingup | 399 |
| Yunderup | 1 751 | Eneabba | 389 |
| Gelorup | 1 728 | Williams | 384 |
| Moora | 1 664 | Dowerin | 378 |
| Mt Barker | 1 648 | Quindalup | 365 |
| Donnybrook | 1 635 | Ravensthorpe | 354 |
| Norseman | 1 516 | Wyalkatchem | 349 |
| Wagin | 1 337 | Koorda | 348 |
| Coolgardie | 1 258 | Mukinbudin | 347 |
| Capel | 1 258 | Carnamah | 338 |
| Kambalda | 1 200 | Walpole | 337 |
| Dunsborough | 1 154 | Jerramungup | 332 |
| Southern Cross | 1 147 | Tambellup | 325 |
| Augusta | 1 087 | Kondinin | 322 |
| Boddington | 1 043 | Hopetoun | 319 |
| Kojonup | 1 035 | Dardanup | 314 |
| North Pinjarra | 1 011 | Mingenew | 313 |
| Furnissdale | 997 | Cranbrook | 283 |
| Pemberton | 994 | Kulin | 271 |
| Binningup | 859 | Burekup | 265 |
| Kellerberrin | 855 | Darkan | 265 |
| Northampton | 842 | Dumbelyung | 264 |
| Wongan Hills | 813 | Green Head | 245 |
| Beverley | 787 | Peppermint Grove | 245 |
| Pingelly | 756 | Wooroloo | 244 |
| Brunswick Junction | 752 | Cuballing | 241 |
| Gnowangerup | 737 | Northcliffe | 239 |
| Cunderdin | 715 | Tammin | 236 |
| Quairading | 706 | Bremer Bay | 221 |

## CASE STUDY

## The urban network of the Avon region

The Avon statistical area is located around 50 kilometres inland from Perth, in the central wheat belt. It is a rectangular area approximately 200 kilometres by 100 kilometres in size. An extensive survey of the urban network of the region, undertaken in June 2002, forms the basis of this case study. Information was gained from field surveys, local business directories, local government statistical data, ABS data and earlier research documents. The region was chosen because it illustrates many of the processes and influences that apply to the urban network of the South-West as a whole. These have been summarised below and will be discussed in greater detail in the body of the case study.

- Evolution of the urban network—growth and decline
- Rural land use and central place functions
- Central places and specialist places
- The nature of the urban hierarchy
- The future of the urban network

### The evolving urban network

The development of central places within the Avon region conforms to the premise that the population of a certain amount of production land will set up a sufficient demand for non-rural goods and services to justify the establishment of business to provide these facilities, and that these functions will most effectively be provided for at a central node within the region.

As the map of central places in the Avon (figure 7.17) indicates, the network mainly developed in line with the pattern of railway construction. There were, however, two distinctive phases to the growth of towns within the region. The first was the pre-railway phase. The towns of Toodyay, Northam, York and Beverley were established prior to the 1890s and used the Avon River as their focus. There were several reasons for this. The river and its tributaries provided the early settlers with a way of travelling into an unknown region. While not suitable for boats, it was a route along which settlers travelled to and from Perth. The river valleys were also areas of fertile alluvial soils and provided permanent water supplies. As farms were established within the Avon and Mortlock valleys, towns developed to serve these rural communities.

After the 1890s, there was a long period of railway development and associated rural and urban settlement. The railways were not constructed all at one time. As each section was constructed, the farmland was allocated to settlers and in time a town developed at the rail head. This was then followed by further extension of the rail into the region. The town of Goomalling to the east of Toodyay illustrates this process.

The land around Goomalling was first settled in 1853, when a Mr George Slater secured a pastoral lease around Goomalling Spring. With the discovery of gold at Coolgardie in 1887, the Slater homestead became a stopping point for travellers on their way to the goldfields and more people began to pass through the district. In addition, there was an increase in the number of farming pioneers and, in 1885, the Goomalling Roads Board was established to oversee the construction of roads within the district. The town of Goomalling did not come into existence until 1903 and was the result of the completion of the railway to this point (the railway continued past this point as it developed north towards Wongan Hills and Dalwallinu, and east to Dowerin). The town commenced with an initial release of 30 blocks. Some of the first buildings to be constructed included the Goomalling Hotel and the railway station. Closely following this was a blacksmith, saddlery, general store, bakery, bank and a fruit and vegetable shop. Several years later, services including a police station, medical practice and several churches were established. The first setback for the young town was World War I, when a large percentage of the district's young men enlisted, with almost 25 per cent killed in action. Following the war, the district grew rapidly with a large rural population being established. The use of horse-drawn equipment and small-scale machinery (see figure 7.18) at this time required a large workforce to be present during seeding and harvesting. In addition, the small farm sizes (100–200 hectares) encouraged a closely settled rural community. Towards the end of the 1920s, the effect of the Great Depression and the massive problems posed by rabbits brought this growth to a halt. The town's growth faltered, with many businesses and farmers leaving, and the population declined. In some respects the district did not fully recover from this setback. After World War II, the 1950s brought high wool prices and the introduction of myxomatosis finally brought the rabbit numbers under control. This was to be the last significant period of expansion in the town, with a number of new services and clubs being added to the facilities. By the mid-1960s the region's population reached 1500 people, but the boom times had ended and the town started a slow decline. The shire population fell to around 1000 by the early 1990s and a number of businesses closed. Particularly significant was the closure of two banks and the district high school. Today, the town continues to survive and look for ways to bring people and services into the district. The spread of hobby farms along its western boundary, near Toodyay, offers some possibilities. Like many other small towns in the Avon region, it encourages activities that will bring people to the town. One such event is the re-establishment of the 'round the houses' car racing event, similar to the 'flying 50' conducted in the town of York, which is located in the south-west of the Avon region.

In recent times, several centres within the study area have actively pursued policies of growth aimed at attracting both people and services. The survey of the region in June 2002 revealed evidence of marked growth in the following towns: Dalwallinu (where a number of houses and a new motel were under construction), Dowerin (with the construction of a number of houses as well as several new light industrial buildings) and

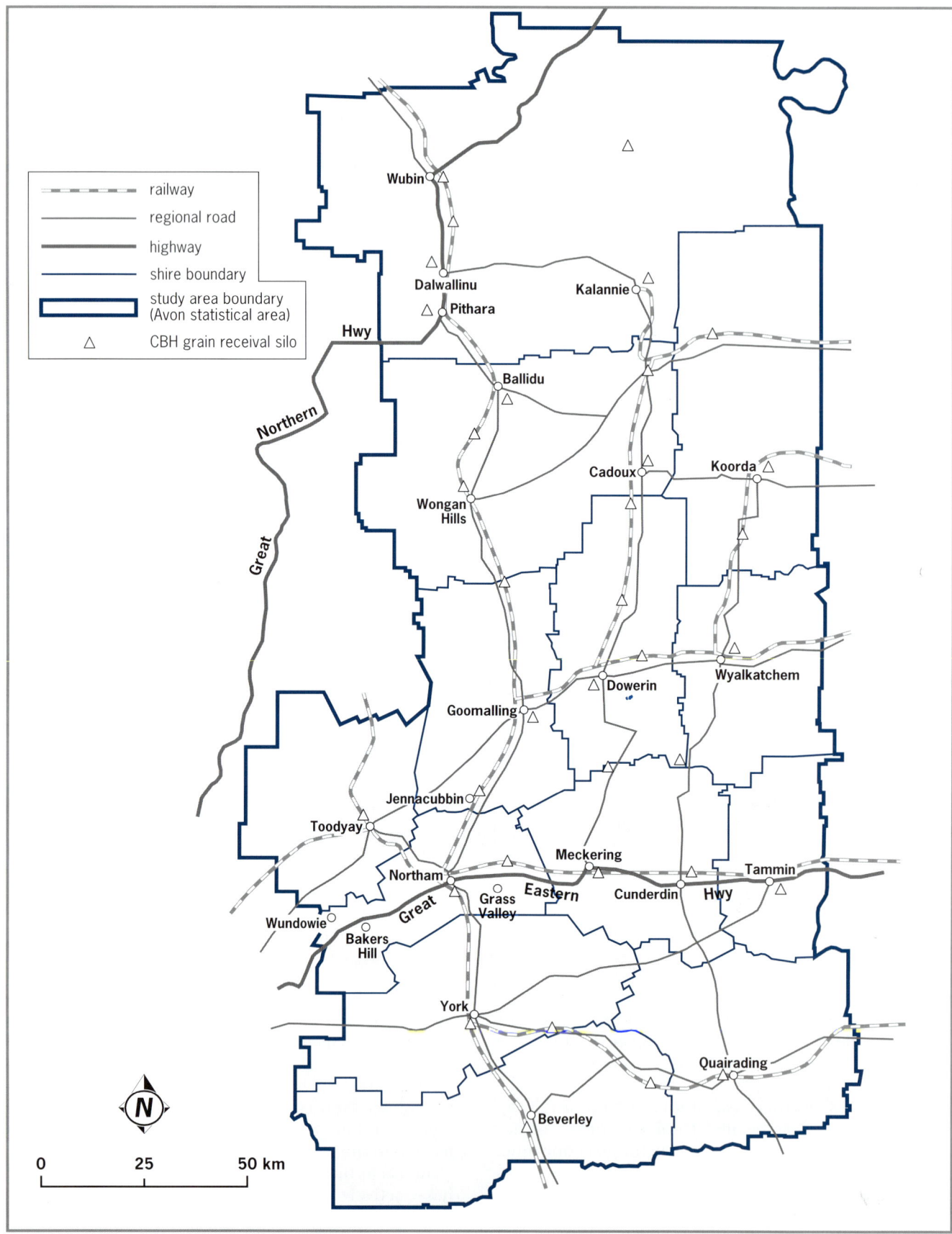

**Figure 7.17** Urban network of the Avon study area

# SETTLEMENT PATTERNS AND URBAN NETWORKS

**Figure 7.18** Early horse-drawn farm equipment

Northam (where several new housing and industrial estates have recently been created).

Other towns such as York, Quairading and Toodyay all indicated economic stability, while the towns of Wongan Hills and Cunderdin showed evidence of losses of functions as well as some new ones being introduced. One town that had undergone significant change was Meckering.

In 1968, Meckering was the centre of a major earthquake in the region. Virtually all of the buildings in the main street were destroyed. These included substantial commercial and residential properties. The effect of this event can be seen in figure 7.21, a buckled section of the trans-Australia rail line which runs past the town. The fault that opened up caused the land to drop about a metre along its length. While the town was rebuilt after the event, it had lost a lot of its accumulated building capital and, unlike other towns which still retain many of their older buildings, the town had little building stock to house its functions. In June 2002, all functions with the exception of a roadhouse had closed and about 25 per cent of the 100 houses were up for sale. The event of 1968 and the expanded influence of Northam to the west and Cunderdin to the east have had a dramatic effect on the town's future survival.

## Central place functions

The Avon region illustrates a distribution of central place functions in keeping with the basic concepts of threshold population and range. Tables 7.3 and 7.4 show the frequency and types of function found in different size settlements within the region. A preliminary study of these functions will reveal a number of characteristics of the urban network.

### Community functions

Both local and state governments have been major contributors to the provision of services to people in the central wheat belt. This is clearly evident in many of the towns, where recreational, educational, health, fire and police services are to be found. Community

**Figure 7.19** New motel at Dalwallinu

**Table 7.3** Functions of the Avon region—smaller central places

| Services | Jennacubbin | Ballidu | Pithara | Wubin | Kalannie | Cadoux | Meckering | Grass Valley |
|---|---|---|---|---|---|---|---|---|
| *Accommodation* | | | | | | | | |
| Caravan park | | 1 | | 1 | 1 | | | |
| Hotel | 1 | 1 | 1 | 1 | | | | 1 |
| Motel/units/B&B | 1 | | | 1 | | | | |
| *Agricultural services* | | | | | | | | |
| Farm machinery sales | | | | | 1 | | | |
| Farm supplies general | | 2 | 1* | 1 | 1 | 1 | 1* | |
| Seed/stock food | | | | | 2 | | | |
| Wheat silos | 1 | 1 | 1 | 1 | 1 | 1 | | 1 |
| *Community services* | | | | | | | | |
| Church | 1 | 2 | | 1 | 1 | | 1 | |
| Club/association | 1 | | | | | | 1 | |
| Halls | 1 | 1 | 1 | 1 | 1 | | 1 | 1 |
| Leisure/sporting facilities | 1 | 3 | 2 | 2 | 3 | 4 | 2 | 3 |
| Museum/art gallery | | 1 | | | | | 1 | |
| *Educational* | | | | | | | | |
| Primary and preschool | | 1 | | 1 | 1 | 1 | 1 | 1 |
| *Manufacturing* | | | | | | | | |
| Engineering–fabrication | | 1 | | 1 | 1 | | | |
| *Motor vehicles* | | | | | | | | |
| Bulk fuel depot | | | | | 1 | | | |
| Fuel sales retail | | | | 2 | 1 | 1 | 1 | |
| Tyre sales | | 1 | | | | | | |
| *Other services* | | | | | | | | |
| Fire and emergency | | 1 | | | | | | 1 |
| Post office/agent | | 1 | 1 | | 1 | | 1* | |
| Telecentre | | | | | 1 | | | |
| *Retail–consumables* | | | | | | | | |
| Butcher | | | | | | | 1* | |
| General store/deli | | 2 | 2* | 1 | 1 | 1 | 1* | |
| Newsagent | | 1 | 1 | | | | 1* | |
| *Transport/storage* | | | | | | | | |
| Airstrip | | 1 | | | | | | |
| Haulage–truck | | | | | 1 | | | |
| Rail–freight | 1 | 1 | 1 | | 1 | 1 | | |
| Rail–passenger | | | | | | | 1 | 1 |
| Total services | 8 | 22 | 8 | 14 | 20 | 10 | 9 | 9 |
| Town population 1996 | 8 | 80 | 15 | 110 | 150 | 110 | 150 | 45 |
| Shire | Goomalling | Wongan Hills | Dalwallinu | Dalwallinu | Dalwallinu | Wongan Hills | Cunderdin | Northam |

Note: *indicates not in use

**Table 7.4** Functions of the Avon region—larger central places

| Services | Koorda | Wyalkatchem | Tammin | Bakers Hill | Wundowie | Toodyay | Goomalling | Wongan Hills | Dalwallinu | Dowerin | Cunderdin | Quairading | York | Beverley | Northam |
|---|---|---|---|---|---|---|---|---|---|---|---|---|---|---|---|
| *Accommodation* | | | | | | | | | | | | | | | |
| Caravan park | 1 | 1 | 1 | 1 | | 3 | 1 | 1 | 1 | 2 | 1 | 1 | 1 | 1 | 2 |
| Hotel | 1 | 1 | 1 | 1 | | 3 | 1 | 1 | 1 | 1 | * | 1 | 3 | 2 | 6 |
| Motel/units/B&B | 1 | 1 | 1 | | | 5 | | 3 | 2 | 2 | | 2 | 14 | 1 | 6 |
| *Agricultural services* | | | | | | | | | | | | | | | |
| Farm machinery | 1 | | | | | 1 | 3 | 4 | 2 | 2 | 5 | 2 | 1 | 1 | 3 |
| Farm supplies–general | 4 | 2 | 2 | 1 | | 1 | 1 | 3 | 3 | 3 | 6 | 3 | 3 | 4 | |
| Livestock sale yards | | 1 | 1 | | | | | 1 | | 1 | 1 | 3 | | | 2 |
| Seed/stock food | | | | | | | | 1 | | 2 | 1 | 1 | 1 | | 2 |
| Veterinary practice | | 1 | | 1 | | 1 | 1 | | | | | 1 | 1 | 1 | 2 |
| Wheat silos | 1 | 1 | 1 | 1 | | 1 | 1 | 1 | 1 | 1 | 1 | 1 | 1 | 1 | 1 |
| *Building* | | | | | | | | | | | | | | | |
| Architect/builder | | | | | | | 1 | | | 1 | | | | 2 | 2 |
| Earthworks | | | | | | | | | | 1 | | | | 1 | 5 |
| Electrician/plumber | | | 1 | | | 1 | | 1 | 1 | 1 | 1 | 1 | 2 | 1 | 5 |
| Garden supplies/plants | | 1 | 1 | 1 | | 1 | | 1 | 1 | 1 | 1 | 1 | 2 | 1 | 3 |
| Painter/decorator | | | | | | | | | | | 1 | | 2 | | 2 |
| Salvage | | | | | | 1 | | | | | | | | | 1 |
| *Community services* | | | | | | | | | | | | | | | |
| Cemetery | 1 | 1 | | | | 1 | 1 | | | 1 | | 1 | 1 | 1 | 1 |
| Church | 2 | 1 | 3 | 2 | 2 | 4 | 4 | 3 | 5 | 3 | 2 | 3 | 6 | 4 | 4 |
| CWA/club/associations | 1 | | 1 | | 1 | 1 | 2 | 1 | 2 | 2 | 2 | 6 | 2 | 4 | 4 |
| Halls | 2 | 1 | 1 | | 1 | 2 | 2 | 1 | 1 | 3 | 2 | 4 | 2 | 2 | 4 |
| Library | | 1 | 1 | | 1 | 1 | 1 | 1 | 1 | 2 | 1 | 1 | 1 | 1 | 2 |
| Leisure/sporting facility | 7 | 7 | 8 | 4 | 8 | 6 | 6 | 8 | 8 | 12 | 12 | 11 | 13 | 14 | 20 |
| Museum/art gallery | 1 | 2 | | | | | 1 | 1 | | 1 | 1 | 1 | 5 | 2 | 3 |
| *Educational* | | | | | | | | | | | | | | | |
| District high school | | 1 | | | | 1 | | 1 | 1 | 1 | 1 | 1 | 1 | 1 | |
| Higher education | | | | | | | | | | | 1 | | | | 2 |
| Primary/preschool | 1 | | 1 | 1 | 1 | 1 | 2 | | 1 | 1 | | | | 1 | 4 |
| Senior high school | | | | | | | | | | | | | | | 2 |
| *Financial services* | | | | | | | | | | | | | | | |
| Accountant/ins. broker | | 1 | | | | 1 | 1 | 2 | 2 | 1 | 1 | 1 | | 4 | 8 |
| Bank or agency | | | 1* | | | 1 | 1 | 3 | 2 | 1 | | | 2 | 1 | 5 |
| *Government & admin.* | | | | | | | | | | | | | | | |
| Govt dept | | 1 | 2 | 1 | | | 1 | 2 | 1 | 1 | 2 | 1 | | | 11 |
| Shire admin/depot | 1 | 1 | 1 | | | 1 | 1 | 1 | 1 | 1 | 1 | 1 | 1 | 1 | 2 |
| Court house | | | | | | | | | | | | | | | 1 |
| *Health services* | | | | | | | | | | | | | | | |
| Alternative health | | | | | 1 | | | | | 1 | | | | 1 | 3 |
| Ambulance depot | 1 | 1 | 1 | | 1 | 1 | 1 | 1 | 1 | 1 | 1 | 1 | 1 | 1 | 1 |
| Dental | | | | | | | 1 | | 1 | | | 1 | 1 | 1 | 2 |
| Doctor | 1 | 1 | | | 1 | 1 | 1 | 1 | 1 | 1 | 1 | 1 | 2 | 1 | 5 |
| Hospital | | 1 | | | | | | 1 | 1 | 1 | 1 | 1 | 1 | 1 | 1 |
| Infant health | 1 | 1 | | | 1 | 1 | | 1 | 1 | 1 | 1 | 1 | 1 | 1 | 1 |
| Pharmacy | | | | | | 2 | 1 | 1 | 1 | | | | 1 | 1 | 2 |
| Physiotherapy/other | | 1 | | | | 1 | | | | | 1 | 1 | 1 | | 1 |
| *Manufacturing* | | | | | | | | | | | | | | | |
| Building products | | | | | | | | | | 1 | | 1 | 4 | | 10 |
| Engineering–fabrication | 1 | 1 | 1 | | | | 2 | 2 | 1 | 2 | 2 | | | | 3 |
| Furniture/cabinet maker | | | | | | | | | 1 | | 1 | 2 | 2 | | 2 |
| *Motor vehicles* | | | | | | | | | | | | | | | |
| Bulk fuel depot | 1 | 1 | | | | | | 2 | 1 | 1 | 2 | 1 | | | 2 |
| Car sales | | | | | | | 1 | 2 | | | | 1 | | | 7 |
| Fuel sales–retail | 1 | 2 | 1 | 1 | | 3 | 2 | 1 | 3 | 2 | 3 | 2 | 2 | 2 | 7 |
| Mechanical repairs/parts | 1 | 2 | 2 | | | 1 | 4 | 2 | 3 | 3 | 5 | 2 | 2 | 1 | 10 |
| Tyre sales/repairs | 1 | 1 | 1 | | | 1 | 1 | 2 | 1 | 1 | 1 | 1 | 1 | | 2 |
| Motor wreckers | | | 1 | | | | | | 1 | | | | 1 | | 1 |

*(continued)*

**Table 7.4** Functions of the Avon region—larger central places (continued)

| Services | Koorda | Wyalkatchem | Tammin | Bakers Hill | Wundowie | Toodyay | Goomalling | Wongan Hills | Dalwallinu | Dowerin | Cunderdin | Quairading | York | Beverley | Northam |
|---|---|---|---|---|---|---|---|---|---|---|---|---|---|---|---|
| *Other services* | | | | | | | | | | | | | | | |
| Beautician | 1 | | | | | | | | | | | | | | 1 |
| Community newspaper | | | | | | 1 | | | | 1 | 1 | | 1 | | 1 |
| Department store | | | | | | | | | | | | | | | 2 |
| Electrical repairs | | 1 | | | | | | | 1 | 1 | | | | | 3 |
| Employment agency | | | | | | | | | | | | | | | 3 |
| Fire and emergency | 2 | 1 | 1 | 1 | 1 | 1 | 1 | 1 | 1 | 1 | 1 | 1 | 1 | 1 | 1 |
| Florist | | | | | | 1 | | 1 | | | | | | 1 | 2 |
| Funeral director | | | | | | | | | | | | | | | 2 |
| Hairdressser | | | | 1 | | 1 | 1 | 2 | 1 | 1 | 1 | | 1 | | 4 |
| Photographic developing | | | | | | | | | | | | | | | 1 |
| Police | 1 | 1 | | | 1 | 1 | 1 | 1 | 1 | 1 | 1 | 1 | 1 | 1 | 1 |
| Post office/agent | 1 | 1 | 1 | 1 | 1 | 1 | 1 | 1 | 1 | 1 | 1 | 1 | 1 | | 1 |
| Printer | | | | | | | | | | | | | | | 1 |
| Real estate agent | | | | 1 | | 4 | | 1 | | | 1 | | 5 | 1 | 5 |
| Radio station | | | | | | | | | | | | | | | 1 |
| Solicitor | | | | | | | | | | | | | | | 1 |
| Telecentre | 1 | 1 | | | | 1 | 1 | 1 | 1 | 1 | 1 | 1 | 1 | 1 | |
| Travel agent | | | | | | 1 | | | | | | | | | 1 |
| Video hire | | | | 1 | | 1 | | 1 | | | | | 1 | 1 | 2 |
| *Retail–consumables* | | | | | | | | | | | | | | | |
| Bakery | | | | | | 1 | | 1 | | | 1 | | 1 | | 2 |
| Butcher | 1 | 1 | | | | 2 | 1 | 1 | 1 | 1 | 1 | 1 | | 1 | 2 |
| Café/fast food store | 1 | 1 | | 1 | | 4 | 2 | 2 | 1 | 1 | 2 | 1 | 7 | 2 | 8 |
| Fruit and vegetable shop | | | | | | | | 1 | | 1 | | | 2 | | 3 |
| General store/deli | 1 | 1 | | 1 | 2 | | | | | 1 | 1 | | 1 | 1 | 5 |
| Liquor store | | | | | | 1 | | | | | | | 2 | 1 | 3 |
| Newsagent | 1 | 1 | | 1 | 1 | 1 | 1 | 1 | 1 | 1 | 1 | 1 | 1 | 1 | 2 |
| Office supplies store | | | | | | | | | | | | | 1 | | 2 |
| Restaurant | 1 | | | | | 1 | | | 1 | | 1 | | 4 | | 6 |
| Supermarket | 1 | | 1 | | | 1 | 1 | 1 | 1 | 1 | 1 | 1 | 1 | 1 | 3 |
| *Retail–durable* | | | | | | | | | | | | | | | |
| Art & picture framing | | | | | | | | | | | | | | | 2 |
| Bicycle sales/repairs | | | | | | | | | | | | | | | 2 |
| Books | | | | | | 1 | | 1 | | | | | | | 1 |
| Building supplies | | | | | | | | 1 | | 1 | 2 | | | | 5 |
| Clothing | 1 | | | | | 1 | 1 | 3 | 1 | 2 | 1 | | 2 | | 6 |
| Computer/office equip | 1 | | | | | 1 | | 1 | | | | | 1 | | 3 |
| Electrical–domestic | | | | | | | | 1 | 1 | 2 | | | 1 | | 5 |
| Furniture–new | | | | | | 1 | | | | | | | 1 | | 3 |
| Furniture/ornaments–used | | 2 | 2 | | | 2 | 2 | | 1 | 2 | | 1 | 6 | 1 | 4 |
| Hardware | | 1 | | 1 | | 2 | 2 | 2 | 1 | 1 | 1 | | 2 | 1 | 2 |
| Home furnishings/drapery | | | | | | | | 1 | | | | | | 1 | 2 |
| Jewellery | | | | | | | | | | | | | | | 1 |
| Souvenirs/gifts/crafts | 2 | 1 | | | | 6 | | | 2 | 1 | | | 12 | | 1 |
| Toys | | | | | | | | | 1 | | | | | | 2 |
| *Transport/storage* | | | | | | | | | | | | | | | |
| Airstrip | | | | | | | 1 | 1 | | 1 | 1 | 1 | 1 | 1 | 1 |
| Bus–passenger | | | | | | | | | | | | | | | |
| Haulage–truck | | 2 | | | | | 1 | 3 | 2 | 1 | 1 | 1 | | 1 | 3 |
| Rail–freight | 1 | 1 | 1 | | | 1 | 1 | 1 | 1 | 1 | 1 | 1 | 1 | 1 | 1 |
| Rail–passenger | | | 1 | | | 1 | | | | | | 1 | | | 1 |
| Warehousing | | | | | | | | | | | | | 1 | | 2 |
| Total services | 50 | 54 | 40 | 25 | 24 | 90 | 63 | 90 | 74 | 85 | 90 | 72 | 148 | 78 | 291 |
| Town population 1996 | 348 | 349 | 236 | 160 | 613 | 674 | 482 | 813 | 697 | 378 | 715 | 706 | 1923 | 787 | 6300 |
| Shire popn (rural) 1996 | 300 | 290 | 260 | Northam** | Northam** | 3200 | 620 | 646 | 785 | 460 | 655 | 494 | 1250 | 800 | 2250 |

Note: **indicates towns located within the shire of Northam. See Northam entry to find the rural population

**Figure 7.20** New housing development at Dowerin

functions do not react as quickly to changes in threshold populations as they are not driven by purely commercial decisions. As a result, they tend to remain in place even after the population has declined. However, when the demand eventually becomes too small they are withdrawn. Conversely, these services tend to lag behind when a central place grows and are often seen as inadequate in these circumstances. The closure of Goomalling district high school in the early 1990s is one example of the loss of a community function.

**Figure 7.21** Buckled section of the trans-Australia standard gauge rail near Meckering

### Commercial functions

In contrast to public or community functions, commercial functions are more closely affected by financial considerations. These functions will often indicate the type of economic activity of the region. The emphasis is on wheat and sheep in the Avon, so numerous functions associated with this can be found within the central places of the region. Co-operative Bulk Handling (CBH) has a great number of silos or wheat bins spread throughout the region and this is one of the most common services to be found. It will even survive after all other commercial functions have disappeared from a town in decline (see figure 7.17). Another rural-based service is the seed cleaning companies that provide seed to the farmers in order to plant the next year's crop. Figures 7.23 and 7.24 illustrate the wheat storage and seed supply functions of the study area.

Commercial functions in the study area indicate a relationship between the size of the central place and the type of service being provided. Low-order functions such as a general store or small supermarket tend to be found in most centres, while the production of a weekly newspaper is limited to towns such as Northam, York, Toodyay and Cunderdin and the only solicitor in the region is located in Northam.

One consideration in providing a service that is often overlooked in the study of thresholds and central place functions is the importance of existing building stock. For example, hotels are found in a number of locations,

**Figure 7.22** Empty shops at Meckering

**Figure 7.23** Wheat silos at Tammin

**Figure 7.24** Seed supplies at Quairading

both small and large. In nearly all cases these were constructed when the wealth generated by the gold rushes and good rural commodity prices encouraged people to invest in the Avon region. They are often one of the most substantial buildings in town. It might be concluded that they are low-order functions. Cunderdin does not have a hotel—it was destroyed by fire in 1999. The cost of rebuilding is estimated to be around $900 000 and so far no developer has come forward to undertake the project. If this function were to be introduced into many of the towns today it would probably not happen. In reality, then, it is a higher order function than it appears to be.

Multi-functional outlets illustrate a complication in calculating the threshold population for a service. In small centres there is often a business that has a number of different services that would be provided by specialist functions in the larger centres. For example, a video library as a separate entity is only found in Northam and York. Figure 7.25 shows that the small, first-order town of Cadoux is serviced by a general store that provides food, farm supplies, newspapers, fuel, a bank agency, video hire and general hardware.

Over time the types of services found within central places will change according to a range of factors. As already discussed, changes in threshold populations can produce significant changes in the types of services available. Other agents of change include technology, and social economic and political changes. Technology sees older functions replaced by newer ways of satisfying people's needs. It will also see the introduction of previously unknown services into the urban network.

While there is an extensive rail network within the region, it is no longer used as a means of public transport—its prime function is to move wheat. People now either use private transport or a regional bus service provided by the state government.

A new function to appear in the region is the telecentre. Only recently introduced into a number of towns, the centre is a joint project undertaken by local and state government and local businesses to provide access to online education at the secondary and tertiary level. Photocopying and secretarial services and business planning are provided, in addition to high-speed Internet access for the general population. Figure 7.26 shows that this service can be provided from an existing

**Figure 7.25** Cadoux general store

shop or, alternatively, it may be attached to the local shire library.

The range of central place functions and the shape of the hinterlands in the Avon region indicate a degree of compliance with the central place theory. Hinterlands for the different orders of central places vary in size, with the smaller centres having limited ranges compared to the higher order centres. The higher order centres such as Cunderdin, Northam, Quairading and York have hinterlands that include the populations of the smaller surrounding settlements. Northam's influence extends over settlements such as York, Goomalling, Dowerin, Quairading, Beverley and Meckering given its top position within the urban hierarchy of the Avon region. Along with the differences in the sizes of the hinterlands there is also a corresponding increase in the number and variety of central place functions offered by the towns in the highest orders. These have been discussed previously.

One consequence of the decline in rural population has been the disappearance of a number of smaller first order towns. These developed around the transport nodes created by the construction of wheat silos and sidings on the railway lines and spurs. Here they benefited from the activity generated when the harvested wheat was brought to the railway for transportation to Perth. They offered a small number of low order functions to the farmers, rural labourers and railway workers. With the changes in transportation, large scale mechanisation and farm amalgamation the rural population declined and fell below the threshold needed to support the services. In addition the growth of the larger towns resulted in the 'capture' of their hinterlands. As a consequence these centres have declined and all but disappeared. Today abandoned and derelict buildings mark the location of these former centres, even though they will often still have an operational wheat silo. Manmanning, just south of Cadoux, is one such town. Figure 7.27 shows the abandoned general store that once served the local community. Recreational facilities in this centre have also been left to fall into disuse.

With increasing ranges and smaller populations, the urban network of the Avon region has undergone significant readjustment. There are fewer towns spread further apart. The majority of the small first-order centres have disappeared. Increasing ranges have also seen the loss of functions from existing towns. Several towns with reasonable numbers of people, such as Ballidu, do not have a service station. In this instance people need to drive to Dalwallinu in order to buy petrol.

### Specialist centres

While the majority of towns in the Avon can be classed as central places, there are several exceptions and this is reflected in the type and range of services offered. York and, to a lesser extent, Toodyay have evolved into tourist centres. While they both commenced as central places, tourism has seen a major shift in services.

Prior to the mid-1970s, York was a town in decline. The rural population it relied on was insufficient to maintain the type and variety of services that it provided. However, its proximity to Perth and its range of historic buildings were attracting weekend visitors and, over the last 30 years, it has grown into one of the South-West's larger tourist centres. Much of the tourist activity is based on historical buildings and other attractions including 'The Residency', motor museums, the courthouse, town hall and numerous bed and breakfast accommodation outlets. York also organises major tourist-related events, such as vintage car rallies, the Winter Festival and the York Jazz Festival, and is promoted as a conference centre for professional and academic associations. The number of tourist visits increased from 7000 in 1975–76 to 47 253 in 1987–88. As a result, York can now be considered a specialist tourist centre, although it is still fulfilling basic central place functions. A comparison of both York and Toodyay's functions will reveal marked differences to the region's other central places. In addition to the large range of accommodation, the presence of a number of real estate agents indicates the entry of the hobby farmer into the community. Both towns actively market small rural lots and weekend retreats.

Wundowie is another specialist centre. In contrast to York and Toodyay, it did not commence as a central place. Wundowie was built to accommodate workers

**Figure 7.26** York telecentre

**Figure 7.27** Abandoned general store at Manmanning

and their families for a nearby steel works. This operation produced specialist steels using charcoal smelting methods and, at its peak, employed over 400 workers. Today, the steel works no longer operate and the town functions as a dormitory centre for commuters as well as a place to retire. Surrounded by state forest, it has little rural population to serve and while its population is still around 600 it has a very narrow range of services.

## The urban hierarchy

The Avon region today has an urban network and hierarchy shaped by a number of factors that have been discussed previously. Significant features of the network include the spatial arrangement of the central places and the numbers of centres within each of the orders.

### The spatial arrangement

Transport systems and the biophysical environment have had a strong and ongoing influence on the size and location of towns. The maps in figures 7.17 and 7.28 illustrate these influences. As in other regions in the South-West, the transport systems have produced a linear pattern of settlement. In particular, the gradual development of the railway system determined both the location and the spacing of the towns. In turn, where the railway was constructed was affected by the landforms of the region. In numerous situations the line followed streams and rivers, avoiding higher areas.

In order to get onto the plateau, the railway system follows the valley of the Swan and Avon river systems, through the Darling Scarp to where it reaches Toodyay. It continues along the Avon Valley to Northam and then heads south to York and Beverley. Branching of the line occurs at Northam, York, Toodyay, Goomalling and Dowerin, making these towns significant transport nodes. With the development of road systems those towns that also had important positions, such as Northam, retained their locational advantage. The more recent expansion of Dalwallinu may be accounted for by this influence. It is on the Great Northern Highway as well as being at the termination point of several large regional roads, so is well positioned to serve both local people and passing traffic.

Variations in soils and topography have also influenced the spatial arrangement. The southern part of the region contains more centres than do the northern and eastern parts. Soils in the southern areas are relatively well developed and fertile, while towards the north-east they become lighter, with mainly yellow duplex types. In addition, primary and secondary salinity becomes more apparent in these areas. Along the western margin of the study area, the land is more undulating and the farms sizes are smaller. Here, urban settlements are more closely spaced. In contrast, the eastern areas are either flat or gently undulating, with farms being much larger and the towns more widely spaced. In addition, the northern and eastern areas have experienced a much larger degree of farm amalgamation, with

# 172 LANDSCAPES AND LAND USES

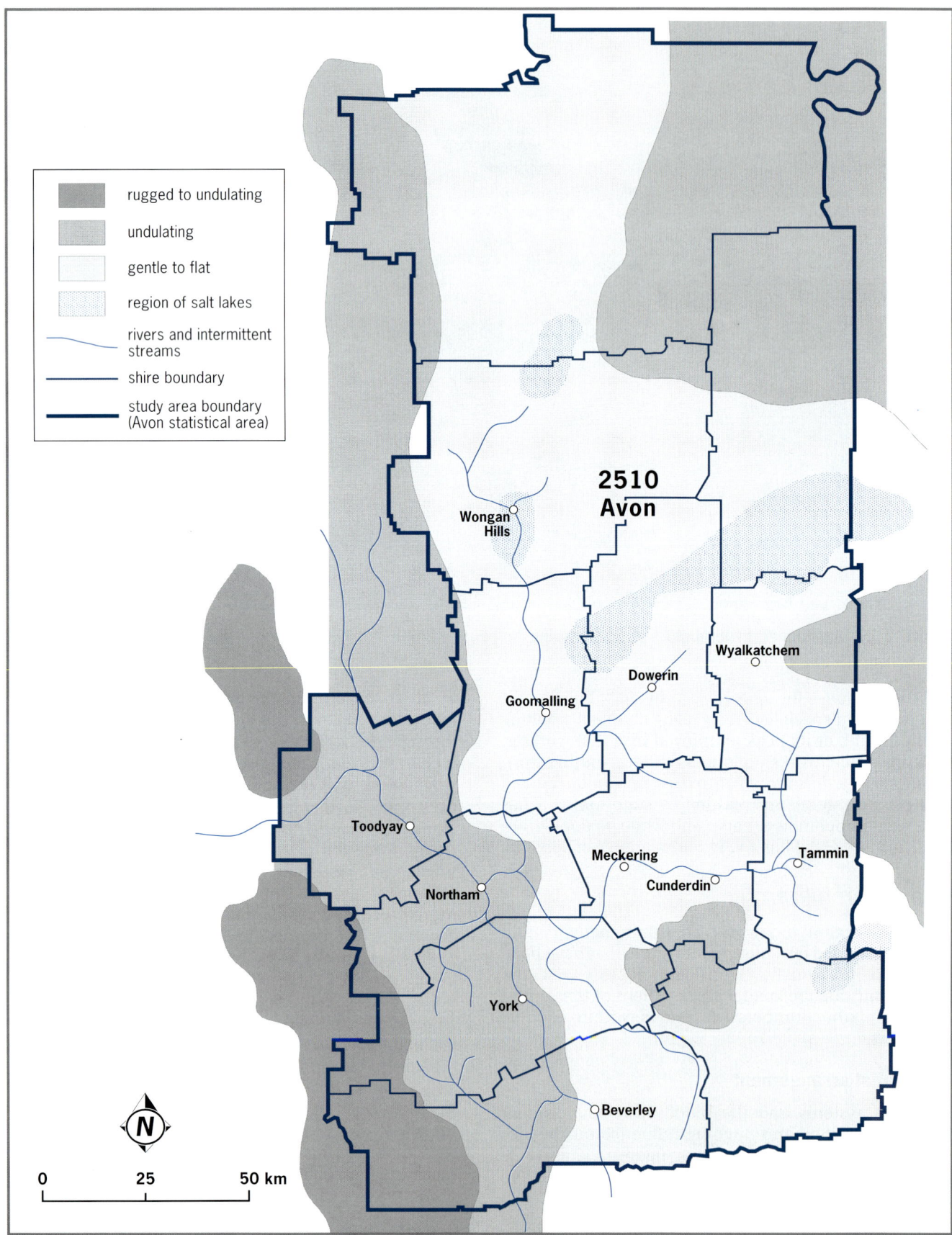

**Figure 7.28** Physical features of the Avon study area

average farm size increasing from several hundred hectares following World War II to over 2500 hectares today.

## Changing central place orders

Scattered throughout the region are remnant first-order centres. Towns such as Manmanning, Jennacubbin, Kondut and Konnonopping fall into this category. Many of these localities still have operational wheat silos and a rail siding. In some cases they may have a shire hall or a church, but all commercial central place functions have disappeared. The location of these former towns corresponds to the grain receival silos shown on figure 7.17. The substantial loss of first-order central places from the urban hierarchy of the study area has had a significant influence on the number of central places within each order. The number of first-order central places is disproportionately low and nearly all of the remaining ones are in rapid decline. One interesting exception is the centre of Grass Valley, 13 kilometres east of Northam. This appears to be growing slightly and functioning as a dormitory suburb for Northam.

The significant feature of the hierarchy, therefore, is one of structural re-adjustment, with fewer but larger centres. Towns such as Dowerin, Goomalling, Beverley and Quairading are absorbing the population of the surrounding smaller centres. Their local governments are actively pursuing a policy of retaining the regional population. One obvious feature of this is the construction of retirement centres and units within the towns to service the ageing population and reduce their need to move into Northam or to Perth.

A further change has been the turnaround of Northam. For a number of years it was in slow decline, but over the last four years it has grown and its population has increased from about 6500 in 1996 to 7000 in 2001. The town is actively pursuing its role as the major centre within the region, despite the drift of trade and functions to Perth. As tables 7.3 and 7.4 show, there are currently four distinct orders within the study area, with Northam operating as the fourth order.

### EXERCISES

1 Using the data on central place functions provided for the Avon region, carry out the following activities.

 (a) Rank all of the central places identified in tables 7.3 and 7.4 according to their population. Graph them in rank order, with population on the *y*-axis and the rank on the *x*-axis. Describe and account for the pattern and order.
 (b) Rank and graph the centres according to their number of functions. Compare and contrast this pattern with the one for population.
 (c) Placing functions on the horizontal axis of a graph and population on the vertical axis, graph the number of functions and the population for each town. Use dots to show the position for each centre. There should be a corresponding increase in functions in conjunction with greater populations. Identify any significant variations and suggest reasons for these.
 (d) Identify and list the most typical function offered by towns within each of the four orders.
 (e) For a primary school, district high school and a senior high school, calculate the approximate threshold population. Do this by counting the number within the region and then dividing it by the regional population (the total population for all the listed towns as well as the rural populations for each shire). Note that a district high school also includes a primary school.
 (f) Calculate the threshold populations for a community newspaper and a caravan park.
 (g) Using the map of the Avon study area, draw a map showing the draw areas for the CBH grain receival silos and calculate their approximate range.

2 Draw a map of the Avon study area and on it place the location of the study towns as well as the shire boundaries. Using the information on town and shire populations, construct a dot distribution map for population. Refer to chapter 9 on mapping techniques to complete this activity.

3 Using the information on central places and their orders, draw a map of the region and then construct hinterlands or draw areas for each central place. In doing this consider the following:

- each centre will be served by all four levels
- no area can be outside a hinterland
- the shape will be affected by roads, spacing of centres and topography
- some towns with few services may be serviced by more than one centre.

4 Analyse the different orders within the urban hierarchy of the study area and comment on the numbers of centres within each order, their functional complexity and the probable extent of their hinterlands.

5 Using symbols and a map of the Avon study area show the location of churches, telecentres, district high schools, libraries and real estate agents. Calculate the approximate range for each.

6 Compare the services offered by York, Toodyay and Wundowie with those offered by the other centres within the study area and comment on any apparent differences. Discuss the extent to which these towns may be classed as specialist centres.

7 Identify how environmental, economic and technological factors might change the spatial characteristics of an urban network, using examples from the Avon study area.

# Chapter 8
# Urban Morphology

> The purpose of this chapter is to outline the concepts and theories of urban morphology, and to apply these to the study of Perth and Northam. It includes the study of processes and problems associated with urban growth and change.

Within all but the smallest urban centres, the spatial arrangement of services and functions tends to present an observable pattern or layout. This pattern takes the form of distinctive functional zones or areas. Given the degree to which this pattern is organised it is reasonable to conclude that the arrangement of different urban functions is not a random event but rather the result of different forces that act to shape the land use patterns of towns and cities. This observation has given rise to the study of urban geography and, in particular, the study of urban morphology. 'Morphology' in broad terms refers to elements of shape, function and change. Hence, urban morphology may be defined as the ways in which the external and internal land use patterns of a city are arranged and change over time.

## External morphology

The appearance of an urban centre can be studied by observing the division that occurs between the urban and the *rural* land use zones. At the edge of a city the urban land uses give way to either rural activities or the biophysical environment. This region may be one of gradual transition or, alternatively, the boundary may be well defined. In general, the larger the city the less clear the boundary and the broader the transition zone. In addition to locating the edge of the city, the geographer is also concerned with identifying the shape that the boundary produces and seeking reasons for the pattern of growth or spread of the urban area, or in some cases the pattern of contraction. These observations become significant when predicting future changes and accounting for their impact on the surrounding regions.

### *Factors affecting external morphology*

The external morphology of a city is subjected to a number of biophysical and cultural forces. Biophysical influences such as landforms or relief, vegetation, geology and soils, and microclimatic conditions make up the site characteristics of a city and act to either encourage or discourage the direction of urban development. Cultural factors are many and varied and include technology, and socioeconomic, political and historical factors.

### Biophysical or site influences

Elements of site such as the patterns of drainage systems, coastlines, soil types, conservation zones, aspect, prevailing wind patterns, slope, altitude, relief and natural harbours may all influence the layout and external shape of an urban centre. In figure 8.1 the location of a natural harbour with a river flowing into it, and the restrictions on outward growth imposed by the topography have all influenced the shape and direction of growth. The built-up area follows the coastline in a narrow band, only extending inland along the river valley. Future growth is planned along existing roads on the lower ridges of the area of high land to the east of the river. The location of beaches in sheltered bays along the coast has attracted urban development.

### Cultural influences

In addition to the influence of site on the shape of urban centres, a number of other factors will interact with each other and with the biophysical environment to produce the final pattern of development. Transport arteries often encourage linear development and produce areas of accelerated outward growth. For example, the opening up of new industrial and residential areas has closely followed the north–south extensions of the major freeway system of Perth. The presence of national parks, conservation zones and water catchments within the Darling Escarpment to the east of Perth has exerted a controlling influence on urban development within this region. The competing interests of orderly and cost-effective urban planning versus uncontrolled growth will have a significant effect on where and how the city expands.

The capacity to build in certain environments is a consequence of the types of technologies available.

URBAN MORPHOLOGY 175

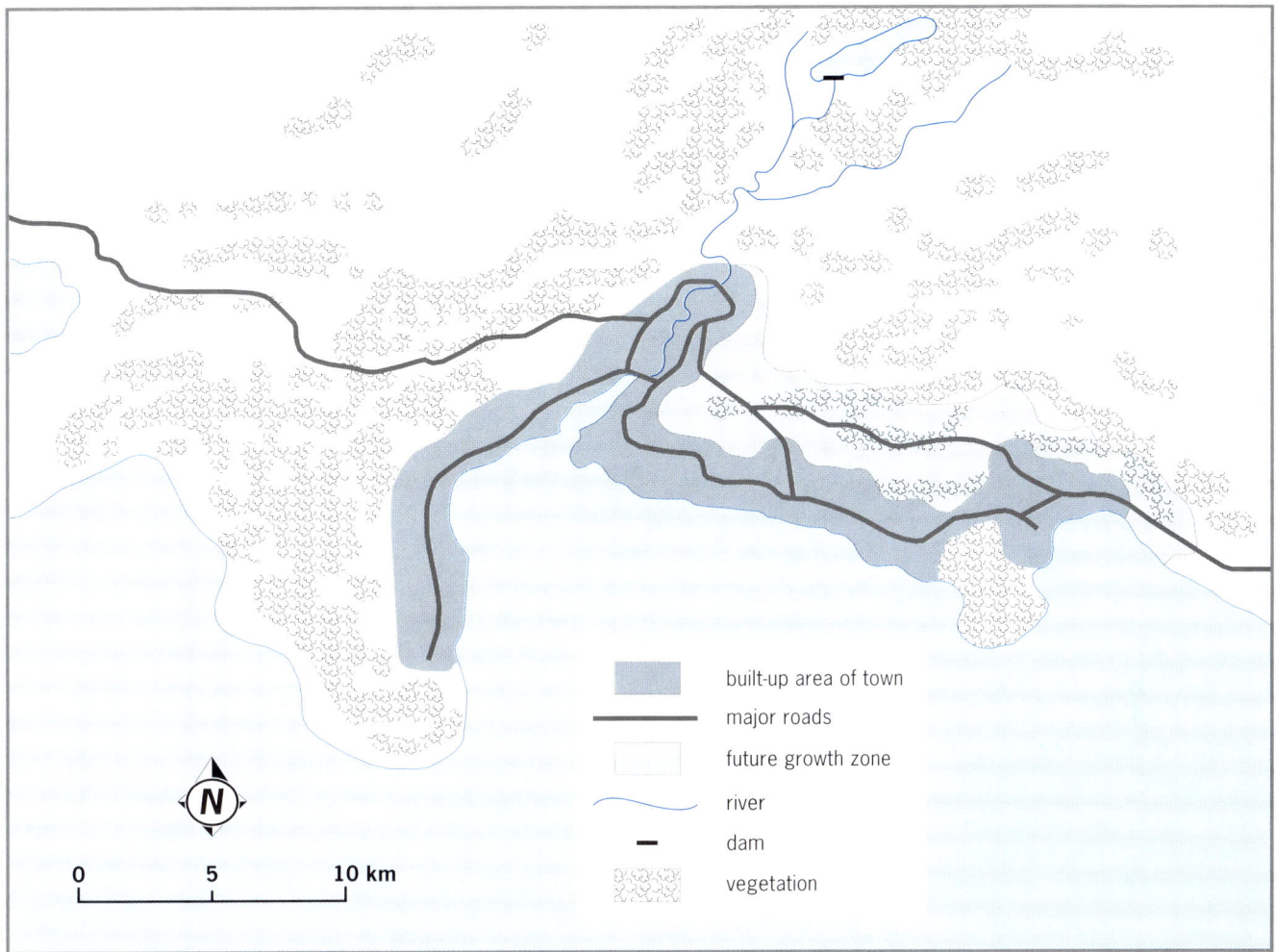

**Figure 8.1** External morphology of a hypothetical urban centre

These include the ability to reshape the site to allow for construction, as well as the capacity to provide infrastructure to support the growing city. Infrastructure in the form of roads, power, gas, water and sewerage is an essential component of a large, modern city. In figure 8.1, the arrangement of the road system has helped to shape the growth and, in turn, the external shape of the hypothetical urban centre. The retention of areas of vegetation and the presence of a dam and water catchment indicate the influence of planning and conservation considerations. This is further reinforced by the presence of a **deferred urban** or future growth zone, which has been located in an area away from these features.

### EXERCISES

1. Explain external morphology and identify the difficulties in defining the edge or fringe of a large city.
2. Using sketches, illustrate the following external morphology patterns:
   (a) linear
   (b) rectangular
   (c) stellate
   (d) circular
   (e) irregular.
3. Using figure 8.2, sketch the possible shape of a city constructed to the east of Gregson's Bay. Note the location of the planned roads and farmland as well as the different site factors. Describe the external morphology of your planned city and identify the factors that influenced the final shape. What problems would the city face as it expanded?

## Internal morphology

The study of the internal morphology of an urban centre involves the examination of distinctive land use or functional zones within the city boundaries. Such an analysis can be structural and functional in nature. The constructions or buildings (structures) that house and assist the variety of urban activities (functions) within the city provide evidence of the type and location of different land uses. The study (structural–functional analysis) of the spatial arrangements of urban land uses has given rise to several theoretical patterns, each of which will be discussed later in this chapter.

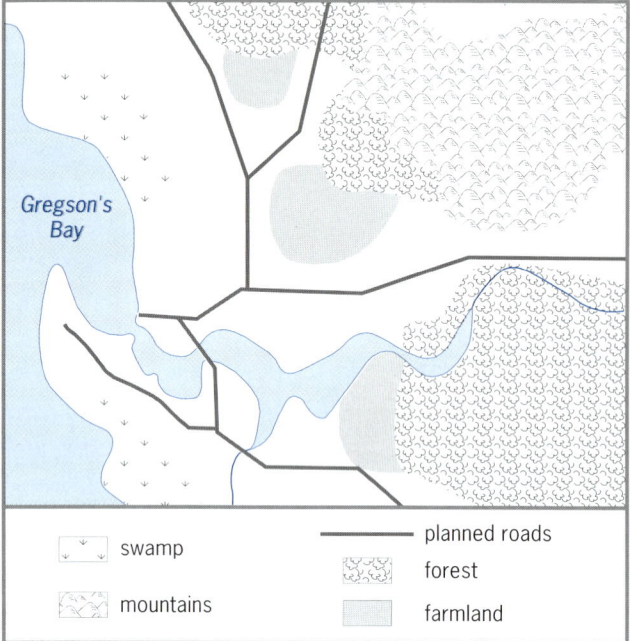

**Figure 8.2**

## Functional zones

Distinctive urban land uses based on similar functions tend to be arranged in recognisable regions within an urban centre. These zones can be identified by their appearance and dominant activities.

### Commercial and business zones

The largest and most important of the commercial and business zones is the central business district (CBD). It usually occupies the most accessible location within the urban centre and, in many cases, is the initial point of settlement from which the city has subsequently expanded. The CBD has a number of distinctive characteristics:

- *Pedestrian scale.* The outward growth of the business district is controlled by the need to maintain a size convenient to high levels of pedestrian movement. This pedestrian scale limits the spread of the commercial zone and encourages vertical growth.
- *Vertical zonation.* As the city develops vertically it exhibits specialisation within the multi-storey buildings. The ground floors house functions that need to have maximum exposure to passing pedestrian traffic, while the upper floors often house functions that are associated with administration or professional services. Across the CBD similar functions tend to group together and this in turn produces distinctive precincts.
- *Horizontal zonation.* Areas such as the **primary** and **secondary** office or financial zones and the uptown (primary) and downtown (secondary) shopping precincts can be identified within the CBD and illustrate the horizontal variation in land use patterns within the city centre. The primary shopping area is characterised by the retailing of high-value or prestige products while lower value goods and less profitable services will be relegated to the secondary shopping areas. In addition, there may be cultural and government sectors as well as historic and recreation precincts which contribute to the horizontal zonation of the CBD.
- *Dynamic character.* Economic, political and social forces will operate to continually shift the locus of specialist precincts within the CBD, resulting in areas of growth and decline. A shift in the types of transport systems used by commuters may influence the points at which they enter the city centre and, in turn, affect the location of shopping facilities. What were once prime shopping areas may decline, producing a region of gradual deterioration and urban blight. Processes of city centre renewal, on the other hand, can encourage increased patronage and bring new commercial activity to areas that were once derelict. The skyline of many large city centres is in a state of constant change, reflecting this dynamic character.
- *Population densities.* Typically, the CBD will have a very high day-time population density and a low night-time one. It is a region of office and service workers who commute from the residential or dormitory zones and who vacate the city by night. Some night-time population associated with city apartments and hotels offsets this daily exodus.
- *Transport node.* Given its significance as a major destination for urban commuters, the CBD is a major transport node within an urban centre. Both public and private transport facilities can be identified. Rail and bus services often radiate outwards from the CBD, while freeways and extensive car parking facilities are a common characteristic of the zone. Other transport systems, including river or sea ferry services and helipads, offer alternative forms of transportation for commuters in some large cities.

Beyond the CBD of a large city, a number of regional and local commercial and business functions can be identified. These often group together into **outer business districts** (OBD). In a way that is similar to the smaller central places in an urban network, these zones offer a smaller number of lower order services. They are characterised by the following features:

- an emphasis on retailing functions, selling everyday consumable items as well as durable products
- financial or banking services focusing on standard customer requirements
- higher-use professional services such as accountants, real estate agents, business brokers and standard health services
- limited or little vertical and horizontal zonation of functions.

Finally, small suburban shopping centres specialising in low-order, low-range and low-threshold services make up the lowest level of the commercial structure of a large urban centre. These include the corner store though, like the smallest central places in an urban network, these are in decline due to the centralisation of functions within the larger shopping centres.

## The inner mixed zone

The **inner mixed zone** (IMZ), as the name suggests, is a zone of mixed urban functions. Its location adjacent to the CBD influences its functional characteristics. It is often one of the oldest areas within the city and contains a range of building stock of varying age and quality. The main characteristics of the IMZ include:

- *Zone of invasion.* As competition for the limited sites in the CBD occurs, some functions that do not consider a central city location to be essential will locate on the fringes of the CBD, in the IMZ. This action will displace other functions. Over time this process of invasion will eventually alter the functional mix of the zone in a process known as succession.
- *Industrial and wholesale functions.* One traditional focus of the IMZ has been to provide services that are required by CBD functions. Light industrial activities such as food preparation, printing, office equipment repair and servicing, and storing of stock and materials are often found in the zone.
- *Secondary retail.* In association with the warehouse function, it is often a zone of secondary retailing. Furniture, whitegoods, clothing and bulk or cheap food outlets are examples of these functions.
- *Transport services.* Given its proximity to the CBD, it is also a major zone of car parking as well as housing public transport depots.
- *Horizontal zonation.* Like the CBD, the IMZ can also have distinctive areas or zones. These may include an emphasis on entertainment, retailing, residential or business activities.
- *Urban blight.* Given the age of some building stock, the zone is also distinguished by the poor quality of some of these structures. Derelict and vacant buildings reduce the attractiveness and amenity of the location and can contribute to further deterioration in a process of urban decay or blight.

## The industrial zone

Throughout a large city manufacturing and associated functions concentrate in selected areas. These industrial zones may vary from those concentrating on warehousing and transportation, or light industrial manufacturing and fabrication, through to general and heavy industrial activities. They can also vary in age and standard of presentation. Distinguishing characteristics of the industrial zone include:

- *Agglomeration economies.* Industrial activities will often locate where they can use the products of related businesses or take advantage of the specialised infrastructure that is provided for this function. Some industrial zones may specialise in transportation and warehousing of goods. They will be well served by road and rail freight terminals or, in some cases, port facilities. Other businesses that service transport and warehouse equipment may be attracted to the location.
- *Industrial variation.* Factors such as past development and planning decisions will often produce different industrial zones. Heavy and general industrial areas may contain chemical and mineral processing functions and heavy or large-scale fabricating activities. The activities in these areas often cause pollution of the soil, air and water as well as generating high levels of noise. Consequently, they must be confined to limited locations within a city. Light industrial areas, on the other hand, are far less intrusive and have often been located near residential zones. They are far more widely dispersed within the urban landscape.
- *Industrial estates.* New industrial areas often show a much higher degree of planning than older areas. They include landscaping, buffer zones and good quality building stock and are generally located on large sites to accommodate future growth as well as efficient movement and storage of materials. In contrast, older industrial areas are restricted in their future growth and have building stock that was designed around older technologies. They can be hemmed in by surrounding residential development and have limited or non-existent buffer zones. The original locational advantages of these sites may no longer exist.
- *Industrial inertia.* Individual functions as well as whole industrial zones may often outlast their original locational advantages. The considerations that were important in selecting and developing a site will, over time, become less so. Customers may move away, accessible locations may become congested as traffic increases, and changing regulations may cause compliance difficulties. High relocation costs make moving unattractive, but a business will generally reach the point where the choice is to move or go out of business. This industrial inertia is sometimes a characteristic of industrial activities that remain within the IMZ.

## Residential zones

The residential zones of an urban centre occupy the greatest percentage of its site or area. Part of the reason for this is the generally low density of the majority of residential development. In addition, the zone contains a number of recreational and ancillary services that add to its overall area. Like the other major zones of a city, the residential zone has a number of distinguishing characteristics:

- *Residential density.* Structures in residential areas vary from high-rise apartments through to detached bungalows. This difference often corresponds to their location within the urban centre. Inner suburban areas tend to contain higher density accommodation, while larger building sites are located within the established and newer residential areas.
- *Residential age.* The age of housing stock closely corresponds to the growth of the city, with older buildings located in the original suburban areas and new development on the city fringes. An example of the latter is the newer growth zone (NGZ), on the

outer edge of the established residential areas, with its new housing estates as well as planned industrial areas. Exceptions to this pattern occur where previously undeveloped sites within established areas are built on, and when older towns or villages are overtaken by the outward growth of the major city. In addition, processes of demolition and redevelopment of existing housing stock will also result in differences in the age of residential areas or sites.

- *Gentrification and renewal.* Redevelopment of large areas of existing land uses often results in new housing estates within the city. This process of urban renewal is seen where old industrial or warehousing zones are demolished or refurbished to provide a residential function. Dockland areas in many large cities provide good examples of this form of residential development. Alternatively, an influx of younger home purchasers may rejuvenate the population of old, inner-city areas and can lead to gentrification of a suburb. In this case, heritage housing is renovated and the quality and property values of the affected areas increase. The Paddington area of Sydney provides good examples of gentrification.
- *Suburban services.* While the primary objective of residential areas is to provide a dormitory function, a number of services are provided to cater for the needs of the population. These include recreational facilities, suburban shopping centres, health and financial services, and social and cultural outlets and services.

### The rural–urban fringe

At the edge of the city, where urban and rural functions meet, there is often a zone of transition in which the two land uses mix. The **rural–urban fringe** (RUF) contains a mixture of rural and urban functions. It is different to the NGZ in that the NGZ represents the earlier RUF of the city. It has already undergone the transition from rural to urban. This indicates that the process of invasion has been largely completed, while in the RUF it is still under way. The existence of the processes of invasion and succession points to some similarities in processes between the RUF and the IMZ. In addition, the RUF has several distinctive features of its own:

- *Invading urban functions.* Functions seeking cheaper land or larger areas in which to develop will often relocate to the RUF. These include industrial as well as recreational and some residential activities. The relatively quiet and natural environment offered by the area will often attract residential functions. In turn, the increase in new residential subdivisions can produce conflict with the existing rural functions. Market gardeners and orchardists may find an increase in theft and vandalism as the urban population increases, while owners of livestock are sometimes affected by domestic dogs attacking their animals.
- *Changing rural functions.* The RUF is also a zone in which competition for land has changed the rural land use activities. In keeping with the concept of the economic rent mechanism, the most intensive and profitable land uses will be located in this region, given its proximity to a major market. As the city expands, less intensive rural land uses are displaced by more intensive ones and these in turn give way to the newly arrived urban functions. One such function, which is semi-rural in nature, is the hobby farm. The owners earn their primary incomes from urban occupations, but engage in rural activities as a hobby or form of recreation.
- *Urban shadow effect.* As with the IMZ, the expectation of invading urban land uses often sees the decline of the original activities and structures. Dilapidated packing or processing sheds and barns, along with fencing in a state of disrepair, provide examples of this form of land use blight.

### EXERCISES

1. Define the following terms:
   - vertical zonation
   - inertia
   - functional zone
   - urban blight
   - urban shadow effect
   - urban sprawl
   - urban renewal
   - gentrification
   - invasion and succession

2. List the main functional zones in a city and, next to each one, list some of the activities that you would find there.

3. Compare and contrast the inner mixed zone and the rural–urban fringe.

4. Draw a model of a 15-storey office block and locate the following functions where you consider that they would be found, in line with the concept of vertical zonation. You may include more than one activity on each floor and some activities may occupy more than one floor. Account for your pattern of vertical zonation.

   - café
   - dentist
   - insurance broker
   - penthouse
   - financial institution administrative offices
   - medical bookshop
   - florist
   - eye specialist
   - gift shop
   - Department of Trade (federal government agency)

5. Place the following retail functions in the primary and secondary retail zones and then explain why these two zones exist in the CBD.

   - jewellery store
   - secondhand book store
   - department store
   - body piercing
   - pet supplies
   - warehouse-type clothing outlet
   - specialist camera shop
   - army surplus sales
   - imported formal wear
   - French perfumery
   - video and computer games—new and secondhand

6. Identify and explain different examples of horizontal zoning that might be found in a typical CBD.

## Models of city structure

There are three basic models of internal city structure. Each one emphasises different forces in accounting for the apparent order that occurs in the spatial arrangement of urban land uses.

## Concentric zone model

This model is based on a single, central commercial district which occupies the most accessible location within an urban centre. Arranged around it in a series of **concentric** circles are other urban land uses. These include an inner mixed or transition zone, established residential zone of semi-skilled and unskilled workers, a better residential zone and a newer residential growth fringe. The model draws partly upon the economic rent model discussed in chapter 4. It is also identified as being **monocentric**, with all commercial and retail functions located in the centre. The pattern is one that assumes the existence of a centrally placed uniform site. One contradiction in the model is the location of low-income groups near the centre, where land values are high. This is partially explained by the presence of high-density residential properties, which provide a large number of rental units per hectare and allow for cheaper rents per person. Another explanation is the presence of substandard housing, which is kept by landlords with the intention of future redevelopment and is therefore rented in the meantime at lower rates. Figure 8.3a shows the main characteristics of this model.

## Sector model

Like the concentric zone model, this pattern is monocentric in character. However, it incorporates the effects of increased accessibility provided by major transport arteries, as well as the possible influences of site factors. This model is basically a refinement of the concentric zone model and suggests a similar arrangement of functions as you move away from the centre. Figure 8.3b illustrates the ways in which a railway system and site factors have shaped the internal morphology.

## Multiple nuclei model

This model identifies a variety of influences on the internal morphology of most large cities. It suggests a **polycentric** pattern, with a number of major and minor commercial districts as well as multiple industrial and residential zones. It takes account of site factors in explaining the growth and development of the city. Other considerations may include the influence of socioeconomic factors, historical antecedents and cultural values, and political and planning decisions. As figure 8.3c indicates (next page), this model is basically a response to a number of forces and, as such, it could be drawn in a number of different configurations.

## CASE STUDY

# Perth

## Perth's external morphology

Perth is the state capital of Western Australia and is located in the south-west corner of the state. As discussed in chapter 7, it exhibits a strong degree of primacy and far outranks all other settlements in the region in terms of its population, economic activity and political importance.

### The shape of the built-up area

The shape of Perth as indicated by the boundary of the built-up area is shown in figure 8.4. The city has a compact core area extending along the Swan River. Major extensions of the built-up area occur along the coast to the north and south, and out to the east and south-east. These corridors of urban development result in what can be described loosely as a star-shaped or stellate pattern. There are pockets of separate urban development on the Darling Escarpment and in the Darling Range.

Perth covers an area of approximately 1500 square kilometres, extending away from the coast by some 25 kilometres and along it for more than 70 kilometres. The city is bisected by the Swan River, with approximately half located to the north and the other half to the south of the river.

### Factors affecting Perth's shape

The external form of a large urban centre develops in response to a combination of a number of different physical, economic, social, political and historical factors. The main factors that have influenced the distinctive shape of Perth are the city's site, situation (accessibility) and functions, the social values of its inhabitants, urban planning, technology and time.

*Site factors*

The characteristics of the land on which an urban settlement is built have significant control over the form or shape of the settlement. Positive site factors, such as relatively flat land, good drainage and assured water supplies, favour uniform urban development. Many urban centres are located initially around specific favourable features, such as deep-water harbours and river crossings. In time, they may expand outwards from this starting point. Negative site factors that prohibit urban development include steep slopes and low-lying marshy land.

The site of Perth, selected in 1829 by Captain James Stirling, the founder of the original colony, was 17 kilometres upstream from the mouth of the Swan River. The settlement was sited on the northern banks of the Swan River estuary, on a ridge of relatively high land, bounded to the west by a 70 metre high ridge of limestone known as Mount Eliza (now Kings Park). A series of shallow lakes and swamps bounded the colony to the north. This site was chosen largely because it was close to good agricultural soils and because sandbars in the river near Heirisson Island prevented further navigation upstream. Building materials and fresh water were also readily available. Figure 8.5 shows the physical factors that have influenced the way in which Perth has grown from 1829 to the present. Most of the built-up area of the city is contained within the 25–35 kilometre

### (a) Simple concentric zone model

Key

- CBD—core of administrative, retail and financial land uses
- Inner mixed zone—'blight' zone of wholesale, light industrial and poorer residential uses with little sorting
- Medium class residential—established zone of better residences and some blocks of industry
- Peripheral growth zone—new workers' and some high-class residences (commuters) and new industry

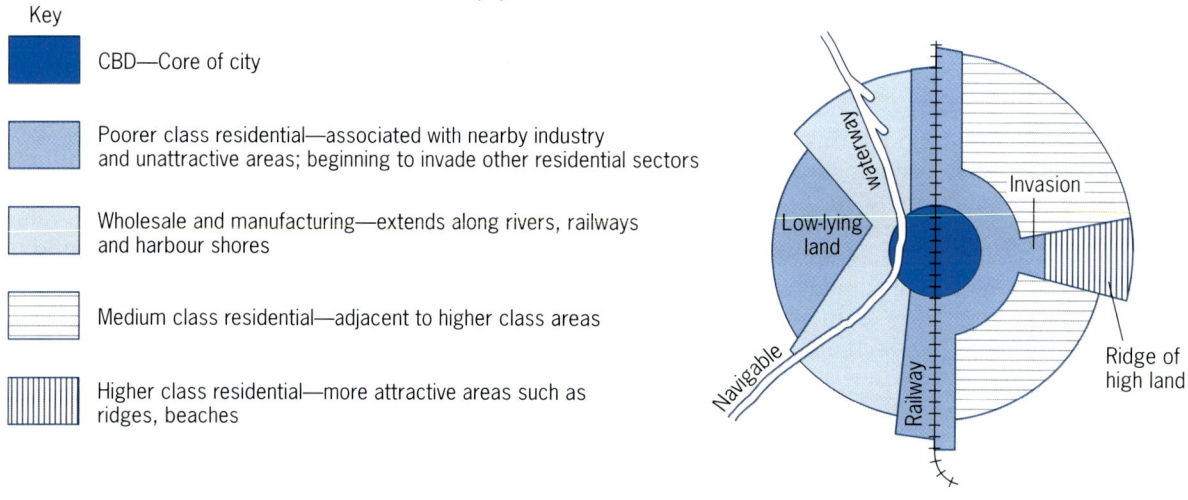

### (b) Sector model

Key

- CBD—Core of city
- Poorer class residential—associated with nearby industry and unattractive areas; beginning to invade other residential sectors
- Wholesale and manufacturing—extends along rivers, railways and harbour shores
- Medium class residential—adjacent to higher class areas
- Higher class residential—more attractive areas such as ridges, beaches

### (c) Multiple nuclei model

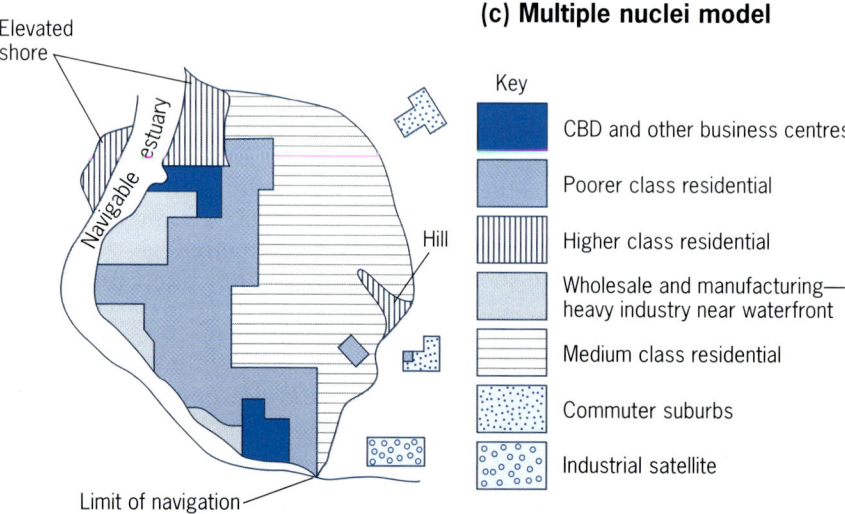

Key

- CBD and other business centres
- Poorer class residential
- Higher class residential
- Wholesale and manufacturing—heavy industry near waterfront
- Medium class residential
- Commuter suburbs
- Industrial satellite

**Figure 8.3** Models of city structure

## Situation factors

Perth's external morphology has been greatly affected by its location in relation to its surroundings and its transport connections to the settlements within its hinterland. Growth along the major transport links that converge on Perth has, therefore, influenced the city's external form.

The extension of the city outwards along major transport links has resulted in sectors of development, producing a stellate urban form. Figure 8.4 demonstrates the radial development that has occurred along the major highways and railways, particularly to the north-east, east, south-east and south-west. Development has occurred, for example, along Albany Highway to the south-east. This major transport route links Perth with agricultural centres such as Kojonup, Mount Barker and the important regional centre of Albany. Extensions to the Mitchell and Kwinana freeways have resulted in growth along the north–south axis of the city and the amalgamation of separate coastal settlements into the metropolitan area as new

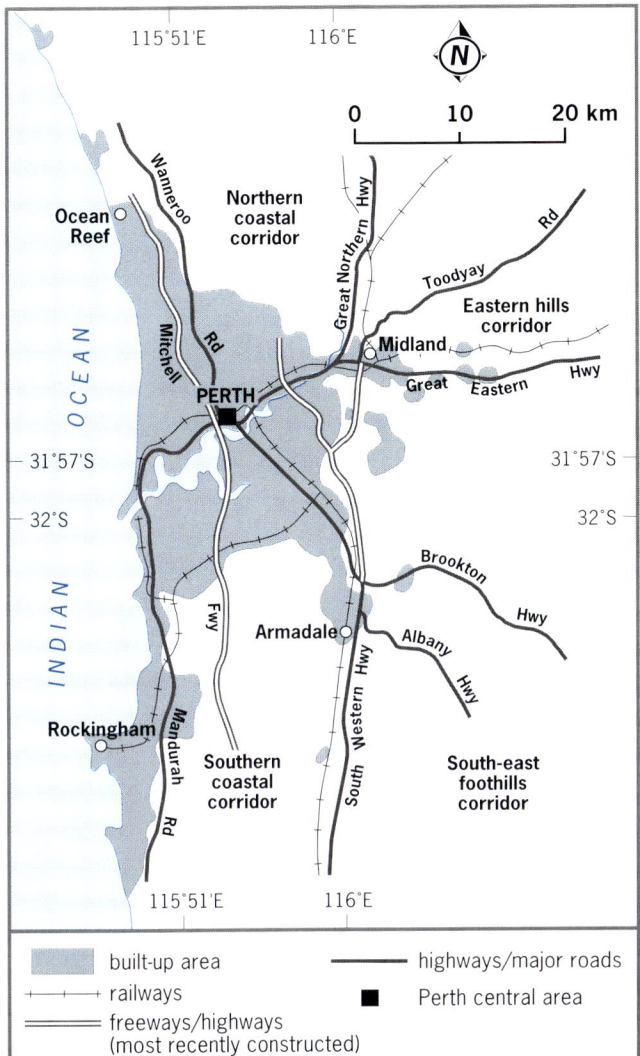

**Figure 8.4** Perth's external morphology

wide undulating coastal plain, bounded to the west by the Indian Ocean and to the east by the Darling Scarp, which rises to a plateau of over 300 metres above sea level. The western part of this plain, which rises up to 60 metres above sea level in places, consists of a series of sand dunes of different geologic ages running parallel to the coast. This results in a landscape of ridges and depressions (many of which contain wetlands). The eastern half of the coastal plain is flatter and relatively low-lying, and is dominated by the floodplains of the Swan, Helena and Canning rivers and the foothills of the Darling Scarp. The only significant physical barriers to development have been the existence of some areas of recent unstable dunes adjacent to the coast, and some pockets of poorly drained land, especially to the south of the Swan River. In some cases negative site factors have been overcome. For example, steep slopes in the area have been levelled for residential development, and some swampy interdunal areas have been reclaimed. Part of the Swan River estuary was reclaimed to enable the construction of the Narrows Bridge interchange, Langley Park and the Burswood entertainment complex.

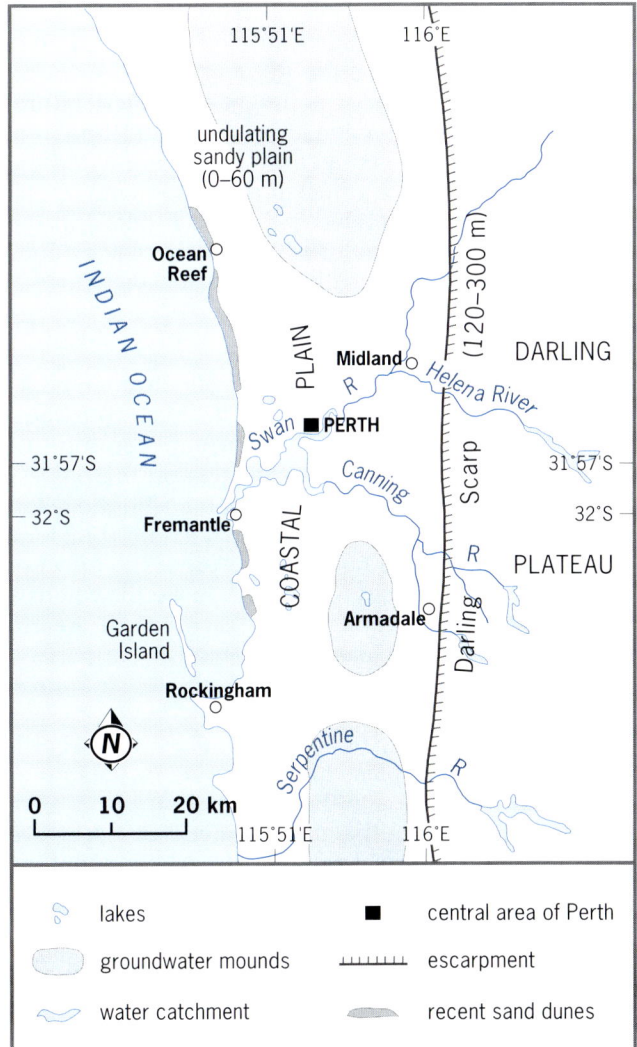

**Figure 8.5** Site features of the Perth area

suburbs have sprung up in the rural land between these settlements.

## Functional influences

An urban centre's functions may have an influence on its shape or configuration. In the case where one function is dominant the centre's shape will strongly reflect that function. Kalgoorlie, for example, has been shaped by the location of the major mineral deposits and more recently by the development of the Super Pit, a large open-cut gold mine. Perth, however, like many other Australian cities, is multifunctional. It is a major port and an important transport node. The city rose to its dominant position during the latter half of the nineteenth century, when there was a great influx of people seeking gold on the Kalgoorlie goldfields and when considerable agricultural expansion was taking place. Because of the large distances to the major urban centres in the eastern states, and the sparse nature of agricultural settlement in Western Australia, Perth attracted most of the new secondary industries, **tertiary** (service) industries and the expanding governmental functions of the developing state.

The various functions of Perth that have developed since the early history of the city have helped to shape its external form. The initial development of the city occurred around Perth as the administrative centre and Fremantle as the port. Opening up the agricultural areas of the city's hinterland led to the establishment of transport links that focused on the city centre and the port. These links have attracted urban settlement. The city's function as an industrial centre resulted in the establishment of an industrial corridor that extends south of Fremantle along the coast to Rockingham.

## Cultural factors

Perth's residential function has also clearly affected its shape. Urban morphology may be affected by the traditions and social values of the urban community. Australians tend to place a great value on the possession of a home on its own piece of land, with the result that the Australian urban area is characterised by low-density housing spread over a large area. The effect of this social value is very noticeable in Perth, where a population of just over 1 million occupies an area of approximately 1500 square kilometres. (In comparison London, with an area of only 1800 square kilometres, has a population of 8 million.)

Another cultural factor that affects the shape of Perth is the value people place on the private ownership of vehicles. A willingness to travel long distances to work has contributed to the development of extensive freeways and major road networks. Development of the city has occurred along these links, reinforcing the stellate pattern of development. The 'suburban dream' of ownership of a house on a large block of land, combined with the high degree of car ownership in Perth, has led to the largely uncontrolled outward residential development of the city—a phenomenon known as urban sprawl. Sprawl is a characteristic of all Australian cities and a major problem for city planners as it is very expensive to provide services to a continually expanding city.

A love of outdoor living and the perceived need to conserve unique natural features have also affected the external morphology of Perth. Areas of the city have been set aside for parks and public open space. Reserves, national parks and water catchment areas around the hills and dams have restricted residential development in some areas of the Darling Scarp.

## Urban planning

Urban planning attempts to channel the growth of a city in a particular way, largely in an effort to solve the problems that result from uncontrolled urban sprawl. The influence of planning on the external morphology of Perth is clearly seen in figure 8.6. The Corridor Plan for Perth, introduced in 1970 and adopted in 1973, had as its major aim the encouragement of even growth of the city along four major transport arteries radiating from the city centre to large regional subcentres. Development occurred to the north-west towards Joondalup, to the east towards Midland, to the south-east towards Armadale and to the south to Rockingham. More details about the Corridor Plan and current alternative plans are provided later in this chapter.

## Historical factors

The external form of an urban centre changes over time, largely as a result of population increase and improved transport technology. Figure 8.7 shows how the shape of Perth has changed since 1921. Until the coming of the railways the development of Perth was restricted by limited transport technology, with the population relying on muscle power for general transport and on river transport for travel between Fremantle, Perth and Guildford. Railway lines were opened in the 1880s, linking Perth to Fremantle and Guildford. Linear development then occurred and new suburbs such as Cottesloe, Claremont and Bayswater grew around stations along the railway. The use of trams allowed infilling of areas between the railways up until the early 1940s.

After World War II, Perth felt the impact of the motor vehicle. The city began to expand outwards in a number of different directions, as growth was no longer restricted to areas served by railway lines. Suburbs began to be established nearer the coast and in the hills. Perth expanded rapidly as its population was boosted by post-war emigration. The preference of the majority of home owners for a single detached dwelling on its own block of land, and for private car ownership, extended the built-up area even more dramatically. Improved transport technology also allowed the development of industrial estates at some distance from port and rail facilities, so reinforcing the pattern of outward growth.

Today, the city exhibits a combination of outward growth, with **greenfields developments** or new housing estates, and urban intensification, with redevelopment within existing suburbs. Pressure for development between the urban corridors is resulting in infill growth in these green areas. Current urban planning takes into

# URBAN MORPHOLOGY 183

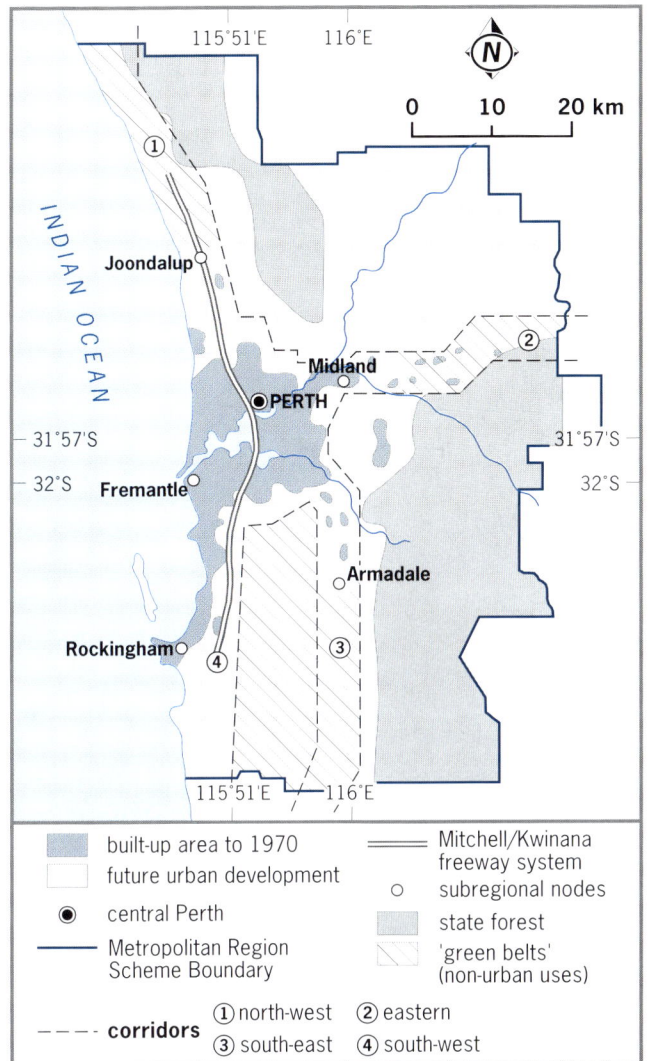

Figure 8.6 The Corridor Plan for Perth, 1973

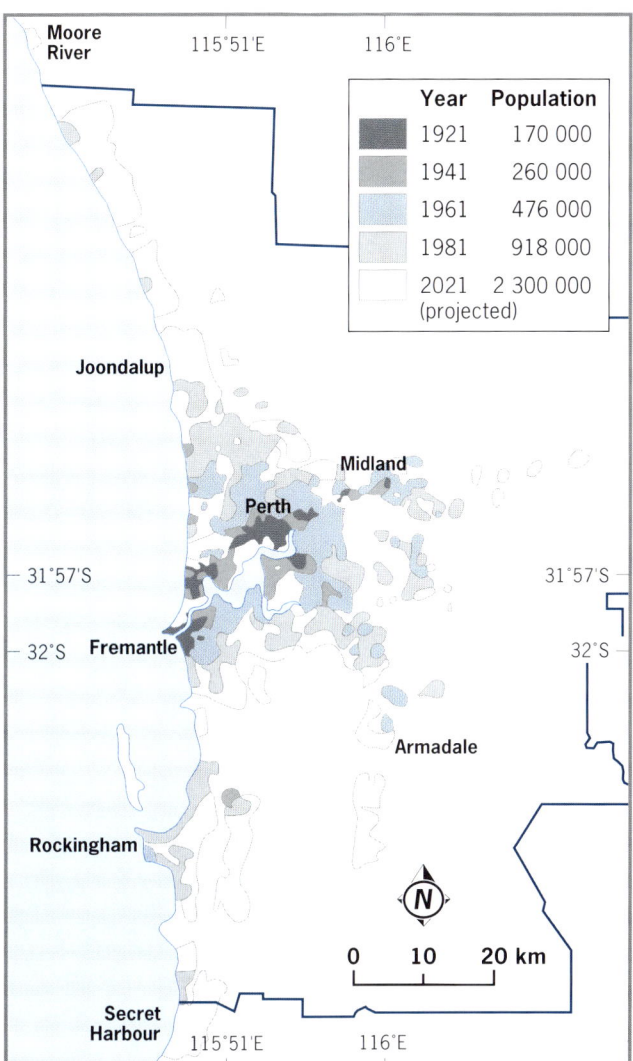

Figure 8.7 The growth of Perth

account these trends and this will be discussed later in the chapter.

## Perth's internal morphology

As in any large city, Perth's metropolitan area contains a large number of functions aimed at meeting the requirements of the residents as well as the population within its hinterland. These include commercial, industrial and residential functions as well as transport infrastructure and recreational facilities. The arrangement of these functions conforms to many of the characteristics discussed earlier in the chapter under functional zones. Some examples of the type and variety of functions are illustrated in table 8.1, while the functional zones can be seen in figure 8.24.

### The central business district (CBD)

Perth's central business district is located on the site of the first settlement. It occupies a relatively small area of the city and has a very compact form. As shown in figure 8.8, it overlooks a broad stretch of the Swan River between Heirisson Island and the Narrows Bridge. This area is laid out in a rectangular, grid-shaped fashion with St Georges Terrace and Hay, Murray and Wellington streets running from west to east, and Milligan, King, William, Barrack, Pier and Victoria streets running from north to south. The alignment of the CBD to the Swan River is illustrated in figure 8.9.

The CBD is the commercial, business and administrative centre for Perth. It is the zone of maximum accessibility and the focus of the communication network for the city and for the whole state. It therefore contains the highest land values, particularly in the core area bounded by Hay, William and Barrack streets and St Georges Terrace. There is keen competition for this limited, valuable land, so only business functions that can operate in a small area for high profit can afford to occupy it. As all available space is used, this zone has the most intensive land use of any zone in the city.

The intensity of the land use in the core area is shown in figure 8.10. A careful study of the photograph shows the large number of multistorey buildings that have been constructed in response to the limited amounts of

### Table 8.1 Perth's functions

| Function | Institutions/Facilities | Example | Location (suburb) |
|---|---|---|---|
| Administration | • Parliament<br>• Law courts<br>• Government departments<br>• Councils | • State Parliament<br>• Supreme Court<br>• Water Authority<br>• Stirling City Council | • West Perth<br>• Central Perth<br>• Leederville<br>• Stirling |
| Business and commerce | • Finance<br>• Business offices<br>• Professional offices<br>• Retail stores<br>• Warehousing<br>• Wholesale | • ANZ bank<br>• Qantas Airways Ltd<br>• Chapman, Glendinning and Associates<br>• Myer<br>• Metropolitan markets | • Central Perth<br>• Central Perth<br>• West Perth<br>• Central Perth<br>• Kewdale |
| Residential | • High density<br>• Medium density<br>• Low density | • Over 3000 } people per<br>• 1000–3000 } square<br>• Less than 1000 } kilometre | • Fremantle<br>• Morley<br>• Kalamunda |
| Secondary industry | • Light industry<br>• Heavy industry | • WA Matchbox Co.<br>• Alcoa of Australia Ltd's alumina refinery | • Osborne Park<br>• Kwinana |
| Transport and communications | • Roads<br>• Railways<br>• Port<br>• Airport<br>• Postal/Telecommunications services | • Albany Highway<br>• Perth–Fremantle<br>• Fremantle<br>• Jandakot<br>• GPO | • Victoria Park to Armadale<br><br><br>• Jandakot<br>• Central Perth |
| Community and welfare | • Health<br>• Education<br>• Recreation<br>• Entertainment<br>• Welfare services | • Child Health Centre<br>• Alexander Library<br>• Superdrome<br>• Gloucester Park (trotting)<br>• Red Cross Society (WA) | • Bullsbrook<br>• Northbridge<br>• Claremont<br>• East Perth<br>• Central Perth |

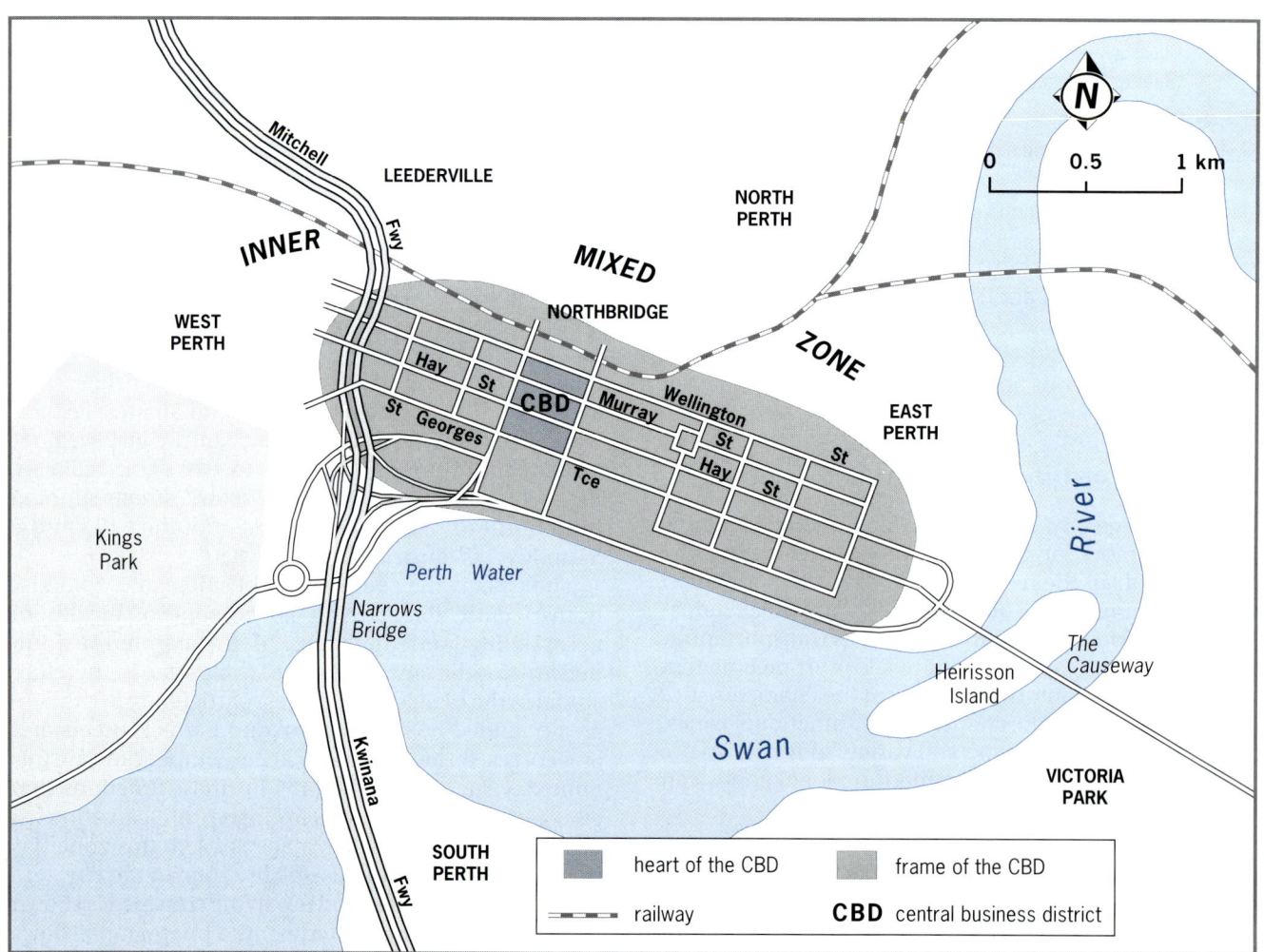

Figure 8.8 Perth's central business district

# URBAN MORPHOLOGY 185

**Figure 8.9** St Georges Terrace looking eastwards from Barrack Street

**Figure 8.10** Perth CBD skyline

extremely valuable land in the CBD. These buildings display vertical zonation of functions. Retail firms and service activities tend to occupy the ground floors of buildings. Business firms, which need less direct contact with customers, tend to locate their offices on higher, less accessible floors. The top floors of office blocks on the south side of St Georges Terrace are prestigious business addresses, as they command views overlooking the river.

Horizontal zonation of functions also occurs in the CBD, as shown in Figure 8.11. In the core zone, financial institutions such as banks, insurance firms and finance companies tend to concentrate in the high land-value area of St Georges Terrace. New office developments have occurred at the freeway end of the terrace, where the higher land improves the river views. Hay and Murray streets, however, are given over to retail functions, especially at ground level. This retail area contains department stores and smaller specialised shops selling high-order comparison goods. The internal structure of the CBD caters for pedestrian access—the Hay and Murray street malls are closed to normal traffic movements, as is Forrest Chase. Pedestrian overpasses and escalators carry people from one part of the retail precinct to the other and connect the shopping areas to the cultural precinct in Northbridge via the central railway station (see figure 8.12). The further development of elevated walkways within the retail areas brings pedestrian traffic to the first floor levels and extends the retail zone vertically.

Traditionally, the CBD has not been an area of residential development. However, in recent times and with a stagnant market for office space, there has been a growth in luxury apartments in this zone. This is a move supported by the Perth City Council and the state planning commission, as it works to encourage a revitalisation of the CBD as the shopping and entertainment heart of Perth. Even so, the CBD still retains the characteristic pattern of very high day-time population density and low night-time densities, with some 85 000 workers drawn to the city each day.

The CBD is a very dynamic zone because of the competition for the most desirable locations within the small area that it occupies. Older buildings are constantly being replaced by newer developments, particularly in St Georges Terrace, where businesses that can afford the location compete for river views. Figure 8.13 shows the demolition of the old BP building taking place in the foreground while a new 29 storey office block at the rear is being built on the northern side of the terrace. This competition and change also applies across the city, with processes of invasion and succession constantly altering the horizontal zonation patterns. **Space-extensive** activities such as bus terminals, railway stations, car parks and civic gardens occupy the fringes of the CBD and often mark the boundary between this area and the inner mixed zone.

While the forces of change tend to dominate the city, this does not mean that all functions will be affected. Perth's CBD also includes important heritage precincts

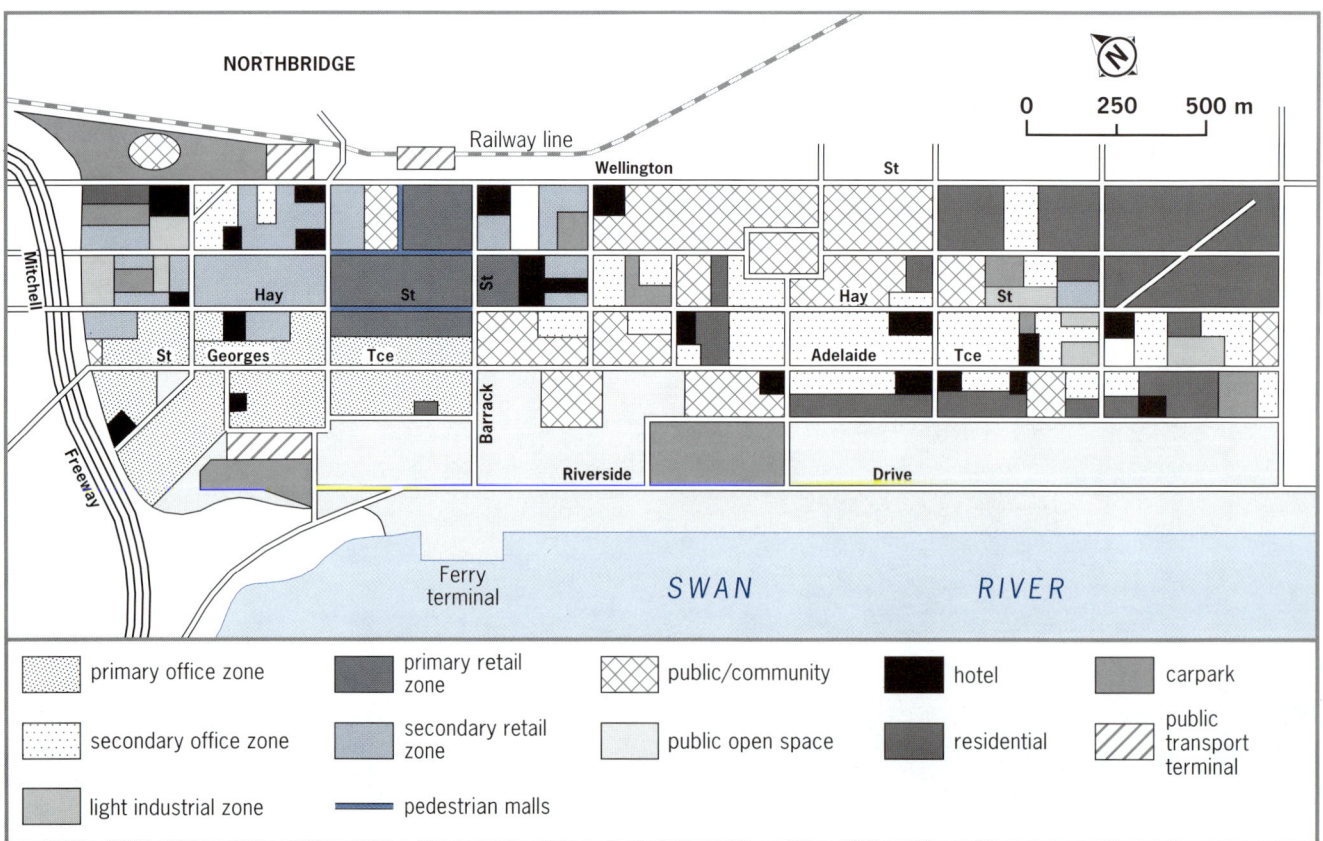

**Figure 8.11** Perth CBD land uses

**Figure 8.12** Murray Street Mall with Forrest Chase pedestrian overbridge

**Figure 8.13** Dynamic nature of the CBD—office development on St Georges Terrace

of a social, cultural and religious nature. The zone contains a number of important religious buildings and these continue to exist amongst the commercial skyscrapers. The growth of an active heritage movement has also seen the facades of older buildings preserved as a condition of redevelopment. Evidence of this influence can be seen in figures 8.14–8.16.

### Outer business districts (OBD)

As a consequence of Perth's size, a hierarchy of business districts and precincts operates beyond the CBD. This is a characteristic of all large cities and reflects the influence of increasing distances between the CBD and some suburban areas. These commercial districts range from the single corner store, which can still be found in some of the older suburbs, to the suburban shopping centre and on up to the regional shopping centre. Underlying this pattern is a broad planning strategy used by local and state planning authorities in determining applications for new developments. The state planning commission imposes a strict limit on retail floor space within the metropolitan area, and especially in the CBD and the regional centres. This has in fact produced a retail hierarchy with local, district, regional and citywide levels apparent.

The regional shopping centre is the largest of the suburban commercial precincts. These multifunctional centres provide retail, business, professional, administrative and community services to a number of suburbs. They offer a range of low- and high-order goods and services including both durable and non-durable items. The Westfield Carousel centre, located

**Figure 8.14** Facade retention at Bankwest Tower

**Figure 8.15** King Street historic shopping precinct

Figure 8.16 St Georges Anglican Cathedral, St Georges Terrace

on Albany Highway some 12 kilometres from the CBD, forms part of a regional city centre within the Canning **local government area**. This location was designated as a major regional commercial district under the metropolitan planning scheme, Metroplan, in the 1980s. The shopping centre, which provides 75 000 square metres of retail shopping space, is one of the largest in the metropolitan area. It houses around 100 businesses and forms the nucleus for a commercial precinct covering some 100 hectares. In contrast, the Riverton Village shopping centre provides about 500 square metres of retail space containing convenience goods and services including a liquor store, supermarket, take-away foods and hairdresser (see figures 8.17 and 8.18).

While most centres operate as functional nodes with defined draw areas or hinterlands, some services achieve a citywide customer base due to their high level of recognition as a supplier of certain goods and services. The aggregation of new and used car yards along Albany Highway and Scarborough Beach Road gives these two locations high exposure throughout the metropolitan area. Their arrangement also points to the **ribbon** or linear pattern that occurs with some OBDs.

### Inner mixed zone (IMZ)

Perth's inner mixed zone is made up of the older inner-city suburbs such as West Perth, Leederville, North Perth and East Perth and is situated beyond the frame of the central business district, particularly to the north (see figure 8.8). While the river to the east provides a distinctive break to the central Perth region, evidence of invasion can also be seen in the suburb of Victoria Park, where IMZ characteristics occur. This zone is one in which a variety of land uses (heterogeneity) may be found. It is also referred to as a zone in transition, where the mixed pattern of land use is constantly changing as central business district functions encroach on the older residential areas.

This inner-city transition zone displays a semi-intensive mixture of land uses, shown by its dissimilar building types. Factories, warehouses, flats, terrace houses, transport terminals, restaurants and other structures housing specialist functions are found close to one another along transport links. Old building forms, such as semi-detached terrace houses, are found alongside new commercial buildings. Many functions are carried out in buildings not originally designed to house them. A restaurant, for example, may now occupy an old house (**change of function**).

Some parts of the inner mixed zone show visual evidence of deterioration or decay. This decaying process, or **urban blight**, is attributable in part to the reluctance of property owners to renovate sites that may be worth more if taken over by central business functions (see figure 8.19). The absence of open space, and the higher density development and congested transport networks are also a feature of this zone. Offsetting these is the locational advantage of the zone in terms of its proximity to the CBD. As a result, Perth's IMZ has in recent times attracted significant levels of urban renewal and redevelopment. This has added to the older residential buildings that are a legacy of a past era. This zone was the original residential area in the walk-to-work era before modern transport and population growth led to residential development in the outer-city areas. The residential character of the IMZ is now one of great variation. There are areas of poor quality properties, older high-density flats, terrace housing which shows evidence of gentrification and

**Figure 8.17** Westfield Carousel, Albany Highway, Cannington

**Figure 8.18** Riverton Village shopping centre

new small and larger scale redevelopments. There is also evidence of variations in the socioeconomic status of the inner suburbs, with some expensive apartments being newly constructed in the older prestigious areas of West Perth and Kings Park.

The transport links of the inner mixed zone channel commuter traffic into and out of the CBD. Because of this, and also because the grid-shaped road network was not designed to handle the large volumes of traffic associated with the outward growth of the city, congestion is a characteristic and a problem of this area. Ribbon development of commercial enterprises along the major transport links has also added to the traffic congestion problem. Commercial enterprises, for example, have wholly replaced houses along Hay Street in West Perth, and in Subiaco as far as the railway line.

Although the inner mixed zone as a whole is characterised by a mixture of land uses, there are some areas in which functions display horizontal zonation. The West Perth region shows the invasion of CBD functions in this transition zone. Located on a ridge of high land overlooking the CBD, this suburb now has specialised offices belonging to the legal, medical, accounting and architectural professions, along with a number of restaurants. It is also an area in which government departments have located in close proximity to the parliamentary precinct. In recent times there has also been a move back towards apartment blocks, with a number of these being constructed along Kings Park Road on the south-western side of the zone. Figures 8.20 and 8.21 show the invasion of this zone by office functions, as well as the re-establishment of residential functions.

East Perth has undergone a dramatic change in the last 10 years. Originally an area of poor quality housing, warehousing and industrial functions, it was the subject of large-scale urban renewal initiated by the state government. Under special legislation the East Perth Redevelopment Authority was established, charged with the responsibility of revitalising this area. Significant pollution of soil and water had occurred here and this had to be remedied. Soil was removed, new roads were constructed and the Swan River foreshore was landscaped and cleared of its old industrial buildings. The significance of the local Aboriginal people and of the original environment has been recognised in the public art works and parks that have been developed in the area. A large number of small blocks were created and sold. They have zero front and side setback, in keeping with the planned and uniform appearance of the area. As shown in figure 8.22, the medium-density dwellings have been built up to the lot boundaries on each side, as well as the front, rather than being set back in the style of the detached bungalow found in most suburban areas. In addition,

**Figure 8.19** Urban blight, Northbridge

developers have built high-density apartments and the area is now one that commands high property prices and is attracting a growing number of people wishing to live there. In some respects the region is no longer a zone in transition, as the older land uses have been largely replaced by the newer ones.

**Figure 8.20** High-density residential and office complexes, West Perth

**Figure 8.21** Office redevelopment site, West Perth

Figure 8.22 Urban renewal, East Perth

To the south-west of the suburb of North Perth, on the northern boundary of the CBD, lies the Northbridge area. This locality is characterised by a concentration of high- and medium-density residential functions, entertainment facilities, restaurants and cultural centres. Perth City Council, in association with the Northbridge Chamber of Commerce, initiated Northbridge's development as an entertainment and restaurant precinct. It is recognised as a high profile location within the metropolitan area and attracts patrons from all areas of Perth. The precinct also illustrates, through its variety of restaurants, the impact of the cultural diversity of the people of this area. Migrants from Greece, Italy, Vietnam and China have all made a contribution to the international character of the Northbridge area. Figure 8.23 illustrates a typical street scene within the restaurant strip in the area.

### The residential zone

Perth's residential zone, as demonstrated in figure 8.24, is by far the most space-extensive area within the city's boundaries, extending from the inner mixed zone to the rural–urban fringe. It consists of a mosaic of dwellings of different ages, sizes, styles and quality, the transport network that connects them, and the buildings that provide goods and services for the residents. It is a dynamic zone in which change is regulated by population movement. As the city's population has grown and more residential land has been required, growth has occurred outwards along major highways towards the city's periphery as well as in the areas between the corridors. Urban sprawl has created a city with a very low residential population density compared with cities in other nations.

Although the residential zone consists of a mixture of different types of dwellings spread over a large area, it is possible to identify distinctive patterns within the overall zone. Distinctions may be made on the basis of socioeconomic characteristics, the age of the suburbs, housing density, and location or position within the metropolitan area. (See chapter 6 for an analysis of socioeconomic characteristics, including the levels of home ownership and the proportions of rental

Figure 8.23 Restaurants in James Street, Northbridge

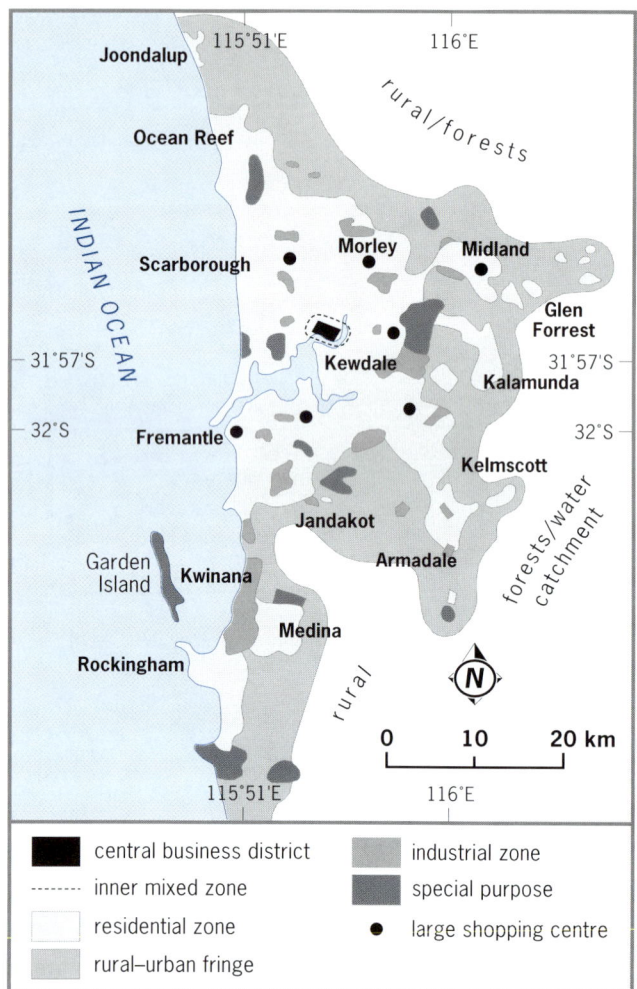

**Figure 8.24** Functional zones of Perth

centre, hydrotherapy pool and leisure pool as well as spa, sauna and physiotherapy rooms. In addition, it is attached to a large public library.

Disruption due to new developments is relatively minimal, though there may be some new construction in areas where land has been rezoned to allow for higher density construction. In addition the desire to live in some of these established suburbs can also result in small-scale redevelopment. Some suburbs are found on desirable sites and therefore command higher land values than in other areas. They have a turnover of population as new families buy into the established zone. Figure 8.26 shows small-scale redevelopment within the established residential zone.

While the housing stock is older than in the newer growth zone, the quality is generally good and houses are well maintained, with established gardens and mature street trees. Figure 8.27 shows houses typical of many in the established zone. Suburbs such as Willetton to the south of central Perth, Bayswater to the east, Morley to the north and City Beach to the west are examples of this residential zone.

### The newer growth zone (NGZ)

Lying beyond the established residential zone is the newer growth zone—an area of greenfields subdivisions and recently constructed houses. Figure 8.28 is typical of this type of development. Developers have generally provided a high degree of amenity to attract the home owner. This includes special road treatments, underground power, established parklands, attractive road design and covenants to ensure a quality standard of construction. Where this development has occurred between the corridors, such as at Canning Vale, Alexander Heights and Cockburn, the areas are relatively close to the city. On the other hand, the NGZ along the coastal regions to the north and south is located some distance away (suburbs such as Quinns Rock to the north and Secret Harbour to the south are some 40 to 50 kilometres from the CBD). These outer suburbs rely heavily on private transport and are generally not as well served with public transport as the established zone. In addition, the NGZ does not have the same range and availability of community services as the established zone. It may have to wait some time

properties within different suburbs.) The age of a suburb will indicate marked differences between various residential areas. In addition to the age of the housing stock, there will also be differences in the population demographics and the type and variety of services available. The division of the residential zone into established residential and newer growth zone is one approach to studying the different characteristics within this functional zone.

### The established residential zone

Beyond the old and new residential properties of the CBD and the IMZ lies a large region of established suburbs. They represent the past fringes of the city and substantially developed following World War II, though development of some of the inner areas precedes this point in time. The established suburbs are located in areas that enjoy good public and private transport facilities. They are generally well served by a range of recreational, cultural, commercial and professional services. Figure 8.25 illustrates the high level of recreational services available within the established zone. This complex includes a 50 metre indoor pool, function centre, children's gymnasium, adult fitness

**Figure 8.25** Riverton recreation complex

URBAN MORPHOLOGY 195

**Figure 8.26** Small-scale redevelopment in established residential zone, Shelley

**Figure 8.27** Residences in the established zone

for both local and state governments to provide schools, community halls, recreation centres and public libraries.

People are attracted to the NGZ by the availability of cheaper land or houses and the opportunity to live within a new suburb. In addition, the lower population densities associated with the zone reduce the impact of traffic congestion and noise that is associated with inner-city and established zones. In living within the NGZ, people accept the disadvantages of greater distances and fewer services. In time, this zone will become part of the established residential region as the city expands further.

A variation on the newer growth zone is the process of urban infill or brownfields development. This occurs on medium to large parcels of land that were left vacant as the city expanded outwards, or in areas that may have contained old industrial activities. Some of these areas would have been owned by government, while others were held by private individuals or companies, or religious or educational institutions as land endowments. One such development is the Mt Henry estate being constructed near the Kwinana Freeway and overlooking the Canning River. Figure 8.29 illustrates this form of residential development.

## The industrial areas

The industrial areas of Perth have evolved as the city has grown and reflect changing technologies, community attitudes and expectations, and planning decisions. Older industrial functions will often have **non-conforming rights** that allow them to continue to operate even when surrounded by residential development. Small-scale industrial activities can be found dispersed throughout the city. Light and heavy industrial activities are allocated to different zones, depending on their potential impact on other land uses. Newer industrial estates show a higher standard of planning and presentation than the older ones.

### Inner-city industries

Small-scale and **space-intensive** industries that are closely associated with CBD functions have a tendency to locate close to the city centre. Stationery and printing firms, clothing manufacturers, speciality food production or processing and jewellery producers fall into this category. These types of industries may be found in suburbs such as Leederville, East Perth, Victoria Park and North Perth. Older industrial areas within the inner-city and established residential zones are often associated with rail transport. This enabled the movement of products by train and allowed the workers ready access to this form of public transport. This type of development also occurred in more recent times along major roads. The development of industrial zones within the suburbs of Maylands, Welshpool, Bassendean and Bayswater illustrates the importance of the railway systems, while the more recent growth of Osborne Park and Kewdale indicate the change in emphasis to road transport.

**Figure 8.28** Residential land development in the NGZ, Canning Vale

**Figure 8.29** Urban infill, Mt Henry

*Planned industrial estates*

The impact of industrial functions on other land uses often produces negative effects. Higher environmental standards and stricter planning standards have given rise to the large industrial estates. Within these, the siting of different industries is closely considered with a view to creating buffer zones and meeting environmental laws. Thus, industries that deal with hazardous materials or have emission problems are sited to minimise their impacts on other industrial activities as well as other land uses. The location of a cement works, as illustrated in figure 8.30, involves a different set of standards to those for a warehouse function, as illustrated in figure 8.31.

While the Canning Vale industrial estate illustrates the newer estates, the Bentley Technology Park, situated near Curtin University, is an example of a high technology estate. Here, pharmaceutical companies, computer and electronic equipment producers, and biotechnology enterprises engage in **quaternary**-type industries.

*Heavy industry*

**Heavy industries** usually agglomerate around a port or major railway freight terminal system. Due to the noise and pollution associated with these industries they are separated from residential areas by relatively wide buffer zones of natural vegetation or rural land use. The Kwinana heavy industrial area, for example, was built on an unused part of the coastal plain about 40 kilometres south of Perth. The sheltered waters of Cockburn Sound, though shallow, provided the opportunity to construct wharves that could take large tankers and bulk carriers. Early industries included the BP refinery and the BHP steelworks. It is also an area in which Western Mining and Alcoa operate refineries. Further south towards Rockingham, the CBH grain silos illustrate another use of the port facilities in Cockburn Sound.

### The rural–urban fringe

The rural–urban fringe is the growing outer edge of a city located between the built-up area and the rural countryside. This area, being the space into which the city extends, is only partly assimilated into the growing urban complex and therefore retains a lot of its rural characteristics. The uneven development of the city outwards, often in a haphazard manner, results in a mixture of land uses. The rural–urban fringe, like the inner mixed zone, is a transitional zone. One similarity is the deterioration (urban shadow effect) in structures that occurs as the landowners wait to redevelop their rural holdings or sell to housing estate developers.

Intensive farming is the principal rural land use in the wide zone on the outskirts of Perth. Market gardens are found to the north (in Wanneroo, for example) and to the south-east (in Spearwood), and vineyards are found to the north-east in the Guildford area. In the hills area to the east, fat lamb farming and stone fruit

**Figure 8.30** Cement works, Canning Vale industrial estate

**Figure 8.31** Warehouse and manufacturing functions, Canning Vale industrial estate

production occur. Other rural land uses in this zone include horse studs, dog kennels, poultry farms and nurseries.

The rural–urban fringe is also home to a variety of specialised, space-extensive land uses, some of which are relegated to these areas due to their incompatibility with other urban functions. These land uses include industrial estates (such as that in Wangara), prisons (Canning Vale Prison), sewage treatment plants, water catchment areas, and some quarrying.

Increasing numbers of commuters moving into the rural–urban fringe have caused the older, semi-rural settlements to become more urban in function. Kalamunda, Gooseberry Hill and Mundaring were formerly small country towns along the Darling Scarp. They are now multifunctional urban centres, though they still have a predominantly rural character. New housing estates also occur frequently throughout the zone. As in the newer growth zone, the provision of amenities and services to these estates may lag behind settlement.

The rapid encroachment of residential and other urban land uses has resulted in a partially developed urban landscape, which often has an untidy appearance. This has occurred in particular in the north of the city, the area experiencing the most rapid growth. Many of the characteristics of the rural–urban fringe are shown in figure 8.32. While the suburb of Ocean Reef has been completely developed, Iluka and Kinross are characterised by incomplete residential development, with subdivided lots yet to be built upon and some transport links yet to be completed. The **curvilinear pattern** of minor suburban roads is a characteristic of recent residential development (compared with the grid system used in established parts of the city). The city has developed further northwards since the photograph in figure 8.32 was taken. This includes the newly developed suburbs of Clarkson, Mindarie and Merriwa. This northern region also illustrates the uneven nature of the rural–urban fringe, with suburban development leaping over undeveloped areas as each individual developer achieves planning approvals for their own piece of land.

### Special purpose areas

Land use activities that cannot be classified into one of the recognised land use functions, or that are too few or scattered in their distribution, are designated as 'special purpose areas'. Parks and reserves, railway yards, airports, sewage treatment areas and military camps are some of the special purpose areas found in urban settlements.

Examples of relatively space-extensive land uses in Perth include Karrakatta Cemetery, Perth and Jandakot airports, Sir Charles Gairdner Hospital and various large tertiary educational institutions, such as the University of Western Australia Murdoch and Curtin University. Often, these land uses are simply included as a part of the residential area in which they are located.

### EXERCISES

1. Study the map of the horizontal zonation of Perth's CBD and then compare and contrast the changes in land use that occur as you proceed south-east along Wellington Street, Hay Street and St Georges Terrace from the Mitchell Freeway to Barrack Street.
2. Identify and describe the different types of new residential development that you would find in Perth's inner and established residential zones.
3. What would some of the issues be in developing new residential subdivisions on the outskirts of Perth in areas that were used for farming or areas that are still covered in natural woodlands?
4. What are the advantages and disadvantages of low-density urban development or urban sprawl?
5. What are the main differences between the rural–urban fringe and the newer growth zone?
6. What would some of the differences be between the established residential zone and the newer growth zone?
7. What are the main differences in the location and characteristics of the different industrial areas of Perth?
8. Study figure 8.33 and then identify and describe the different land uses found within the urban landscape. Note that the picture includes the tip of a plane wing on the right.

## Urban processes affecting Perth's development

As with other urban areas, Perth's appearance and growth is not a matter of chance. A number of important factors and processes have acted over time to produce its distinctive urban landscape.

### Processes that group similar functions

Processes of agglomeration and aggregation can be used to account for the clustering of similar functions. **Agglomeration** was initially used to account for the grouping of industrial functions within a given area. Individual industries within an industrial zone will often use the products of their neighbours, or supply semi-finished products to other industries for further processing. The benefit of being in close proximity to each other is reflected in savings in transport costs or in the ability to integrate their production processes. They can also take advantage of common infrastructure and ancillary services that have been established to support the industrial activities. For example, the heavy industrial zone of Kwinana has well-developed power, water and gas supplies and extensive road, rail and port facilities. Transport companies seek to locate here to move the products that are manufactured, and companies that supply component parts to other businesses for further assembly or fabrication are also found within the industrial strip. Businesses thus use the by-products or semi-processed products of other companies in the manufacture of their goods, while others supply specialist components to the producers of larger items. They also operate within an environment that allows for levels of noise and air pollution that would be unacceptable in other parts of the city. In

**Figure 8.32** Perth's rural–urban fringe (north coastal)

this way they take advantage of the zoning of the area and the economies of operation that flow from the industrial agglomeration.

Within the CBD, the agglomeration of financial, commercial, professional and retail functions allows for close contact between providers and consumers of these types of services. The Australian Stock Exchange, located in St Georges Terrace, is in close proximity to a number of major publicly listed company offices as well as banks, stockbrokers, lawyers and accountants. The grouping of these functions illustrates a form of commercial agglomeration.

**Aggregation** occurs where like functions seek a common location. The presence of medical specialists in West Perth is one example of this, while the location of car yards in close proximity to each other in Victoria Park is another. Within the suburban areas the development of fast-food outlets next to each other is not a matter of chance. The advantage of aggregation lies in the tendency for customers or clients to associate certain locations with certain services. This then leads them to seek out these locations should the need arise. It may also allow each individual business to be associated with a prestige location, especially if this location has a history of high quality service or standards.

While these factors are socioeconomic forces, they are often formalised through the planning process. An important element of urban planning is the grouping of

**Figure 8.33** Urban landscape

compatible land uses and the establishment of buffer zones around incompatible ones. Industrial areas will often have warehousing or wholesaling between them and residential zones as a land use buffer. Heavy industrial areas may be separated by transport or green belt buffer zones. These decisions form part of the planning process.

The Kwinana industrial area was zoned for industry in the 1950s. Industry was provided with large areas of coastal land, relatively cheaply, and the necessary ancillary services and infrastructure were also provided. Once the main **location leader** industries (such as the BP oil refinery) had been established, agglomeration occurred as other industries moved to the area to take advantage of the economies that would accrue from a location adjacent to customers, suppliers and service firms in the area.

## Processes that produce the patterns of land use zones

The actual arrangement of Perth's functional zones (their position in relation to one another) can be attributed largely to the effects of land use competition. **Accessibility** is the dominant force behind this important process, as it influences land values, which in turn determine the use that is made of particular urban sites. (It also determines the use made of agricultural land, through the operation of the land rent mechanism first discussed in chapter 3.)

The most desirable sites for many urban land uses lie close to the city centre, where maximum accessibility is provided by converging transport links. As central locations are the most limited in number, the competition between functions requiring them forces the land values up. Only functions that can achieve high returns or profit from a small land allotment can afford a location in the most central part of the urban area. Those functions that need larger parcels of land in order to produce a profit will be forced to operate further away from the centre as they cannot compete for expensive land.

Figure 8.34 shows the effects of accessibility and land use competition. The graph demonstrates that higher land values encourage a higher intensity of land use close to the city centre and that the lower land values of outer areas attract the more space-extensive land uses, which cannot compete for the more expensive sites. Land-use competition accounts for the location of retail and commercial organisations, for example, within the central business district. The profits that they are able to make from a very limited floor space far outweigh the excessively high cost of their inner-city site. (High-rise development and the vertical zonation of land uses within this area occur in response to the demand for limited central business district sites.) Figure 8.34 shows that more intensive industrial users of the land will tend to occupy land further out from the central business district. The residential areas, which have an even smaller return from a much larger area of land, are therefore located in the outer urban zones where land values are even lower.

The pattern of urban land use in Perth conforms to the model to a certain extent, but there are differences caused by the existence of other high-cost, accessible and desirable sites, as shown in figure 8.35. Although peak land values in Perth occur in the block bounded by Hay, William and Barrack streets and St Georges Terrace, there are other accessible and desirable areas that attract space-intensive commercial, retail and some industrial functions. Major intersections, for example, attract retail and commercial development. These land uses are also located along major transport links, creating ribbon development of central business-type functions such as that along Albany Highway, Victoria Park. Planning can also create accessible sites, as in the case of some industrial estates and regional shopping centres. Particular site factors such as high land and proximity to rivers and the ocean also create relatively high residential land values.

## Processes that act to limit change

Many urban functions, once established in an area, will resist changing their location even if the original reasons for their establishment in the area have long since ceased to be important. Known as **inertia**, this resistance to change particularly affects functions that have long been established in or near the central urban zones. Inertia also affects industries that have been established as a consequence of industrial agglomeration, especially when a firm's major suppliers or purchasing businesses leave the industrial zone. Relocation is often not feasible, primarily because the costs of re-establishing the industry somewhere else would be too high.

The Royal Perth Hospital was built in the 1950s on a spacious site in the East Perth residential area. Now the hospital is almost within the CBD (because of the CBD's growth outwards), and its buildings occupy the whole site. While the location has lost some of its original advantage, the high cost of relocation has caused the hospital to remain on its original site. In order to continue to perform its function, the hospital has had to replace the original buildings with multistorey structures. St Mary's Cathedral, located next to the hospital, is also affected by inertia. The reluctance to relocate this institution stems from the church's cultural significance, even though there is no longer a large local congregation and parishioners have difficulty accessing the church due to a lack of parking.

## Processes that produce changes in location

While inertia acts in many cases to keep Perth's pattern of land uses static, there are a number of important processes that, in combination, are responsible for the dynamic nature of the city. Some of the forces that cause the relocation of functions are demonstrated in figure 8.36. There are processes responsible for attracting functions to the centre of a city, while others result in an outward movement or decentralisation of functions.

**Centripetal forces** are forces that attract functions to the central, most accessible parts of the urban area. This attraction results in a concentration of activities, particularly retail and commercial functions, within the central region of Perth. The accessibility of inner-city areas to both clients and workers is one of the most significant reasons for the concentration of functions in the central region. Functions will also be attracted to the inner city by the benefits that accrue from associating with other centrally located functions of a similar type. A high degree of public awareness of the central

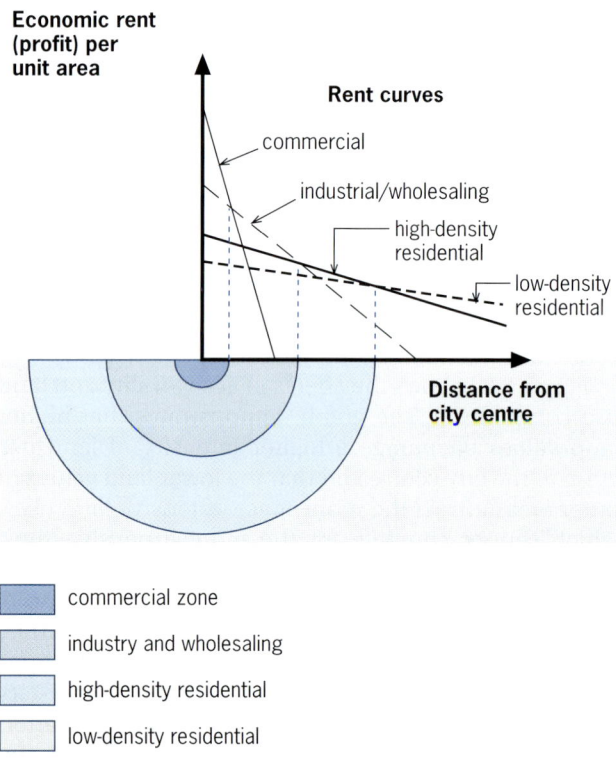

**Figure 8.34** The effect of land use competition and profitability on urban land use patterns

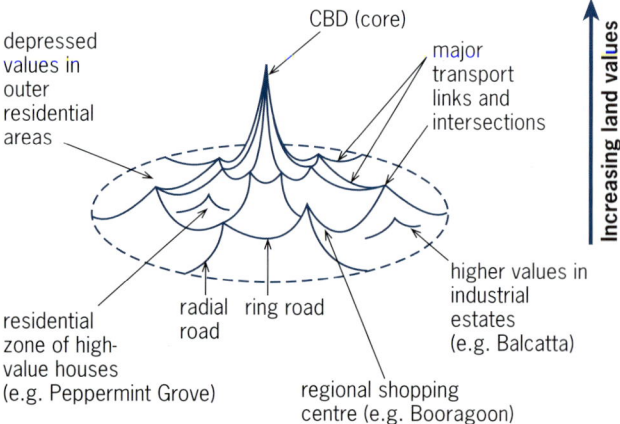

**Figure 8.35** Land value surface model

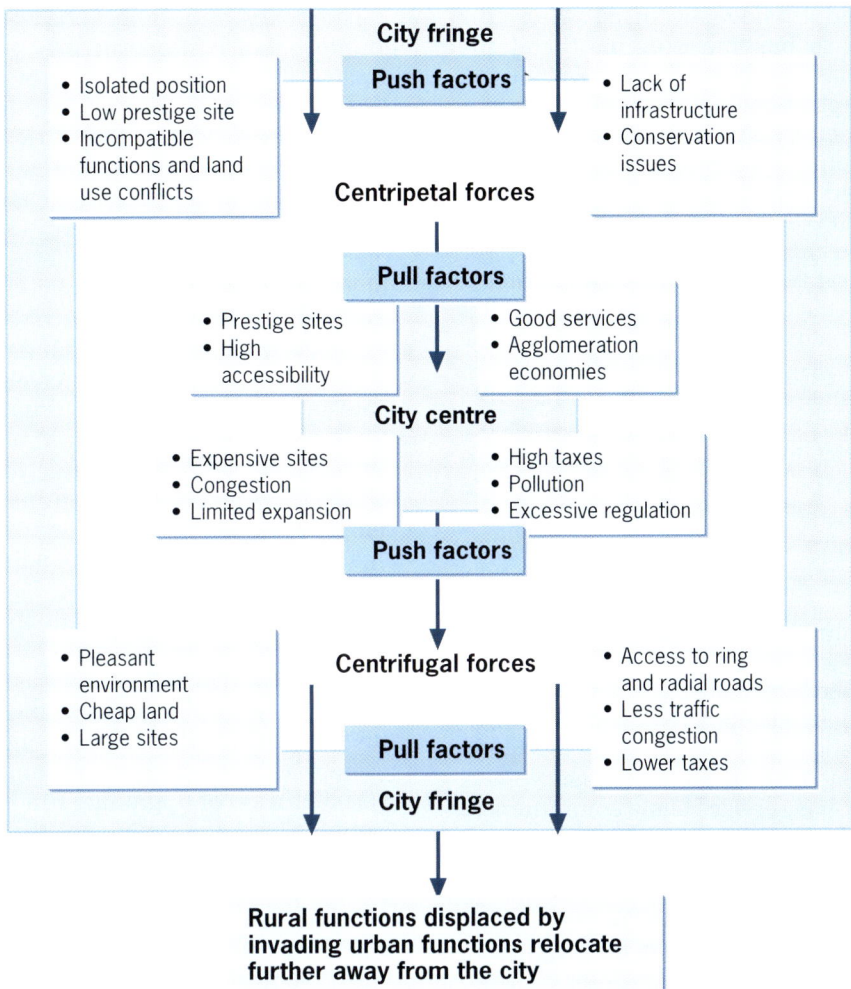

Figure 8.36 Dynamic forces affecting urban patterns

business district also attracts business functions, as does the area's established business reputation.

**Centrifugal forces** are forces that cause functions to move away from central areas towards the periphery of Perth. The process of dispersal (decentralisation) of activities towards the rural–urban fringe results from such factors as the prohibitive cost of land and the high rates and land taxes in more accessible areas. Traffic congestion, parking problems, local government restrictions on land use, lack of space for expansion and the rising cost of such public utilities as power and water can also be disincentives. Functions that gravitate towards the periphery are attracted by such factors as lower land prices and taxes, decreased traffic congestion and room for expansion.

Centrifugal forces help to account for the continual development of the residential growth zone and the decentralisation of industry to those outer areas that have become zones for industrial development. The location of manufacturing establishments in Perth since the mid-1950s has changed significantly. For example, the inner-city suburb of West Perth experienced a decline in the number of industrial establishments (from 697 in 1954 to only 21 in 1986). In the north-west industrial suburb of Osborne Park, the number of industrial establishments rose from 82 to 408 in the same period. Many outer suburbs that had no industrial function in 1950 (such as Jandakot and Balcatta) have since developed a significant industrial capacity. This decentralisation process has largely occurred in response to the increasing diseconomies associated with operating in congested inner-city locations.

The processes of **invasion and succession** help to explain the changes that are taking place in the transition zones of Perth as the city grows outwards over time. Invasion occurs in the inner mixed zone when retail and business functions, unable to afford the high land prices in the CBD, buy older and cheaper premises in the neighbouring inner mixed zone. These buildings are renovated and converted to suit their new function. After a time, the increased flow of new businesses into the zone causes land prices to rise. Property developers begin to buy up the old places, demolish or modify the buildings and construct new premises more suited to the incoming functions. Invasion is also a process that occurs within the rural–urban fringe, where the most frequent invaders are residential and industrial functions.

Succession is said to have occurred when the invading function has become dominant and the original land use has disappeared. In Hay Street, West Perth, for example, retail and business functions have replaced a residential area. In parts of Karrinyup and Balcatta, residential areas now exist in locations once occupied by market gardening.

The influence of Perth's population increase and changing transport technology on the growth of the residential zone outwards has already been outlined in this chapter. Much of the change that has occurred has been simple expansion, triggered by the growth of the urban population. Invasion (and subsequent succession) has led to urban sprawl as growth has occurred along the road and rail systems that radiate from the central business district. Infilling between the transport links has then taken place. Some leap-frogging has occurred as industrial and residential estates have been established on attractive sites in the rural–urban fringe before land nearer the city centre has been fully developed.

Since the Australian urban community as a whole prefers the most space-extensive type of residential development (detached houses on large blocks), **urban sprawl** is a prominent characteristic of Australian towns and cities. Changes and improvements in transport systems have encouraged greater urban sprawl. For instance, express bus services allow commuters to travel further to work, so many will settle in outlying districts where they find the environment more satisfying. Private ownership of motor vehicles has had an even greater effect on urban sprawl. People are no longer restricted to living within walking distance of bus routes or railway stations so they can live further away from the major transport routes and still have good access to other areas of the city. Faster movement along upgraded transport networks has had the effect of 'reducing' distance. People and goods can be transported much further in a shorter amount of time (functional distance), so the city has expanded rapidly into the surrounding countryside.

The process of invasion, while acting to change land uses of transition zones and helping to spread the city outwards, is also responsible for the deterioration of both the inner mixed zone and the rural–urban fringe. In the inner mixed zone this phenomenon, known as **urban blight**, occurs because property owners are reluctant to upgrade properties in view of the probability of invasion by central business functions. Landlords lease rundown houses with inadequate facilities to people in low-income groups or to firms requiring cheap storage space.

The **urban shadow effect** is a process similar to blight. This is experienced in parts of the rural–urban fringe of Perth. Because farmers are uncertain of the future of their farms as the urban area encroaches, they do not feel inclined to invest money in their properties. Competing residential and industrial land uses are likely to force the land value (and land taxes) up and farmers are likely to be paid more for their land than would have been offered before encroachment. As they wait, deterioration of the farmland and its buildings occurs, producing a rundown, degraded landscape.

There are a number of processes that tend to counterbalance the processes of blight and urban shadow in the transition zones. **Urban renewal** and **urban redevelopment** result in an upgrading of blighted areas in the inner mixed zones, while in the rural–urban fringe the upgrading and intensification of rural land use may occur where owners or investors are more certain of the long-term future of the property. The retention of the Swan Valley wine growing region of middle and West Swan through zoning regulations is one example of this resistance to urban invasion and the loss of good agricultural land.

Urban renewal in the inner mixed zone is a deliberate process involving the rebuilding of parts of an urban area by demolishing old buildings and erecting new ones. Renewal occurs largely as a result of the need for space as functions develop and expand. Urban redevelopment usually involves improving or upgrading an area by retaining existing buildings and remodelling them to suit their new uses. The design of any new buildings to be introduced into the redeveloped area is complementary and sympathetic to existing structures. Renewal and redevelopment has occurred in the East Perth area. This involved the transformation of rundown, polluted industrial land into a sought-after residential, business and education region.

The process of upgrading rundown old housing and warehouses in the inner suburbs of Perth is known as **gentrification**. In recent years, there has been widespread restoration and refurbishment of these old buildings, many of which were abandoned and some of which the community regards as having historical worth. Many gentrifiers are young professional people who prefer inner-city living, with its greater access to employment, services, social life and the entertainment opportunities provided by the city. Gentrification brings changes to the visual appearance of the area through renovation, rehabilitation and restoration of old residences and factory warehouses and the commercial redevelopment of service areas.

The area adjacent to the port in Fremantle provides a very good example of the process of gentrification. In the early 1980s, industrial and port activities in the area declined as many functions relocated to Kwinana and Rockingham. The tourism and entertainment functions in the generally attractive, historic area were increasing. People in professional occupations began to move into the area and renovate dwellings formerly occupied by factory and dockside workers. The staging of the 1986–87 America's Cup yachting defence in Fremantle and the associated influx of tourists accelerated the gentrification process. Today, Fremantle is a bustling community and attracts large numbers of people to its cafes, restaurants and retail outlets.

### EXERCISES

1. Study the model of dynamic forces affecting urban patterns illustrated in figure 8.36 and explain how push and pull factors affect the location of urban functions.
2. Using examples, explain aggregation and agglomeration.

3 Using examples, illustrate how urban land uses may be affected by centripetal and centrifugal forces.
4 Identify difficulties associated with the location or siting of Perth's CBD and discuss ways in which these are being overcome.
5 Using examples, compare and contrast the dynamic residential processes of urban renewal, urban redevelopment and gentrification.
6 What types of functions appear to have invaded West Perth and how are the inner mixed characteristics of this suburb different to East Perth?

## Perth: urban problems and planning solutions

Urban problems develop as a consequence of changes within an urban centre. These problems may be the result of urban growth or decline. These problems highlight a range of economic, social and environmental issues. Table 8.2 identifies the principal urban problems associated with urban growth.

Urban problems are often overcome through formal planning processes. The Western Australian State Planning Commission has the responsibility of overseeing the broad planning strategies throughout the state, while local councils are responsible for the implementation and regulation of town planning schemes and building regulations at the local level. Recent changes in the structure of government departments have resulted in issues of public and private transport being assigned to the Minister for Planning, while there has been a growing involvement of the Department of Environmental Protection in urban planning and land use matters. The Minister for the Environment has responsibility for a range of environmental issues—an important element of urban planning—and oversees heritage issues, which also has important planning and development implications. Figure 8.37 identifies the main planning authorities in Western Australia.

A local council wishing to implement a new town planning scheme involving the location of industrial, residential and recreational functions would need to make sure that it met the broad intentions of the Metropolitan Region Scheme. It would seek approval from the state planning commission, as well as advertising for submissions from the public. Where the new plan involved bushland redevelopment or new industrial zones it would also need to seek approvals from the Department of Environmental Protection. The new town plan would then need to be approved by the Minister for Planning and passed by the state parliament to give it statutory authority. In all, the process might take up to two years to be completed.

### Issues of urban growth

The population of the Perth region is forecast to increase from 1.3 million in 2002 to 2 million by the year 2029. This increase will mainly be accommodated in the coastal regions to the north and south, thus stretching the city out in a broad corridor from Alkimos in the north to Mandurah in the south, a distance of some 150 kilometres. During the same period it has been estimated that the average daily journey for each resident will grow from the current 25 kilometres to 30 kilometres (in 1986 this was 14.1 kilometres). These two factors alone will add to some of the problems of urban growth already facing the city—finding enough suitable land for future subdivision, providing adequate transport infrastructure and services, meeting the environmental expectations of the population, while at the same time minimising land use conflicts.

**Table 8.2** Problems created by city growth and sprawl

| Economic | Environmental | Social |
|---|---|---|
| Traffic congestion:<br>• peak-hour delays<br>• lack of parking<br>• stress<br>• pollution<br>• time/money lost<br>• accidents | Pollution (water, air, noise, visual):<br>• inadequate waste-disposal facilities<br>• build-up of noxious fumes, effluent<br>• unsightliness<br>• incompatibility of industry | In blighted areas:<br>• poverty<br>• unemployment<br>• overcrowding<br>• high crime rates<br>• homelessness<br>• stress<br>• loneliness |
| Lack of services<br>Lack of amenities/<br>high cost of community services in outer areas | Waste disposal (liquid, solid) | In outer areas:<br>• isolation<br>• unemployment<br>• boredom<br>• lack of services |
| Depletion of resources (water and biotic), e.g. ground water | Climatic variations<br>• increased temperatures in CBD<br>• wind tunnels | |
| Incompatibility of land uses (land-use conflict), e.g. industrial vs residential, urban vs rural | Lack of open space | |
| Urban blight<br>• substandard buildings<br>• inadequate facilities<br>• congestion<br>• social problems<br>• land-use conflict<br>• pollution<br>• low prestige | | |

### Land for urban development

The low-density nature of the city has produced a number of concerns. It imposes demands on the planning authorities to approve sufficient new land releases each year to satisfy the housing market. As the city grows outwards transport, infrastructure and service demands grow and natural environments are removed to make way for urban land uses. The monetary cost of this can be significant. Each new residential, commercial or industrial lot must have access to roads, footpaths, power, gas, water, telephone, sewerage and drainage. In addition, the population needs to have reasonable access to services such as shopping centres, schools, public open space and medical facilities. The total cost of all development and servicing in 2002 could be estimated to exceed $50 000 per lot within a new residential subdivision.

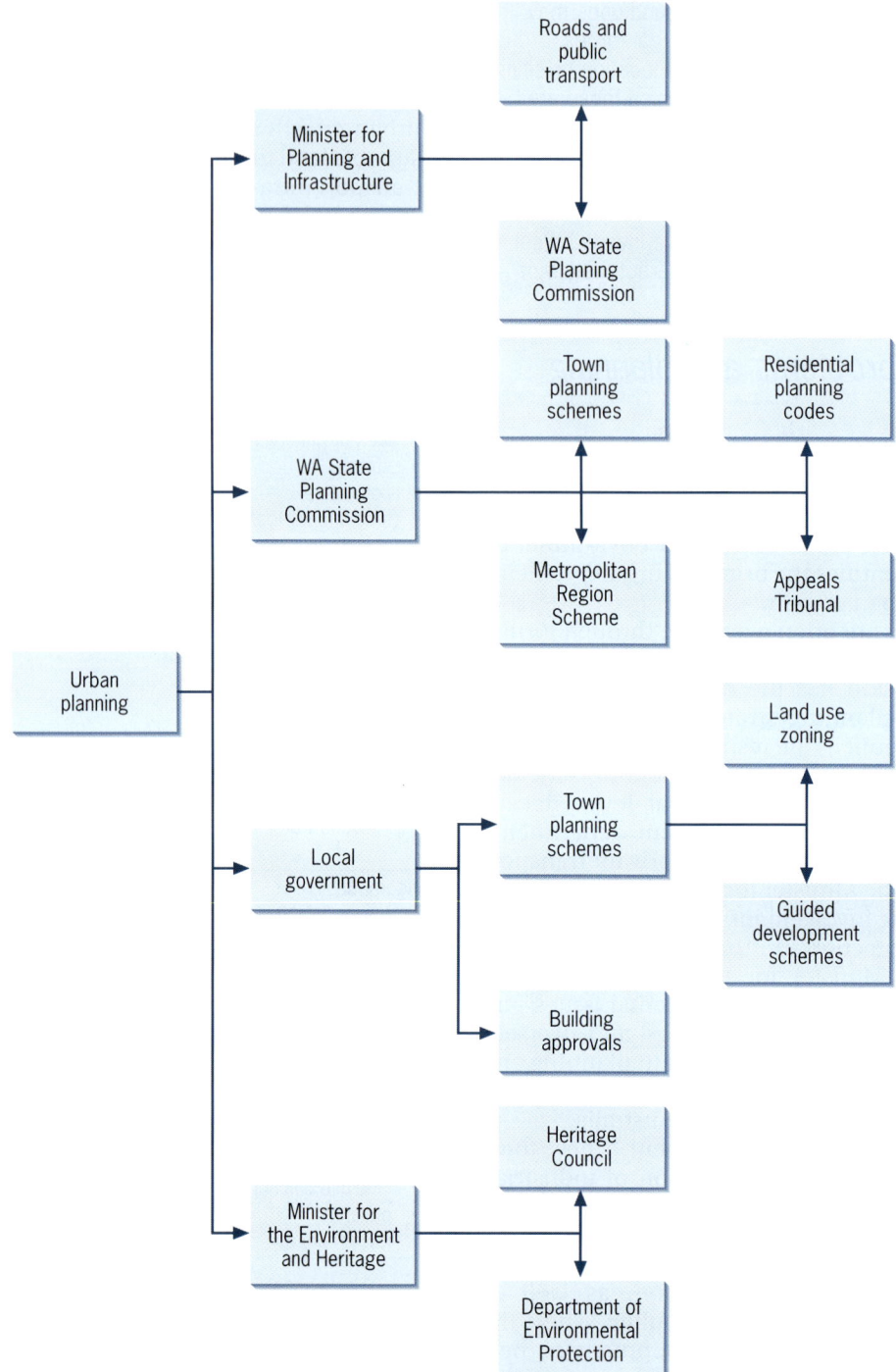

**Figure 8.37** Urban planning authorities in Western Australia

*Transport*

The economic advantage of any large city is largely determined by its capacity to efficiently meet the demands for goods and services that are generated both within and beyond its boundaries. This requires an effective transport system, both for people and goods. The outward growth of Perth has increased journey distances and at the same time has reduced the number of people using public transport. Figures 8.38 and 8.39 dramatically illustrate this problem. The increasing reliance on the motor vehicle as the main means of moving people and goods around the city will bring rising levels of traffic congestion and increasing environmental and economic costs. Patronage of public transport is falling and the number of single passenger car journeys is rising. Nine out of 10 cars have only one person in them, making this a very inefficient use of the road system and adding to the level of traffic congestion for the Perth region.

The number of kilometres travelled by car each day in Perth is growing faster than population growth, with the increasing dependence on the car as the preferred form of personal transport indicated in the changes that have taken place in getting students to school. In 1986, only about one-third of children were taken to school by car. Most walked, cycled or caught public transport. In 1998, 62 per cent of children were being driven to and from school.

*Regional employment and services*

The shift of population into the outer residential areas of the city has not been accompanied by a corresponding shift in employment opportunities. Jobs have largely been located in the inner city and established zones of Perth. This has contributed to the problems of commuting to work and the rise in the length of daily journeys. This issue will affect the future planning of the city, with consideration given to land use zoning for adequate industrial, retail and commercial precincts within the outer suburban areas. It will also influence the decision of governments as to the location of major public service departments and their workforce. The relocation of the Department of Land Administration from central Perth to the regional centre of Midland illustrates one solution to this problem.

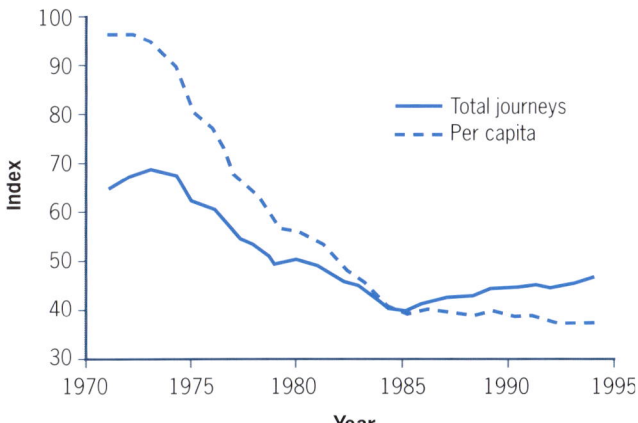

**Figure 8.38** Public transport journeys per year, 1970–95

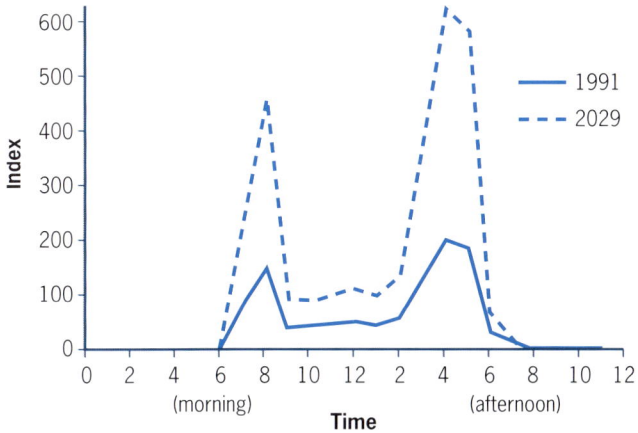

**Figure 8.39** Estimated traffic congestion for the Perth region

*The biophysical environment*

Urban development has the potential to produce negative impacts across a range of biophysical elements. These include the loss of remnant bushland, increasing levels of greenhouse gas emissions and fossil fuel consumption, pollution of streams and ground water reserves, degradation of soils, destruction of wetlands and the use of increasing amounts of land for waste disposal.

The great emphasis on the use of private motor vehicles within the Perth region, and the increasing distances travelled by people on a daily basis, are contributing to the rising levels of emissions of carbon monoxide, sulfur dioxides and unburnt hydrocarbons. Related to the levels of vehicle emissions and energy consumption is the energy use within the city associated with industrial and domestic activities. Increasing emphasis on domestic airconditioning and heating, as well as the use of other electrical products, is increasing levels of electrical consumption. This in turn requires the generation of extra power. In Perth, the supply of most of the electricity consumed is via the power stations located on the Collie coalfields. These coal-burning stations produce nitrogen oxides as well as sulphur dioxide, while the consumption of electricity in the city contributes to localised heating of the environment in a process known as the **heat island effect** Some large cities have temperatures around 6°C above the surrounding countryside, because of the heat generated by urban activities.

The expansion of the Perth Metropolitan Area has come at the expense of the local flora and fauna. The coastal plain contained a wide variety of plant and animal species unique to the South-West. Significant areas have been cleared to make way for urban development, with small amounts of remnant bushland left in some areas as the city expanded. Wetlands and other areas considered to have low economic or ecological value were used as waste disposal sites. These included significant areas of floodplain along the Swan and Canning rivers. The use of the sandy areas of the Perth coastal basin for waste disposal has raised another issue—ground water pollution. Rubbish disposal sites contain organic and inorganic waste, which generate both greenhouse gases and organic acids. The acids are leached into the ground water system, often taking a variety of heavy metals and other toxic materials with them. They spread out beneath the sand and are carried slowly by ground water towards the lowest part of the landscape. There they may be brought back to the surface via domestic bores or they may enter the deeper aquifers. These aquifers are fast becoming a major source of water for the city, as the dams in the Darling Scarp fail to keep pace with growing water demands.

*Cultural issues*

A variety of issues have been identified within this general area that require the attention of planning authorities. Aboriginal heritage is one such issue. The development of the city has in the past paid little attention to this aspect of the cultural environment.

As a consequence, traditional sites and areas of cultural significance to Aboriginal people have been overlooked or destroyed. Other issues include the protection or retention of European heritage sites and buildings, and the redevelopment of areas of poor quality housing or old industrial sites, while meeting increased expectations for better public amenities and catering for the cultural diversity of the population.

## Planning strategies and solutions

Central to the planning solutions for Perth's current and future problems is the determination of how the land will be used. The development of a series of regional and local plans and policies over the last 50 years illustrates the ways in which the city's problems have been addressed. The planning process involves a number of state and local government agencies, previously identified in figure 8.37.

### Metropolitan regional plans

The Stephenson–Hepburn Plan of 1955 represented the first regional planning approach to the long-term planning of Perth's growth. The Corridor Plan of the 1970s evolved out of this plan and has been the basis for planning the city's development up until the mid-1980s. It identified four main corridors of development, previously discussed, with areas of green belt falling between them. It tackled the prospect of endless urban sprawl by aligning the growth of the city along transport corridors which, it was considered, would also provide a means of providing urban infrastructure economically to the growing population. While it was not the prime intention of the plan, it also fortuitously protected areas of natural bushland and kept development away from the ground water mounds of Jandakot and Wanneroo, which were to become significant as new sources of water were sought for Perth. Even today, much of the shape of the city's urban development can be attributed to the Corridor Plan.

In 1987, after a period of review, the first major change to the Corridor Plan was endorsed by the government of the day. Metroplan brought some important changes to the concept of corridor development. One of its prime objectives was to slow down the outward growth of the city by encouraging an intensification of urban land use. This was to be achieved through urban consolidation. In addition to the traditional forms of detached housing, there was to be an increasing emphasis on well designed medium-density housing located close to centres of activity and public transport facilities. Urban renewal and urban redevelopment was also encouraged and this was emphasised to local councils when they reviewed their town planning schemes. The result was widespread increases in lot densities across the metropolitan area. The older 1000 square metre lots were often rezoned to allow the construction of three or four new residences.

Larger areas of **public housing** were either demolished or refurbished and sold to long-term tenants in an attempt to break up the state housing commission suburbs, while large inner-city industrial sites were rezoned and redeveloped as medium- to high-density residential precincts. These included areas in Wembley, Subiaco, East Perth and Fremantle. This was often done in conjunction with private developers.

A further consideration of the plan was to identify strategic regional centres of industrial and commercial development as a means of reducing commuting distances and shifting employment and purchasing opportunities into the suburbs. These centres can be seen in figure 8.40. This figure also identifies the areas of potential urban consolidation and existing and future growth zones. Metroplan addresses falling public transport patronage by emphasising the need for higher-density housing along the existing rail transport links and around the strategic regional centres. The city centre plan for the Canning regional centre illustrates the response of the local authority to this push. Residential densities within this area are set at R40 (40 units per hectare), compared to the city average of R17.5.

Since the introduction of Metroplan, there has been no major new planning strategy. However, a series of planning initiatives have been developed in response to the urban problems identified previously. Many of these have been developed in line with a planning review known as 'Future Perth—a vision for Perth in 2029'.

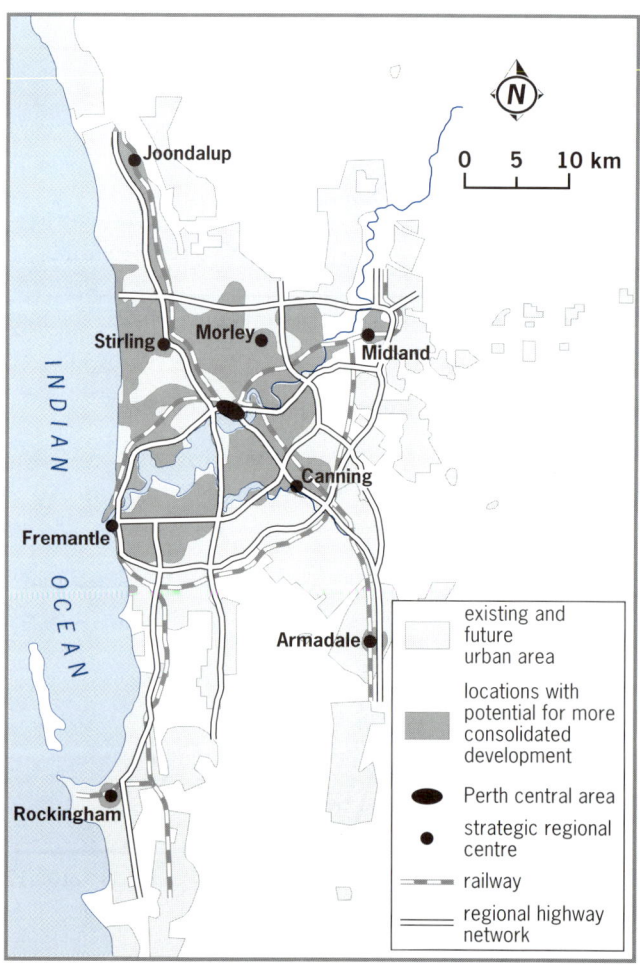

**Figure 8.40** Metropolitan urban development strategy

One such planning initiative is the 'better cities, liveable neighbourhood' strategy, which aims to address social, economic and environmental issues. This initiative includes urban designs that emphasise compact structures in which neighbourhood community and commercial centres provide a strong focal point. It argues that development should be restricted to a range of some 400 to 450 metres around these facilities to encourage walking. The liveable neighbourhood strategy also encourages the retention of significant environmental and cultural features, and the use of urban designs that will promote community interaction and personal safety. Housing that uses energy-efficient design would also be encouraged through the use of building design guidelines controlled by the local authorities (including allowing higher building densities where designs use energy efficient techniques).

Another significant policy development has been the 'Bushplan' or 'bush forever' program, which promotes the identification and retention of significant areas of remnant bushland within the urban areas. This plan reflects the growing community concern at the loss of native vegetation and habitats. Developers planning new housing subdivisions must conduct flora and fauna surveys to identify significant areas of remnant vegetation and to put conservation measures in place to ensure their retention.

The increasing reliance of the population on the motor vehicle as the preferred means of transport is creating a number of problems. Planning for the future transport needs of the city focuses on two major approaches:

- *Developing a new road system*. Extension of the freeway system to the north and south, and the development of highways encircling the outer areas of the city, represent some of the ways of dealing with the problem of road congestion. Metroplan identifies the completion of the Roe Highway from Welshpool through to Fremantle as the final link in a major ring-road system that includes the Reid and Tonkin highways. This will allow movement from Stirling in the north to Midland in the east, and then to Fremantle and Kwinana in the south, without passing through the CBD. It could be argued, however, that this approach is simply adding to the problem by encouraging greater use of motor vehicles.
- *Providing a public transport system that is flexible, with services frequent enough to encourage people to leave their cars at home*. The construction of bus lanes along freeways, and the extension of the rail system, represent major investments in public transport.

Another approach being considered is to encourage people to car pool, which would reduce the number of single occupant vehicles.

## Local government

The role of local governments is to develop town planning schemes in line with the regional planning schemes. These local plans are highly detailed and take into account a number of considerations including the local environment, the need for an equitable spread of commercial and community services, the history of past development and the concerns of the local population. An example of the last factor is the changing attitudes of local communities towards planning for higher-density residential areas. While these have been accepted in some locations, in other areas they have been opposed by the community, which has led to modification of the plans.

Local government also oversees the construction of public and private buildings within its area. The local government authority (council) controls building standards, and the size and position of buildings on sites. Applications to build must be in line with the land use zoning and with the land use density. They must also follow building by-laws and policies relating to such things as building height, issues of privacy, retention of vegetation, use of approved building materials and use of energy-efficient designs.

Many local councils work with community groups to preserve remnant ecosystems. This includes providing technical, material and financial assistance, supporting funding applications to state and federal departments, rezoning land to achieve conservation status and encouraging preservation through community education programs. A number of councils are closely involved in programs such as Bushplan, ribbons of blue (protecting river and riparian vegetation), drains to living streams and local wetland rehabilitation.

They also oversee aspects of the Aboriginal Heritage Act and work with local Aboriginal communities to identify culturally significant sites before undertaking rehabilitation or urban development activities. Often, this process will involve site registration and conservation or community education through the erection of plaques and other signage.

### EXERCISES

1. Define or briefly describe the following:
   - urban sprawl
   - bushplan
   - urban consolidation
   - greenfields development site
   - residential density code
   - corridor plan
   - strategic regional centre
   - Metropolitan Region Scheme

2. Why is urban sprawl considered to be a major urban problem and how does Metroplan address this?
3. Discuss the changing pattern of public and private transport use in Perth and suggest reasons for this.
4. Identify what you consider to be the main urban problems facing people living in inner-city suburbs and those living on the fringes of the city. How are these being addressed by the various planning authorities?
5. How would increased construction of higher density housing solve some of the problems facing Perth and what might be the disadvantages of this solution?
6. Conduct a survey of students in your class to determine the numbers using public and private transport. Discuss your results in terms of any transport problems in your locality.

# Morphology of an Australian country town: Northam, Western Australia

Northam is located in the Avon statistical area (see chapter 7). It is the largest centre within this region and occupies the highest order within the urban hierarchy. As such, it fulfils an important role within the central wheat belt, providing higher order functions to the surrounding towns and rural communities. While the town of Northam is the largest inland centre not based on mining in Western Australia, it has been through a period of declining population which has only been reversed in the last five years. In 1991 the population was 6560, dropping to 6300 in 1996. In 2002 the town had a population of some 7000 people, with another 3500 in the surrounding shire of Northam. The separation of local government responsibilities between an urban town council and a rural shire council is a characteristic of some settlements in Western Australia, including Northam. This arrangement is sometimes described as a donut pattern (because the town is surrounded by the shire).

Even though Northam occupies the highest order within the Avon statistical area, its relatively close proximity to Perth (95 kilometres) has resulted in a degree of **trade capture** by the larger city. As a consequence, there has been some restriction on the range of services offered within the town. Despite this, Northam still provides goods and services not found in any of the other central places within the local urban network.

Compared to other towns within the region, Northam has a number of locational advantages. Its initial siting (in the 1850s) on the junctions of the Avon and Mortlock rivers made it a natural focal point for settlers opening up the farming land to the north, south and east of the town. It won the battle of the railway routes in the 1850s, when it was chosen over York and Toodyay as the junction for railway lines constructed to the north, east and south. In the 1890s, the Great Eastern Highway to the eastern goldfields was constructed through Northam and the goldfields water supply pipeline was also routed through the town, further enhancing its position as a regional centre.

## Northam's external morphology

The shape and growth of the town has been in response to a variety of physical, historical, economic and cultural factors and processes. The urban area is trapezoid in shape with the base, which is located on the southern side of the Avon River, formed by Throssell Street. To the north of the river, the town site is mainly confined to a bend in the Avon, which has tended to restrict its northern and eastern growth. The construction of a railway line to the west, and the more rugged relief, marks the limit of urban growth in this direction. A further limitation on the growth of the town is the extensive granite outcrops on the edges of the valley that have prevented the construction of roads and buildings. Some of these granite outcrops can be seen in the Northam street map (figure 8.41).

The initial focus of the town's development was the valleys of the Avon and Mortlock rivers. It followed the main channel of the Avon on either side, with development primarily contained within the narrow floodplain and much of it falling between the 150 and 180 metre contours. The topographic map extract of Northam (figure 8.42, see p. 212) illustrates the shape and size of the town in 1977 and its relationship to the rivers and contour patterns.

The urban morphology of the town is very much a product of past planning decisions. The grid layout of the town parallel to the Avon has been important in creating the present shape. In addition, the development of the railway system through the town on the southern side of the river was significant in determining the current external morphology. While the railway system was important in shaping the development of the town in the past, the road system has played an increasingly important role since the early 1900s. Yilgarn and Mitchell avenues (formerly the Great Eastern Highway), Goomalling Road and the Northam–York Road illustrate this, with linear development extending outwards along these transport routes.

Recent new residential and industrial growth has occurred to the south-east and north-east of the town and this has given these areas an indistinct boundary, in contrast to the more defined borders to the north-west and south-west. The opening of the Great Eastern Highway bypass (June 2002), which runs to the north of the town, will in time play a part in changing the town's external morphology. This development, in conjunction with the standard gauge railway system, is already attracting storage and warehouse facilities and some limited industrial development towards the north. Figure 8.44 shows the external morphology of Northam in 2002. By comparing the aerial photo (figure 8.43, see p. 213) with the topographic map (figure 8.42), the areas of expansion can be clearly identified.

## Northam's internal morphology

Northam's size and position in the urban hierarchy of the Avon region has influenced the land use patterns or internal morphology of the town. It is a multifunctional centre, offering a variety of low and higher order goods and services to its residents and to its hinterland. As it is a central place within a well-established agricultural region, some of these services reflect the demands of the wheat and sheep farmers who live within the region. Given its relative proximity to Perth, the type of services will be limited by the impact of the larger city as a provider of central place functions. This has affected the shape of its hinterland or draw area. To the north, east and south its influence extends for some 100 kilometres. However, on the western side its influence is limited to some 30 kilometres, with the population in the western parts of the shire focused more on Perth to satisfy their needs.

A study of the internal morphology of Northam will reveal processes operating in the town similar to those in Perth, but on a much more limited scale. Thus it is

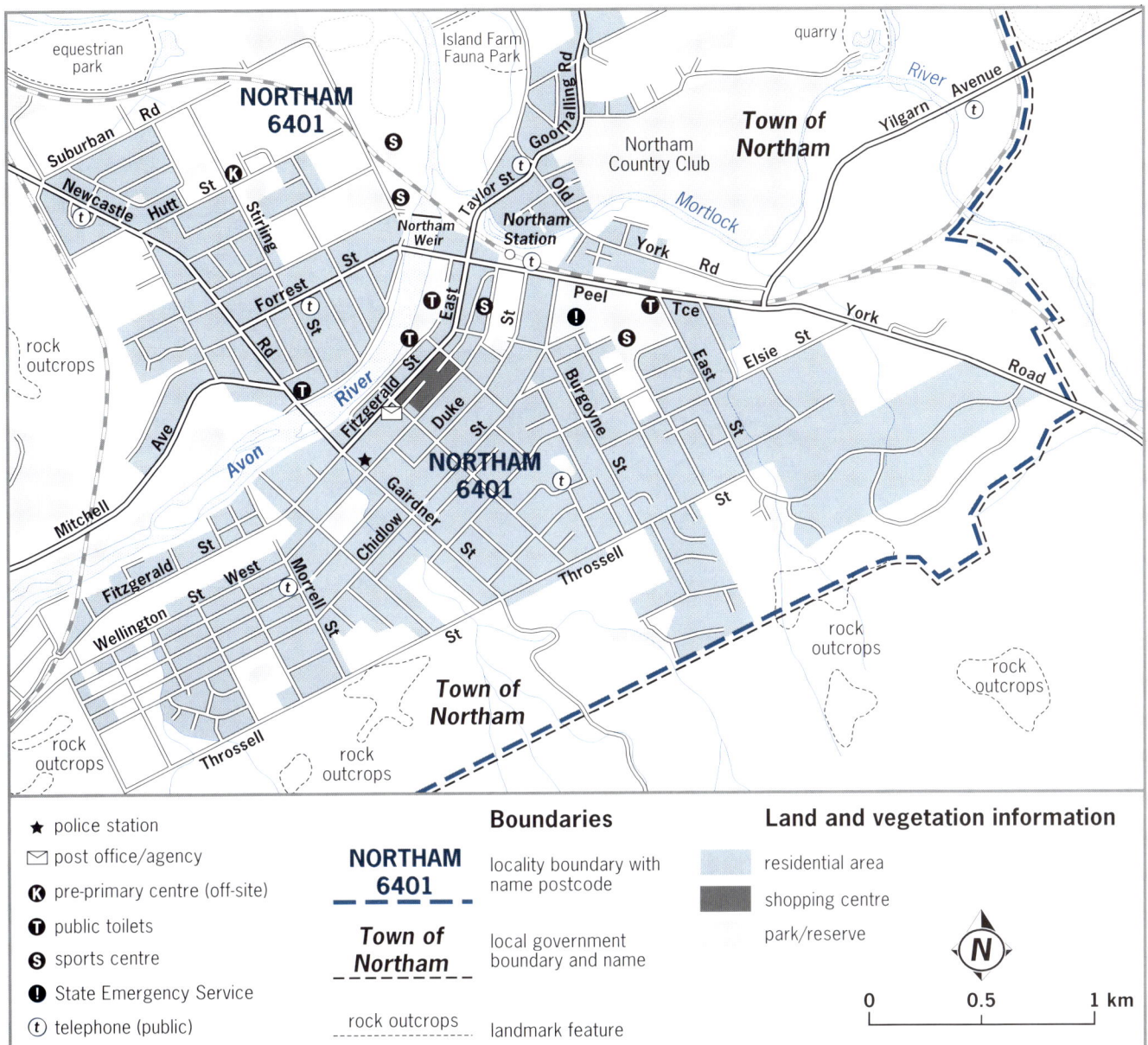

**Figure 8.41** Street map of Northam

possible to identify the main functional zones: the CBD, IMZ, industrial, established and newer residential, and RUF. Processes such as the economic rent mechanism, invasion and succession, and agglomeration and aggregation can also be identified. Figure 8.45 (page 215) illustrates the size and location of the main functional zones of Northam. It represents a simplified view, based on generalised land use analysis. More detailed studies would indicate a greater mixture of services and functions than is shown on the map.

## The central business district

Most of the retail functions are located along the eastern half of Fitzgerald Street, extending south to Wellington Street. Figure 8.46 (page 216) illustrates the types of services that can be found in the town. Northam is the only centre within the region that has a radio station, and one of the few with a local newspaper. Aggregation of like functions can be seen in the location of the five banking outlets within the same section of Fitzgerald Street. This is an area in which a number of the town's accountants and insurance brokers are also found.

Within the Fitzgerald Street commercial zone there has been a gradual shift of business activity eastwards as a consequence of the construction of the new railway station in the 1960s, as well as the rerouting of traffic along Forrest Road. The old railway station was located at the western end of Fitzgerald Street and prior to its closure in the 1960s it acted as an important focus for retail functions. This area of the commercial zone subsequently went into decline, with shops closing and a change in land use to the storage of building materials or simply being left vacant. This area at the western end of Fitzgerald Street illustrates the process of urban

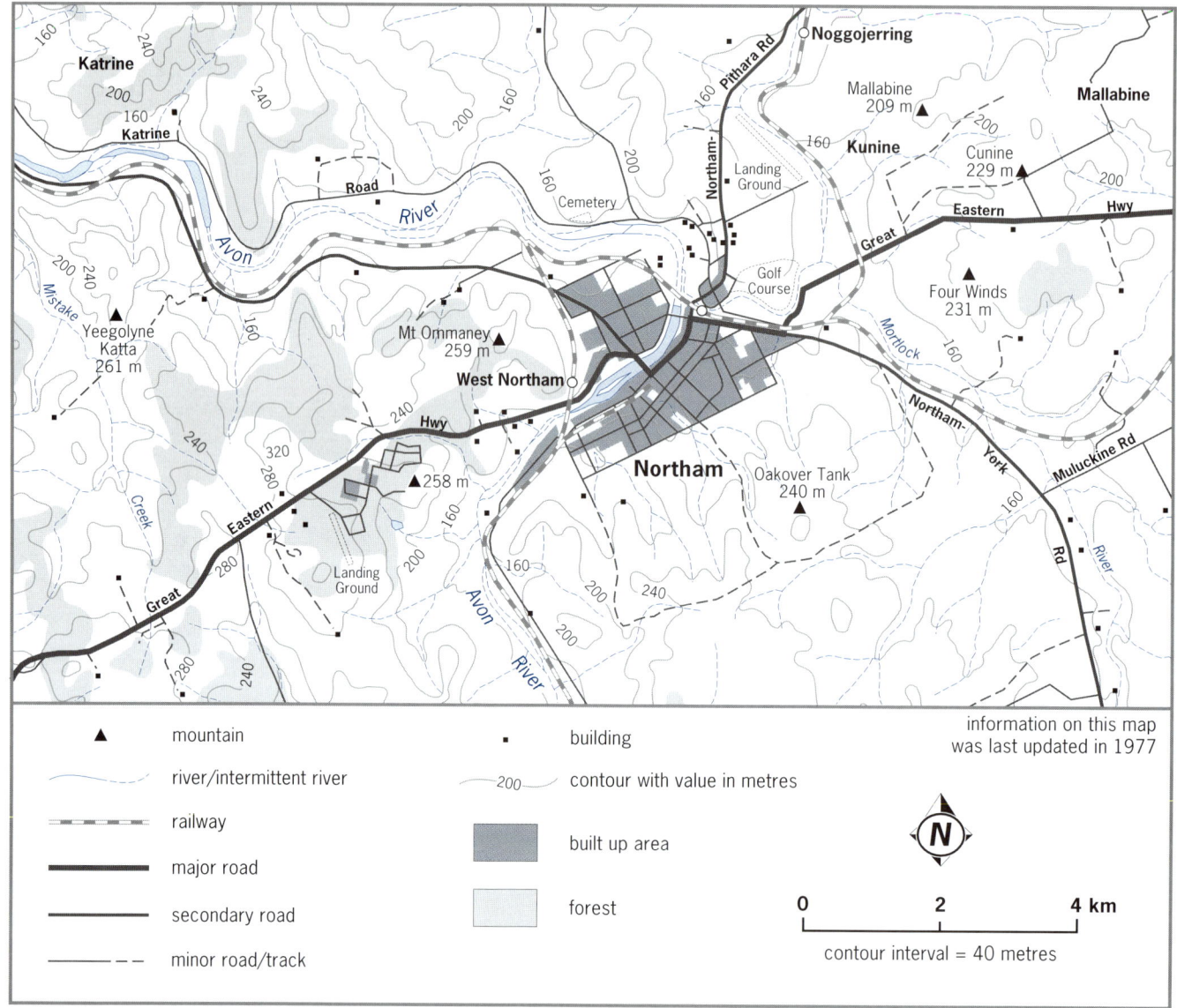

**Figure 8.42** Topographic map of Northam SOURCE MAP COURTESY GEOSCIENCE AUSTRALIA, CANBERRA. CROWN COPYRIGHT ©. ALL RIGHTS RESERVED. www.ga.gov.au/nmd

blight or decay, though on a much smaller scale than that seen in Perth. In general the area is now one of poor quality and derelict buildings (see figure 8.48 on p. 217).

The intersection of Peel Terrace with Fitzgerald and Taylor streets and Forrest Road is the major traffic node for the town (the only point where there is a set of traffic lights, which indicates the high level of traffic passing through this intersection). It is also the centre of much of the current commercial activity. Focused around this point are car sales yards, fast-food outlets, supermarkets, shopping malls and financial institutions. Immediately to the west of this location the emphasis changes, with a predominance of private and public (state government) office functions and local government services.

While virtually all of the retail outlets within Northam are concentrated within the CBD, the town is large enough to support the operation of several small corner stores or delicatessens within the residential areas.

In contrast to larger cities, there is very limited vertical growth within the commercial areas of the town. The tallest building is in fact the old regional hospital, located to the south of the commercial district and currently vacant. Most buildings along Fitzgerald Street are one or two storeys high and show little evidence of vertical zonation. This illustrates the limited range of higher order functions compared to Perth's CBD, and the lower value of the land. The land value between the centre of the town and the inner mixed zone does not change greatly. The demand for prime central sites is correspondingly much lower, with limited redevelopment occurring.

### Inner mixed zone

While it is possible to identify sections adjacent to the CBD that have elements of mixed functions, this is neither extensive nor continuous. There are small sections along Taylor Street and at the far western end

**Figure 8.43** Aerial photograph of Northam

of Fitzgerald Street where **light industry** and residential functions mix. Small sections of mixed land use may also be seen along the streets that run off Fitzgerald and Wellington streets. Here, a mixture of light industrial, residential and retail functions occurs.

## Industrial zones

As indicated in chapter 7, there are approximately 15 industries associated with the manufacture of building and engineering products, which makes these fabricating industries an important feature of the zone. However, most of the activities in Northam's industrial zones are associated with vehicle repairs, transportation and storage. Some activities reflect the agricultural activities of the region, with sale and stockyards as well as wool brokers and grain buyers and sellers. The storage and sale of farm machinery and equipment is also a feature of the zone.

## The residential zone

As with all other towns and cities in Australia, the residential zone of Northam occupies the largest area within the town's boundaries. It can be classified into the older, inner residential areas (which include the buildings that make up the original town site), a large established residential zone and a newer growth zone. The original town site extended from the junction of the Mortlock and Avon rivers along the south-eastern edge of the Avon River. Fitzgerald, Wellington and Duke streets were the first streets to be surveyed. Along the north-western edge of the Avon, Broome Street also contained a number of early buildings.

Much of the town was constructed during and after World War II, when it was the site of an important army base constructed about 5 kilometres to the west. In addition, the town functioned as the centre for administration and maintenance of the railway network that fanned out from it into the central wheat belt. Much of this residential development was public housing constructed by the state government. In recent times the town has grown towards the north-east and the south-east, where the main subdivisions for new estates are found. These areas are not as rugged as the land to the west and are away from the railway systems. They occupy land previously used for farming and represent an invasion of the rural fringe.

## Public functions

This land use zone contains a mixture of educational, medical and recreational functions. The concentration of primary, secondary and tertiary educational institutions in the north-east of the town is one example of this zone. Another example is the Northam racecourse, which can be found to the east of the town, surrounded by agricultural land.

## The rural–urban fringe

This zone is discontinuous. Often the line between the rural and urban functions is a boundary road, while in other areas a broader transition zone is more apparent. The northern and eastern fringes of the town have the most pronounced rural–urban transition zone. Here, it is possible to identify a variety of space-extensive urban activities, such as ovals, trotting and horse racing

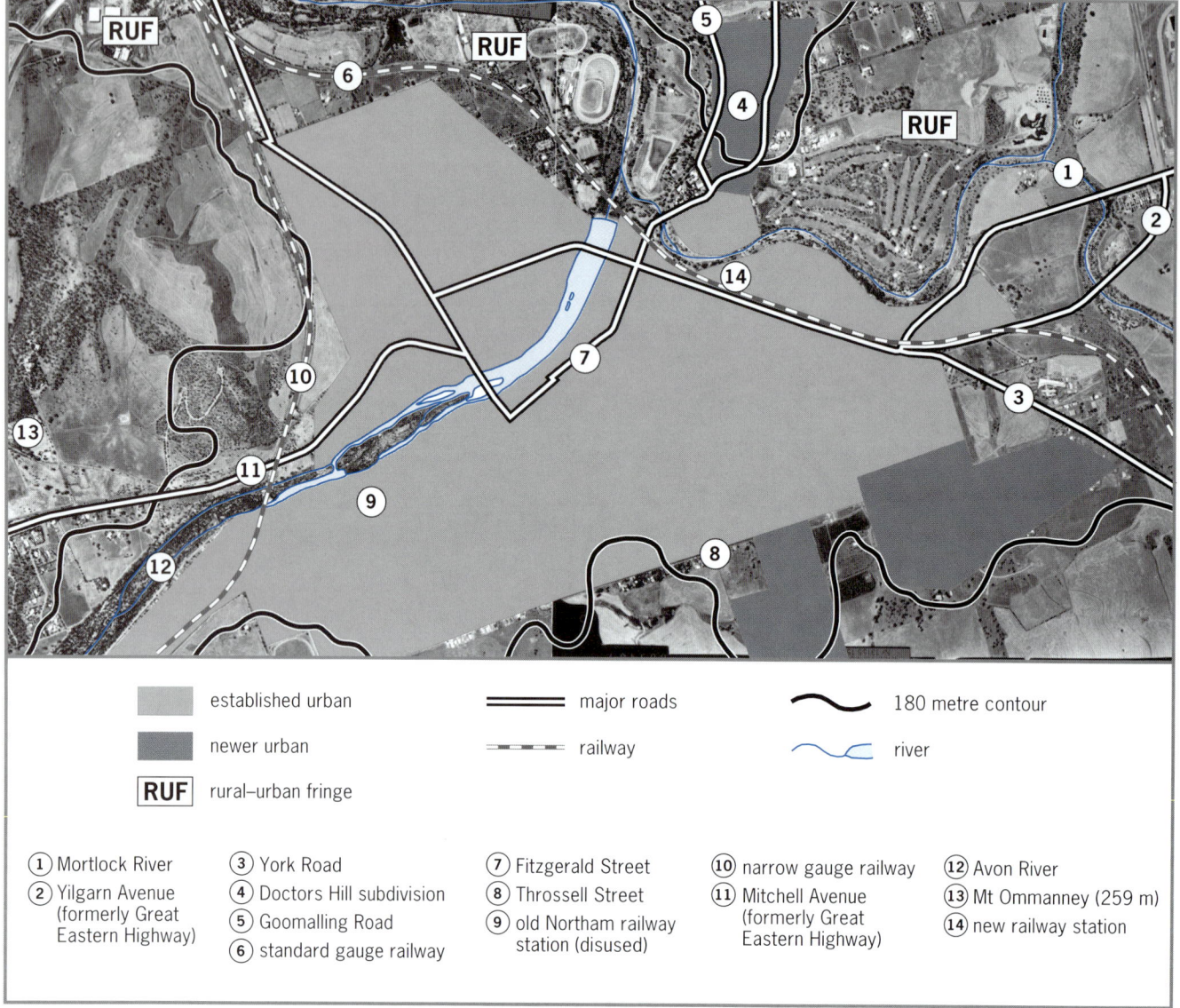

**Figure 8.44** External morphology of Northam, 2002

tracks, a golf course, plant nursery, caravan park and warehouses. Transport facilities, such as the new rail freight yard and road train assembly points, occur to the north and west in association with the standard gauge railway and the Great Eastern Highway bypass. Rural functions within this zone include stables, hobby farms and the farm paddocks of adjoining crop and livestock properties.

## Elements of change

Northam has seen a number of changes associated with changing economic fortunes, government policy and technology over the time that it has been in existence. Since its inception in 1850, the town has grown significantly. Its early development was closely tied to the growth of the surrounding agricultural region, with early services including an inn, a blacksmith's shop, clothing store, church and a steam-driven flour mill.

The prosperity created by the gold rushes resulted in the growth of the town's population from about 400 in 1890 to over 2000 in 1901. The population grew steadily until the 1950s, when rapid growth was largely brought about by the influx of migrants into the town (approximately 1600 between 1948 and 1954). This led to an expansion of the residential area. Adding impetus to the town's development was increased service trade resulting from the establishment of the iron and steel plant at nearby Wundowie. A further contribution to the prosperity of the 1950s was a period of buoyant agricultural activity and prices enjoyed by the surrounding farming community.

Major changes in the town were brought about by the completion of the transcontinental standard gauge railway link between Kalgoorlie and Perth in 1966. This new line approaches the town from the east, beside Peel Terrace, but then proceeds straight ahead across the Avon. The Great Southern Railway Line was diverted so that it now crosses the river a little south of the town,

URBAN MORPHOLOGY 215

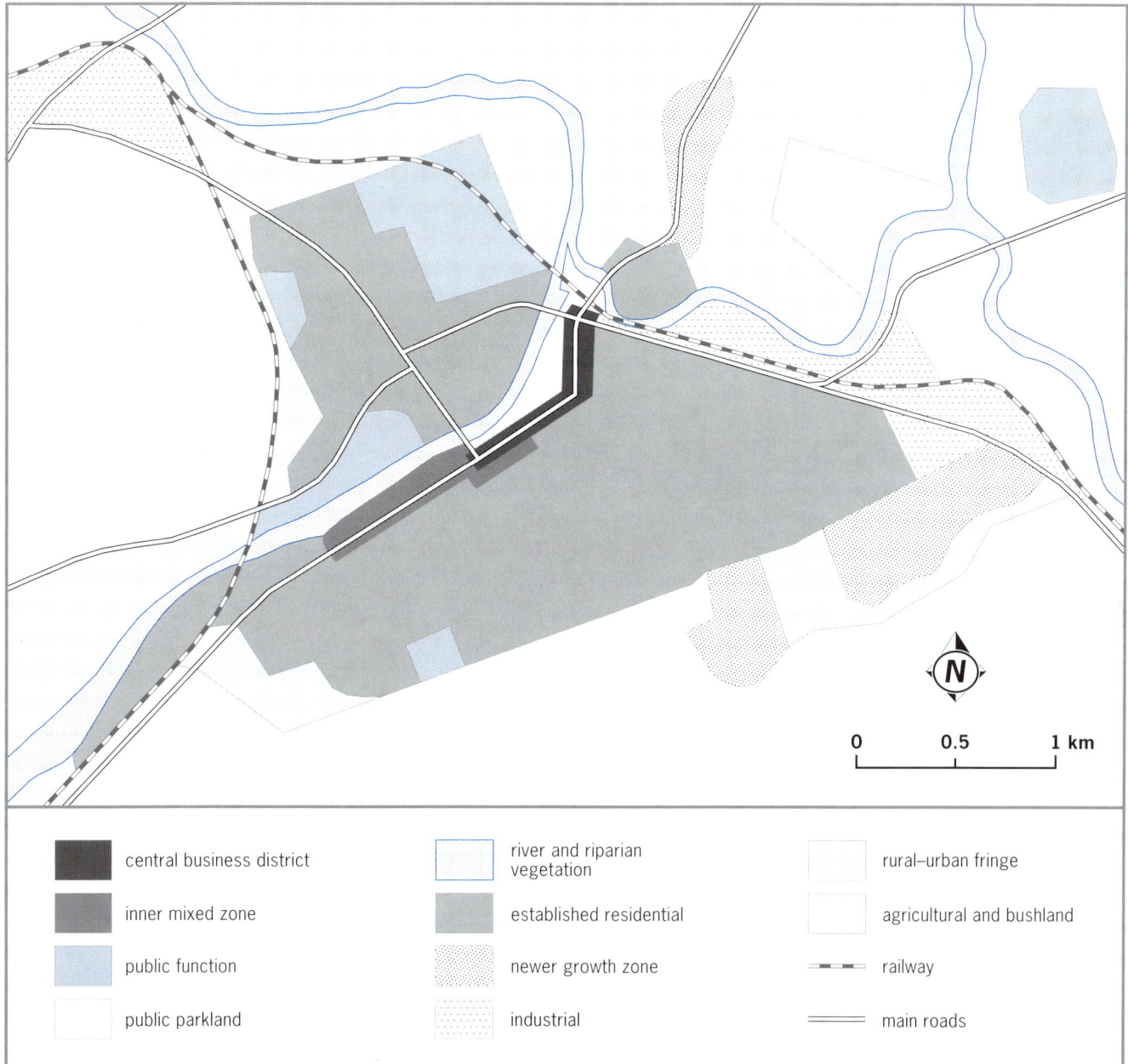

**Figure 8.45** Internal morphology of Northam, 2002

skirts the foot of Mount Ommanney and joins the standard gauge line. In addition, the old railway route through the centre of Northam was closed, the new Avon railway yards were opened, a bulk grain receiving installation was established and the old Northam station was replaced by one at the eastern end of town (see figure 8.44 on p. 214). The rerouting of the railway line and the construction of the new station has had a considerable effect on the western end of town, as discussed earlier under functional zones.

Modernisation of many aspects of the old town has occurred since the 1960s. Many of the small community centres (such as schools and halls) that were dotted throughout the Northam area have disappeared, while other older functions have been replaced by new buildings. Some early residences have disappeared with the demolition of the rickety Grey Street Bridge, which was superseded by a new bridge at Peel Terrace. Old residences that have remained, especially those along Fitzgerald Street, have been remodelled. Many of the changes within the town may be attributed to the high level of motor vehicle ownership that has become common within Australian society. This increased mobility has reduced the necessity for services to be located within walking distance of the population.

The growing efficiency of road transport has decreased the functional distance between Northam and Perth, so that trade has tended to bypass Northam in favour of the greater attractions offered by Perth. To date, industrial development has not occurred at the rate hoped for, because of the obvious advantages of Perth as a centre for industrial location. Other factors

**Figure 8.46** Fitzgerald Street, Northam

**Figure 8.47** Post Office and newspaper production, Fitzgerald Street

**Figure 8.48** Urban blight, Fitzgerald Street

that inhibited growth until the mid-1970s included the high rate of movement of residents of retirement age to coastal localities, and the poor position of the town in terms of the development of a major tourist industry. A decreasing rural population in the hinterland also contributed to the slow growth rate.

Since the mid-1970s, Northam has generally had a positive growth rate and there is a slow but steady expansion of the town. Land has been zoned and regulated according to the town planning scheme, which has been in place since 1976–77. Compared with the scheme for Perth, this scheme is relatively simple and flexible. It was determined that growth of the town would occur mainly to the north-east and south-south-east. The slopes of Mount Ommanney are considered unsuitable for residential development because of their steepness and isolation from the town (this isolation is partly due to the separation caused by the railway and the river). In the late 1980s some 40 to 50 new residences were being constructed each year, mainly in the vicinity of Doctors Hill. There are, however, some constraints to development in the area. The cost of blocks is relatively high, as is the cost of their development. Some of the more suitable land had already been subdivided into 4000 square metre blocks, prohibiting more intensive development. The city was allowed to expand outwards during the 1980s and 1990s, resulting in inefficient use of existing infrastructure. In the late 1990s the emphasis shifted to urban consolidation and growth within existing town boundaries. Urban renewal or the use of existing vacant blocks would make better use of the road, electricity, water and sewerage systems that were in place, without the need to construct and maintain new infrastructure.

A joint initiative between the western-most councils of the Avon region and the state government is now addressing the need to retain the population by developing a broader economic base within the region. The Avon Arc Program aims to support new industrial and agricultural processing enterprises within the region, as well as looking at ways in which mining engineering works could be undertaken in this location to service the eastern goldfields. Other initiatives include the relocation of government departments into the region and the improvement of public transport for people commuting from Northam and Toodyay to Perth.

### EXERCISES

1  Draw a sketch map of Northam's external morphology and then identify and label the main biophysical and cultural elements that have influenced the town's shape.
2  Study the street map provided for Northam and complete the following activities:

   (a) Compare the aerial photo and the map of Northam's internal morphology and give examples of streets that are found in the CBD, IMZ, industrial areas and NGZ.

(b) List examples of urban activities located within the rural–urban fringe and suggest reasons why these functions are found there.

(c) Construct a table and then tally the different functions that can be identified within the town and surrounding areas. Use the following classification system:

- educational institutions
- health services
- government services
- public communication services
- religious institutions
- public utilities
- transport services
- recreational services

(d) Compare and contrast the area of Northam found to the north of the Avon with that found to the south.

(e) On a sketch map, identify the streets that have been referred to in the description of the external and internal morphology of Northam provided in the text.

(f) Write a description of the land use characteristics and landform features encountered as you move from the railway crossing at the junction of Peel and Yilgarn roads, along Peel Terrace to Fitzgerald Street east and then across the Avon Bridge to the western end of Forrest Street.

3 Study the topographic map and aerial photo of Northam and then complete the following activities:

(a) Using the aerial photo, draw a sketch map and on it identify the major recreational facilities and suggest their use.

(b) Study the map and photo and determine the direction of flow for the Avon River. Identify map and photo evidence that supports your answer.

(c) Using your sketch map, identify and mark the changes that have occurred in Northam since the map was published. Discuss the impact that these changes have had on the external morphology of the town.

(d) Identify and discuss the ways in which the industrial, residential and commercial zones of Northam can be identified from the aerial photo. In other words, how does their appearance differ?

4 Describe the location and shape of the CBD of Northam and suggest reasons for its present location.

5 What factors appear to have contributed to the growth of the town towards the north-east and south-east?

6 List and analyse the main differences between the CBD of Northam and the CBD of Perth.

7 Outline the evidence that points to Northam being a higher order centre within the Avon region.

# Chapter 9

# Practical Mapping and Research Skills

> The purpose of this chapter is to describe a range of geographic skills associated with the analysis of patterns and processes within the biophysical and cultural landscape. Students will be shown how to apply geographic tools such as maps, statistics, photographic and sketch methods, and primary research or fieldwork methods to the study of these patterns and processes. A summary of web sites relevant to the different topics covered in the text has also been included for further reference.

## Introduction

The study of geography introduces students to a wide range of skills and understandings. While many of these have applications beyond the subject, there are some that form an important foundation to understanding the geographic approach. A number of these skills are introduced in this chapter but, while they are discussed separately, they should not be treated in isolation. Rather, they should be applied to the areas of study within the earlier chapters of this book. It is envisaged that students will use the various skills discussed to develop their understandings of this content.

The geographic approach is one that emphasises the concepts of location, spatial arrangements and geographic processes. The skills that are covered pay special attention to these concepts. The regional and systematic approaches are discussed within the chapter and applied to various examples to further illustrate the geographic approach to the study of spatial information and data. In addition a variety of mapping, graphing, statistical, fieldwork and computing skills are discussed to show the ways in which these can be applied to the study of geography.

## Regionalism

Regional classification is one of the fundamental processes utilised in the geographic approach to the study of patterns and processes in the biophysical and cultural landscape. In any science, the process of classifying provides order and understanding to information that at first appears complex and chaotic. Once information is organised, it provides a basis for investigation and analysis. Cause and effect relationships and linkages between different features or variables can be observed and explored.

In geography, the process of regional analysis is based on the observation that areas in close proximity to one another have strong connections or similarities, and that these diminish with increasing distances.

Regions can be used in a formal and a functional way to illustrate geographic processes. They can be illustrated by the use of maps and models and are an important geographic tool for the analysis of spatial data.

### Formal regions

The uniform, static or formal region is used to classify spatial data on the basis of similarity of features or functions. In this way the classification of areas of the Earth's surface according to similar rainfall and temperature characteristics results in a formal regional map of world climates. Other types of biophysical data that can be dealt with in this way include vegetation, landforms, soils and habitats. The same approach can be applied to cultural features. Formal or uniform regions can be mapped for agriculture, urban land use zones and population densities.

The numerical range from one category to another will influence the number of individual classifications found on a map. Hence, a rainfall map for Australia using a numerical range of 5 millimetres per category would show a great number of different zones compared to one where the categories are based on a 50 millimetre range. Likewise, a range of 4000 millimetres is of limited value as there would only be one category shown on the map. As a general rule, when developing formal region maps, the number of categories falls between five and 10.

Once a system of classification has been devised, the next step is to divide the study area into the different groupings and determine their boundaries. The classification system should be such that all areas are included. Often, areas with a mixture of features will be identified in the construction of the formal regions and these are classified as transitional zones. On completion of the mapping of formal regions the final step is to devise a series of descriptive titles for each region that reflects their dominant characteristics. The steps in region building are illustrated in figure 9.1, where land use types have been identified and then classified.

Major formal regions containing a variety of natural and cultural features were often used by geographers as the basis for studying the different parts of the globe. This gave rise to regional geography as a field of study. These broad regional classifications often used one or two outstanding features as the basis for classifying the region. Thus, a study of Monsoon Asia would often involve looking at climate, vegetation, soils, agriculture, settlement and people of this part of Asia, even if the means of identifying the region was based primarily on certain climatic characteristics. The initial identification and location of the Monsoon Asia region involved the steps illustrated in figure 9.1. The classification was based on climatic data, with the boundary set by the limits of the monsoonal influence, and the name that it was given described the major characteristic that gave the region its distinctive identity.

## Functional regions

The functional, nodal or dynamic region identifies areas on the basis of movements to and from a nodal point or nucleus. The region is defined not by the uniformity or similarity of its features, but by the length and frequency of movements between the nucleus and its sphere of influence. Functional regions are dynamic because their extent is defined by movement rather than by static objects. So, while the classification of urban land uses as illustrated in figure 9.1 is based on building structures and is thus formal, a study of the functional regions within the same area would look at the movement of people or goods between these. A factory unit within the study area would have customers, raw materials, finished products, workers and communications systems all producing flows or

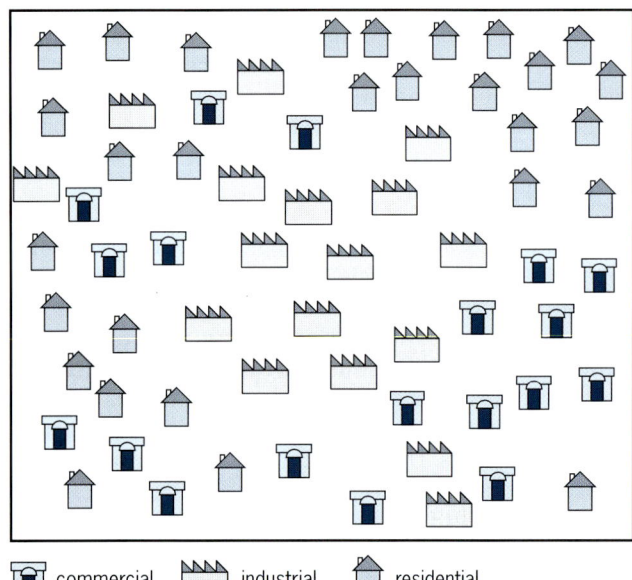

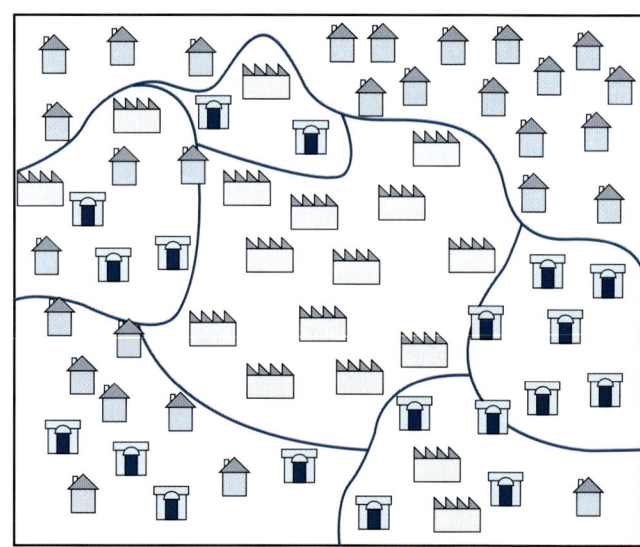

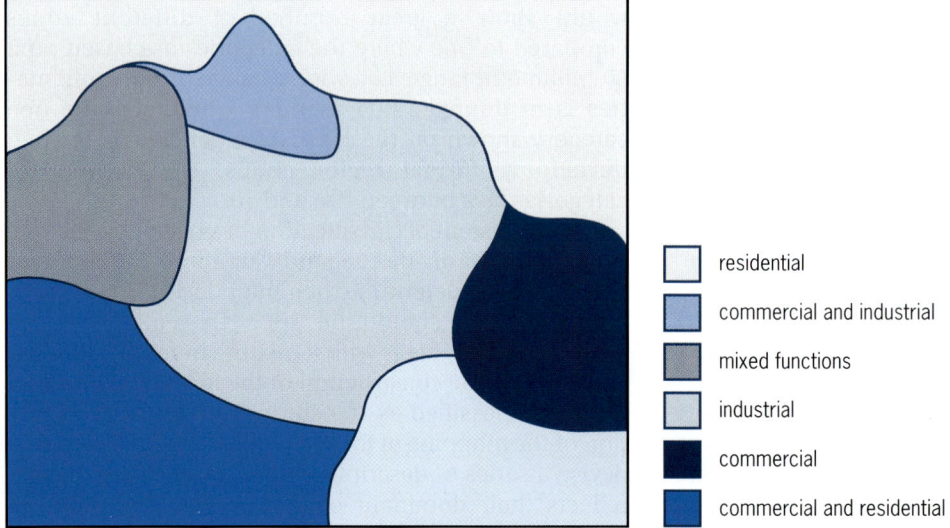

**Figure 9.1** Steps in region building

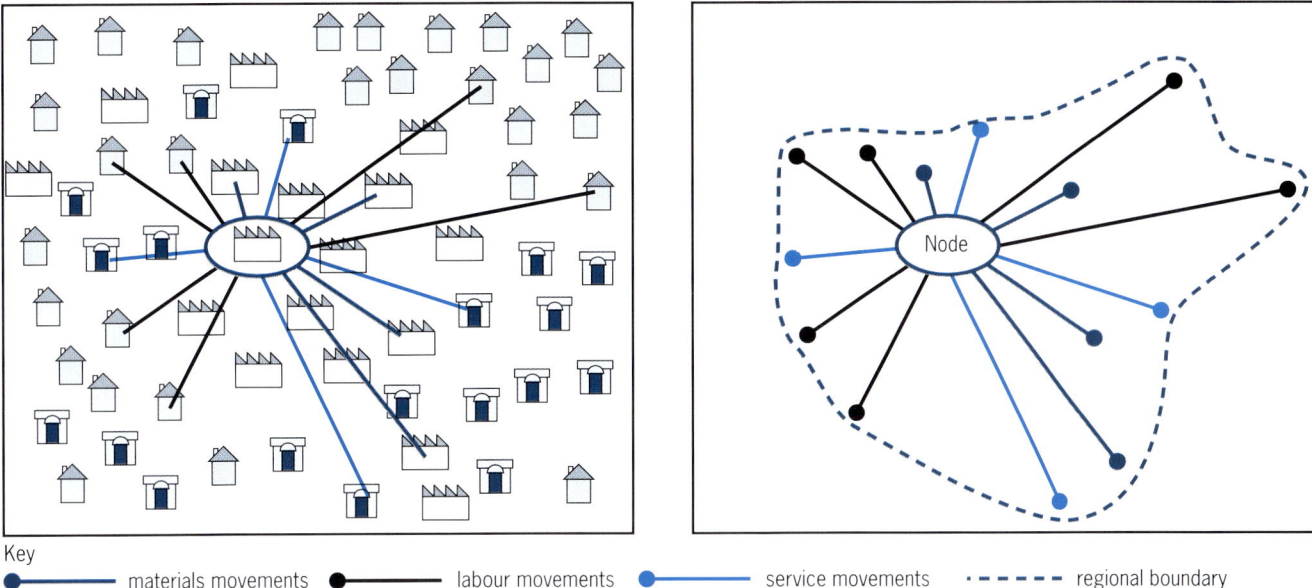

Key
●———— materials movements   ●———— labour movements   ●———— service movements   ·----- regional boundary

**Figure 9.2** Functional regional model

movements. Figure 9.2 illustrates a functional region within an urban landscape.

Within the functional region the frequency and number of movements is greater near the nodal point and diminishes towards the boundary of the region. In this way it is possible to see and map variations in the intensities of this movement. Around the node will be the core of the functional region (a zone of high intensity activity), while the outer part of the region will form a zone of transition or fringe (a zone of low intensity activity). If this variation in frequency and movement is plotted, the result is a formal regional map. This is illustrated in figure 9.3.

Examples of functional regions can be found in the study of urban networks. Functional analysis and regional mapping of central place functions will reveal information such as where, when and how frequently people will move to use a given service. The analysis of the functional regions created by these interactions will illustrate important differences in these movements, which will in turn reveal hierarchical structures. In this way functional regions can be superimposed on each other and further understandings of the patterns of movements can be discovered. Thus, the study of a large metropolitan area will show a complex network of nodes and movements in which roads, rail systems, telecommunication networks and energy systems are all utilised to support the operation of the functional regions.

Within the biophysical environment functional regions could also be identified. The movement of animals to and from a waterhole provides one example, while the cycling of nutrient to and from a plant is another. Within a river basin, erosion of soil and its transportation to a delta could be seen as the operation of a nodal region. However, as the soil could not be returned naturally along the same transport routes, this would limit the application of the concept of the functional region to this example.

# Maps

Maps provide the geographer with one of the major means of plotting and studying the Earth's spatial data. There is a vast array of maps and methods of representing these data. These different representations and functions are discussed in further detail in the following sections.

## Statistical maps

Statistical data on demographics, biophysical characteristics, land use and economic statistics are all capable of being plotted on maps to reveal patterns of frequency or density and arrangements or distributions. Time-interval data can also be plotted to show

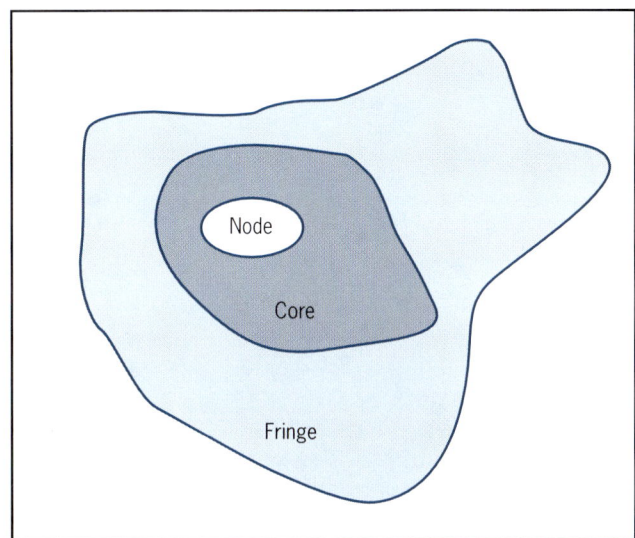

**Figure 9.3** Formal region developed from a functional region

growth or **decline** rates and trend patterns. Isopleth, choropleth and dot distribution maps are all forms of statistical maps and the methods used to produce these maps is similar to that illustrated in the section on formal regions.

## Isopleth maps

The isopleth or isoline map illustrates variations in the patterns of spatial data by the use of lines that connect points with the same values. These values may relate to frequencies, densities or ratings. The pattern of the lines shows distributions as well as rates of change. Common maps using this method include contour maps, synoptic charts, rainfall and temperature maps and soil maps illustrating acidity, salinity or fertility levels. This method of mapping is used where the data is continuous within the region and allows for sampling at an infinite number of points. Data that is discontinuous is mapped using other methods.

Figure 9.4 shows the construction method for isopleth or isoline maps. In this case, soil pH data are collected at a series of sampling points. Lines are then drawn at an interval of one on the pH scale. The lines are plotted so that they pass between points with higher and lower value numbers and closest to the values that are almost the same as the isoline. If the point value is the same as that of the isoline, then the line passes directly through that point. The map can now be left as it is or the areas between the values can be coloured to indicate zones with the same values.

## Choropleth maps

Where isoline maps are shaded to show zones with similar values they illustrate one form of choroplethic map. In figure 9.4 there are three different pH zones (less than 5, 5–6 and over 6). When shaded, each becomes a formal region and the type of map produced is a choropleth map. Choropleth maps can also be constructed using information for existing statistical areas. The volumes, frequencies or rates of the feature being mapped are classified using the methods previously illustrated in the section on formal regions. They are then mapped by colouring the different statistical areas according to the amount of the feature found there—light colours representing low amounts and darker colours higher amounts. This process allows the geographer to identify and analyse variations in the mapped information from one statistical area to the next. The socioeconomic data maps for Perth (see chapter 6) show the use of the choroplethic technique to map data for statistical areas. In these examples the categories range from light to dark depending on the density, frequency and rates for the data being mapped. Where the information is not based on statistical variations such as these, then it is not essential to use colour grading. This is the case with choropleth maps that illustrate different climatic, land use or vegetation zones.

Figure 9.5 illustrates the method of constructing choropleth maps. In this example, frequency data are represented, with the statistical areas being coloured according to a four-point scale. The map in its finished form does not show individual variations for each area but rather a broad pattern of variation. As with the methods used to classify formal regions, the cut-off point between one grouping and the next is set by the map maker. However, the general rule is to show variation while not producing maps that are unnecessarily complex.

## Dot distribution maps

Statistical information such as population density and distribution, livestock density and distribution, and building density and distribution can be mapped using dot distribution techniques. The data from different

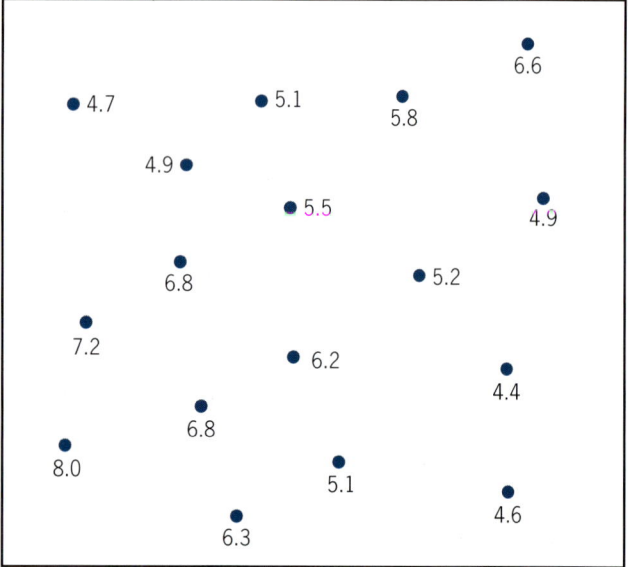
**(a)** Soil pH data are collected at sampling points

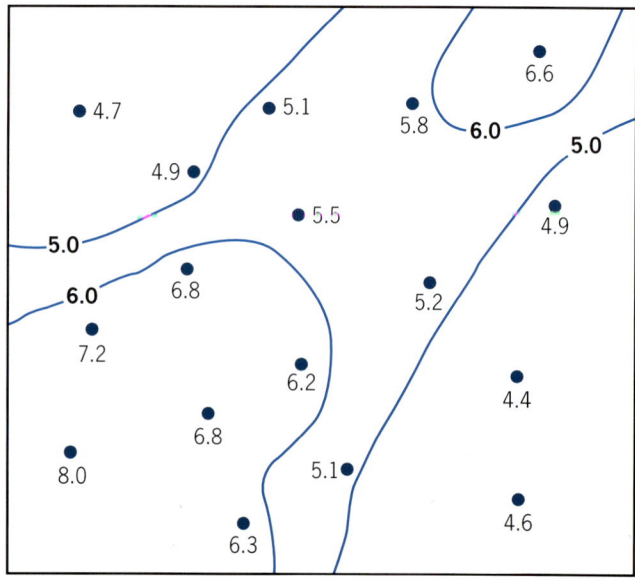
**(b)** Isolines drawn to show pattern of pH levels

**Figure 9.4** Soil pH isoline map

statistical regions can be represented by determining the number of units per dot and then plotting these dots on a map of the study area. In doing this, the following issues should be considered:

- *The most appropriate scale.* This should not result in too many or too few dots. Ideally, the pattern should be one that provides a clear indication of the variations in density of the data being mapped.
- *The pattern of distribution.* Where obvious elements such as landforms or settlement patterns could influence the distribution of dots within a statistical area then these need to be considered when placing the dots on the map.
- *Variations at the statistical boundaries.* The boundaries between areas should not, as a rule, show distinctive changes in densities; rather the transition should be gradual. This is because the statistical areas are arbitrary while the actual data will tend to blend gradually from locations of high density to locations of low density.

In figure 9.6 the number of computers within each statistical region has been counted and the data represented as a dot distribution map. Higher numbers of computers within small regions show a higher density than the same number within a larger region.

## Map types and scale

The size or scale of a map largely depends on its purpose and the amount of detail required. Thematic maps showing global patterns of elements such as population distribution, climates, landforms and national political boundaries will tend to be small scale, while maps showing city streets, house designs, local relief and land uses will tend to be large scale.

### Small-scale maps

Small-scale maps are those that represent large areas of the Earth's surface. They have been reduced many times from the largest size (full size or full scale) and hence show features as being small on the actual map. For example, a map of Australia would show the capital cities as small dots, while in reality they cover hundreds of square kilometres. Small-scale maps are generally those that have been reduced by more than a million times and include maps of countries, states, world regions and the world. In an atlas, maps of Australia may be found drawn at scales between 10 and 40 million times reduced from full size.

### Large-scale maps

Large-scale maps are those that represent small areas of the Earth's surface. Compared to small-scale maps their reduction is much less. This allows the cartographer to show much greater amounts of information (detail) compared to the small-scale map. Large-scale maps are generally those that have been reduced by less than a million times. They include topographic maps with scale reductions between 25 000 and 250 000 times, street directories with reductions of 20 000 times and house plans, which may only be reduced 100 times.

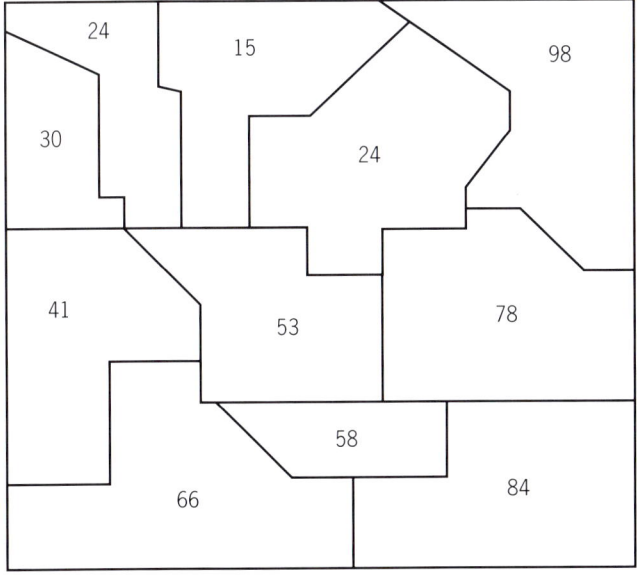

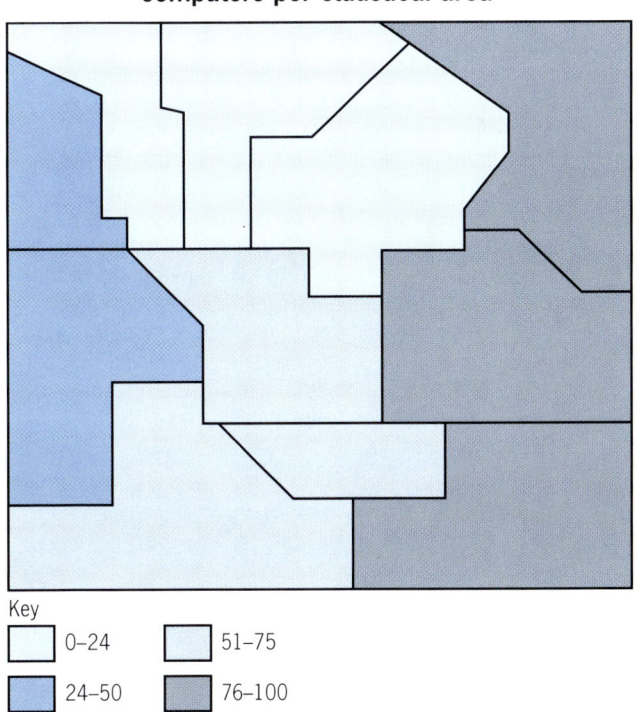

**Figure 9.5** Percentage of household computers per statistical area

## 224 LANDSCAPES AND LAND USES

(a) **Number of personal computers**

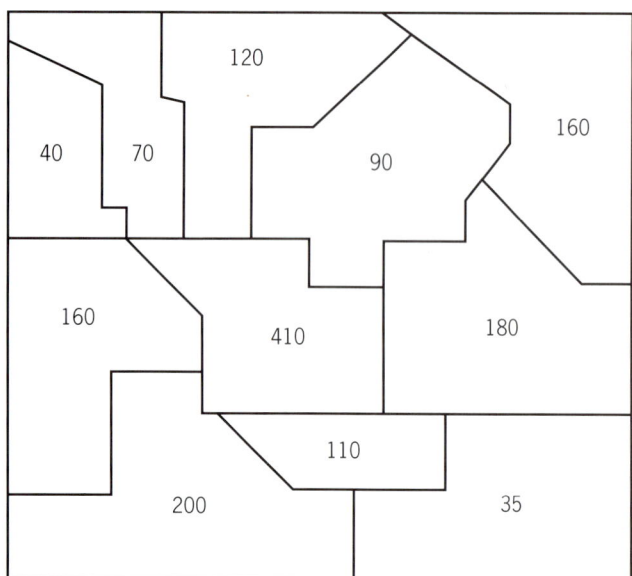

(b) **Dot distribution map of personal computers per statistical area**

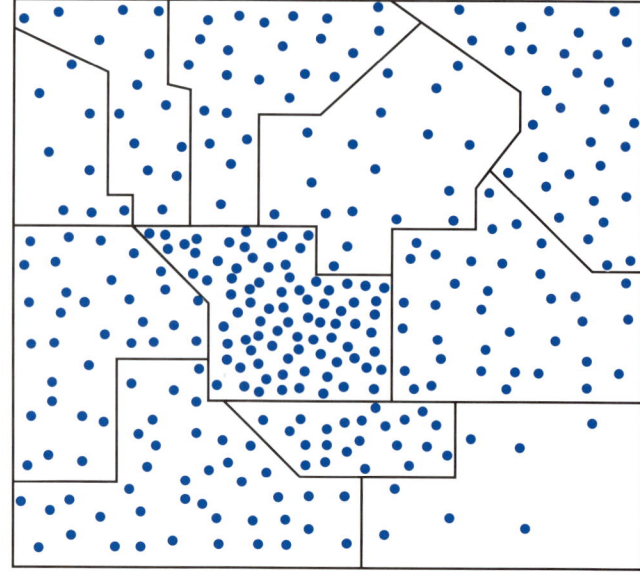

• = 5 personal computers

**Figure 9.6** Number of personal computers

### EXERCISES

1. Using a variety of examples, explain why maps are important geographic tools.
2. Discuss the differences between isopleth, choropleth and dot distribution maps.
3. Outline the steps required in the construction of isopleth, choropleth and dot distribution maps.
4. Conduct a class survey of students attending your school and then plot their movement to and from school on a map. Describe the pattern in terms of directions, distances and densities. Explain why the map illustrates a functional region.
5. On a sheet of paper, draw up a rectangle measuring 20 cm × 10 cm. Divide it into 2 cm squares (10 × 5) and label the columns A, B, C, D, E and the rows 1 to 10, so that the data in the table can be transferred to the squares. Now, using the data in the following table, construct isopleth, choropleth and dot distribution maps. When constructing the isopleth maps the value for each square is assumed to fall in the centre of the area. For example, square 2C is 180 and this value is plotted as a dot in the centre with the number written next to it. Briefly describe the pattern revealed by each map by referring to direction, density, distribution and changes from location to location.

|    | A   | B   | C   | D   | E   |
|----|-----|-----|-----|-----|-----|
| 1  | 80  | 110 | 150 | 210 | 210 |
| 2  | 120 | 140 | 180 | 230 | 220 |
| 3  | 170 | 180 | 210 | 240 | 240 |
| 4  | 190 | 210 | 230 | 240 | 250 |
| 5  | 220 | 240 | 250 | 260 | 270 |
| 6  | 190 | 200 | 220 | 250 | 260 |
| 7  | 180 | 180 | 210 | 230 | 250 |
| 8  | 170 | 170 | 170 | 220 | 240 |
| 9  | 130 | 130 | 140 | 210 | 230 |
| 10 | 100 | 90  | 110 | 190 | 220 |

6. Using the information provided in figures 9.5 and 9.6, calculate the number of houses in each statistical area. Assume that there is only one personal computer per household. This data can be found by using the following formula:

$$\frac{\text{Number of computers}}{\text{Percentage of computers}} \times \frac{100}{1} = \frac{\text{Number of houses}}{\text{per statistical area}}$$

Sketch a map of the statistical areas and, using either the choropleth or the dot distribution mapping technique, show the number of houses per area. Compare this map to figures 9.5 and 9.6 and identify the main variations.

## Mapping skills

The construction and interpretation of maps requires a level of knowledge of the various processes and features associated with these activities. Map reading and map making skills are discussed in the following sections and cover such elements as the use of scales, directions, coordinates, legends and relief.

### Symbols

Map symbols are simplified illustrations of biophysical and cultural features found on the Earth's surface. They are generally not drawn to scale. On Australian topographic maps they are referred to as conventional symbols. So that they are more easily recognised and remembered, they have been standardised. Conventional symbols are grouped into distinctive categories based on their purpose. Table 9.1 lists different symbols and their related groupings and colours.

**Table 9.1** Conventional map symbol groups

| Linear features | Point features | Area features |
| --- | --- | --- |
| Roads (black and red) | Waterhole (blue) | Vegetation (green) |
| Railway lines (black) | Rock, bare or awash (black) | Perennial or intermittent lake (blue) |
| Piers, wharves (black) | Buildings (black) | Quarry (black) |
| Fences (black) | Mine (black) | Rock shelf (black) |
| Windbreaks (green) | Spot height (black) | Built-up area (light red) |
| Streams (blue) | Railway station (red) | Land subject to inundation (blue) |
| Contours (brown) | | |

Conventional symbols are found in a legend or key on a topographic map. There will often be several symbols on the same line in the key, or multiple symbols linked together (such as with the symbols related to railway lines). The name or descriptor for each symbol is separated by a semicolon and it is essential that these names be matched in the correct order to the appropriate conventional symbol.

## Direction

Map directions are based on the points or degrees of a compass. Virtually all maps are drawn with north at the top and can be orientated by aligning the map in this direction. Topographic maps often have three references for north, as illustrated in figure 9.7. Magnetic north (MN) indicates the position of the magnetic north pole at the time that the map was drawn. This position shifts slightly each year (magnetic declination) and its current position would have to be calculated and appropriate adjustments made if the topographic map were to be used for cross-country navigation. True north (TN) is marked by the intersection of the meridians of longitude and represents the northern end of the Earth's axis or north pole. Grid north (GN) may vary from left to right of true north as illustrated in figure 9.7, based on the topographic map's location on the Earth's surface. In other words, when a map of an area is drawn, it may have TN on the left or right depending on where it is positioned in relation to the nearest line of longitude. This difference is due to the construction of the grid as a parallel system rather than as a true representation of the meridians, which are curved and intersect at both poles while being widest apart at the equator. When using directions on a topographic map in the classroom, north is represented by grid north.

The four major points of a compass are referred to as the cardinal points—north, south, east and west. Between the cardinal points are the four intermediate points—north-east, south-east, south-west and north-west. It is possible to divide these angles further to show eight more points and then, between these, another sixteen. The traditional compass rose used for early sea navigation has 32 points. Figure 9.8 illustrates an 8-point compass.

## Scale

Map scales are reduction scales and indicate the number of times that an area of land has been reduced from full size when the map was produced. As already stated in the section on map types and scales the further a map is reduced from full size, the smaller the features and the smaller the scale. Any map is drawn at a fraction of an area's actual size and the reduction may therefore be expressed as a fraction of true size. For example, a scale of one-half is a 50 per cent reduction in actual distances, while one-fifth is 20 per cent of the distance. Scales are expressed in three ways, with each having certain advantages:

- *Representative fraction scale.* The expression of a scale as a representative fraction (RF) identifies the ratio between distances on the map and actual distances on the ground. The numerator is always a unit of one, while the denominator is the number of units on the ground or the number of times that the map has been reduced. The following examples illustrate the scale as an RF and as a ratio.

$$\frac{1}{20\,000} \text{ or } 1:20\,000 \qquad \frac{1}{500\,000} \text{ or } 1:500\,000$$

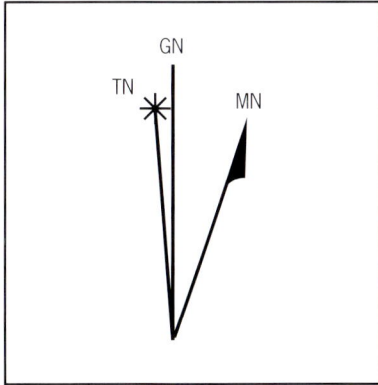

**Figure 9.7** North points

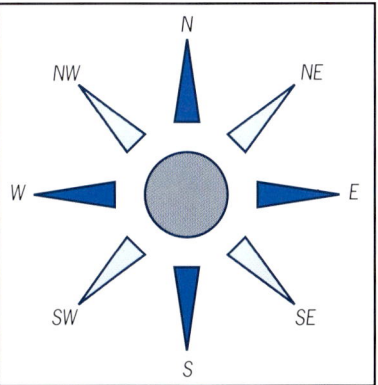

**Figure 9.8** A compass rose

Both the numerator and the denominator must be expressed in the same unit of measurement. Hence, if the 1 is in centimetres then the 20 000 must also be in centimetres.

- *Statement in words.* A further method of expressing a scale is as a word statement. This has the advantage of providing the map user with an easily understood statement on map distances. The ratio 1:20 000, as already indicated, could be expressed in words as '1 centimetre represents 20 000 centimetres'. However, when expressed in a statement, the denominator is converted to the largest practical unit of measurement, in this case metres. Thus the scale expressed in words would be '1 centimetre represents 200 metres'.
- *Line scale.* The construction of a scaled line on a map allows the user to mark or measure distances and then compare them to the graduated line. The line is accurately drawn to represent the map ratio or RF. It normally expresses ground distances in the largest practical unit and generally starts at zero. Sometimes the zero point will be placed in from the left-hand end of the line, with smaller units of measurement being found to the left and the larger ones to the right. The two different methods of constructing a line scale have been shown in figure 9.9.

## Measuring distance

Straight line distances can be measured on a map by marking the start and finish points on the edge of a piece of paper and then placing this against the line scale and reading the distance. Alternatively, the map distance can be measured by using a rule and then calculating the distance by multiplying the measurement and the RF. A distance of 6.5 centimetres on the rule, multiplied by 100 000 (RF 1:100 000), would result in a total distance of 650 000 centimetres or 6.5 kilometres. Where the linear distance being measured is not a straight line then the method shown in figure 9.10 can be applied. In this case the length of the road is plotted in segments on the edge of a piece of paper, which is then placed against the line scale to determine the total distance.

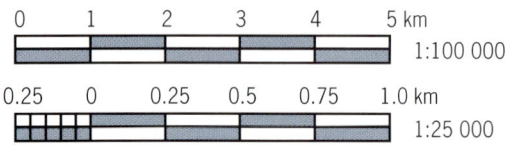

**Figure 9.9** Line scales

## Measuring area

The calculation of area using the map scale involves finding the length and breadth of a map feature and then multiplying these dimensions to produce an answer in square metres, square kilometres or hectares. Where the area to be measured is irregular in shape, several methods can be used to determine the actual size:

- *Estimation.* Many metric topographic maps have a grid system based on the map scale. A map with a scale of 1:100 000 will have a grid pattern measuring 1 kilometre by 1 kilometre. Count the number of map grids that cover the feature being studied, making adjustments for part grid squares, and then calculate the total area. In the case of the scale 1:100 000, four grid squares would cover 4 square kilometres.
- *Area reproduction.* A more accurate method is to trace the map feature onto a piece of paper and then superimpose a grid pattern over this, with a size set by the map scale. Once this is done, the full and part squares are counted and multiplied by the area of one square. Part squares can be counted either as half squares or, for a more accurate result, as a fraction of 10, expressed as a decimal (0.1 to 0.9). In figure 9.11 the squares have been drawn at 2 centimetre intervals and the map scale is 1:50 000. The size of a grid square is 1 square kilometre, 1 000 000 square metres or 100 hectares. The area of the feature is calculated to be 6 square kilometres or 600 hectares.

### EXERCISES

1. Explain the difference between true north and grid north.
2. Write the correct bearings or degrees on an 8-point compass rose. Note that north is both 0 and 360 degrees.
3. Find an example of a 32-point compass rose and calculate the bearings for each point. Note that these can be found on the Internet by conducting a search using the term 'compass rose'.
4. Rank the following scales from largest to smallest and rewrite all of them as ratio scales.

   - 5 centimetres represents 100 metres
   - 1:500
   - 1:100
   - 1 centimetre represents 1 kilometre
   - 1 metre represents 1000 metres
   - 1:250 000

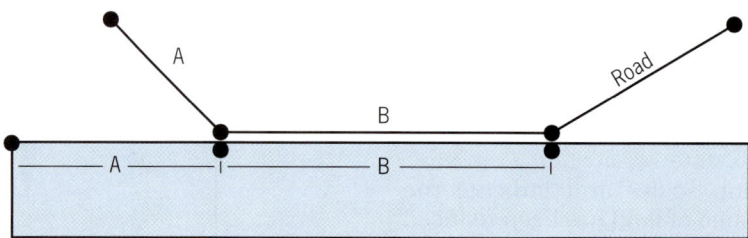

**Figure 9.10** Measuring irregular lines

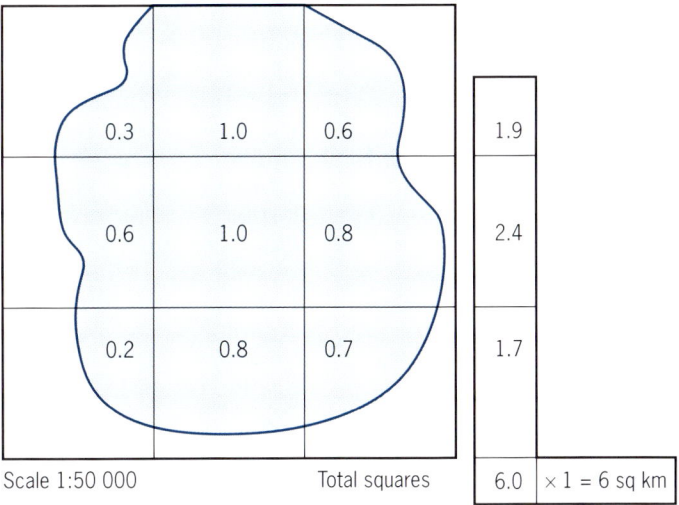

**Figure 9.11** Measuring area

5 Draw a map of a classroom or something similar at a scale of either 1:50 or 1:100. Identify the correct position for north and include a 16-point compass rose. Use appropriate symbols or colours to indicate the main features of the room.
6 Using a topographic map legend, find and draw the symbols identified in table 9.1.
7 Using the aerial photo of Northam (figure 8.43) measure the distance between the two road bridges that cross the Avon River.
8 Measure the distance from the junction of Fitzgerald Street and Peel Terrace, along Fitzgerald Street to the Avon River bridge (south-west bridge).
9 Draw a grid system with 10 columns and 10 rows with 1 centimetre intervals. Use a map scale of 1:200 000. Superimpose on the grid pattern a lake that is twice as long as it is wide, with an irregular shoreline and covering an area of 32 200 hectares.
10 Using the topographic map of Northam (figure 8.42) and the aerial photo (figure 8.43), calculate the size of the built-up area in 1977 and then in 2002.

## Coordinates

Maps are two-dimensional representations of the Earth's surface and in order to locate a position on this surface a grid pattern of lines can be used. The intersection of these lines provides a reference point or coordinate. Globally, the grid system used is latitude and longitude, while on topographic maps a system of local grids called eastings and northings is employed.

### Latitude and longitude

Latitude is depicted as a series of parallel lines running around the Earth and measured as an angle from the equatorial plane. The position of these lines is expressed in degrees and minutes north and south of the equator, with the highest value of 90 degrees corresponding with the North and South poles. In figure 9.12 the method of locating latitude is illustrated. Navigators would use a sextant to measure the angle between the horizon and the noonday sun on a given day to find their latitudinal position. This method has largely been replaced by Global Positioning Systems (GPS), using satellite broadcasts.

Longitude is depicted as a series of circles, called meridians, passing through the North and South poles. The position of these lines on the globe is measured in degrees and minutes east and west from the plane of the prime meridian. The highest value, 180 degrees, corresponds with much of the International Date Line. (The International Date Line follows the 180 degree meridian of longitude except where it passes through land. Here it deviates around the area of land to prevent a location having two different days.) Unlike the lines of latitude, meridians of longitude are not parallel and they are all of equal length. The method of calculating a meridian of longitude is illustrated in figure 9.13.

A coordinate in latitude and longitude is expressed as: latitude degrees and minutes N or S and longitude degrees and minutes E or W. Topographic maps often

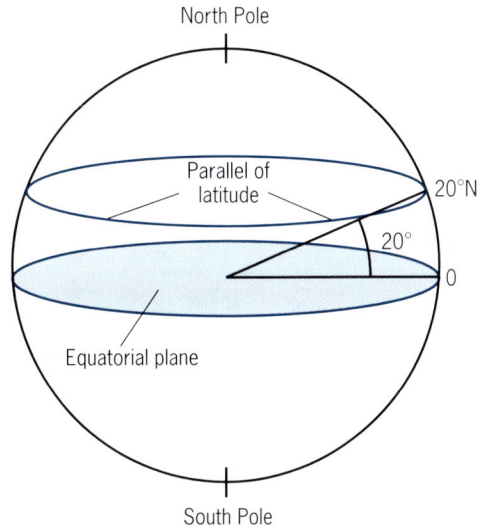

**Figure 9.12** Degrees of latitude

# 228 LANDSCAPES AND LAND USES

cover less than one degree of latitude and longitude and giving a coordinate involves calculating the number of minutes. The minutes on a topographic map are marked with small ticks on the map margin, or with a black and white border. In figure 9.14 the position of several coordinates is illustrated using latitude and longitude.

## Grid references

The grid lines on a topographic map are used to locate areas or points. They are numbered from 00 to 99. These numbers are based on a referencing system that, in the case of Australia, applies to the whole continent. The sequence is repeated once the numbers reach 99. The lines that run vertically on a topographic map have numbers that increase in value to the east and are therefore called eastings. The horizontal lines increase in value to the north and are known as northings. Unlike readings in latitude and longitude it is the vertical lines or eastings that are referenced first and the horizontal or northings second.

Figure 9.15 illustrates the method of giving a four-figure or area reference and a six-figure or point reference. Six-figure references use a decimal system to determine the position of a feature falling between two grid lines. Thus, a reference of 125434 has the first two numbers based on an easting (12) with the third number (5) being half way between easting 12 and easting 13. The fourth and fifth numbers (43) correspond to the northing with the sixth (4) being four-tenths of the way between northing 43 and northing 44. Location A in figure 9.15 falls within a grid square with a four-figure area reference of 1445. Location B has a six-figure reference of 120420 and location C's grid reference is 117457. Note that the size and positioning of the grid lines is set by the scale of the map and bears no relationship to latitude and longitude. The two systems are separate and should not be confused.

### EXERCISES

1 On a blank sheet of paper, draw a rectangle measuring approximately 20 cm × 15 cm. On it, mark the equator and the prime meridian so that they cross at the centre of the rectangle. Mark in lines of latitude and longitude at intervals of 20 degrees. Name the northern, southern, western and eastern hemispheres. Using an atlas, find the degrees of latitude and longitude that mark the start and finish of the perimeters of the following countries and then mark their location on your chart by means of a rectangle. The top and bottom of the rectangle will be set by the degrees of latitude while the left and right hand sides will be set by the degrees of longitude.
- The United Kingdom
- South Africa
- Australia
- Spain
- Greece
- United States of America
- Brazil
- Afghanistan

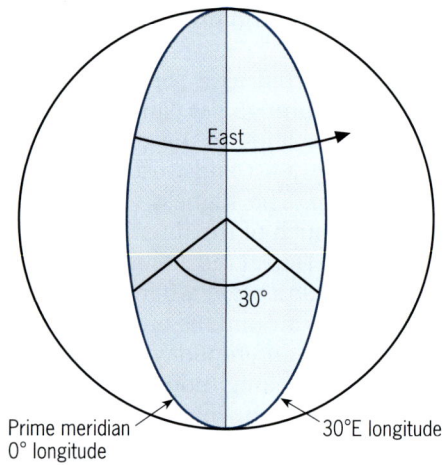

**Figure 9.13** Degrees of longitude

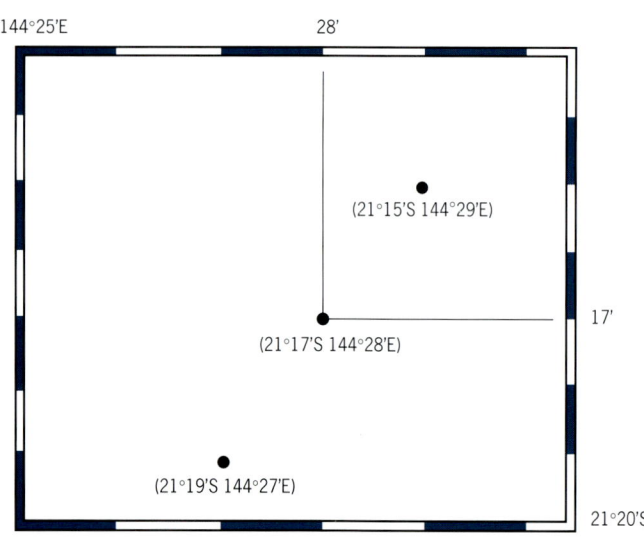

**Figure 9.14** Latitude and longitude on topographic maps

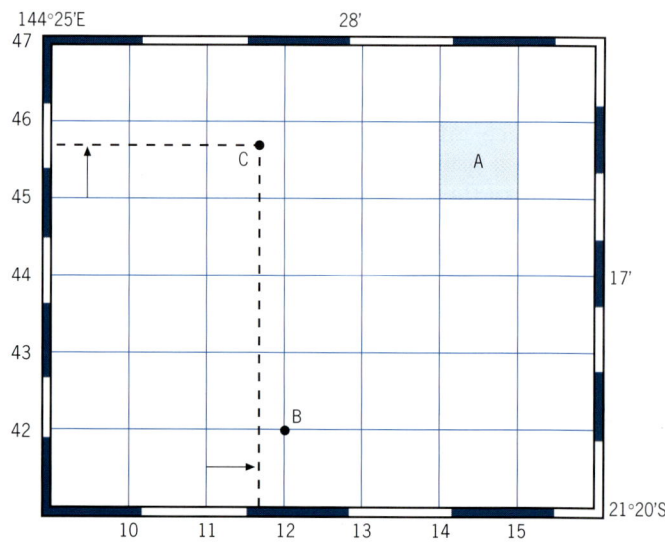

**Figure 9.15** Grid references

2 Using a diagram, explain the equatorial plane and how it is used to measure degrees of latitude.
3 Identify the ways in which the lines of longitude are different from the lines of latitude.
4 Research the ways in which latitude and longitude are found when navigating. Note that one uses the angle of the sun to the horizon, while the other uses Greenwich Mean Time.
5 Draw figures 9.14 and 9.15 and mark the following locations:

- 129430
- 0944
- 151461
- 21° 15′ S 144° 28′ E
- 21° 19′ S 144° 26′ E

## Relief

The shape of the land (its relief) is represented on topographic maps by contours, spot heights shading (shadow) and colouring. Contours are isolines that connect points that are at the same height above sea level. They provide a great deal of information on landforms, slope, drainage patterns, erosion and deposition features, and tectonic processes for an area of land. When analysed correctly, contours provide a three-dimensional perspective of the mapped area. The heights of successive contours are measured using a set vertical scale known as the contour interval. In figure 9.16 the vertical interval is set at 5 metres and the feature being mapped is a conical hill 24 metres high. Where the slope of the hill is steep the contours are close together, while they are further apart where the slope is flatter.

## Cross-sections

The profile or shape of the land may be illustrated as a line graph constructed by plotting the heights taken from contour patterns. The scale of the *x*-axis (horizontal axis) is taken from the map scale, while the scale of the *y*-axis (vertical axis) is based on the contour heights. The relationship between the two scales is expressed as a ratio. If the two scales are the same, then the ratio is 1:1. If the vertical scale is larger than the horizontal scale then the ratio will have a denominator that is greater than one. In figure 9.17 the contour pattern is based on a map scale of 1:20 000. The vertical scale is 1:1000 and the vertical exaggeration is 1:20. The vertical exaggeration of a cross-section is calculated by using the formula:

$$VE = \frac{\text{Vertical scale}}{\text{Horizontal scale}} = \frac{\frac{1}{1000}}{\frac{1}{20\,000}} = \frac{1}{1000} \times \frac{20\,000}{1} = 20$$

## Gradient

The slope of the land can be referred to in descriptive terms such as 'steep', 'moderate', 'gentle' or 'flat'. These terms indicate the type of angle or gradient that might be expected when traversing a landform. The spacing of contour lines on a topographic map gives some indication as to the type of slope that could be expected in any particular area. Closely spaced contours illustrate a rapid increase or decrease in height and a steep slope, while widely spaced contour lines indicate a more gentle slope. For example, in figure 9.16 the slope at the top of the hill is gentle and becomes gradually steeper as you progress down the hill.

The slope of the land can be calculated either as an angle or as a ratio. The angle of a slope is measured from the horizontal, while the ratio is a comparison of the horizontal distance and the change in altitude or height experienced over that distance. In figure 9.18 the distance between points A and B is 1000 metres, while their difference in height is 100 metres. Thus, for every 10 metres travelled the height increases or decreases by 1 metre. This gradient is expressed as 1 in 10 or 1:10. The horizontal distance is found by using the map scale, while the difference in height is calculated by reading the contours and finding the heights for point A and for point B.

The formula for gradient is expressed as:

$$\text{Gradient} = \frac{\text{Differences in height in metres}}{\text{Distances in metres}} \quad \frac{100}{1000} = 1:10$$

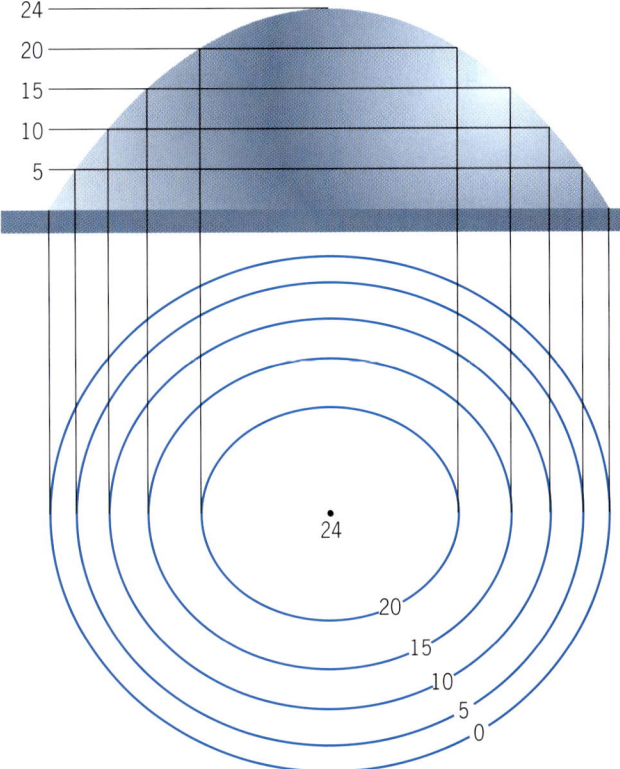

**Figure 9.16** Rounded hill or knoll contour pattern

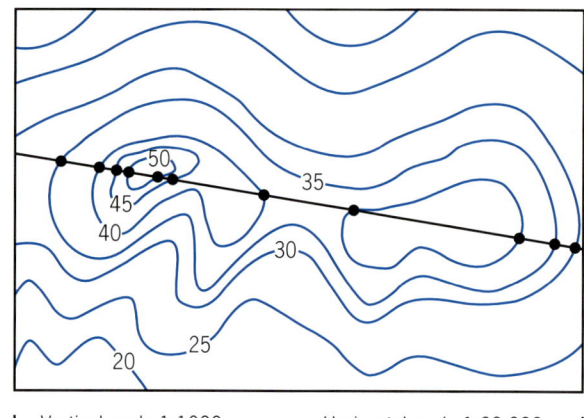

**Figure 9.17** Cross-section

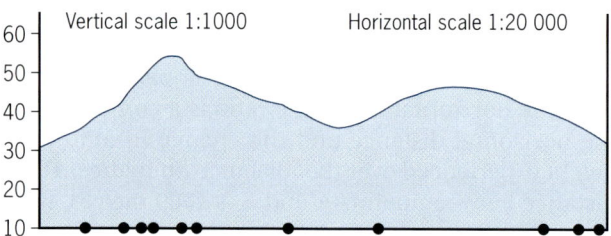

**Figure 9.18** Calculating gradient

# Interpreting topographic maps

The interpretation of the landscape represented on topographic maps requires an understanding of the ways in which the biophysical and cultural landscapes are presented. In this section a number of the more common features of topographic maps are identified and discussed.

## Interpreting the biophysical landscape

### Landforms

Topographic maps contain a great variety of landform features that are illustrated by contours, colours, symbols and labels. Some of these are summarised in the following list and figure 9.19 illustrates a number of landforms with their associated contour patterns.

- *Ridge.* A series of summits or high points roughly aligned along an axis. It has lower land on either side.
- *Plateau.* An extensive, raised, relatively flat surface whose boundary may be marked by an escarpment or range of hills or mountains.
- *Slope and gradient.* The angle of a landform. The gradient or steepness of a slope is indicated by the closeness of the contours. The average gradient of a slope is the ratio of the vertical height to the horizontal distance. Thus, a 1 in 5 slope is one that increases in height by 1 metre for every 5 metres travelled horizontally.
- *Cliffs.* A very steep or vertical slope.
- *Escarpments.* Associated with faulting and marks the boundary between two regions of different heights, such as a lowland plain and a plateau.
- *Col* (saddle). An area of lower land between two higher regions such as two summits. Often seen as a dip in the horizon along a ridgeline.
- *Spur.* An area of land extending out from a region of high land. It is surrounded by lower land on three sides and can often be seen between two valleys.
- *Valley.* An eroded section of land formed on slopes, with higher land on three sides.
- *Knoll.* A low hill with a relatively wide summit that gently slopes upwards to the highest point.
- *Summit.* The highest point on a hill or mountain, forming a distinctive apex.

### Fluvial features

The action of water flowing across the surface of the Earth produces a number of distinctive features. These fluvial features can be identified on topographic maps by their shape, location or contour pattern. The main fluvial features have been summarised in the following list and they are also illustrated in figure 9.20.

- *Meander.* As a river crosses an area of relatively flat land it will erode and deposit soil laterally. This creates exaggerated bends or curves in the channel called meanders.
- *Oxbow lake.* When a meander is separated from the main channel by a process of erosion and deposition, then it results in a semi-circular lake on the river floodplain known as an oxbow lake.
- *Gorge.* On a raised plateau or surface, weaker rock is often eroded rapidly downwards producing a narrow, steep-sided valley. This valley or gorge is also referred to as a chasm or canyon.
- *Lake.* These inland water bodies can vary in size and may be either temporary (intermittent) or permanent (perennial) in nature. This feature is illustrated by both the oxbow lake and the coastal lake in figure 9.20.
- *Estuary.* Where the lower sections of a river valley have been invaded (inundated) by a rise in the sea level, then the body of water formed by this process is an estuary. Estuaries are characterised by salt water and tidal movement.
- *Delta.* These may be formed at the mouth of a river where it enters a lake or sea. Deltas are depositional features and are dependent on the rates at which alluvial soil is carried down the river and deposited. This rate must exceed the rate of removal by the action of the sea or lake. In figure 9.20 the islands that make up a delta have occurred where the river enters the estuary. Here, the river has broken into a number of channels called distributaries.
- *Floodplain.* This is an area of flat land on either side of a river channel that may be subject to flooding or inundation and is formed by the deposition of alluvial soil.

## Coastal features

A great variety of coastal features and landforms may be found on topographic maps, some of which have been described in the following list and illustrated in figure 9.21.

- *Lagoon*. This term is applied to a variety of water features. It may be used to describe a permanent lake, fully or semi-enclosed bay or region of protected water surrounded by a reef.
- *Beach*. Where the processes of deposition are greater than those of erosion along a coastline then a build-up of sand or pebbles occurs, producing a beach. These are often marked on topographic maps by the symbol for sand or mud.

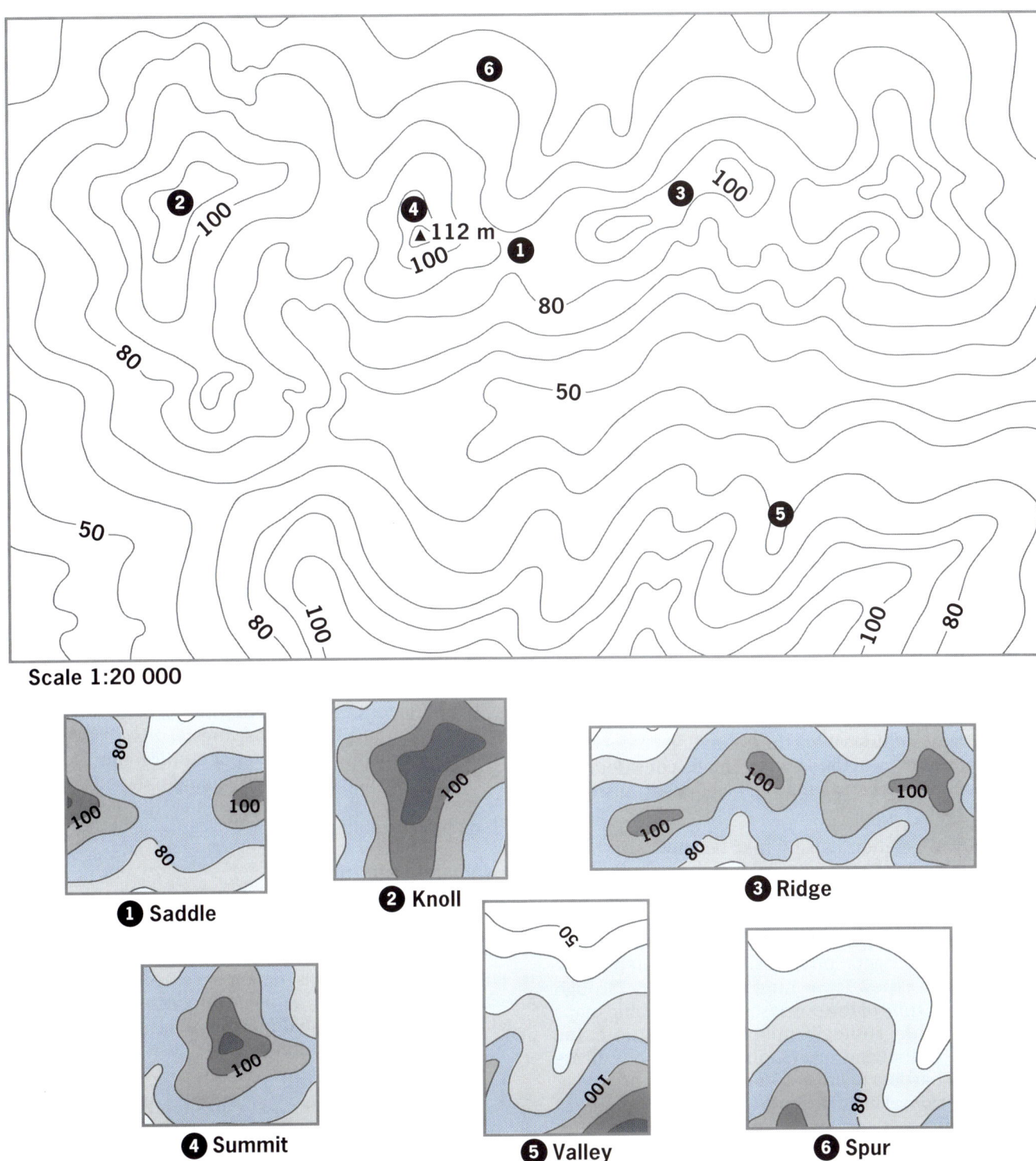

**Figure 9.19** Contour patterns

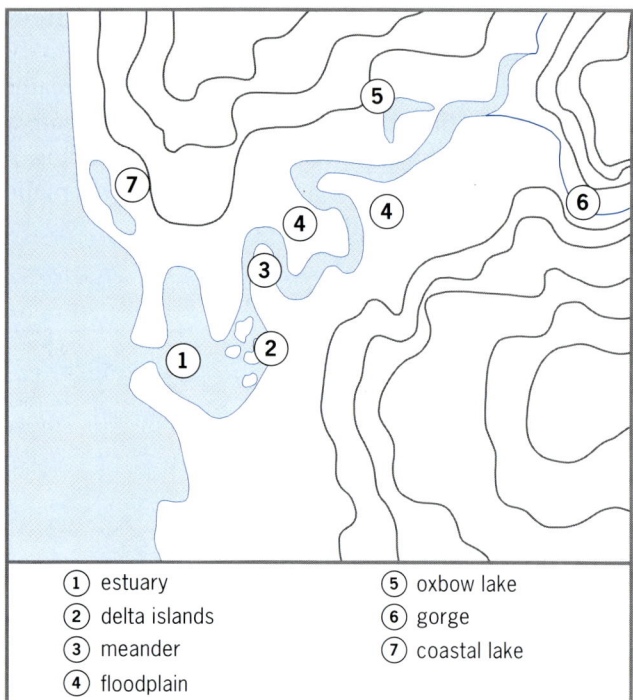

**Figure 9.20** Fluvial features

1. estuary
2. delta islands
3. meander
4. floodplain
5. oxbow lake
6. gorge
7. coastal lake

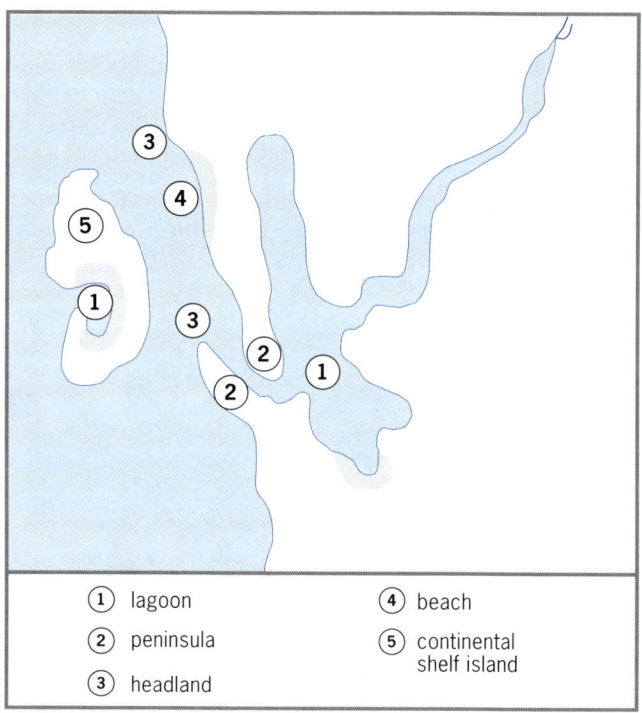

**Figure 9.21** Common coastal features

1. lagoon
2. peninsula
3. headland
4. beach
5. continental shelf island

- *Peninsula.* Coastal landforms that project out into the sea form a peninsula, with a headland occurring at the end.
- *Headland.* Where the coastline is exposed to the erosional forces of the waves, more resistant areas will eventually protrude as rocky features such as wave-cut platforms, cliffs, stacks and sea caves. These features are typical of headlands.
- *Island.* Oceanic and continental shelf islands are landmasses surrounded by water and in the case of coastal (continental shelf) islands are much smaller than the adjacent mainland.

## Vegetation patterns

In observing and describing the vegetation characteristics illustrated on a topographic map, consideration should be given to elements such as density, distribution and type.

- *Density.* Vegetation density may be shown on topographic maps as either dense, medium or scattered. This often reflects the degree of land clearing that has taken place or the influence of soils, local climate variations and topography. The density and general height of the vegetation may also be indicated by the variation in types. These range from forests through to scrub. As one of the purposes of topographic maps is to provide strategic military information, the purpose of showing vegetation density is to indicate possible barriers to cross-country movement.
- *Distribution.* The distribution of vegetation is of special interest to the geographer as it illustrates variations and the possible influences of cultural variables and other biophysical factors. The extent of clearing for rural and urban land uses is one such influence. Areas of intensive agriculture or broad-acre farming will have a greater degree of clearing than areas of extensive grazing. The retention of ecosystems in national parks or the controlled use of vegetation through the setting aside of state forests will also shape patterns of distribution.
- *Types.* While formal vegetation zones are not shown on topographic maps, it is possible to identify some variation in the vegetation and then draw conclusions about the overall patterns. Open forest, rainforest, shrubland, mangrove and swamp vegetation represent some of the common types of vegetation that can be seen on topographic maps. Introduced vegetation in the form of orchards, vineyards and coniferous forests (plantations) can also be seen on different topographic maps. Students should be careful not to include the introduced vegetation types if asked to identify and discuss natural vegetation patterns.

## Drainage patterns

Surface water movement produces a variety of distinctive drainage features that can be identified on topographic maps. They develop in response to the shape of the landforms, which have been produced by various tectonic and geologic processes. In addition climatic factors play an important role, as does the type and density of the vegetation.

- *Perennial and intermittent.* In humid climates with regular rainfall patterns streams tend to flow all year round. These drainage systems are perennial or permanent in nature. Major rivers or tributaries are more likely to be perennial than smaller ones, which will often only contain water after rainfall and then only for a limited time. Small tributaries or creeks therefore tend to be intermittent. Whether a stream is perennial or intermittent indirectly illustrates elements of the region's climate.

- *Basins and watersheds.* The area of land drained by a river system forms a drainage basin. Its perimeter is marked by a zone of high land (summits and ridges), which indicate the **watershed** or boundary between one basin and another. Drainage basins can be identified on topographic maps by locating the summits and ridges between streams. In figure 9.22, the watersheds between three partial drainage basins (marked 1, 2, 3) have been shown with a dashed line. Where a river enters a lake or the sea then that is also the point at which the basin commences, with the watershed converging to meet the mouth of the stream.
- *Radial, trellis and dendritic patterns.* The streams or channels that make up a river system can often form distinctive patterns or shapes influenced by a region's shape and geology. Where streams flow outwards from an isolated area of high land the pattern is radial in appearance. Streams flowing down slopes towards the main channel of a river can produce a **trellis** pattern. This is especially apparent where there is a series of ridges with valleys between. Trellis drainage patterns are common in areas of tectonic folding. One of the most common patterns is the **dendritic** drainage system. This tree-shaped pattern has a main trunk, with a series of branches that divide into smaller and smaller tributaries in a process termed **bifurcation**. In many cases this splitting of streams appears to conform to a set ratio. For example, a drainage network with a bifurcation ratio of 3 will have one large stream, three secondary streams and nine minor streams. In figure 9.23 the three types of drainage patterns have been illustrated.

## Interpreting cultural landscapes

The cultural landscape on topographic maps can be interpreted by using direct evidence as well as by looking for clues from which land uses can be inferred.

Land uses may often be identified by the use of symbols, labels and place names. The most common types of cultural landscapes include rural or farming landscapes, mining landscapes and urban landscapes.

### Rural landscapes

Direct evidence of the presence and types of rural activities can be found in such features as processing, storage and transport facilities. Packing sheds, dairies, shearing sheds, stockyards, piggeries and timber mills may be labelled on a map. Silos indicate the storage of wheat and other grains, while railway sidings may have depots labelled for different rural produce. In some instances, such as with vineyards or orchards, the map may have specific symbols for this type of land use.

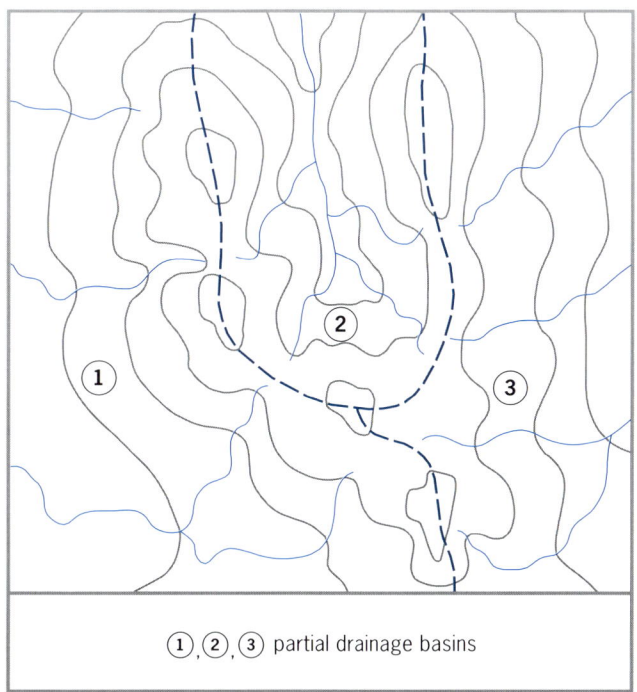

**Figure 9.22** River basin and watershed

 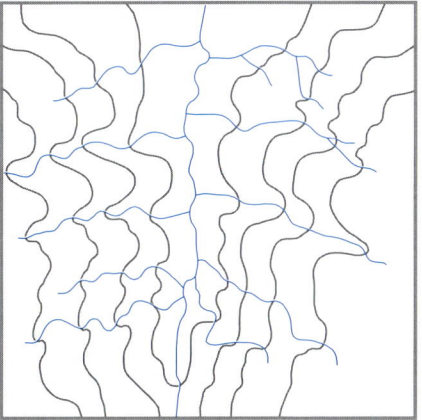 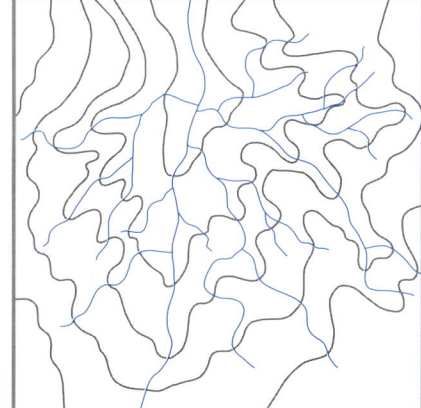

**Figure 9.23** Radial, trellis and dendritic drainage patterns

Indirect evidence of rural activities includes the estimated size of rural landholdings, the presence of livestock watering points such as windmills and small dams and evidence of intensive irrigated farming in the form of irrigation channels. Other types of indirect evidence to look for are the degree of clearing of the natural vegetation, fencing, and the size of any paddocks or fields. Activities such as intensive grazing or grain farming will have significant land clearing, while extensive grazing will have less clearing and very large paddocks. The type, pattern and frequency of transport systems are also significant in interpreting the rural landscape. Intensive forms of farming will have a well-developed transport network, often with sealed roads. Extensive livestock ranching on the other hand will have few roads, and these will often be unsealed. It is important to use the map scale and have a good understanding of typical farm sizes for different land uses when drawing conclusions based on indirect evidence.

### Mining landscapes

This landscape may have identifiable features such as quarries or open-cut pits and the site will often be marked by the conventional symbol for a mine. Storage and processing facilities may be labelled on the map and there might also be distinctive transport systems, such as conveyor belts, in addition to the presence of spur lines and mine roads. In contrast to agriculture or forestry, mining occupies a relatively small site and is therefore quite localised in its pattern.

### Urban landscapes

Topographic maps provide a number of opportunities to interpret urban landscapes. This includes the investigation of urban networks and the external morphology of urban centres, and the interpretation of internal morphology as well as site and situation. The study of urban networks would include the numbers and distributions of different urban centres, how they are linked and any evidence of central place services. These services could be identified by conventional symbols and might include schools, post offices, churches and police stations. The spacing of the urban centres would also provide information on the potential sizes of the hinterlands and the ranges of the different centres.

The external morphology of an urban centre is shown by the boundary of the built-up area. This shape indicates whether it conforms to rectangular, linear, radial, circular or irregular patterns, which are typical of many urban centres. In addition to the external pattern it is also possible to identify elements of the internal morphology of a town or city. Zones such as the rural–urban fringe, newer growth zone, established residential zone and main commercial zone can often be seen on topographic maps. For example, the location of space-extensive activities such as sporting facilities, sewage treatment plants, rubbish tips and cemeteries on the outskirts of towns indicates the position of the rural–urban fringe. In addition, this zone will often have individual buildings and small, semi-rural properties marked. Where a number of individual services are concentrated in the built-up area and marked with conventional symbols (such as the post office, police station, public hall, fire station and hospital) then it may be inferred that this area represents the main commercial zone of a town.

## Interrelationships between the biophysical and cultural landscape

The study of biophysical and cultural features on topographic maps can reveal a number of 'cause and effect' relationships. These provide the basis for interpreting the landscape in order to identify significant geographic patterns and processes. Some common interrelationships are discussed in the following sections, along with typical examples that can be seen on topographic maps.

### *Vegetation, drainage and relief*

Variations in vegetation may be accounted for by the different landforms found within a region. The influence of coastlines, valleys, steep slopes and areas of highland may be associated with changes in vegetation types. The landforms may in turn have different soil types, which will also produce variations in vegetation. Sandy coastal soils and sand dunes will often have shrub or heath vegetation while mountainous regions and valleys may contain different types of forest. The influence of drainage patterns can also account for differences in vegetation. Rivers, estuaries and lakes will tend to have distinctive vegetation patterns that can often be seen on topographic maps. This fringing or riparian vegetation will often contrast with the vegetation patterns on land away from the water body. Intermittent rivers in arid regions will have permanent trees, which survive on the water that soaks into the ground after rainfall. Areas of poorly drained land such as freshwater swamps or mangrove swamps are also associated with specialised vegetation types.

Relief and drainage are closely related. As already discussed, landforms can produce a variety of drainage patterns. The dominant slope of the land will direct the surface water flow, while the permeability of the soils and geologic structures will affect the number and shapes of the drainage channels that can be seen. Rocky, impermeable, steep landforms will have a large number of surface drainage features, while sandy, flat landforms will have few surface channels as any rainfall will tend to infiltrate rather than run off. Low-lying sand regions may also have extensive areas of swamps or wetlands if they are located in humid climates. These features are often surface expressions of the watertable.

## Settlement, landforms and drainage

Features such as transport systems, rural land use and urban settlements are often strongly influenced by landforms and drainage systems. Valleys can be attractive locations for the development of agriculture and the siting of towns. Here the presence of a water supply and alluvial soils support these activities. Valleys also provide a cost-effective way for transport systems to pass through areas of highland or to connect areas of lowland and plateau. Their contour patterns reduce the gradients of slopes and provide better access for roads and railways.

The shapes or external morphology of towns may be affected by landforms, which either encourage or restrict the growth of settlement. This would include the effects of coastlines, rivers, lowlands or basins, and areas of highland. Coastlines may encourage settlement, with attractive views and potential for economic and recreational activities to occur. Rivers often provided suitable sites for towns and cities, but in some cases the potential for flooding can also be seen as a disincentive. Mountains and steep slopes are costly to build on and to service and areas such as these tend to be avoided.

Transport systems will reveal distinctive patterns when constructed on different landforms. In steeply contoured regions, the roads and railways will often closely follow the shapes of the contours, while in flat or undulating areas the transport network will tend to produce a grid or straight linear pattern. Transport systems and land uses will also be closely linked. Intensive forms of rural land use will have numerous interconnected roads compared to the absence of transport networks in areas of extensive agriculture. Large urban settlements will have a variety of transport links radiating out into their hinterlands, while small settlements will have few connections.

## Sketch mapping

Sketch mapping involves the interpretation and pictorial representation of topographic map information. As such it is has several purposes:

- It tests a person's ability to correctly interpret information from a map.
- It is a valuable aid to describing landscapes.
- It assists in developing a general overview of the complex information contained on a topographic map.

Where a sketch map focuses on one particular element of the landscape then it may be referred to as a précis map. Précis maps include maps of settlement types and density, vegetation, landforms and rural land uses. Sketch maps can also be produced to highlight the relationships between significant features of the landscape. This might include landforms and road patterns or rural land use and vegetation. When a sketch map is drawn it should be presented clearly, with a legend or labels to identify the main elements. While it is not necessary to maintain an exact scale, the map should retain the proportions and general shapes of the elements being mapped. Where small exceptions to the general pattern occur, these can be ignored in order to maintain an emphasis on simplicity and clarity. In figure 9.24 landforms, land uses, major transport systems and settlements have all been mapped. The area of forest may have had several small zones of intensive grazing, but it is shown as a continuous band in order to illustrate its general appearance and location. Note that the map has been divided into quadrants. This helps to retain the proportion, shape and position of the features.

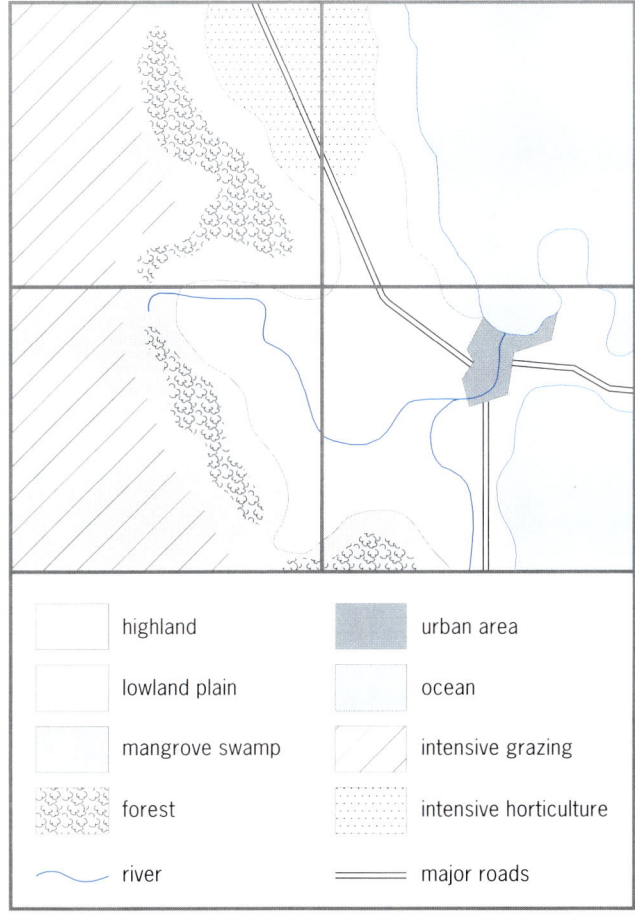

**Figure 9.24** Hypothetical sketch map

### EXERCISES

1  Calculate the vertical exaggeration for the scales provided in the following table.

| | Vertical scale | Horizontal scale |
|---|---|---|
| A | 1:50 000 | 1:100 000 |
| B | 1 cm represents 50 m | 1:25 000 |
| C | 1:100 | 1 cm represents 50 m |
| D | 2 cm represents 100 m | 5 cm represents 2.5 km |

*Answers VE × 5, VE × 10, VE × 2, VE × 50*

2 Using figure 9.19, construct a cross-section from point 6 to point 5 with a vertical exaggeration between 5 and 10. The map scale is 1:20 000.
3 Using figure 9.19, calculate the gradient between point 6 and the spot height at point 4.
4 Calculate the gradients for the information provided in the following table.

|   | Altitude | Map scale and distance |
|---|---|---|
| 1 | A is 139 m and B is 290 m | 1:20 000. 5 cm on the map |
| 2 | Difference in height = 400 m | 1 cm is to 1 km. 2 cm on the map |
| 3 | A is 700 m and B is 100 m | 1:50 000. 4 cm on the map |

*Answers approximately 1 in 3, 1 in 7, 1 in 5*

5 Draw the contour patterns for a saddle or col, summit, steep slope, valley and a spur. For each draw cross-sections to show how they would appear in profile. You would need to draw two of these for the saddle, valley and spur in order to show their shape—one cross-section would go along the feature while the other would cut across it.
6 Either use a topographic map or draw your own contour pattern to produce a map that has the following fluvial features:
   - Watershed and basins
   - Dendritic drainage patterns
   - Radial drainage patterns
   - Oxbow lakes
   - Meanders
   - Flood plain
   - Gorge
7 Using a series of topographic maps, identify and describe the ways in which the road and rail systems appear to have been influenced by the contour patterns. Also suggest ways in which the construction of these transport systems may have been accomplished by the alteration of the landforms (cuttings and embankments).
8 Identify and describe the biophysical factors that either encourage or discourage the development of the cultural landscape as illustrated on topographic maps.
9 Draw a sketch map of the Northam area provided in the topographic map (figure 8.42) and include the following:
   - major road and rail systems
   - urban centres
   - major drainage patterns
   - areas of medium forest or vegetation cover
   - areas of sparse vegetation cover

# Statistical analysis

Statistical analysis and measurements provide important tools for geographers in the study of patterns and processes. The basic statistical measures used by geographers and their particular advantages are discussed further in the following sections.

## Measures of central tendency

The **mean**, **mode** and **median** are measurements of central tendency. They provide information on the central point within a range of data. Each one is a slightly different way of providing a summary of this point.

The *mean* is an average and is often used to illustrate similarities and differences between sets of data. The average life expectancy of men (76) and of women (80) in Australia is one such application. The mean suggests that the majority of cases will be found around this number and it is therefore considered to be reasonably accurate in predicting or representing an outcome. The mean is calculated by totalling all the scores or numbers in a sample and then dividing this by the number of scores. A mean is considered to be most appropriate when the sample has an even distribution above and below the average.

Where the sample is **skewed** then it is sometimes more appropriate to use the *median* score. For example, rainfall in Australia for different locations can vary significantly from year to year. Ten years of exceptionally heavy rainfall can distort the results from 50 years of recordings. The result will be skewed towards a higher result. The use of an average in this case will result in a score that is not indicative of the true situation. The median on the other hand takes the score where 50 per cent of the results fall above and 50 per cent below. Where the number of scores is even (for example, eight) then the median falls halfway between (in this case, between scores four and five). This then produces a more reasonable prediction. Medians are therefore used where individual results can significantly distort the sample.

The *mode* is another measurement of central tendency. It is the most commonly occurring score within a sample. For example, in the monthly temperatures recorded for a locality over the course of a year, five may be the same, while three are higher and four lower. In this case the mode will provide a reasonable degree of accuracy in predicting the type of temperature that might be experienced in any particular month for the location. Thus the mode is used when there is little variation in the sample data.

## Measures of variability

The spread of scores or data either side of the measure of central tendency (mean, median, mode) illustrates the variation in the statistical information. A common measure of this information is the range—the difference between the maximum and minimum numbers or scores within a set of numbers. Thus, the range of temperature for a day is the difference between the daily maximum and the daily minimum temperature. This is also referred to as the diurnal range.

Other measurements of variability include the decile and the standard deviation. These measurements provide an indication of how widely spread the sample data are on either side of the mean. In understanding the rainfall variability for a given climate the calculation of the standard deviation will show whether a climate has a large or small variation from one year to the next. This will determine the degree to which rainfall can be predicted. The use of deciles is a relatively simple way of indicating the variation in a set of numbers. The 5th decile is the median—50 per cent of the scores fall below this decile and 50 per cent are greater than it. The 1st decile has 10 per cent of the scores below and 90 per cent above, while the 9th decile has 10 per cent of the scores above and 90 per cent below. In table 9.2 the two locations have a different spread of scores and the 1st, 5th and 9th deciles indicate these. In location A there is a 90 per cent chance that the temperature will exceed 20°C and a 10 per cent chance that it will exceed 30°C, while for 80 per cent of the time it will fall between 20°C and 30°C. The range or spread of data between the 1st and 9th deciles is 10 degrees. Location B has a much greater spread or range at 20 degrees and is therefore a location with a greater expected temperature variation.

**Table 9.2** Temperature deciles

| Location | 1st decile | 5th decile | 9th decile |
|---|---|---|---|
| Location A | 20°C | 25°C | 30°C |
| Location B | 5°C | 15°C | 25°C |

## Measurements of correlation

An important undertaking for geographers is the study of interrelationships. Correlation statistics are a useful tool in determining the strength of the relationship that exists between two or more variables. A standard statistical method used to test the strength of the relationship between two variables is the rank order (rank difference) correlation coefficient. This method is very useful in testing the relationships between functional complexity and population in central places, or the size of farms and total annual rainfall within different localities. The comparison of two sets of data using this method may reveal a positive or a **negative correlation**. For example, the increase in the distance from a market point may correspond to a decrease in the value of rural land. Thus a negative relationship has been observed and tested. Alternatively, an increase in the value of urban land may relate to an increase in the density of housing. Both are increasing thus indicating a **positive correlation**. The rank order correlation ranks the data from two variables and then sees how closely the rank order of the first sample matches the rank order of the second sample. A perfect match produces a score of +1 for a positive correlation and −1 for a negative correlation. The formula for rank order correlation is:

$$r = 1 - \frac{6 \Sigma d^2}{n^3 - n}$$

In this formula the rank order correlation is **r**. The sum of the differences squared is multiplied by six. The number of items in the sample is represented by **n**. The result will be a value between −1 and +1, which is then deducted from 1 to give the final result. In table 9.3 the strength of the relationship between two variables is tested using the rank order formula or coefficient.

In this information there are nine towns ranked and the sum of the differences squared is 10. The correlation coefficient is 0.917. This is very close to 1 and represents a very high positive correlation between the population of the towns and the number of functions. The worked example is provided below.

$$1 - \frac{6 \times 10}{729 - 9} = 1 - 0.083 = 0.917$$

The relationship between the two variables can be graphed, with the strength of the correlation being illustrated by the closeness of the individual plots to the **line of best fit**. In figure 9.25 a positive relationship or correlation is indicated by the slope of the line of best fit upwards to the right. If the line were sloping down to the right then it would be illustrating a negative correlation.

## Tables and graphs

The use and interpretation of tables and graphs is an important geographic skill. Tables provide a range of ways in which statistical data can be organised and presented, while graphs allow this information to be converted into visual or pictorial representations. Tables can provided raw data as well as refined or standardised data. Graphs represent this data by the use of various scales and techniques.

### Tables and data

Data contained in tables may be presented in a variety of ways, depending on its purpose. The most basic form of presentation is as raw data. The total number of people within a statistical area is one example of raw

**Table 9.3** Size and functional complexity of selected central places

| Central place | Functions Number | Rank | Population of central place Number | Rank | Difference in rank | Difference squared |
|---|---|---|---|---|---|---|
| Slittenton | 15 | 7 | 120 | 5 | 2 | 4 |
| Braybrook | 25 | 4 | 156 | 4 | 0 | 0 |
| May | 10 | 8 | 45 | 9 | 1 | 1 |
| Terry | 17 | 5 | 87 | 7 | 2 | 4 |
| Durack | 42 | 2 | 520 | 2 | 0 | 0 |
| Martin | 30 | 3 | 300 | 3 | 0 | 0 |
| Willett | 5 | 9 | 50 | 8 | 1 | 1 |
| Lane Cove | 123 | 1 | 1125 | 1 | 0 | 0 |
| Robinsonville | 16 | 6 | 98 | 6 | 0 | 0 |
| Total | | | | | | 10 |

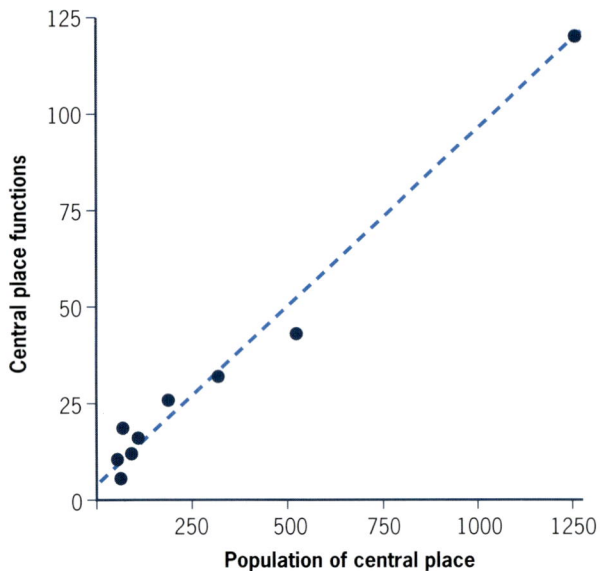

**Figure 9.25** Scatter graph of central place population and functions

data. Sometimes this type of data may be abbreviated or rounded and then summarised by being expressed in hundreds, thousands or millions. The conversion of raw data to percentages provides a method of comparing proportions or rates where the raw data from two or more samples are of significantly different amounts. Relative or comparative sizes are more clearly illustrated. An example is the number of adults classed as literate in Australia and India. There is a large population difference between the two countries (988 million in India and 19 million in Australia in 2001) and the comparison has little value until converted to a percentage. Numerically, more adults in India can read. However, in percentage terms, Australia has 100 per cent literacy compared to 52 per cent in India.

A further refinement may be a comparison of population numbers to regional area to produce a density ratio in which the number of people per common unit of area is calculated. The ratio may compare the average number of people per hectare or square kilometre. Another form of ratio is the rate per common unit. Birth and death rates are often expressed as a rate per 1000 people. In this way, by using the same standard, vastly different populations can be compared. Another type of refinement is to compare the value or amount of some item per person or per capita in a process of averaging. Examples of this method include gross domestic product (GDP) per person or motor vehicles per person. Rates of change in data can be illustrated using percentages or alternatively indexes. With the use of an index a base year is given a value (usually 100) and movements above and below the base are then calculated. The consumer price index (CPI) is calculated from a base of 100. Different stock exchanges also use an index to describe movements up and down in the value of shares. In table 9.4 different types of statistical data have been presented for three countries. By processing this information it would be possible to calculate the area of each country in square kilometres (population divided by density), find the total births for 1998 (population divided by 1000 times the birth rate) and the number of people living in urban areas (population times the percentage divided by 100).

**Table 9.4** Types of data by country, 1998

| Country | Population (raw data) | Density (average per sq km) | Births (rate per 1000) | Urban population (percentage) |
|---|---|---|---|---|
| Australia | 18 500 000 | 2.4 | 13 | 84.7 |
| Japan | 126 300 000 | 334.3 | 10 | 78.2 |
| Indonesia | 206 400 000 | 107.5 | 24 | 36.4 |

## Line and scatter graphs

Line graphs are used to illustrate continuous data such as temperature, population growth, air pressure changes and levels of output or production. A single line graph shows the changes for one variable while comparative lines illustrate a number of related variables. Compound line graphs are used where the different components of a total need to be illustrated. In figure 9.26 various types of line graphs have been constructed using the hypothetical data provided in table 9.5. The data used have been based on a series of sample points (in this case years) and the lines connect these points. The construction of a scatter graph uses a similar process with points being plotted against the $x$ (horizontal) and $y$ (vertical) axes.

**Table 9.5** Population data by region, 1900–2000

| Region | 1900 | 1920 | 1940 | 1960 | 1980 | 2000 |
|---|---|---|---|---|---|---|
| Region A | 555 | 1 235 | 2 345 | 3 324 | 8 900 | 10 230 |
| Region B | 345 | 987 | 3 456 | 4 460 | 7 800 | 13 480 |
| Region C | 456 | 789 | 1 200 | 3 490 | 3 450 | 4 500 |
| Region D | 890 | 2 350 | 4 540 | 7 890 | 12 800 | 17 890 |
| Region E | 245 | 487 | 567 | 987 | 1 200 | 2 580 |

## Bar graphs

Bar graphs are generally used to illustrate discrete data. This is data that are non-continuous or based on separate times or places. Rainfall totals for each month of the year can be shown by a bar graph. Unlike temperature data, these data are not continuous. Likewise populations for different countries, or agricultural output from different farms, can be shown by the use of bar graphs. The bars may be vertical (columnar) or horizontal (lateral) in their construction. Like line graphs they can be single, comparative or compound representations. In figure 9.27 (p. 240) various types of bar graphs have been constructed using the hypothetical data provided in table 9.6.

**Table 9.6** Crop production by region (tonnes)

| Region | Wheat | Barley | Canola | Field peas |
|---|---|---|---|---|
| Region A | 45 000 | 23 000 | 12 340 | 4 500 |
| Region B | 18 000 | 15 600 | 34 000 | 2 400 |
| Region C | 14 500 | 22 000 | 4 500 | 6 570 |
| Region D | 21 500 | 13 450 | 12 340 | 1 250 |
| Region E | 33 500 | 19 500 | 4 400 | 14 500 |
| Total | 132 500 | 93 550 | 67 580 | 29 220 |

PRACTICAL MAPPING AND RESEARCH SKILLS 239

(a) Population of region C, 1900–2000 (Single line graph)

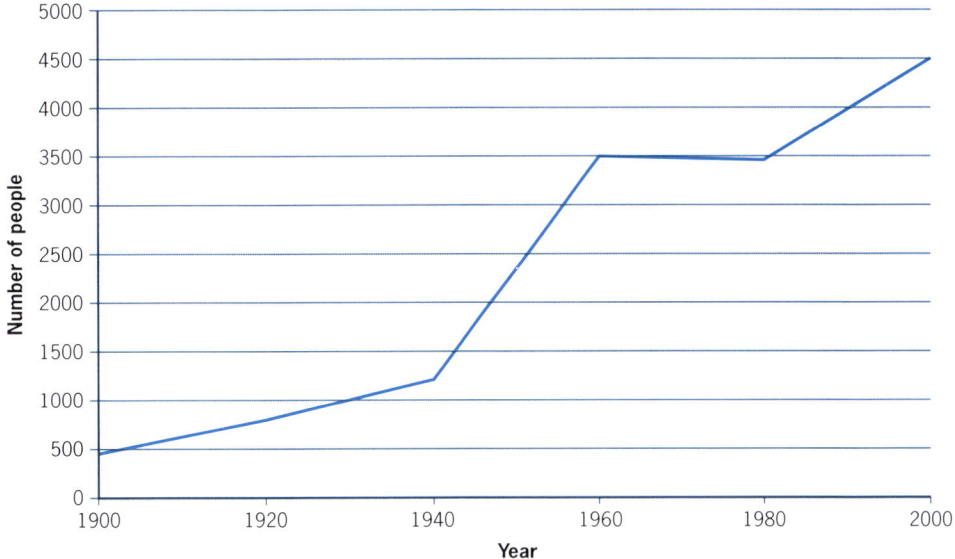

(b) Population by region, 1900–2000 (Comparative line graph)

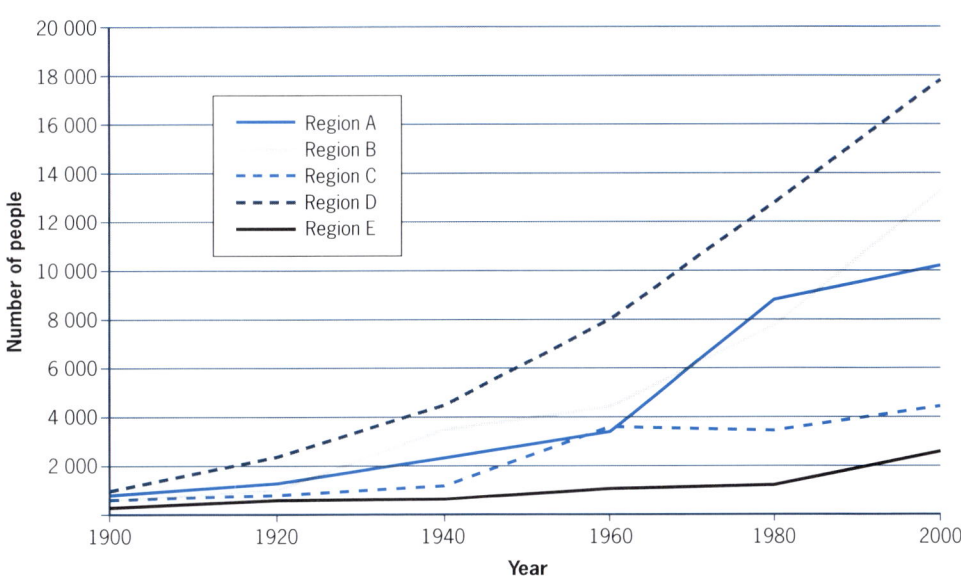

(c) Total population, 1900–2000 (Compound line graph)

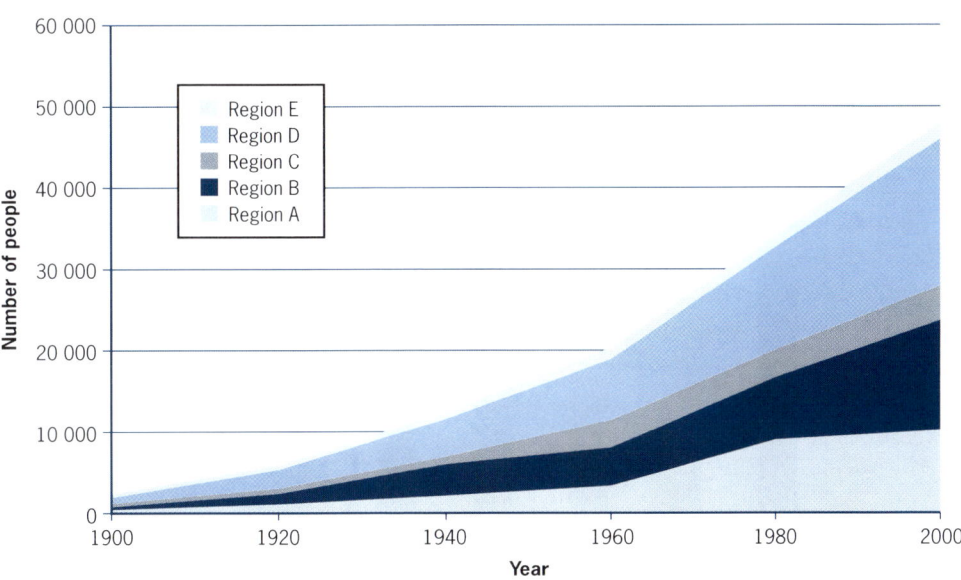

**Figure 9.26** Types of line graphs

## Proportional circle graphs

Proportional circle graphs are constructed by using the area of a circle to represent data. In the case of populations for different locations, the radius of the circle is set by the square root of the total population. Hence a population of 25 has a square root of 5. A scale that is appropriate for the square roots of all the different data is then selected and used to set the radius of the circles. In table 9.7 the size of a set of hypothetical regions is given in square kilometres, with the square root of each. The scale is 1 cm to 10 units (square root). A series of proportional circles is shown in figure 9.28.

**Table 9.7** Regional areas

| Region | Size (sq km) | Square root | Radius (cm) |
| --- | --- | --- | --- |
| Region A | 1 500 | 38.7 | 3.87 |
| Region B | 2 450 | 49.5 | 4.95 |
| Region C | 870 | 29.5 | 2.95 |
| Region D | 345 | 18.6 | 1.86 |
| Region E | 1 720 | 41.5 | 4.15 |

## Pie graphs

Pie graphs are also known as divided circle or pie charts. They illustrate the relative proportions or segments of a known total. The total of a circle chart is equal to 100 per cent or 360 degrees. Each segment of this total is therefore a part of 100 per cent or 360 degrees. In drawing a pie graph the first step is to convert the data to degrees in order to plot the segments using a circle protractor. The starting point is a line drawn from the centre to the top of the circle. The data are then plotted cumulatively to the right or in a clockwise manner. As a rule the smallest segments are plotted first, working progressively up to the largest. This allows for any small inaccuracies caused by the rounding of degrees to be absorbed by the largest segment. If one of the categories is 'other', this should be left until last. In figure 9.29 the data for a set of hypothetical regions (provided in table 9.8) has been plotted to show the finished result.

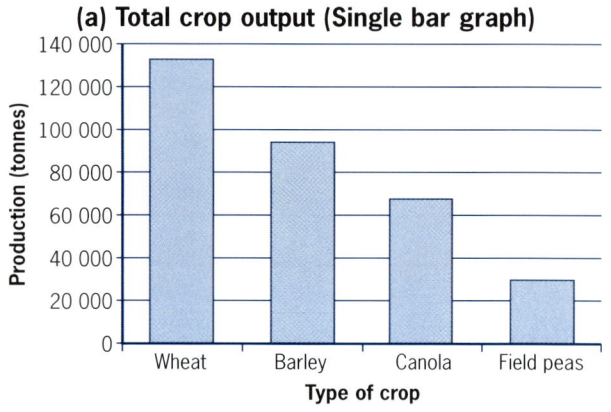

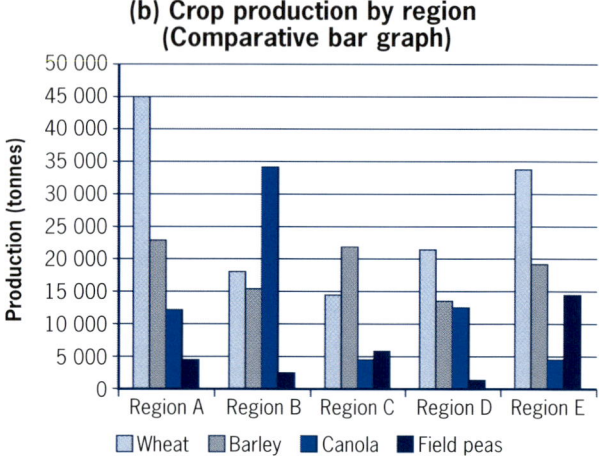

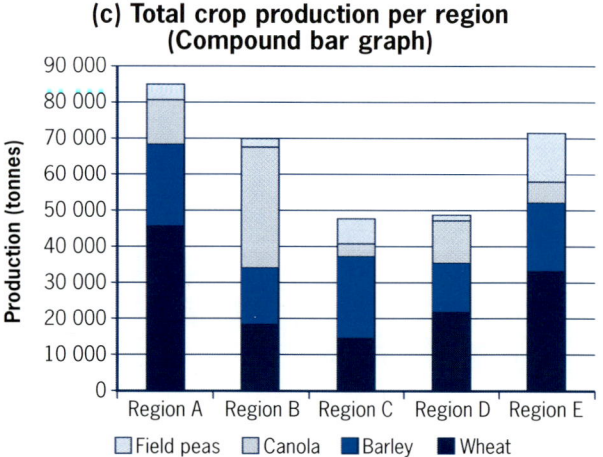

**Figure 9.27** Types of bar graphs

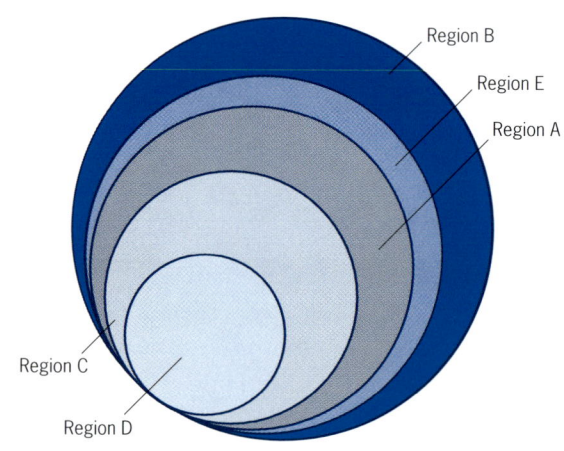

**Figure 9.28** Regional area proportional circles

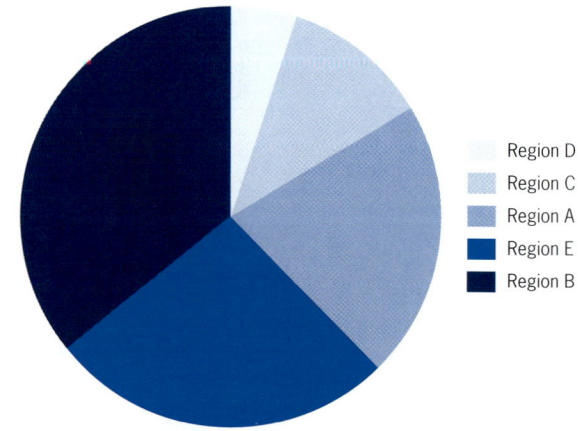

**Figure 9.29** Regional area pie graph

**Table 9.8** Regional areas ranked by size, percentage and degrees

| Region | Size (sq km) | Percentage | Degrees |
|---|---|---|---|
| Region D | 345 | 5 | 18 |
| Region C | 870 | 12 | 43.2 |
| Region A | 1 500 | 22 | 79.2 |
| Region E | 1 720 | 25 | 90 |
| Region B | 2 450 | 36 | 129.6 |
| Total | 6 885 | 100 | 360 |

### EXERCISES

1. Using the rainfall and temperature information provided in table 9.9, complete activities (a) to (f)). Note that this data can be entered into a spreadsheet program and then completed on a computer.

   **Table 9.9** Selected climatic data

   | Locality | Distance from the sea | Annual temperature range | Annual median rainfall |
   |---|---|---|---|
   | Rottnest | surrounded by sea | 8 | 678 |
   | Fremantle | 1 | 9 | 766 |
   | Kalamunda | 20 | 11 | 1 061.2 |
   | Northam | 90 | 14 | 436.9 |
   | Cunderdin | 150 | 14 | 371.4 |
   | Southern Cross | 345 | 14.5 | 282.1 |
   | Leonora | 570 | 15.7 | 217.6 |
   | Warburton | 675 | 15.2 | 169.3 |

   *Note:* the distance from the sea has been measured from the nearest coastline and all stations fall between 26°S and 32°S latitude

   (a) Calculate the mean and the median for both sets of data.
   (b) Find the mode for the annual temperature range.
   (c) Find the maximum and minimum and the range for both sets of data.
   (d) Construct scatter graphs for rainfall and distance, and temperature range and distance, and plot the line of best fit.
   (e) Convert the rainfall and temperature data to percentages with the highest in each being 100 per cent. Plot these as comparative bar graphs and comment on any apparent relationships between the two sets of data.
   (f) Using the rank order correlation coefficient, compare rainfall amounts and distance from the sea by calculating the correlation.
   (g) Calculate the rank order correlation coefficient for temperature range and distance from the sea. Compare this result with the one for rainfall and comment on the type and strength of each correlation.
   (h) Using an atlas, plot these locations on a map of Western Australia and then see if latitude has had an influence on the data provided.

2. Using the population data provided in table 9.10, construct a proportional circle and a segmented circle (pie) graph.

   **Table 9.10** Population of Australia's states and territories, 2001

   | | Population, Dec. quarter 2001 ('000) |
   |---|---|
   | New South Wales | 6 642.90 |
   | Victoria | 4 854.10 |
   | Queensland | 3 670.50 |
   | South Australia | 1 518.90 |
   | Western Australia | 1 918.80 |
   | Tasmania | 473.3 |
   | Northern Territory | 199.9 |
   | Australian Capital Territory | 322.6 |

# Photographs

Photographs provide geographers with a variety of information on the biophysical and the cultural landscapes. Photographs record spatial arrangements, changes over time and variations from place to place in geographic features as well as the way in which these features interact. Used in conjunction with other research techniques they help to develop insights and understandings of geographic patterns and processes. Photographs can present a variety of views or perspectives, ranging from the vertical aerial photo to the oblique and the ground level view.

## Aerial photographs

The vertical aerial photograph is taken at 90 degrees to the horizontal and produces a plan view (see figure 9.30). A camera mounted beneath a plane takes a series of photographs in runs, which overlap to produce a mosaic. As the plane passes over the landscape it maintains a constant height and the photographic images are recorded at a scale set by this altitude. Aerial photographs will have the scale recorded on the margin of the image.

Orthophoto maps combine the elements of a topographic map with an aerial photograph. Names, contours and grid systems may be superimposed on the photo. Elements of the photo may also be enhanced to highlight certain features. These could include drainage systems, road types, land uses and different vegetation formations.

### Scale and measurement

There are two methods to calculate the scale for a vertical aerial photo. One is to compare the **focal length** of the camera to the height at which the photograph was taken. For example, a plane flying at 3000 metres above sea level carrying a camera with a focal length of 150 millimetres (15 centimetres) would produce an image at a scale of 1:20 000. This scale is calculated by converting the height to centimetres and then dividing it by the focal length of the camera. Alternatively, the distance between two objects on the ground being

**Figure 9.30** An aerial photograph of Sydney CBD and Harbour Bridge

photographed can be measured and compared to the distance between the same two features on the image. When the distance on the ground is divided by the distance on the image it will produce a representative fraction. Both methods are shown in the following formulas.

$$RF = \frac{\text{Camera focal length (cm)}}{\text{Altitude of plane (cm) ASL}} \qquad RF = \frac{\text{Distance on photo point A – B (cm)}}{\text{Distance on ground point A – B (cm)}}$$

### Interpreting aerial photographs

The interpretation of information on aerial photos requires access to a number of different landscape images. The distinctive features common to a number of different landscapes have been summarised below and should be studied in conjunction with appropriate aerial photographs.

*Urban landscapes*

- Closely developed transport systems with numerous roads, often forming grid patterns.
- Residential development is the most common feature. The spacing of these indicates the housing density.
- Features such as parks and open space can often be found.
- Commercial and industrial development may occur as distinctive nodes within the residential pattern.
- Special use features such as schools and hospitals often form distinctive clusters of buildings surrounded by car parking and open space.

*Industrial landscapes*

- Large rectangular buildings often feature in these images.
- Open storage space and stockpiles or containers may be identified.
- Transport systems are complex in nature and often show a relationship between different forms such as road, rail and sea. Extensive terminals and transshipment points may be found.
- Vegetation may be limited, with large areas of cleared ground.

*Rural landscapes*

- Fields often form a patchwork of colours and shapes, with their size regulated by the type and intensity of farming practised.
- Seasonal variations may be apparent, such as planting or harvesting activities.
- Permanent forms of plantings such as vineyards or fruit trees often form clearly defined rows.
- Farm buildings may occur as well-defined clusters within one part of the farmland.
- Natural vegetation will indicate varying degrees of modification depending on the land use. Extensive grazing will have a relatively large amount of vegetation left, while broad-acre grain farming will have extensive clearing with some remnant vegetation along roads, fences and streams.

## Oblique and ground level photographs

Oblique and ground level photographs provide a three-dimensional view of landscapes. The oblique image is taken from a high vantage point, or from an aircraft. It shows features looking down at an angle less than 90 degrees to the horizontal. Ground level photos provide a viewpoint looking towards the horizon. In both the oblique and the ground level view, objects in the foreground appear larger than those further away. This provides the observer with a perspective view of the landscape in which the most distant objects lie near the vanishing point or limit of a person's range of sight.

In both the oblique and the ground level image the observer can describe and interpret the landscape by referring to the foreground, middle ground and background. This can be further broken up into the centre, right and left, as illustrated in figure 9.31.

### EXERCISES

1. Find and list examples of aerial, oblique and ground level photographs provided in the text. Write down their figure number and caption.
2. Describe the main differences between the three different types of photographs.
3. Calculate the scale of an aerial photograph where the camera focal length is 100 mm and the altitude of the plane is 1000 m.
4. Find examples of urban and rural landscapes illustrated on aerial photographs. Sketch and label the cultural features that give the landscapes their distinctive appearance.

# Fieldwork

The application of geographic knowledge and principles to fieldwork case studies involves an understanding of different methodologies. Collecting spatial data, field sketching, photographic records, survey and sampling techniques and questionnaires are some of the methods used by geographers to carry out fieldwork activities.

## Spatial data

Spatial data are collected by observing and recording the location and frequency of objects within a study area. Grid recording systems can be used to count and locate various features, which can be mapped and categorised using the region-building steps discussed earlier in this chapter. A variation on this method is the use of a **transect** (where data are sampled along a predetermined axis and subsequently mapped to show interrelationships, frequencies and locations). In studying the variation in vegetation on the coastline an area on the beachfront may be marked out using a grid pattern and then analysed. Inside the grid area, vegetation features such as density, spacing, variety, height and common distinguishing characteristics can be observed. This method is then repeated at several locations further away from the beach until a pattern of change is built up. Alternatively the sampling of an area can be done using a band transect. In this instance a band of vegetation is marked out using two tapes or lines set at a predetermined width. Within this band, changes in the vegetation or other data are recorded along the length of the transect.

## Field sketching

The field sketch is used both to record information for further analysis and to illustrate significant geographic elements within the landscape. As such it is not meant to be a detailed reproduction, rather a visual record of these important elements and in particular how they might be interacting. In producing a field sketch the following steps should be followed:

- From a predetermined vantage point, observe the landscape and decide on the major elements to be recorded. In doing so, consider the purpose of the sketch—is it illustrating cultural or biophysical processes?
- Divide your paper into nine sections with lightly drawn lines. This is to locate the foreground, middle ground and background; right, centre and left. This will also help to maintain the spatial arrangement of the landscape.
- Focus on the most significant elements and sketch these first. In a rural landscape this may include farm buildings, rural roads, fences and stock watering points.
- Use labels or colours to identify the sketched elements, as well as brief notations on any processes that you can observe.
- Date your sketch and note any important factors that may have influenced the scene. Factors such as weather conditions, the day of the week and the season can all have a bearing on the landscape that is being observed.

## Photographic records

Photographs taken in the field provide a visual record of geographic patterns and processes. Their range of applications has been summarised as follows:

- *Evidence of change.* Changes occur within a variety of time frames. The action of waves on a beach is a relatively short time frame. In this instance the series of photographs could include the wave of translation, surf zone, swash and backwash. In the study of urban landscapes, changes in the pedestrian traffic over the course of a day will illustrate the variation between night and day. In the longer term the changing seasons and agricultural activities represent another example of photographic evidence of change.

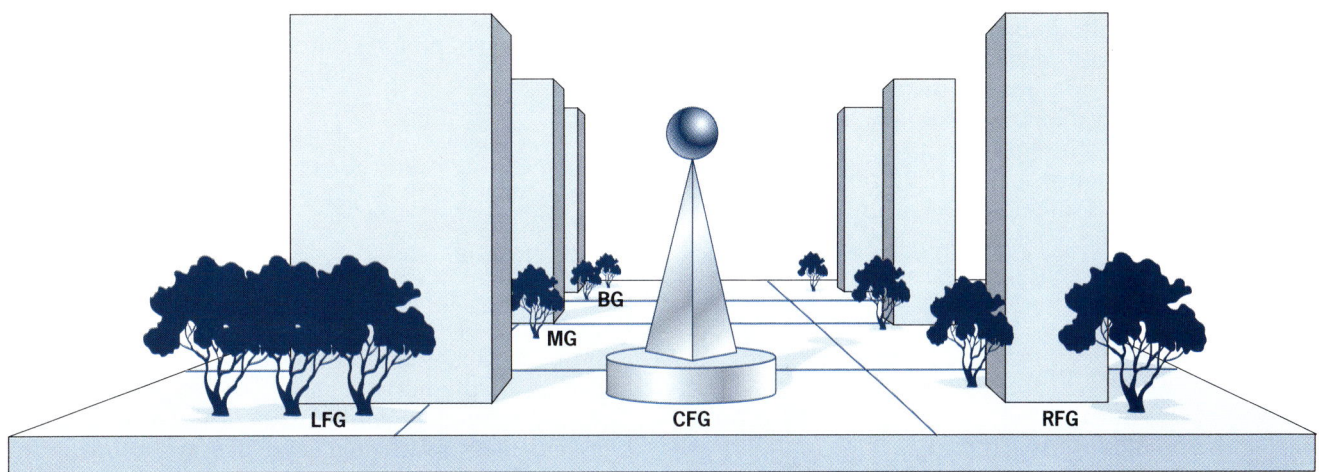

**Figure 9.31** Viewing three dimensional images

Figure 9.32 Oblique view of Sydney CBD and Harbour area

Figure 9.33 Ground level view of Sydney Harbour Bridge and CBD

- *Structures and functions.* Structural characteristics of built features in the landscape provide evidence as to the types of functions that are being fulfilled by these structures. Within the urban landscape photographic information can show the variations in land use zones such as industrial, residential, commercial and recreational functions. Urban blight and functional change are shown by changes or modifications to structures.
- *Spatial arrangements.* The relationship between different elements in the biophysical or cultural landscapes can be recorded photographically. The impact of landforms on roads, settlement patterns and agricultural land uses is just one area in which pictures can illustrate spatial patterns. Photographs can also provide visual records of transitions in land use activity. The patterns of the rural–urban fringe or the change from the central business district to the inner mixed zone are an example of this application.
- *Observing systems.* Systems are comprised of inputs, throughputs and outputs. Images illustrating these components and arranged in a model or diagrammatic form are a valuable means of illustrating fieldwork observations. Agricultural land uses are one example of this approach. In grain farming the inputs include biophysical and cultural factors, while the throughputs include the farming practices. Outputs such as the harvested grain crops and their transportation away from the farm illustrate the final process in this system.

## Analysis

Geographic analysis of information collected during fieldwork activities involves the application of theories, concepts and principles to the field observations and data. Analysis illustrates your ability to see processes and patterns and account for interrelationships between different factors. Analysis can be classified as spatial, trend or systematic.

### Spatial analysis

The observation and analysis of pattern on the land is central to the study of geography. This can involve the identification of regional patterns at a variety of scales. The differences in vegetation along a transect is one example, while the changing climatic patterns of a continent is another. Once a pattern has been observed and described the next step is to account for the variations. Such an approach requires the identification of causal factors. In the study of climatic variations, the identification and application of climatic controls illustrates this approach. Hence factors such as latitude, distance from the sea, altitude and landforms can be referred to in seeking to explain climatic variations.

Spatial analysis can also be achieved by the application of geographic theory. The land rent mechanism discussed in chapter 3 is one such example. It accounts for the variation in agricultural and urban land uses on the basis of the land user's ability to pay the rent or price for the land. The changes in land use that occur as you move away from the point of maximum land rent illustrate changes in the spatial patterns.

### Trend analysis

Geographic changes occur both spatially and over time. The direction of these changes can at times be identified. This then reflects a trend. Trends allow observers to predict changes based on past patterns. The long-term change in climate or the move towards higher density residential land use are trends. They allow predictions to be made as to the future possible landscapes. Spatial trends occur when a pattern of change is observed between two points. Thus the outward growth of the central business district in a large city is one such spatial trend. The shift towards regional shopping centres is both a spatial and a *temporal* trend.

Trends can be observed statistically. Where changes can be recorded then the pattern of the change will be reflected in the rate and direction in which the data move. These data when graphed will illustrate the trend. Rates of change can increase uniformly or *exponentially*. They may decrease slowly or at an increasingly more rapid rate. Figure 9.34 illustrates the different trends that can be observed in statistical information.

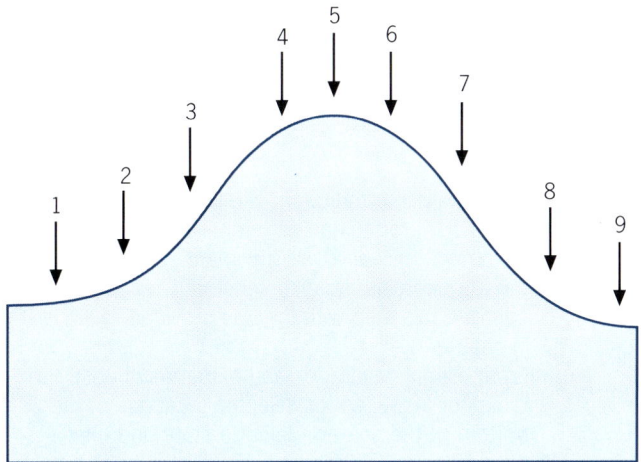

1 increasing slowly   2–3 increasing at an increasing rate
3–4 increasing at a decreasing rate   5 static
6–7 decreasing at an increasing rate
7–8 decreasing at a decreasing rate
9 decreasing slowly

**Figure 9.34** Rates of change—trends

### Systems analysis

As discussed previously, systems involve the identification of inputs, processes (throughputs) and outputs. Analysis of a system requires the observer to identify the three main elements and then identify the interactions or relationships between them. Analysis would involve identifying the relationships between the processes and the types and relative importance of the inputs as well as the outputs. For example, in an agricultural system, the application of technology to the various farming activities and processes will influence the type of equipment, materials and knowledge that are the input into the system. In turn, the processing that occurs will determine the nature of the outputs.

> **EXERCISES**
>
> 1 Discuss the importance of fieldwork in geography.
> 2 Identify the major methods of collecting and recording information during fieldwork.
> 3 Prepare a plan for carrying out a fieldwork study in either an urban, mining or agricultural investigation. In doing so include information on the following:
>   - the types of data that you could record and the methods of collecting this information
>   - the types of statistical analysis that could be done using this data
>   - the ways in which the landscape could be recorded and described
>   - the types of maps that could be produced or used
>   - the major types of geographic analysis that could be carried out with the information gained from the fieldwork study

### Web sites

The Internet is a valuable resource for the student of geography. It contains a number of highly relevant Australian and international sites and the following short list will provide a useful supplement to the information available in this book.

- *Australian Bureau of Statistics (ABS)*
  www.abs.gov.au
  The ABS provides a very wide range of statistical information on population, industry, agriculture and other social and economic trends. A large amount of information can be freely accessed, however some links can only be made for Auststat subscribers. In addition to statistics there are also summaries, fact sheets and thematic maps. Much of the statistical information can be downloaded and pasted into Excel or similar programs. This allows the user to produce graphs or carry out different statistical processes. Some useful activities include constructing population graphs and graphs showing the significance of different agricultural outputs by state and nationally.
- *Bureau of Meteorology*
  www.bom.gov.au
  The Australian Bureau of Meteorology web site contains wide ranging information on climate and weather at the national and regional level. Statistical data covering a number of climatic variables can be downloaded for hundreds of sites throughout Australia. This is formatted as comma separated values (CSV) files, which can be automatically pasted into Excel for further processing or graphing. The site also covers a number of climatic factors such as the El Nino effect, the southern oscillation index, climatic controls and long- and short-term weather patterns and climatic trends.
- *National land and water resources audit*
  www.nlwra.gov.au
  This site is a federal government initiative and is an excellent source of information on Australian agriculture and related environmental issues. It contains a variety of maps in a resource atlas that can be used as an interactive program. The information is very recent and is being constantly updated. Students wishing to seek information on the state of the nation's soils or the nature of individual farming systems should make good use of this site.

- *CSIRO plant industry*
  www.pi.csiro.au
  This site is quite technical. The information it presents covers a range of cutting-edge technologies in agriculture, and from that point of view is well worth visiting to find details on the application of genetic engineering, developing new plant varieties, sustainable use of soils and water resources and the packaging and preparation of produce for movement to the market.
- *Department of Petroleum and Mineral Resources, Western Australia*
  www.dme.wa.gov.au
  This site is maintained by the Western Australian Department of Petroleum and Mineral Resources. It is well laid out with an extensive range of information on the location and extraction of minerals within the state as well as details on the various legislative requirements in areas of environment, infrastructure development and land use. Students should also visit the Chamber of Minerals and Energy at www.mineralswa.asn.au/cme. This site is maintained by the industry representative body and contains some very informative fact sheets on mining activities within the state.
- *New South Wales Department of Agriculture*
  www.agric.nsw.gov.au
  This site covers a number of issues with regard to agriculture in New South Wales, including farming activities, environmental issues, new technologies and the biophysical environment as well as providing links to other related web sites.
- *Population planning in Western Australia*
  planning.wa.gov.au/publications
  This web site contains a large range of demographic information on Western Australia in portable data format (PDF) files that can be downloaded for use in class or for further research. Information covered includes population distributions, densities and projections, regional statistics and age and sex distributions. Detailed explanations and analysis of population trends are included. There is also information on the statistical methodologies and their relative degrees of accuracy.
- *Regional Development Council, Western Australia*
  regional.wa.gov.au
  This site contains wide ranging information on the local government areas (LGA) of Western Australia. Information provided includes population, employment and unemployment, mineral and petroleum production. PDF files can be downloaded as hard copy.

### EXERCISES

1. Access the ABS web site home page and click on 'Statistics'. Under latest population statistics go to population for states and territories. Using this information, complete the following activities:

   (a) Rank the states and territories by population growth rates from slowest to fastest.
   (b) Account for the variations in rates of growth.

2. Using the Bureau of Meteorology web site complete the following activities:

   (a) From the home page, choose climate and then climatic averages and extremes. Now click on 'Climate zones' and select climate zones based on the Koppen classification scheme. Compare and contrast this map with the one provided in the text in chapter 1.
   (b) From climatic averages and extremes choose the tables of averages for selected locations at the bottom of the screen. Select a state and then between 5 and 10 actual locations within the state. Using the statistical data provided for these stations, construct graphs for monthly median rainfall and monthly maximum and minimum temperatures. The stations should provide a spread of information to illustrate the climatic differences of the state. Compare this information with the text and web site climate maps and then describe and account for the climate characteristics of each location.

# Glossary

**acacia**  group of trees and shrubs belonging to the pea family (leguminosea); also known as wattles

**accessibility**  ease of access to a location

**acid soil**  soils deficient in bases such as calcium or sodium; with a pH less than 7

**acidity**  describes substances having a pH less than 7; not alkaline

**agglomeration**  the grouping together of different or related land use functions that benefit from each other's operations or in the use of shared infrastructure

**aggregation**  the grouping together of similar land use functions that benefit from the higher profile that comes from sharing a common location

**air mass**  a relatively large body of air in the atmosphere that is characterised by a uniformity of temperature and moisture

**alkaline**  material with a pH more than 7; not acid

**alluvial**  material such as soil or particular minerals carried and deposited by water on the land's surface

**alumina**  the mineral found within bauxite ore; a white powder, when refined

**aluminium**  the metal produced from the smelting of alumina

**Anglo-Celtic**  people of British, Irish, Welsh or Scottish descent

**aquifer**  rock occurring within sedimentary layers that is porous enough to hold ground water; often associated with artesian basins

**arable**  land that is able to be cultivated or farmed

**aridity**  describes regions that are arid in nature; having predominantly low unreliable rainfall

**average**  a measure of central tendency, found by dividing the total by the number of individual scores within a sample

**banksia**  member of the protea family, native to Australia; often characterised by large flowers and seed cones

**basalt**  igneous rock formed from the flow of lava on the Earth's surface or beneath the sea; fine grained as a consequence of its relatively rapid cooling

**bifurcation**  the separation of a stream into two or more branches or tributaries

**biome**  a major plant and animal community, usually corresponding to a climatic region

**brackish**  containing salt levels somewhere between fresh water and sea water; applies to surface or ground water

**butte**  small section of flat-topped higher land or remnant of an older plateau surface, with steep sides; smaller than a mesa

**calcareous**  applied to soils containing high levels of calcium; includes beach sands

**casuarina**  type of tree or shrub with thin jointed leaf structures; includes the desert oak, she-oak, forest oak and swamp oak but is not related to the European oak

**central business district**  major central commercial and business zone within a city

**central place theory**  model of urban settlements occupying discrete orders or positions within a hierarchy of settlements (urban network); also central place model

**centrifugal forces**  forces that move elements towards the edge. In the case of cities, means by which functions are pushed or pulled to the outer zones

**centripetal forces**  forces that move elements towards the centre. In the case of cities, means by which functions are pushed or pulled to the inner zones

**change of function** where a building or site undergoes a change in its activity or use

**concentric** describes a series of circles one inside the other

**continental block** a large region of crust which has a similar geologic structure or history. Continents are often composed of a number of these blocks

**conurbations** large urban areas formed by the amalgamation of two settlements

**craton** alternative name for a continental block

**cultural landscape** landscape containing evidence and elements of human land uses

**curvilinear pattern** when applied to settlements and road patterns describes their appearance if they form circular or semi-circular shapes along curved lines

**decile** unit of measurement based on the division of a sample into tenths

**decline** sloping downwards at an angle; opposite to incline

**deferred urban** describes land that is still rural but is set aside for future urban development

**demographic transition** changes that occur over time in the birth and death rates in a region; the transition has four distinctive stages

**dendritic** tree shaped. When applied to rivers it describes drainage patterns that have a main trunk and then branching tributaries

**diastrophic** tectonic forces causing folding and faulting of the Earth's crust. Also termed diastrophism

**diurnal** over the course of a day; for example the pattern of temperature from day to night

**dolerite** medium-grained basic igneous rock similar to basalt and often found in sills and dykes

**dolomite** rocks formed from the chemical alteration of limestone or calcium

**draw area** area from which central place functions draw customers; also known as the hinterland

**ecosystem** formed by combining living and non-living components of the biosphere into a distinctive community; can be of any size and duration

**endemic** unique to a given location or environment

**ephemeral** of temporary duration; often applied to plants in desert regions that germinate and flower following rainfall

**epicormic** describes how new shoots and leaves grow out of the trunk and main branches of a tree; a characteristic of eucalyptus trees following a fire

**epiphyte** plants that attach themselves to other plants in order to reach sunlight in rainforests; includes orchids and ferns

**Eucalypts** family of plants occurring naturally in Australia. Literally meaning 'well covered' due to the protective seed pod on the flower; also referred to as hard leafed woody plants or sclerophyllous

**eutrophication** the accumulation of nutrients within waterways as a consequence of fertiliser runoff from agricultural land. Leads to algal growth and de-oxygenation of the water, causing decreases in aquatic life forms or death

**exponentially** occurring at an accelerating rate

**farm amalgamation** the joining together of smaller farms to create larger ones

**fiord** long narrow inlet extending inland from the sea; produced by valley glacier

**fiscal policy** government policy of influencing the economy through the use of budget expenditure and taxation levels

**fluvial** associated with surface water movement in the form of rivers or sheet flow

**focal length** the distance between the surface of the film in a camera and the lens (a telephoto lens has a long focal length)

**formal region** an area classification based on similarities in the characteristics being observed e.g. a climatic region; also called a uniform region

**freehold** describes land ownership indicated by the holding of a title deed

**functional distance** the time taken to travel between two points

**functional region** an area classification based on movements of people, goods, energy, information between a hinterland or draw area and a nodal point; also called a dynamic region

**gangue** waste material of little value surrounding a metallic ore body or lode

**general industry** classification of manufacturing and fabricating applied to industrial activity or zone

**gentrification** small-scale renewal of older inner city buildings associated with an influx of younger property owners into the area

**geosyncline** a large depression or trough in the Earth's crust; a syncline on a very large scale

**gneiss** metamorphic rock type with distinctive banding, often found within igneous granites and occurring as a result of further melting or compression of this rock type

**graben** a rift valley formation occurring as a result of subsidence between two or more parallel faults

**gradational** describes the wearing down and movement of surface material by weathering, erosion and deposition

**granite** igneous rock formed from the slow cooling of magma beneath the Earth's surface

**greenfields developments** new housing estates or housing tracts developed on rural land or on land still containing its natural vegetation

**ground water** water found beneath the surface of the land

**heat island effect** increased local temperatures caused by the effects of urban land uses on atmospheric warming

**heavy industries** industries associated with large scale manufacturing, processing and fabrication

**hectare** an area measuring 10 000 square metres; there are 100 hectares to a square kilometre

**hinterland** area of rural population surrounding a central place and using the services that the town provides

**homestead** buildings associated with dispersed rural settlements, such as those found on pastoral leases

**horst** block mountain or section of land forced upwards between fault lines

**hydrophytes** plants with a high tolerance of water; occur in wetlands, estuaries, lakes and in rainforests

**inertia** the tendency for an object to remain stationary if still or to keep moving if in motion. Applied to land uses which tend to remain within an area after the initial advantages have declined or disappeared

**inner mixed zone** zone of mixed land uses around the central business district of an urban settlement

**intertropical convergence zone** zone of convergence between trade winds within the tropics corresponding to the zone of equatorial low pressure

**invasion and succession** the entry into and eventual occupation of a location, such as the entry of central business functions into the inner mixed zone

**isobar** line on a weather map connecting points with the same barometric air pressure

**isohyet** line on a map connecting points with the same amount of rainfall

**isotherm** line on a map connecting points with the same temperature levels

**karst** a limestone region in which most of the drainage is underground in caves and caverns; often appears dry and barren

**leasehold** describes use of property or land gained by a set period of rent. Pastoral leases are applied to Crown lands used for the grazing of animals such as sheep or cattle

**legumes** pea family of plants; have the capacity to convert atmospheric nitrogen to soil nitrate via the actions of bacteria found within the root system

**life expectancy** the age to which a person might expect to live at the time of their birth; set by the age that 50 per cent of the population reaches

**light industry** generally small-scale industry engaged in repair or fabrication; includes storage and warehouse functions

**lignotuber** woody root mass from which branches regenerate after fire or drought

**line of best fit** on a scatter graph the line through a series of points with an equal number of points above and below it

**linear** forming a line or ribbon pattern

**local government area** area under the control of a local shire or council

**location leader** industrial or commercial activity that will encourage other related activities to follow when it is established within a new location; steel and oil refineries are included in this group

**manufacturing milk** milk used to produce products such as cheese, milk powder, yoghurt and butter

**mass wasting** the movement of weathered material down a slope under the influence of gravity; includes landslides, soil creep and earth slump

**mean** the average of a set of scores calculated by adding them and dividing the total by the number of scores

**median** the middle score or fifth decile; found by ranking the scores and then finding the one in the centre

**megalopolis** structure formed by the growth and amalgamation of large urban areas; the best known is along the north-east coast of the USA, where a continuous urban area stretches some 600 kilometres from Boston to Washington

**melaleuca** group of trees or shrubs commonly referred to as paperbarks

**mesa** remnant plateau surface surrounded by steep sides and having a large raised flat surface; also known as a table top hill or mountain

**microclimatic** climatic conditions experienced within a small area; localised variations in elements such as temperature, rainfall, humidity, wind and sunlight

**mineral seam** mineral rich body of ore within a rock body

**mode** the score that occurs most often within a sample

**monadnocks** an isolated hill or body of rock that stands above the surrounding land, due to its greater resistance to weathering and erosion

**monetary policy** the use of interest rates and exchange rates to influence levels of income, output and expenditure within an economy

**monocentric** describes development around a single central point

**monsoons** seasonal reversals of trade winds within the equatorial regions. This is most pronounced between South-East Asia and Northern Australia

**negative correlation** an increase in the level of one variable corresponds to a decrease in the level of a second variable; for example, increasing distances from the ocean and decreasing levels of rainfall indicate a negative correlation

**nitrogenous** describes a soil nutrient derived from atmospheric nitrogen

**non-conforming rights** describes functions that have the right to continue to operate in an area due to their establishment before a change in the land use zoning; for example a transport depot in a residential zone

**orogeny** the forming of mountain systems by the collision of tectonic plates

**orographic** referring to mountain. The uplift of an air mass caused by higher land and resulting in higher rainfall on the windward side and a rain-shadow effect on the leeward side is the orographic effect

**outer business district** suburban or regional business district within a large metropolitan area; lacks the size and variety of the CBD

**palatable** able to be eaten

**pastoral** describes the raising of animals on natural or introduced pastures

**pedalfer** soil group associated with hot wet climates; deeply weathered and heavily leached with concentrations of iron and aluminium oxides in the lower levels

**pedocal** soil group associated with dry climates; characterised by calcium and salts in the upper levels and may be alkaline in nature where these minerals are deposited on the surface by high evaporation rates

**pH** measurement of the degree of acidity or alkalinity of sample material; a pH of 7 is neutral

**phosphorus** soil nutrient derived from rocks

**podsolic** describes soils that develop in cool and wetter conditions and are generally sandy and acidic in character

**polycentric** having many centres; development around more than one focal point

**positive correlation** an increase in the level of one variable corresponds to an increase in the level of a second variable; for example, increasing distances from the ocean and increasing temperature ranges indicate a positive correlation

**primary** first within a sequence. Primary industry is the production, collection or extraction of unprocessed raw materials

**public housing** housing provided by government agencies such as Homeswest in Western Australia

**push–pull factors** factors that repel (push) and attract (pull) land uses or activities

**quartz** igneous or metamorphic rock made up of silica, the most common element of sand; very hard

**quartzite** metamorphic rock formed from the compression of sandstone and containing high levels of silica

**quaternary** the most recent time period within the geologic time scale

**quota** a set amount of something; similar to a ration

**R codes** residential codes; planning codes or rules governing the design and construction of houses

**radial** describes a linear pattern extending out in a series of rays from a central point

**rain-shadow** region of low rainfall experienced on the leeward side of a zone of highland

**range** the distance or difference in values between two points

**refinery** industrial facility designed to concentrate or separate minerals or fuels from ore bodies or other source materials

**regolith** the weathered surface material making up the top layer of the Earth's crust

**rejuvenation** renewal or rebirth; occurs when younger people take up residence in an urban area with an ageing population

**ria** a coastal region which has been inundated or drowned by a rise in sea levels, creating long inlets that have formed in the valleys of rivers and are separated by the ridges and spurs

**ribbon** describes land use development that follows a linear feature; often applied to the development of commercial land uses along major roads in urban areas

**rift valley** a valley formed by the subsidence of land between two roughly parallel normal faults (the great rift valley of Africa has been produced by the separation of two continental plates)

**riparian** describes vegetation along a river bank or around a lake edge

**rural** describes structures and functions associated with primary activities

**rural–urban fringe** a zone of transition from urban to rural land uses around most urban centres

**salinity** the concentration of salts within the top layers of soil or in drainage systems

**sandstone** sedimentary rock formed from the consolidation of layers of sand; may be soft or hard depending on the amount of compression and the types of minerals that have cemented it together

**scats** solid pellets of waste material defecated by animals

**schist** metamorphic rock that has formed into thin layers due to intense pressure and heat; its origin may be sedimentary or igneous

**sclerophyllous** describes hard-leafed or woody plants containing a high number of scleroid cells; applied to most eucalypts

**secondary**  following primary; applied to manufacturing processes in the industrial context

**shale**  a sedimentary rock produced from clay and composed of fine particles; under pressure may be metamorphosed into slate

**siliceous**  describes the mineral silica, commonly found in quartz and as grains of sand

**site**  the physical landscape on which a building or city is constructed

**situation**  the relative position and importance of a locality

**skewed**  describes what happens when a sample that is biased towards a certain position produces a predetermined result and can affect statistical analysis of data

**smelter**  industrial facility processing refined minerals by melting to produce metals

**southern oscillation index**  the variation in air pressure above and below the mean experienced in the region between Darwin and Tahiti. Negative values are associated with the El Nino effect and drought conditions in eastern Australia

**space-extensive**  describes functions requiring large areas of land and often located on the edges of urban areas

**space-intensive**  describes high density land uses

**spatial**  describes the way in which objects are arranged on a surface

**succulents**  type of plants with fleshy leaves or stems, designed to store water; a type of xerophytic adaptation

**sunkland**  area of low land between higher regions, formed by subsidence or downwarping of the Earth's crust

**tariff**  a tax placed on imported goods and services

**tectonic**  describes forces originating from inside the Earth, including vulcanism and diastrophism

**temporal**  referring to time or sequence of events

**tertiary**  third level; tertiary industry is the provision of services

**thermal equator**  a line connecting points of maximum temperature and lying within the tropics. It shifts north and south with the changing seasons and deviates away from the equator where it passes over land

**total fertility rate**  the number of children born to women within the 15–44 age group and then divided by the number of women in that group to give an average

**trade capture**  the movement of consumers from the hinterland of one central place to another central place in response to greater choice or variety of central place goods and services; occurs where a smaller centre is near to a larger one

**transect**  a line along which data is recorded and collected

**trellis**  pattern based on a single main line and a series of minor lines branching on either side

**undulating**  gentle to moderate rise and fall, such as the rise and fall associated with alternating hills and valleys

**urban**  describes structures and functions associated with secondary and tertiary activities

**urban blight**  evidence of poor quality urban structures and their deterioration

**urban primacy**  describes a central place within an urban network that has achieved a high level of dominance; measured by population and number of functions offered

**urban redevelopment**  small-scale redevelopment of urban areas which may have experienced urban blight, or have older land use activities that are no longer in operation; usually initiated by private landowners or developers

**urban renewal**  redevelopment of areas of urban blight, usually initiated by government agencies and occurring within a relatively large area; can involve the redesign of roads and the introduction of new infrastructure

**urban shadow effect**  the deterioration of rural structures in areas adjoining an urban centre resulting from land use conflict and an expectation of an eventual invasion of urban land uses

**urban sprawl**  the low-density outward growth of urban areas into the surrounding rural lands and natural ecosystems

**vegetation formation**  distinctive vegetation zone or biome

**vulcanism**  the external and internal movement and deposition of molten material coming from the Earth's mantle

**watershed**  the boundary between two or more drainage basins; marked by high points on the landscape such as summits, ridges and spurs

**watertable**  the line marking the boundary between the zone of infiltration and the zone of saturation in soils

**whole milk**  white milk and other liquid milk products; minimal processing is involved and there is no loss of bulk as is the case with manufacturing milk

**work complementary**  refers to the scheduling of multiple work tasks so that the quiet time for one activity corresponds to the busy time for another

**xerophytic**  specialised characteristics of plants that have allowed them to tolerate low and unreliable rainfall regimes; these plants are called xerophytes

# Credits—photos, charts, statistical data

**Cover image:** Fred Williams 1927–1982 Australia
*Iron ore landscape* 1979
gouache on paper
56.0 x 76.2 cm
Presented through the NGV Foundation by Rio Tinto Limited, Honorary Life Benefactor, 2001
National Gallery of Victoria, Melbourne

All photos, with the exception of the following, were supplied by the author.

All ABS data is used with permission of the Australian Bureau of Statistics (www.abs.gov.au).

**1.4** WA Curriculum Council Geography Broadsheet 1997, modified by author; **1.8** © Manfred Gottschalk, Lonely Planet Images; **1.9 (a)** and **(c)** D. Lane; **1.11** Devonian Ridge, J. Barnes; **1.12** Australian Picture Library; **1.13** © Diana Mayfield, Lonely Planet Images; **1.13** Tourism South Australia; **1.16** and **1.17** N. Rowell; **1.19** Australian Picture Library; **1.20** D. Lane; **1.28** Australian Picture Library; **1.29** and **1.35** D. Lane

**2.7** Australian Tourism Commission; **2.8** D. Lane; **2.18** © Commonwealth of Australia reproduced by permission

**4.1** Tertiary Entrance Examination, 1998 Geography Broadsheet Side 11. Curriculum Council of Western Australia; **4.4** David Melling; **4.6** John Deere; **4.7** New Holland Australia; **4.8** Australian Picture Library; **4.12** J. Kalajzich; **4.15** John Deere; **4.16** © Mitch Reardon, Lonely Planet Images; **4.20** D. Lane; **4.21** J & N Reeves; **4.22** Australian Picture Library; **4.23** © Mitch Reardon, Lonely Planet Images; **4.29** and **4.30** Photos courtesy of Bega Cheese; **4.33** News Limited; **4.38** GAWA

**5.1**, **5.6** and **5.7** D. Lane; **5.9** Department of Minerals & Energy WA; **5.10** D. Lane; **5.11** S. Lavery; **5.13 (top)** D. Lane; **5.13 (bottom)** Kalgoorlie Consolidated Gold Mines; **5.14** and **5.15** D. Lane; **5.20** Argyle Diamond Mine Pty Ltd; **5.26** R. Green

**7.3** Australian Picture Library/Corbis/L. Clarke

**8.27** GAWA; **8.33** Aerial photography reproduced by permission of the Department of Land Administration, Western Australia, CL77/2002; **8.40** and **8.42** Reproduced by permission of the Department of Land Administration, Western Australia, CL93/2002; **8.41** Copyright Commonwealth of Australia, Geoscience Australia. All rights reserved. Reproduced by permission of the Chief Executive Officer, Geoscience Australia, Canberra, ACT

**9.30** and **9.32** Land and Property Information NSW. Courtesy Map Sales & Aerial Photography; **9.33** Australian Picture Library

# Index

Page numbers in **bold** print refer to main entries. Page numbers in *italics* refer to figures and tables

## A

Aboriginal people
  early evidence of, 10
  farming practices of, 43, 47
  fire stick practices of, 43, 47
  mining activities &, 94, 112
  Perth urban renewal &, 191
  population of, **123**, 126
  urban development &, 191, **207–8**, 209
acacia, 23, 39, 41, 42, 45, 75, 105, 247
accessibility, **201**, 202, *203*, 247
acid rain, 119
acid soil, 27, **56–7**, 58, 63, 92, 247
acid water, **107**
adaptation, 23, 24, **25**, 26, 41, 42, 45
Adelaide, 16, *122*, 123, 150
Adelaide Geosyncline, 6
adiabatic cooling, 17
Aeolian landforms, 31, 45
aerial photographs, **241–2**, *242*
age-sex distribution, **128–9**, *129*, *130*, *133*, 135, **136–8**, *137*, *138*, *139*, 158
ageing population, 128, **133**, *133*, 136, **137**, *138*, *139*, 173
agglomeration, **150**, **177**, **199–200**, 201, 202, *203*, 211, 247
aggradation, *3*
aggregation, 199, **200**, 211, 247
agricultural chemicals, *64*, 65, 66, *66*, **70**, 91
agricultural systems, **47–9**, *47*
agriculture, 94, 95, 103, 110
  case studies in, **59–93**
  development of, **61–3**, **149–50**, 155, 157, 159
  history and politics &, **61–3**, 65, **74**, **85**, **149–50**
  intensive/extensive types of, **49–50**, 51, *51*, *60*, 61, **71–93**, **124–5**, 126, *126*
  issues in, **55–8**, *56*, *57*, **70–1**, **77–9**, **91–3**
  locational influences on, **50–3**, *51*, *52*, **59–63**, *60*, *61*, *62*, **72–4**, *72*, *73*, **74**, **80–4**
  mapping and research skills &, 232, **234**, 235, 242, 244
  population distribution &, 123, *123*, **124–6**, *125*, 131, **134**, 135
  rural cultural landscapes &, **53**, *54*, **55**, *55*, **67–9**, *68*, *69*, **90–1**, *91*, **145–6**, *146*
  rural-urban fringe &, **197**, 199, 204
  vegetation &, 23, 39, 41, 42, 43, *47*, 48, *52*, 57, *57*, **61**, **64**, 68, 73, 75, *75*, **77**, *77*, **86–7**, 91
  *see also* farm amalgamations; farming inputs; farming outputs; farming throughputs; irrigated agriculture
air masses, **17**, 247
air pollution, 49, 105, **107**, **119–20**
Albany, *36*, 134, 135, 155, *157*, 159, *160*, 181
Alexander Heights, 194
algal growth, 55, 56, 91
Alice Springs, 15, *122*, 123, 126, *150*
alkaline soil, 27, 247
alluvial deposits, 5, 8, 31, 32, 33, 34, 45, 48, 50, 81, 97, 123, 235, 247
alluvial diamonds, 108, 109, **110**, 111, 113
alluvial gold, 94, 97, *97*, **99**, 105
alumina, 94, 116, 247
aluminium, 45, 247
alpine vegetation, **26**, 27
altitude, 17, 27, 245
Amadeus Basin, 6
animals *see* cattle; fauna; introduced fauna; livestock; sheep
Antarctica, 1, 29
Archaean era, 3
area measurement (scale), **226**
Argyle diamond mine, 95, **108–9**, *108*, 110, **111–13**, *111*, *112*, *113*
Argyle Diamond Mines Pty Ltd, 110, 112, **113**, *113*
Argyle diamond pipes, **108–9**, *109*, 110, 111
arid landscapes, 4, **5**, **13**, **15**, *15*, **24–5**, 45, 46, 73, 77, 96, 123, 126, 234, 247
Armadale, 135, 182, *183*, **194**
Arnhem Block, 1

Arnhem Land, 16
Arnhem Plateau, 6, **8**
artesian water, 73, 76
  *see also* ground water
ash grey mouse, **43–4**
Ashburton region, 74
Atherton Tableland, 10, **11**, *11*
Augusta, 29, 134, 159, *160*
Australian Alps, 10, **11**, *11*, 16, 17, **26**, 27
Australian Bureau of Statistics web site, **245**
Australind, 159, *160*
averaging process, 238, *238*, 247
Avon regional urban network, **161**, *162*, **163**, *163*, *164–6*, **167**, *167*, *168*, **169–71**, **169**, *170*, *171*, *172*, **173**, 210, 218
Avon River, 159, 161, 171, 210, 213, *214*
Ayers Rock, 6, *7*, 15
azonal soils, 33, 34

## B

Balcatta, 204
Ballidu, *164*, 170
banksias, 22, 23, 39, *40*, 41, 42, 247
bar graphs, **238**, **240**, *240*
Barkly Tableland, 8, 25, 126
barley, *59*, 62, 66, *67*
Barracouta gas field, 95
barrier reefs, 8, *8*
Barrow Island, 95
basalt, 4, 11, 13, *30*, 32, 247
basins, **233**, *233*, 235
Bass Strait, 11, 13, 95
Bass Strait Islands, 25
Bassendean, 31, *32*, 33, 41, 196
batholiths, *2*
bauxite, 94, 95, 116
Bayswater, 182, 194, 196
beaches, 13, *30*, 31, *32*, 174, 232, *232*
Bega, 17, *18*, 80, 81, *81*, *82*, 83, *83*, **85**, **86**, 87, *87*, 88, 89, **91**, *91*, **92**, *92*
Bellenden Ker Range, 11
Beverley, 157, 159, *160*, 161, **165–6**, 170, 171, *172*, 173
BHP Billiton, 95
bifurcation, 233, 247

253

Binningup, 159, *160*
biomass, 20, 45, 55–6
biomes, 20, 247
biophysical environment, 148
    climatic controls &, **17**, *18*, *19*, 244–5
    climatic pattern &, **13**, *14*, **15–17**, *15*, *19*, **34–5**
    introduction to, **1**
    mapping and research &, 219, 221, 224, **230–3**, *231*, *232*, *233*, **234–5**, 244
    physiographic regions of, **1–6**, *2*, *3*, *4*, *5*, *6*, *7*, **8–13**, *8*, *9*, *10*, *11*, *12*
    population distribution &, **123**, 134, 135
    south-west urban network &, 155, **171**, *172*, **173**
    urban development &, **174**, **207**
    vegetation patterns &, **20–7**, *20*, *21*, *22*, *23*, *24*, *25*, *26*, *27*
biophysical interrelationships, **28–46**
biotechnology, 48
birds, **42–3**, *43*
black coal, 114
black swan, **43**, *43*
block mountains, 5, 10
Blue Mountains, 11, *11*, 74, 125
Bluff Knoll, *29*
Boddington, 116, 158, *160*
Bogong High Plains, 11
bolson, 5
Boulder, 132
Bow River diamond mine, 108, *108*, 109, *109*
Boyanup, *160*
Boyup Brook, 114, *114*, 116, *160*
Bridgetown, 158, *160*
Brisbane, 121, *122*, 123, 132
broad acre farming, 60, 70, 232, 242
Broken Hill, 95, 126
Bronzewing mine, 96, *98*, 99
brown coal, 114
brownfields development *see* urban infill
buffer zones, 197, 201
Bunbury, 29, *30*, *36*, 116, 134, 135, 155, *157*, 159, *160*
Bungle Bungles, 6, 8, *8*
Bureau of Meteorology web site, **245–6**
business zones *see* commercial and business zones
Busselton, 134, 159, *160*
butte, 5, 247
butter *see* dairying

## C

Cadoux, *164*, 169, *169*
Cairns, 16, 121, 126, 132
calcareous sands, 26, 33, 41, 45, 247
Cambrian period, 8
Canberra, *122*
Canning Basin, 18
Canning River, 13, 32, 207
Canning Vale, 194, *196*, 197, *198*, 199
canola, *67*
Cape Don, 17, *18*
Cape Leeuwin, 32, 38
Cape York Peninsula, 9, 16, 21
Capel, 95
capital cities *see* state capitals
capital equipment and investment, 47, 48, 49, 50, *50*, 52, **64–5**, *64*, *65*, 71, 74, **75**, *75*, 76, 79, *84*, 85, 86, *86*, **87–8**, *88*, 89, 91, 146, *146*, *163*
carbon dioxide emissions, **119**
carbon monoxide, **207**
Carnamah, 159, *160*
Carpentaria Basin, 9
Carpentaria Lowlands, **9**
casuarina, 247
cattle, 48, **50**, *55*, 59, *60*, *61*, 63, 67, 68, 71, 72, 158 *see also* dairying
caves, *2*, 8–9
Central Australia, 5, 6, *7*, 15, *122*, 123, *126*
Central Australia pastoralism case study, **71–9**
central business district, 136, 140, **176**, 177, *180*, **183**, *184–5*, **186**, *186*, *187*, **188**, 191, *194*, 200, 201, 202, *203*, **211–12**, *242*, **244**, 247
Central Lowlands, 3, 4, 73
central place theory, 148, **151–5**, *152*, *153*, *154*, *163*, **164–6**, **167**, **169–71**, **173**, 247
Central Tablelands, 10, **11**, *11*
central tendency measures, **236**, *237*
centralisation, 124, 126, 151
centrifugal forces, **203**, *203*, 247
centripetal forces, 151, **202–3**, *203*, 247
cereal crops, 57, 58, **59–63**, *61*, **63–7**, 70 *see also* barley; wheat
Cervantes, 31, *31*
Chalice mine site, **103**
Chase Syndicate, 63
cheese, 79, 83, 84, 85, *85*, 89, 93
chemicals *see* agricultural chemicals
Chichester Range, 5
Chicken Creek coal mine, *115*, 117, *117*
choropleth maps, **222**, *223*
Christaller, Walter, **151–5**
cities, **146–7**, *147*
    *see also* state capitals
city structure models, **178–9**, *180*
Claremont, 137, 138, 142, 182
cliffs, *2*, 230
climate
    agriculture &, **47**, 48, 52, **59–60**, 63, *64*, 70, **73**, 75, *75*, **80–1**, **86–7**, *86*
    demographic patterns &, 121, **123**, 148, 155
    fauna interaction &, **46**
    landform interaction &, 11, **17**, **44**, 245
    soil interaction &, 33, **45**
    topography &, **37–8**, 233
    vegetation interaction &, 1, 25, **26**, **44–5**, 232
    *see also* micro-climatic variations
climate change, 1, 12, 26, **27**, 40, 205
climatic controls, **17**, *18*, *19*, 244–5
climatic patterns, **13**, **15–17**, *15*, 29, 33, **34–8**, *35*, *36*, *37*, *38*
cloning, 48
closed forests/vegetation, **20–2**, *21*, *22*, **26**
closer settlement schemes, 85
coal and coal mining, 10, 94, 95, **114–20**, *114*, *115*, *117*, *118*, 207
coal fines, 118, **119**
coastal features, *2*, **232**, *232*, 234, 235
Coastal Lowland landforms, 3, *4*, **12–13**, *12*, *14*, 28, *28*, **30–2**, *30*, *31*, *32*

coastal population distribution, 121, 123, 124–5, 134–5
coastal regions, 149, 150, 155, 159
coastal soils, **33–4**, *33*, 61
coastal vegetation, *20*, 27, **41**
cobalt, 94
Cockburn, 194
Cockburn Sound, 197
Collie, *114*, 115, **116**, *117*, 119, 135, *160*, 207
Collie Coal Basin, **114–15**, *114*, *115*, 116, *117*, 118, 119, 120
cols, 230, *231*
commercial and business zones, **176–7**, 183, *184*, *185*, **186**, *186*, *187*, 188, 189, *190*, 191, *192*, 193, *193*, 200, 201, 202, *202*, 208, 211, **212**, *216*, 220, 234, 242
commercial considerations (agriculture), **51–2**, *52*, 53
commercial functions, *164*, **165–6**, **167**, *168*, **169–70**, *169*, *171*
community functions, **163**, *164–6*, **167**
comparison goods and services, 152, *153*, 154
concentric zone model, **179**, *180*
conservation controls, 53
continental block, 29, 248
continental drift, 9
continental shelf, 12, 31
contours, 229, *229*, 230, *231*, 235
conurbations, 123, **147**, 248
Coober Pedy, 15
Coolgardie, 99, 102, *160*, 161
Coorong, 9, 10, 13
copper, 94, 101
correlation statistics, **237**, *238*
Corridor Plan (Perth), 182, *183*, **208**
Cottesloe, 182
cotton, 58
CRA, **107–8**, 110
Cradle Mountain, 12, *12*, 26
craton, 32, 248
    *see also* continental block
crops, 49, 53, 57, 58, *61*, 62, 65, 66, *66*, 67, *67*, **69–70**, 124 *see also* barley; wheat
cross-sections, **229**, *230*
Crown land, 52, 71
CSIRO plant industry web site, **246**
Cue, 102
cultural influences on urban centres, **174–5**, 182
cultural inputs (agriculture), *47*, **48–9**
cultural issues and urban development, **207–8**
cultural landscape, 49, **53**, *54*, **55**, *55*, **67–9**, *68*, *69*, **77**, *77*, **90–1**, *91*, **95–7**, **102–5**, *103*, *104*, *105*, **110–12**, **116–18**, *117*, 219, 224, 232, **233–5**, 244
Cunderin, *38*, *160*, 163, *164*, **165–6**, 167, 169, 170, *172*
cyclones, 16

## D

Daintree rainforest, 20
Dairy Exit Program, 84–5
dairy factories, **83**, *83*, 85, 91, 92
Dairy Structural Adjustment Program, 84

dairying, 48, 49, 50, 51, 52, *52*, 53, *55*, 58, *60*, 61, *61*, **79–93**, *79*, *80*, *81*, *82*, *83*, *84*, *85*, *86*, *87*, *88*, *90*, *91*, *92*
Dale, 134, *134*
Dalwallinu, *160*, 161, *163*, *164*, *165–6*, 170, 171
Darling Escarpment, **9**, 28, **29–30**, *30*, 31, *32*, 37, 45, 48, 123, 134, 135, 136, 155, 171, 174, 179, 181, 182, 199, 207
Darling Fault, *28*, 29, *32*
Darling Plateau, 4, **9**, **28–9**, *28*, 31
Darling Ranges, 30, 37, 42, 95, 179
Darling River, 9, **10**
Darwin, 16, *122*, 132
decentralisation, 126, 202, 203
deciles, 237, *237*, 248
deferred urban zone, 175, 248
deflation armour, 5
degradation, *3*
  see also land degradation
demographic transition, 127, *127*, **128**, *128*, 248
demography see population
Denmark, *160*
density ratio, 238, *238*
Department of Agriculture (NSW) website, 246
Department of Petroleum and Mineral Resources (WA) website, 246
deposition, *2*, *3*, 6, 12, **13**, 34, 97
deregulation, 48–9, 53, 66, **84–7**, **92–3**
deserts, *2*, **5**, **15**, *18*, *20*, 24, **25**, 39, 61, 102, 126 see also arid landscapes
Devonian period, 8
diamond mining, 94, 95, **108–13**, *108*, *109*, *111*, *112*, *113*
diastrophic movements, 1, *2*, 12, 248
distance measurement (scale), **226**
dolerite, *32*, 97, 248
dolomite, 5, 248
Donnybrook, *160*
Donnybrook Sunkland, 29, *30*
dot distribution maps, **222–3**, *224*
Dowerin, *160*, 161, *165–6*, *167*, 170, 171, *172*, 173
drainage patterns, **233**, *233*, 234, **235**
draw and range area, **152**, 153, *153*, *154*, 163, 248
drought, 15, 24, 25, 38, 45, 50, 70, 73, 74, 76, 80
drowned coastlines, **12–13**
dryland salinity, 56, *56*, **57**, *57*
dugite, **44**
Dumbleyung, 68
Dunsborough, 31
Dunsborough Fault, 29
duplex soils, **34**, 60, 63
Durack Range, 6
dust pollution, **107**
Dwellingup, 37, *160*
dynamic character, **176**, 186, 193

**E**
East Perth, 189, **191–2**, *193*, 196, 204, 208
Eastern Australia intensive pastoralism case study, **79–93**
Eastern Highlands, 1, 3, *4*, 6, 9, **10–11**, *11*, *12*, 15, 16, 22, 27, 123, 126
Eaton, 159

economic development, **126**, 128
economic factors, **160**
economic rent, 51–2, *52*, 61, *202*, 211
ecosystems, 47, 48, 49, 53, *54*, 91, 96, 105, 107, **113**, 209, 248
educational attainment, **141**, *141*
electric power generation, 114, *114*, 115, **116**, **117–18**, *118*, **119–20**, 207
Ellendale, 108, 109, *109*, 110
employment opportunities, **207**
  see also rural labour; urban labour
Eneabba, 63
environment see biophysical environment
environmental controls (agriculture), **50–1**, *52*, 63
environmental impact
  agriculture &, 47, 49, 50, **53**, **55–8**, *56*, *57*, **69–70**, **77–9**, **91–2**
  mining industry &, 94, **96–7**, **99–100**, 102–3, **105**, *106*, **107–8**, **112–13**, 116, 118, *118*, **119–20**
  urban development &, *205*, **207**, **209**
  see also land clearance; pollution
Eocene era, 34
ephemeral plants, 25, 248
epicormic shoots, 22, 248
epiphytes, 20, *20*, 21, 248
erosion, 1, *2*, *3*, 4, 8, 9, 12, **13**, 29, 30, 33, 44, **45**, 46, 48, 53, 56, 58, 63, 64, 68, **69–70**, 76, 77, **91–2**, 107, 113, 120
escarpments, 230
  see also Darling Escarpment
Esperance, *33*, *36*, 38, 59, 63, 70, 134–5, *160*
established residential zone, **194**, *195*, 211, 213, *214*
estuaries, 231, *232*, 234
ethnic population distribution, **142–3**, *142*, *143*, *144*
eucalypts, 1, *5*, 22, 23, 24, 25, 26, *40*, 42, 61, 105, 113, 248
Eucla, 74
eutrophication, 55, 91, 92, 248
evapo-transpiration rates, 15
Ewington mine, 114, *115*, 117, *117*
Exmouth Gulf, 95
export markets, **70–1**, 84, 85, 89, 93, 94, 95, 101, *102*, 149
extensive agriculture, **50**, 51, *51*, *60*, 61, **71–9**, 126, *126*
external morphology, **174–5**, *175*, **179**, **181–3**, *181*, *183*, **210**, *214*, **234**
Eyre Peninsula, 6

**F**
farm amalgamations, 48, 53, 58, 59, 65, 134, 160, 170, 171, 248
farm debt, 71, 78
farm site cultural landscape, **68**, **77**, **91**
farming see agriculture
farming inputs, 47, **48–9**, **63–6**, *64*, 74, **75–6**, *75*, **86–9**, *86*, 90, 244
farming outputs, 47, **49**, *49*, 58, 63, *64*, **67**, 74, *75*, **76–7**, *86*, **89**, 90, 244
farming throughputs, **49**, *64*, **66–7**, *66*, 74, *75*, **76**, *86*, **89**, 90, 244
farmsteads, **145–6**, *146*
faults, 1, *2*, 5–6, 9, 10, 11, 29, 30, *30*, 97, *97*, 114, *115*

fauna, 56
  agriculture &, 77
  climatic interaction &, **46**
  mining industry &, 108
  south-west and, 39, **42–3**, *43*
  topograpic interaction &, **46**
  urban development &, 207
  vegetation interaction &, **46**
  see also introduced fauna
fertilisers, 61, 63, 64, *66*, 86, *86*, 87, 92, 120
field sketching, **243–4**
fieldwork, **243–5**, *245*
fieldwork analysis, **244–5**
Fimiston, 100, 104
fiords, 12, 13, 248
fire, 21, 22, 23, 24, 41, 43, 47
fiscal policy, **53**, 248
fishing, 145, 158, 159
Fitzgerald, 159
Fitzgerald National Park, 32
Fitzroy Crossing, *18*
Flinders Range, 6, *8*
flood, 50, 91, **92**, 235
floodplains, 231, *232*
flora see vegetation
fluvial landscapes, 4, 11, 12, 13, 29, 34, **230–1**, *232*, 248 see also alluvial deposits
fly ash, 119–20
focal length, 242, 248
folding, *2*, 6, 10, 97
forestry, 160
formal regions, 13, **219–20**, *220*, *221*, 248
Fortescue River, 5
freehold land, 52, 248
Fremantle, *36*, 135, 137, 138, 141, 142, 158, 182, *183*, *194*, 204, *208*, 209
Frenchmans Cap, 12
fruit growing, 48, 50, 58, *61*
functional distance, 68, 248
functional influences, **182**
functional regions/zones, 151, *152*, **176–8**, **183–99**, *194*, 200, **211–15**, **220–1**, *221*, 248
future growth zone, 175

**G**
gangue, 99, 248
Gardner Plateau, 6
gas, 95, 116
Gascoyne region, 73
Geelong, *150*
Geikie Gorge, 8
general industry, 248
gentrification, **178**, 190, **204**, 248
geosyncline, 10, 248
Geraldton, 31, *36*, *36*, 155, 159, *160*
gibber deserts, *2*, 5
Giles, 17, *18*
Gingin, 31
Gippsland, 80, 81, *81*, 83
glaciated landforms, *2*, 4, 11, 12, *12*
  see also fiords; ice ages
global position, 148
Global Positioning Systems, 227
global warming, 29, 38, 119
  see also greenhouse effect
gneiss, 4, 29, 30, 32, *32*, 248
gold, 124, 135
Gold Coast, 121, *122*, 126, 131, 132, *150*

## INDEX

gold mining, 94, 95, 96, **97–105**, *97*, *98*, *99*, *100*, *101*, *102*, *103*, *104*, *105*, *106*, **107–8**, 159, 161, 182, 215
Golden Bay, *160*
Goldsworthy, 96
Gondwanaland, 1, 3, 4, 9, 22
Goomalling, *160*, 161, *164*, *165–6*, 167, 170, 171, *172*, 173
Gooseberry Hill, 199
gorges, 5, 230, *232*
government
　agricultural sector &, **48–9**, **52–3**, *52*, 53, 55, **62–3**, **65–6**, 70, *75*, 79, **84–5**, **89**
　mining sector &, 95, 96, 102, **107–8**, **112–13**, 114–15, 116, 119
　urban networks &, *163*, *165*, **169–70**, 173
　urban planning &, **205**, **209**
government web sites, **246**
graben, 5, 30, 114, 248
gradation, 1, 2, *2*, *3*, 28, 31, 46, 248
gradients, **229–30**, *230*, 235
grain, 48, *61*, 67, *67*
　see also barley; wheat
granite, 4, 6, 11, 13, 29, *29*, 30, *30*, 32, *32*, 33, 34, 46, 60, 64, 73, 114, 248
Granny Smith, 96, *98*
graphs and tables, **237–8**, *238*, *239*, **240**, *240*, *241*
Grass Valley, *164*, 173
grasses, 23, **24**, **25**, 26, *26*
grasslands, *20*, *21*, **24–5**, 26, 73, 74, 126
Great Artesian Aquifer, 9
Great Western Plateau landforms, **3–6**, *4*, *5*, *6*, *7*, **8–9**, *8*, *9*, 60
greenfields development, 182, 194, 249
greenhouse effect, 17, 207
　see also global warming
grid references, **228**, *228*
grid systems, 226, **227–8**, 243
Griffin Coal Mining Company, 114, 116, 117, 120
Grimwade, 160
ground level photographs, 242–3, *243*
ground water, 57, *57*, 58, 68, 69, 73, 207, 249
Guildford, 135, 182, 197
Gulf of Carpentaria, 9, 13
Gulf St Vincent, 6
Gutha, 159

## H

Hamersley Range, 5
hamlets, **146**, 159
Hampton Hill, *76*
Harvey, 134, 157, *160*
hay, *61*, *66*, 80, *86*, 87, 89
headlands, 232, *232*
heat island effect, 207, 249
heathlands, *21*, **24**, *25*, 26, 39, **40**, **42**, 45, 234
heavy industry, **197**, 199, 201, 249
herblands see grasslands
heritage sites/issues, 186, 188, *188*, *189*, 205, **207–8**, 209
hinterlands, 148, 149, 151, *152*, **153–4**, *154*, 156, 158, 159, *170*, 181, 183, 210, 249
Hobart, *122*, 148–9
hobby farms, 178
home ownership, **140–1**, *140*, 193

horizontal zonation, 176, 177, **186**, 191
horst, *2*, 5, 29–30, *30*, 249
household income, **139–40**, *140*
human factors and population distribution, **123–6**
human inputs (agriculture), **64–5**, *64*, **75–6**, *75*, *86*, **87–9**
Hunter River area, 92
hydrophytes, 26, 249

## I

ice ages, 5, 9, 12, 29, 30
　see also glaciated landforms
Iluka, 199
immigrants see migrants
Indo-Australian Plate, 2
industrial agglomeration, **150**
industrial estates, 197, *198*, 199
industrial inertia, **177**
industry, 145, 146, 150, 151, **177**, 179, *180*, 182, 183, *184*, *186*, **194**, **196–7**, 201, *202*, 203, 204, 208, 210, 211, **213**, 217, *220*, 242
inertia, **177**, **202**, 249
infrastructure, 48, 50, 51–2, *52*, 53, *54*, 175, 201, 205, 208, 218 see also transport
inner-city industries, **196**
inner mixed zone, **177**, *180*, *184*, *186*, **189–93**, *190*, *192*, *194*, 203, 204, 211, **212–13**, 249
Innisfail, *18*
insects, *46*, 56, 70
intensive agriculture, **49–50**, *50*, *51*, *60*, 61, **79–93**, **124–5**
internal morphology, **175–9**, *180*, **183**, *184–5*, **186**, *187*, **188–9**, *188*, *189*, *190*, **191–4**, *191*, *192*, *193*, *194*, *195*, **196–7**, *196*, *197*, *198*, **199**, **210–15**, *211*, *212*, *213*, *215*, *216*, **234**
International Date Line, 228
Internet, **245–6**
interstate migrations, **129**, **131**
intertropical convergence zone, 17, 249
introduced fauna, 43, 77, *78*, 113, 161
introduced flora, 232
invasion, 177, 191, **203**, **204**, 211, 249
iron ore, 1, 5, 94, 95, 249
ironstone, 5
irrigated agriculture, 48, 52, 58, 80, 85, **86**, 88, 91, 92, 123, 124, 125
islands, 232, *232*
isohyets, 13, 15, 59, 63, 72, 249
isopleth maps, **222**

## J

Jandakot, 208
Jennacubbin, *164*, 173
Jerramungup, 159, *160*
Joondalup, 135, 136, 137, 141, 182, *183*, *194*
Jurien, 159, *160*

## K

Kakadu National Park, 8, 26
Kalamunda, 199
Kalannie, *164*
Kalbarrie, *160*
Kalgoorlie, 15, 95, *96*, 97, 99, *99*, 100, 102, 105, *106*, **107**, 126, 132, 135, 155, *160*, **182**

Kalgoorlie Consolidated Gold Mines Pty Ltd, 104
Kalumburu Joint Venture, 110
Kambalda, 95, 100, *160*
Kanowna Belle goldmine, 98, *98*
Karnet, 37
Karridale, *36*
Karrinyup, 204
karst landscapes, **8–9**, 249
Kata Tjuta, 6, *7*
Katanning, 68, *69*, 157, *160*
Katherine River Gorges, 8, *8*
Kewdale, *194*, 196
Kimberley Diamond Company, 108, 111
Kimberley Plateau, 1, **6**, 8
Kimberley region, 12, *12*, 51, 95, **108–13**, *108*, *109*, *111*, *112*, *113*
kimberlite, 109, *109*
King Leopold Ranges, 6
Kings Peak (Perth), 191
Kinross, 199
knolls, 230, *231*
Kojonup, *160*, 181
Kondut, 173
Konnongorring, 159
Konnonopping, 173
Koorda, *160*, *165–6*
Kosciuszko Plateau, 26
Kosciuszko Uplift, 10
Kununurra, 110
Kwinana, 116, 140, 141, *194*, 197, 199, 201, 204, 209

## L

lagoons, 232, *232*
Lake Alexandrina, 10
Lake Argyle, 113
Lake Eyre, **9**, *10*
Lake Eyre Basin, **9**
Lake Grace, 68, *160*
Lake King, 68
Lake Torrens, 6
lakes, **5**, 25, 29, 31, 34, 57, 73, 155, *172*, 230, *232*
lamproite, 109, *109*, 111
*Land Act 1933*, 52
land clearance, 23, 41, 42, 43, 48, 52, 53, *54*, 57, *57*, **61**, 68, 69, 70, 77, 85, 87, 91, 92, 102–3, **105**, 107, 113, **116–17**, 119, 120, 125, 232
land degradation, *3*, 53, **56–8**, 63, **69–70**, 74, 76, 77, 119
land leases, **52**, 71, 72, 73, 74, 77, 79, 249
land ownership, **52**, *52*, 53, 71, 72
land rehabilitation, **53**, 56, 63, 64, 68, 77
land rents, 245
land use competition, **201–2**, *202*
land values, 51, *52*, 61, 201, 202, *202*, 203, 204, 212
landforms, **1–6**, *2*, *3*, *4*, *5*, *6*, **8–13**, *8*, *9*, *10*, *11*, *12*, **17**, **28–32**, *28*, *29*, *30*, *31*, *32*, **44**, **45**, 46, *123*, 148, 174, 230, **234–5**, 245
lateritic material, 1, 29, 32, 34, 60, 64
latitude, **17**, **227–8**, *227*, *228*, 245
Launceston, 150
leaching, 33, 34, 45, 56, 86, 107
lead, 94, 95, 96
leasehold land, **52**, **71**, 72, 73, 74, 77, 79, 103, 161, 249

Leederville, 189, 196
Leeman, 159, *160*
Leeuwin Counter Current, 38
legumes, *61*, 249
Leonora, 99, 102
life expectancy, 128, 129, 249
lignite, 114
lignotubers, 24, 25, 249
limestone, *3*, 8, *8*, 10, 13, 30, 31, 32, 33, 46, 73, 126
Limestone Creek, 108, 110, 112, 113
line of best fit, 237, 249
line graphs, **238**, *239*
line scales, **226**
lithosols, **34**
Liverpool Range, 11
livestock, 48, 50, 52, 55, 58, **59–63**, *60*, *61*, 62, *62*, **63–4**, 65, **66–7**, *66*, *67*, 68, **69–70**, **71–93**, *71*, 126 *see also* cattle; sheep
local government, 189, 205, 208, **209**, 249
location leaders, 201, 249
locational influences *see* agriculture
longitude, **227–8**, *228*
lupins, *62*, 67

## M
MacDonnell Ranges, 6, *7*
mallee, 24, 25, *40*, 42, 48, 57, 61, 69
mammals, **43–4**, 46
Mandurah, 134, 136, 137, *150*, 155, *157*, *160*, 205
manganese, 94, 95
mangroves, 13, *20*, 26, *26*, 27, *27*, 232, 234
Manjimup, *157*, 158, 160, *160*
Manmanning, 170, *171*, 173
manufacturing milk supplies/products, 79, 80, **83–4**, 85, 87, 89, 93, 249
manufacturing sector, 131, 145, 146, 150, *164*, *165*, 177, *180*, 203, 213
map coordinates, **227–9**, *227*, *228*
map directions, **225**, *225*
map scale, **223**, **225–7**, *226*
map symbols, **224–5**, *225*, 234, 235
mapping skills, **224–30**, *225*, *226*, *227*, *228*, *229*, *230*
maps, **221–4**, *222*, *223*, *224*
Marble Bar, 15
Margaret River, 134, *160*
market gardening, 49, 50, 51, 52, *52*, 55, 61, 74, 83, 124, 155, 197
market threshold, **151–2**, *153*, *154*, 160
markets, *47*, 48, 49, 50, 52, 53, 61, *64*, 67, **70–1**, **74**, 82–3, 84, 89, 93, 94, 95, 101, *102*, 125, 149, 150
marri, 39, *40*, 41, 61
marsupials, 42, 46
Marylands, 196
mass movement, *2*
mass wasting, *3*, 45, 249
McPherson Range, 11
mean score, **236**, 249
meanders, 230, *232*
meat, 59, *60*, 61, 67, 76–7, 89
Meckering, 163, *164*, *168*, 170, *172*
median score, **236**, 249
Medina, *194*
Mediterranean regions, **15–16**, *15*, *18*, 34
megalopolis, 147, 249
melaleuca, 26, 249

Melbourne, 121, *122*, 123, 126, 132, 150
meridians, 228, *228*
Merredin, 68, 135, *160*
Merriwa, 199
mesas and buttes, 5, 6, 249
Mesozoic era, 9, *32*
Metroplan, 189, **208**, *208*, 209
metropolis, **146**
Mettler, *37*, 38
micro-climatic variations, 44–5, 148, 174, 249
Midland, 135, 182, *183*, **194**, 207, 209
migrants, 124, 127, **129**, **131–2**, 133, 136, **142–3**, *142*, *143*, *144*, **150**, 193, 215
Milanna, 63, 64, 65, *65*, 66, *66*, 69
milk industry *see* dairying
milk-quota system, 53, 84, 89, 93
mine site rehabilitation, **96–7**, 99, 103, 104, **105**, **107–8**, **112–13**, 116, 117, **118**, 120
mine sites, **96–7**, **103–4**, *103*, 105, *105*, *106*, 110, **111–12**, *111*, **114**, *114*, **116–17**, **118**, *118*
mine subsidence, **119**
mineral and energy resources
    case studies on, **97–120**
    cultural landscape &, **95–7**, **102–5**, *102*, *103*, *104*, **110–12**, **116–18**, *117*
    development of, **104**, **111**, **116–17**
    environmental impact of, 94, **96–7**, **99–100**, 102–3, **105**, *106*, **107–8**, **112–13**, 116, 118, *118*, **119–20**
    exploration for, **104**, **107–8**, **111**, 116
    history of, **94–5**, 102, **110**, **114–15**, 119, 124, 135
    introduction to, 1, **94**
    location of, **97–8**, *97*, *98*, **108–9**, *108*, *109*, **114**, *114*
    mapping and research skills &, **234**
    mining and processing of, **98–101**, *99*, *100*, *101*, *102*, **104–5**, *106*, **110**, **111–12**, *112*, **115–16**, **117–18**
    population distribution &, 124, **126**, 134, 135
    refining of, **100–1**, **110**
    *see also* open cut mining; underground mining
mineral sands, 94, 95
Mingenew, 159, *160*
mining equipment and structures, 96, 99, 100, *100*, 101, *101*, 102, 104, 105, *105*, 106, 110, 111, 113, *113*, 115–16, 117, **118**
mining settlements, 77, **95–6**, 102, **103**, 110, **111–12**, 15, **116**, *117*, 126, 145, 158, 159, 182
Miocene age, 8
Mitchell Plateau, 6
mixed crop and livestock farming, **59–71**, 125
mode score, **236**, 249
monadnocks, 4, 249
monetary policy, 53, 249
monsoons, *15*, **16**, 17, *18*, 26, 249
Moonie oil field, 95
Moore, 134, *134*
Mortlock River, 210, *214*
Mosman Park, 138
Mt Barker, 157, *160*, 181
Mt Charlotte, 97, 100, 104, *105*
Mt Conner, 6, *7*

Mt Cooke, 30
Mt Dale, 30
Mt Gambier, 10, 83, *150*
Mt Henry, 196, *197*
Mt Isa, 95, 126
Mt Keith, 96
Mt Kosciuszko, 11, *11*
Mount Lofty Ranges, 6, *123*
Mt Magnet, 102
Mt Newman, 96
Mount Ossa, 12
Mt Saddleback, 30
mountain ranges, *2*, 5, 17, 234, 235
Muja coal mine, 114, *114*, 115, *115*, **117–18**, *117*, *118*, 120
multiple nuclei model, **179**, *180*
Mundaring, 41, 142, 199
Murchison region, 73, 74, *98*, 99, 102
Murray-Darling system, 9, **10**, 58, 123
Murray River, 13, 32, 80, 124, 125
Murrumbidgee River, **10**, 125
Musgrave Ranges, 6
Myalup, 159

## N
Narrogin, *36*, 68, 135, 157, *157*, *160*
national land and water resources audit, **246**
native title rights, 71
natural pastures, **73**
Naturaliste-Leeuwin horst-block, 29–30, *30*
Nedlands, 141, 142
New England Tableland, 11
Newcastle (NSW), 121, *150*
Newdegate, 68
newer growth zone, 177–8, **194**, **196**, *196*, 211, 213, *214*, 234
nickel, 94, 95, 101
Norseman, 95, 100, 102, *160*
North Perth, 189, 196
Northam, 38, 68, 159, *160*, 161, 163, *164*, *165*–6, 167, 169, 170, 171, *172*, 173, **210–15**, *211*, *212*, *213*, *214*, *215*, *216*, **217–18**, *217*
Northampton, 59, 61, *160*
Northbridge, 186, *191*, **193**
Northcliffe, 159, *160*
Nullagine, 110
Nullarbor Plain, **8–9**, *9*, 72, 73, 126
numbat, **43**
nutrient leakage, 55, 56, 92
nutrient cycles, 27
nuts, *61*

## O
oats, *67*, 87
oblique images, 242, 243, *243*, *244*
observing systems, **244**
Ocean Reef, 199
oil, **95**
oilseeds, *61*
Olgas, 6, *7*
open cut mining, 96, 97, **99–100**, *100*, 105, 108, **110**, **111–12**, *111*, 114, **115–16**, *115*, **117**, *117*, 118, *118*, **119**, 120
open forests, **22–3**, *21*, **39–41**, 232
Ora Banda, *103*
organic agricultural produce, 48
ornate crevice dragon, **44**

orogenisis, 29, 30, 37, 250
orographic uplift, 16, 17, 250
orthophoto maps, 242
Osborne Park, 196, 203
Otway Ranges, 20
Oudabunna station, 76, 78
outer business districts, **176–7**, **188–9**, 250
oxbow lakes, 230, *232*

**P**
Palaezoic era, 6, 10, *32*
particle emissions, 119–20
pastoral leases, **71**, 72, 73, 74, 77, 79, 103, 161
pastoralism, *60*, **71–93**, *75*, *77*, 124, 125, 126, 155, 159, 250
peat, *32*, 114
pedestrian scale, 176
Peel estuary, 13
peneplains, *2*
Pemberton, 36, *36*, 37, 39, 160, *160*
peninsulas, 232, *232*
Peppermint Grove, 137, 138, 140, *160*
Permian era, 4, 5, 10, 29, 114
Perth
  climate of, 16, 17, *18*, 35, 37, *37*
  history of, **155**, 159, **179**, **182–3**
  population of, *122*, 123, 132, 134, *134*, **135–44**, *136*, *137*, *138*, *139*, *140*, *141*, *142*, 150, 151, **155**, 159, *160*, 182, *183*, 204, 205
  socioeconomic characteristics of, **139–44**, *140*, *141*, *142*, *143*, *144*
  urban hierarchy &, 155, *157*, 159
  urban morphology of, **174–9**, *175*, *180*, **181–3**, *181*, *183*, **184–5**, **186**, *187*, **188–9**, *188*, *189*, *190*, **191–4**, *191*, *192*, *193*, *194*, *195*, **196–7**, *196*, *197*, *198*, **199–209**, *200*, *201*, *202*, *203*, *207*, *208*
  urban problems and planning solutions &, **205–8**, *205*, *207*
  urban processes affecting, **199–205**
Perth Basin, 9, 29, *32*, 116
Petermann Ranges, 6
petroleum, 95
phosphorus, *56*, 92, 250
photographs, **241–3**, *242*, **244**
physical inputs (agriculture), **63–4**, *64*, **75**, *75*, **86–7**, *86*
physiographic regions, **1–6**, *2*, *3*, *4*, *5*, *6*, *7*, **8–13**, *8*, *9*, *10*, *11*, *12*
pie graphs, **240**, *241*
Pilbara, 5, 95, 103, 110
Pingelly, *160*
Pinjarra, 134, 157, 159, *160*
Pinnacles, 31, *31*
Pithara, *164*
Pittwater, 13
planned industrial estates, **197**, *198*
plants *see* vegetation
plateau woodlands, **41–2**
plateaus, **6**, 8, **34**, 230
Pliocene era, 10
Plutonic mine, *98*, 104–5, 108
podsolic soils, 33, *33*, 41, 250
pollution, 49, 105, **107**, **119–20**, 197, 199, *205*, 207

population
  changes in, **127–9**, *127*, *128*, *129*, *130*, **131–3**, *131*, *132*, *133*, *134*, **135–6**
  distribution of, **121**, *122*, **123–7**, *124*, **128–9**, *132*, 134, **135–6**, 145, 148, 150, 151, **155–7**, *156*, *157*
  historical factors &, **123–6**, 135
  introduction to, **121**
  Perth &, **135–44**, *136*, *137*, *138*, *139*, *140*, *141*, *142*, *143*, *144*, 182, *183*, 205
  WA &, 103, *104*, **134–5**, *134*, *160*, 183
  *see also* age-sex distribution; ageing population; rural population
population density, **121**, *122*, 123, **124–5**, *125*, 134, 135, 136, *136*, 155–6, 157, *157*, 176, 193, 208
population growth rates, **127**, *127*, **131**, 132, 133, 135–6, *136*, **150**, 173, 210, 215, 217
population planning in Western Australia web site, **246**
population projections, **132–3**, 136, 205
population pyramids, **128–9**, *129*, *130*
Porongerups, 29, 155
Port Campbell, 13
Port Macquarie, 13
ports, 149, 150
potato farming, 50, *50*, 58
Precambrian period, 1, 6
precipitation *see* rainfall; snow
précis maps, 235
Premier coal mine, 114, *114*, 115, *115*, 117
pressure systems, 17, *19*, **34–5**, 36, *39*
prickly pear, 25
primary index, 148, 149, *149*
primate cities, **149–51**, 155, 156, 159
proportional circle graphs, **240**, *241*
Proterozoic era, 109
public functions, **214**
public housing, 208, 214, 250
Purnululu, 6, 8, *8*

**Q**
Quairading, *160*, 163, *165–6*, *169*, 170, 173
quartz, 97, 250
quartzite, 5, 6, 250
Queenstown, 97, *97*
Quindalup dunes, 31, 33
Quinns Rock, 194
quotas, 48, 53, 84, 89, 93, 250

**R**
rabbits, 161
radial, trellis and dendritic drainage patterns, **233**, *233*, 250
rain-shadow, 16, 250
rainfall, 13, 15, 16, 17, *18*, *19*, 23, 24, 25, 26, 27, 33, 34, **35–8**, *37*, *38*, 39, *40*, 44, 45, 48, 51, 59–60, **63**, *63*, *64*, 66, 70, 71, 72, 73, 75, 80, *82*, 86, *86*, 113, 123, 125, 126, 148, 156, 233, 235
rainforests, 1, **20–2**, *20*, *21*, *22*, 27, 232
range and draw area, **152**, 153, *153*, *154*, 163, 248
range (measure), 236–7, 250
rangeland, 72, 74
rank order correlation coefficient, **237**, *237*
rate per common unit, 238
Ravensthorpe, 68

raw data, 237, **238**, *238*
R codes, 250
regional cities, **146**, *150*, 151, *156*
Regional Development Council (WA) web site, **246**
regional primacy, **149–51**, *150*, *156*
regional shopping centres, **188–9**, *190*
regionalism, **219–21**, *220*, *221*
regolith, *2*, 46, 250
relief maps, **229**, *229*, 235
representative fraction scale, **225–6**
reptiles, **44**, 46
research and mapping skills, **219–46**
residential age, 177–8
residential density, 177
residential zones, **177–8**, 179, *180*, 181, 182, 183, *184*, 186, *186*, 189, *190*, **191–2**, *192*, **193–4**, *194*, *195*, **196**, *196*, 199, 201, 202, *202*, 203, **204**, 205, 208, **209**, 211, **213–14**, 215, *220*, 234, 242
retail functions, 176, 177, **186**, *186*, 187, **188–9**, *188*, *190*, 200, 201, 202, **212**, 213, *216*
ria coasts, 12, 13, 250
rice, 58
ridges, 230, *231*
rift valleys, *2*, 5, 6, 10, 30, 250
riparian vegetation, **26**, 234, 250
river systems, 29, *32*, 56, 57, **233**, 234, 235
river valleys, 13, 31
Riverina, 80, 85
Rockingham, 135, 137, 143, 157, 182, *183*, *194*, 204
rocks, 1, *2*, *3*, 4, 5, **6**, 8, 9, 10, 11, 13, 28–9, *29*, *32*, 33, *33*, 56, 73, 97, 235
Roelands, 159
Rottnest Island, *160*
rural amalgamation, **159–60**
rural cultural landscapes, 49, **53**, *54*, **55**, *55*, **67–9**, *68*, *69*, **90–1**, *91*
rural depopulation, 68
rural labour, *47*, 48, **65**, *75*, **76**, *86*, **87**, 124, 131, 132, 150, 161
rural landscape change issues, 58
rural population, 123, **124–6**, *125*, 131, 132, *132*, **134–5**, *134*, 154, **155–7**, *157*, **159**, **160**, 161, 170
rural settlement(s), 61, **62–3**, *68*, 77, 85, 124, 125, 132, **145–6**, 151, 157, **159**, 160, 235
rural-urban drift, 132, 151, 160
rural-urban fringe, **178**, 193, *194*, **197**, **199**, *200*, 203, *203*, 204, 211, **214–15**, *214*, 234, 244, 250

**S**
saddles, 230, *231*
St Ives gold mine, 98, *98*, 99
saline soil, **34**, 48, 53, 56, *56*, **57–8**, *57*, 63, 64, 68, **69**, 73, 91, **107**, 171, 250
saline water, **107**, 120
salt lakes, **5**, 25, 29, 34, 57, 73, 155, *172*
saltbush, 75
sand dunes, *2*, 5, 26, 30, 31, 32, 45, 181, *181*, 234
sandstone, *3*, 6, 8, 10, 11, 13, *32*, 114, 250
savanna woodlands, **24**
Scarborough, *194*
scatter graphs, 238

schist, 250
sclerophyllous forests, 1, 20, *20*, **22–3**, *22, 23*, **39–41**, *40*, 45, 250
scree, *2*
sea currents, 38
sea-level changes, 1, 9, 12, 30–1, *32*
sea water temperature, 35, 38, 44
secondary retail, 177
Secret Harbour, *183*, 194
sector model, **179**, *180*
sedimentary rock, *3*, 6, 9, 10, 11, *29*, 32
sedimentation, 30, *30*, 32, *32*, **56**, 113
Selwyn Ranges, 8, 9
semi-arid regions, **15**, *15*, *18*, 24, 25, 37, 39, 71, 73, 77, 96, 102, 123
service centres, 146, 159, 210, 215
settlement classes, **145–7**
settlement, landforms and drainage, **235**
settlement patterns, **145–8**, *146*, *147*, 149
sex-age distribution, **128–9**, *129*, *130*, *133*, 135, **136–8**, *137*, *138*, *139*, 158
shale, 5, *32*, 114, 251
Shark Bay, 15, 29, 35, 38
Shatter Belt, 5–6
shearing, 66, *75*, 76, *76*, 77, 78
sheep, 48, 50, 52, 53, 59–60, *60*, **61**, *61*, 62, *62*, 63, *64*, **66–7**, *66*, *67*, 68, 71, *71*, **72–7**, *72*, *73*, *75*, *76*, 125, 155, 157, 158, 167, 210
shrub woodlands, **23–4**
shrubland, 20, *21*, **24**, 25, 39, *40*, **42**, 45, 74, 75, 232, 234
silage, 80–1, 87, 88, 89
siliceous sands, 26, 251
silver, 94, 96, 97, 101
site factors, **148**, **174**, **179**, **181**, *181*, 202, 251
situational factors, **148**, **181–2**
sketch mapping, **235**, *235*
smelters and smelting, 94, **101**, 104, 251
Smoke Creek, 108, 110, 112, 113
snow, 16, 17
social factors, 47, 49, **52–3**, *52*, 193–4, 200, *205*, 209
soil and climate interactions, **45**
soil and topography interactions, **45–6**
soil and vegetation interactions, **45–6**
soil erosion, 45, 48, 53, 56, 58, 63, 64, 68, **69–70**, 76, **91–2**, 107, 113, 120
soils, 1, *2*, 11, 23, 24, 26, **27**, 32, *32*, **33–4**, *33*, 39, **45–6**, *47*, 48, 50, *52*, 53, *54*, 55, **56–7**, *56*, 59, **60–1**, *63*, *64*, 66, **69–70**, **73**, 75, *75*, **81**, **86**, *86*, 92, 113, **123**, 124, 126, 134, 135, 207, 222, *222*, 232, 234, 235 *see also* acid soil; alkaline soil; saline soil
Sons of Gwalia, 97, *98*
South Australian Craton, 29
South-West
  agriculture in, **59–71**
  biophysical interactions &, **44–6**
  case studies of, **59–71**, **155–73**
  climate of, **34–8**, *35*, *36*, *37*, *38*, 39, *39*, *40*, **59–60**, *63*
  fauna of, **42–4**, *43*
  population patterns &, 124, **134–5**, *134*, **155–7**
  primacy network of, *156*
  regional towns of, *156*, 159
  rural-urban drift in, 151

settlement patterns urban network of, **155–73**
soils of, 32, **33–4**, *33*, 39, **60–1**, *63*
special function centres of, **158**
topography of, **23–32**, *28*, *29*, *30*, *31*, *32*, 39, **60**, **63–4**
vegetation of, **39–42**, *40*, *41*, **61**, **64**
Southern Cross, 35, *35*, *36*, 61, *160*
Southern Oscillation Index, 35, 251
space-extensive activities, 186, 204, 234, 251
space-intensive activities, 196, 199, 251
spatial analysis, **244–5**
spatial data, **243**, 244
spatial trends, 245
spatial variations, **139–42**, *140*, *141*, *142*, *143*, *144*, **171**, **173**, 174, 175
Spearwood dune system, 31, *32*, 33
special function centres, **158**, **170–1**
special purpose areas, *194*, **199**
spinifex, **25**
spits, 13
spongelite, 32
spurs, 230, *231*
standard deviation, 237
state capitals, 121, 123–4, 131, **132**, 147, 148–9, *149*, 150, 151 *see also under city e.g.* Adelaide
statistical analysis, **236–8**, *237*, *239*, **240**, *240*, *241*
statistical maps, **221–3**, *222*, *223*, *224*
Stephenson-Hepburn Plan, 208
Stirling, 209
Stirling Range, 29, *29*, 155
subartesian water, 73, 75
Subiaco, 137, 191, 208
suburban areas, 182, 188, **194**, **196**
suburban/regional shopping centres, 176–7, 178, **188–9**, *190*
suburban services, 178, 194, 196
succession, 177, 186, **204**, 211
succulents, 25, 251
sugar, 58, 124
sulfur dioxide, 97, 101, **107**, 119, 207
summits, 230, *231*
sunklands, 12, 29–30, *30*, 251
sunlight, 26, 45
Sunnybank farm, 86, 87, *87*, 88, 89, 91, *91*, 92
Sunshine Coast, 121, 126, 131, 132
Super Pit, 97, *98*, 99, *99*, 107
surface water drainage patterns, **233**, *233*, 235
swamps, *32*, 34, 232, 234, 235
Swan, 137
Swan coastal system, *30*, 31, 123
Swan Hill, *18*
Swan River, 13, 31, 32, *32*, 135, 171, 179, 181, 183, 191, 204, 207
Swan Valley, 48
Sydney, 16, 121, *122*, 123, 126, 132, 147, *147*, 150, *242*, 244
Sydney Harbour, 13
systems analysis, **245**

## T

'3500 Farms Scheme', 62–3
tables and graphs, **237–8**, *238*, *239*, **240**, *240*, *241*

Tamala limestone, 31, *31*, *32*
Tammin, *160*, 165–6, *168*, *172*
tantalum, 94
tariffs, 84, 89, 251
Tasman Fold, 10
Tasmanian Highlands, 10, **11–12**, *12*, 16, 17, 26, *26*, 27
technology, *47*, 48, 49, 50, **65**, **76**, 85, 86, **87–8**, 132, 150, 159, 169, 174–5 , 204
tectonic processes, 1, *2*, *3*, 4, 9, 10, 28, 30, 97, 233, 251
telecentres, **169–70**, *170*
temperate alpine regions, **15**, **16–17**, *18*
temperate maritime regions, **15**, **16**, *18*, 34
temperate rainforests, **20–2**, *21*
temperature patterns, 13, 15, **16–17**, *18*, 25, 26, 33, 34, **35**, *35*, *36*, 39, 44, 45, 48, 60, 75, *75*, 80, 86, 126, 148 *see also* sea water temperature
temporal trends, 245, 251
Tenterden, 61
thematic maps, 223
thermal equator, 16, 17, 251
Thredbo, 17, *18*
Three Springs, 159, *160*
tin, 96
titanium, 95
Toodyay, 159, 160, *160*, 161, 165–6, 167, 170, 171, *172*
topographical maps, **225**, 226, 227, 228, *228*, 229, *229*, **230–4**, *231*, *232*, *233*
topography, **28–32**, *28*, *29*, *30*, *31*, *32*, 35, **37–8**, 39, **45–6**, *47*, 48, 52, *52*, 53, *54*, 55, **60**, **63–4**, 67, **73**, 75, *75*, **81**, *86*, 171, *212*
total fertility rate, 128, 251
tourism, **126**, 145, 146, 158, 170, 204, 217
towns, **146**, *147*, 159
Townsville, 121, 124
trade capture, 210, 251
transitional zones, 174, 179, 189, 191, 192, 197, **203–4**, 214, 219, 221
transport, 50, 51–2, *52*, 53, *54*, 55, 61–2, *62*, 65–6, *66*, **67–8**, **74**, 76, 77, **83**, 84, 96, 114–15, **116**, 124, *124*, 135, 146, 147, 148, 149, 150, 155, **157–8**, *158*, **159**, 161, *164*, *166*, 169, 170, **171**, 174, **176**, 177, 181, *181*, **182**, *184*, 189, **191**, 193, 194, 196, 199, 201, 202, **204**, 205, *205*, **206–7**, *207*, 208, **209**, 210, 214, 215, **217**, 234, **235**, 242
transportation, *2*, 33, 34, 45, 56
trend analysis, **245**, *245*
Tropic of Capricorn, 15, 17
tropical arid regions, **13**, **15**
tropical maritime regions, **15**, **16**, *18*
tropical rainforests, **20**
tropical semi-arid regions, **15**, 71
Tunnel Creek, 8
tuarts, 39, *40*, **41**, *41*, 45, 46

## U

Uluru, 6, *7*, 15
Undara Lava Tubes, 11
uranium, 94, 95
underground mining, 96, 97, **100**, *101*, 102, **104**, 105, *105*, 115, *115*, 116, 119
unemployment, 139, **141–2**, *142*
urban blight, 176, 177, 189, *191*, 204, 205, 212, **217**, 251

urban centres, **121**, 123–4, **131–2**, **135–44**, **174–218**
urban consolidation, 208, 218
urban conurbations, 123, **147**, 248
urban growth issues, 131, **132**, *132*, **205–8**, 217
urban infill, 196, *197*, 204
urban labour, 150, 151
urban morphology
    aerial photographs &, 242, *242*
    external morphology &, **174–5**, *175*, **179**, **181–3**, *181*, *183*, **210**, **234**
    internal morphology &, **175–9**, *180*, **183**, *184–5*, **186**, *187*, **188–9**, *188*, *189*, **190**, **191–4**, *191*, *192*, *193*, *194*, *195*, **196–7**, *196*, *197*, *198*, **199**, **210–15**, *211*, *212*, *213*, *214*, *215*, *216*, **234**
    mapping and research skills &, **234**
    Northam case study, **210–15**, *211*, *212*, *213*, *214*, *216*, **217–18**, *217*
    Perth case study &, **179**, **181–3**, *181*, *183*, *184–5*, *195*, **196–7**, *196*, *197*, *198*, **199–209**, *200*, *201*, *202*, *203*, *207*, *208*, *209*
urban networks and hierarchies, **148–60**, *148*, *149*, *150*, *152*, *153*, *154*, *156*, *157*, *158*, **159–61**, *162*, **163**, *163*, *164–6*, *167*, *168*, **169–71**, *169*, *170*, *171*, *172*, **173**, 221, 234
urban planning, **182**, 188, 197, 200–1, **205**, **208–9**, *208*, 217
urban planning authorities, 205, *206*
urban primacy, 121, **148–9**, *148*, *149*, **159**, 251
urban problems and planning solutions, **205–8**, *205*, *207*
urban redevelopment, **204**, 208, 251
urban relocation, **202–3**, *203*
urban renewal, **178**, **189–91**, *193*, **204**, 208, 218, 251
urban settlements, 53, *54*, **145**, **146–7**, *160*, *157*, *158*, *160*, 235
urban shadow effect, 178, 197, **204**, 251
urban sprawl, 136, 182, 193, 204, 208, 251
urbanisation, **131**, 132, *132*

**V**
valleys, 230, *231*, 234, 235
variability measures, **236–7**
Vasse, 114, *114*, 116, 134, *134*
vegetation, **1**, *5*, **20–7**, *20*, *21*, *22*, *23*, *24*, *25*, *26*, *27*, **39–42**, *40*, *41*, *47*, 48, *52*, *54*, **61**, **64**, 69, 70, 73, 75, *75*, **86–7**, 91, 148, 175, **207**, **209** *see also* introduced flora; land clearance
vegetation and climate interactions, **44–5**
vegetation and fauna interactions, **46**
vegetation and landform interactions, **45**
vegetation and soil interactions, **45**
vegetation patterns, **232**, 233, **234**, 242
vegetation replanting, 53, 56, 64, 68, 69–70, 77, 92, 96–7, 105, 107, 108, 110, **113**, 116, 118, **119**, **120**
vertical zonation, 176, 186, 201, 212
Victoria Park, 137, 189, 196, 202
viticulture, 48, 50, 55, 58, 197
villages, **146**
volcanic activity, *2*, 4, 6, 10, 11, 12, 30, 97, 99, 251

**W**
Wagin, 68
Wallsend Colliery, 114, 120
Walpole, 35, 159, *160*
wandoo woodland, *40*, **41–2**, 45, 61
Wanneroo, 137, 141, 208
warm temperate semi-arid regions, **15**
war service farms, 62, 63, 85, 159
Waroona, 134, 157, *160*
waste products, 49, *64*, 86, 92, 105, **107**, 108, 111, 113, 116, 118, **119**, 205, 207
water *see* artesian water; ground water; irrigated agriculture; river systems
water pollution, 49, **107**, **120**, 207
water quality, **120**
water salinity, **107**, 120
watersheds, **233**, *233*, 251
watertable, 31, 57, *57*, 69, 99–100, 251
Wave Rock, 29
weathering, 1, *2*, *3*, 4, 12, 29, 33, 34, 44, 45, 46
web sites, **245–6**
Welshpool, 196, 209
west coast topography, **31–2**, *31*
West Perth, 189, **191**, *192*, 200, 203, 204
Western Mining Corporation, 95, 116
Western Plateau, 73
Western Power, 119, 120
Western Shield *see* Great Western Plateau landforms
Westfarmers Coal, 114, 115–16, 117, 118, 120
wetlands, **26**, 207, 235
wheat, 23, 48, 52, 53, *55*, 58, **59**, 60, **61**, 62, *62*, 63, *64*, **66**, 67, *67*, 68, **70–1**, *72*, 74, 79, 125, 155, 156, 157, 158, 159, 167, *168*, 169, 170, 210
White Australia Policy, 129
white-tailed black cockatoo, **42–3**, 44
whole milk, 79, *82*, 83, 84, 85, 89, 251
Wiluna, 102
wind, 16, 26, 45, 70, 107, 148
windmills, 73, 75, 76, 77
Windjana Gorge, 8
Winmera, 125
Witchcliffe, 159
Wittenoom Gorge, *6*
Wollongong, 121
Wongan Hills, 68, *160*, 161, 163, *164*, *165–6*, *172*
woodland, 20, *20*, *21*, **23–4**, *23*, 40, **41–2**, *41*, 46, 57, 61, 63, 91
wool, *59*, *61*, *62*, 63, **66**, 67, 68, 72, 73, *73*, 74, *75*, 76, **78–9**, 131, 150, 161
word statement (map scale), **226**
Worsley Alumina Refinery, 114, 116
Wubin, *164*
Wundowie, *160*, *165–6*, *170–1*, 215
Wyalkatchem, *160*, *165–6*, *172*

**X-Y**
xerophytic adaptation, 23, 24, 25, 26, 41, 42, 45, 251
Yalgoo, *36*, *38*
Yarloop, 158, *160*
Yilgarn Block, 9, **28–9**, *28*, 32, 34, 73, 97, 114
Yilgarn goldfields, *98*, 99
York, 61, 68, 146, *147*, 157, 159, 160, *160*, 161, 163, *165–6*, 167, 169, **170**, *170*, 171 *172*
Yunderup, *160*

**Z**
zero population growth, 128
zinc, 94, 95, 96, 97
zirconium, 95
zonal soils, 33, 45
zone of invasion, 177, 191, **203**, **204**, 211